NanoScience and Technology

NanoScience and Technology

Series Editors:
P. Avouris B. Bhushan D. Bimberg K. von Klitzing H. Sakaki R. Wiesendanger

The series NanoScience and Technology is focused on the fascinating nano-world, mesoscopic physics, analysis with atomic resolution, nano and quantum-effect devices, nanomechanics and atomic-scale processes. All the basic aspects and technology-oriented developments in this emerging discipline are covered by comprehensive and timely books. The series constitutes a survey of the relevant special topics, which are presented by leading experts in the field. These books will appeal to researchers, engineers, and advanced students.

Please view available titles in *NanoScience and Technology* on series homepage
http://www.springer.com/series/3705/

Fausto Rossi

Theory of Semiconductor Quantum Devices

Microscopic Modeling and Simulation Strategies

With 125 Figures

Springer

Prof. Dr. Fausto Rossi
Politecnico di Torino
Dipartimento di Fisica
Corso Duca degli Abruzzi 24
10129 Torino, Italy
E-mail: fausto.rossi@polito.it

NanoScience and Technology ISSN 1434-4904
ISBN 978-3-642-10555-5 e-ISBN 978-3-642-10556-2
DOI 10.1007/978-3-642-10556-2
Springer Heidelberg Dordrecht London New York

Cover design: Integra Software Services Pvt. Ltd., Pondicherry

Printed on acid-free paper

Springer is part of Springer Science+Business Media (www.springer.com)

dedicato a Elena, la luce della mia vita

(devoted to Elena, the light of my life)

Preface

Primary goal of the present volume is to provide a cohesive description of the vast area of semiconductor quantum devices, with special emphasis on basic quantum-mechanical phenomena governing the electro-optical response of new-generation nanomaterials. The book covers within a common theoretical framework different types of optoelectronic nanodevices, including quantum-cascade laser sources and detectors, few-electron/exciton quantum devices, and semiconductor-based quantum logic gates. Distinguished feature of the present volume is a unified microscopic treatment of quantum-transport and coherent-optics phenomena on ultrasmall space- and time-scales, as well as of their semiclassical counterparts.

The book, mainly devoted to graduate students as well as researchers working in the field, presents a unified theoretical treatment of semiconductor nanodevices; indeed, the primary goal of this volume is to cover within a common language two different classes of quantum devices, i.e., systems where the quantum nature manifests itself in terms of discrete energy spectra but their dynamics may still be treated within a semiclassical scenario (e.g., infrared laser sources and detectors) and semiconductor devices whose behavior is entirely governed by electronic quantum coherence (e.g., semiconductor-based quantum logic gates).

The field of semiconductor quantum devices is so active and extensive that an exhaustive treatment of the many diverse research areas is nearly impossible; we shall therefore limit ourselves to a discussion of selected theoretical and experimental issues, including recent developments, which have led to fundamental new insights as well as to relevant advances in semiconductor quantum physics and technology. In particular, we shall focus on nonequilibrium carrier dynamics in open quantum devices. Furthermore, we shall not address the vast area of quantum-optics phenomena; the only exception will be the case of carrier-cavity mode coupling in semiconductor microcavities.

The book is organized into 11 chapters plus 4 appendices. In Chap. 1 we shall recall the basic concepts and fundamental properties of semiconductor bulk materials as well as of low-dimensional semiconductor structures like, e.g., superlattices, quantum wells, wires, and dots; in addition, we shall discuss in very general and qualitative terms the link between nanomaterials and corresponding optoelectronic quantum devices. Chapter 2 will focus on the basic assumptions of the conventional semiclassical picture, showing that the latter are not always justified in new-generation semiconductor nanomaterials and nanodevices and therefore fully quantum-mechanical treatments of the problem are imperative. In Chap. 3 we shall recall the fundamentals of the well-known density-matrix formalism applied to the investigation of the electro-optical properties of semiconductor nanomaterials and nanodevices, while in Chap. 4 the same formalism will be extended to quantum systems with open spatial boundaries, corresponding to the case of a generic quantum device inserted into an electric circuit. Chapter 5 will introduce the basic concepts as well as key instruments related to the numerical modeling of semiconductor nanomaterials and nanodevices. In Chap. 6 we shall discuss in very general terms the most effective approaches for the study of unipolar transport in nanodevices; to this end, we shall address separately the low- and the high-field regimes, and for both regimes we shall provide a semiclassical treatment of the problem as well as its quantum-mechanical generalization. Chapter 7 will address the basic physical processes as well as open technological problems related to the design and optimization of new-generation quantum-well infrared photodetectors, focusing on the development of efficient quantum devices for the terahertz spectral region. In Chap. 8 we shall discuss the basic features of quantum-cascade coherent-light sources; to this aim, we shall review a few simulated experiments focusing on the microscopic explanation of the gain regime, both in the mid-infrared and in the far-infrared spectral regions. Chapter 9 will discuss the basic properties and unique features of few-electron/exciton quantum systems, namely single and coupled semiconductor macroatoms, pointing out their potential role in designing a completely new class of optoelectronic quantum devices, like electron-state detectors and quantum logic gates. In Chap. 10 we shall review a few potential implementation strategies for the concrete realization of quantum information processing using specifically designed semiconductor nanostructures, namely quantum dots and wires. In Chap. 11 we shall briefly address two extremely active and stimulating research topics, namely molecular and spin-transport electronics, whose development may lead to completely new paradigms in semiconductor-based electronic and optoelectronic physics and technology.

This volume is the result of about 20 years of research activity on fundamental issues related to quantum-transport as well as coherent-optics phenomena in semiconductor bulk and nanostructures; the latter have been performed at the University of Modena (Italy), at the Philipps University of Marburg (Germany), and mostly at the Polytechnic University of Torino (Italy), involving a number of worldwide collaborations with several leading research groups in the field.

Let me take this unique opportunity to thank a number of people that in many ways and at different stages have contributed significantly to this research effort. First of all, I am grateful to Carlo Jacoboni, the person that opened my young mind to the magic world of quantum mechanics; his enthusiasm, scientific curiosity, and intellectual rigor have strongly influenced my personal as well as professional life. Let me thank Paolo Lugli and Elisa Molinari; their innovative viewpoint on scientific research as well as their continuous help and support played a crucial role during the first stages of my scientific career. I am grateful to Tilmann Kuhn and Paolo Zanardi; they gave key contributions in developing many of the ideas on the theoretical description of quantum devices presented in this book. Let me thank Rita Claudia Iotti for her essential role in setting up and developing our current research group at the Physics Department of the Polytechnic University of Torino, together with a number of Ph.D. students and researchers, including Eliana Biolatti, Remo Proietti Zaccaria, Emanuele Ciancio, Irene D'Amico, Ehoud Pazy, Radu Ionicioiu, Stefano Portolan, Fabrizio Castellano, and David Taj; she has contributed significantly to a large fraction of the research activity reviewed in this book. Let me finally thank Traiano Rossi (my father) for his invaluable help in setting out the manuscript layout. Last but not least, I am profoundly grateful to my family for their never-ending support and patience.

Torino, August 2010 *Fausto Rossi*

Contents

1

Fundamentals of Semiconductor Materials and Devices

In this chapter we shall recall basic concepts and fundamental properties of semiconductor bulk materials as well as of low-dimensional semiconductor structures like superlattices, quantum wells, wires, and dots. In addition, we shall discuss in very general and qualitative terms the link between nanomaterials and corresponding optoelectronic quantum devices.

1.1 An Introductory Overview on Semiconductor Physics and Technology

The following introductory overview by Martin Stutzmann [1] has been written in 2008 for the celebration of the 20th anniversary of the *Walter Schottky Institute (WSI)* of the *Technische Universität München.*

On the 14th of November, 1876, the high school teacher Ferdinand Braun gave a presentation entitled "Experiments concerning deviations from Ohm's law in metallic conductors" in front of the illustrious *Naturforschende Gesellschaft zu Leipzig.* Neither Mr. Braun nor his critical audience were aware of the fact that they were just witnessing the birth of Semiconductor Physics. Indeed, what Ferdinand Braun had discovered by a long series of meticulous experiments, at first sight did not appear too exciting: if one equipped a solid crystal with two metallic contacts and applied an electric voltage, two basic kinds of behavior could be observed. Either no measurable electric current passed through the crystal. Then this crystal obviously was an electric insulator and, therefore, was no longer of interest for further investigation of its electric properties. Or, for the second type of crystals, a sizeable current could be measured. Then this crystal was a metallic conductor and obeyed Ohm's law. If one doubled the applied voltage, also the observed current was doubled. And if one inverted the electric voltage applied to the two contacts, also the current was inverted. At least, that was the way it used to be until 1876...

The deviations from Ohm's law, which Ferdinand Braun reported to the Natural Society at Leipzig, were quite strange, indeed. Instead of using large electric contacts, Mr. Braun had performed some of his experiments with very fine contact needles. In some crystals he then had observed an electric current as in a usual conductor, but when he inverted the applied voltage, the current disappeared and the crystal apparently had converted to an insulator. Similarly unexpected was what happened when Braun changed the magnitude of the applied voltage. Then for these strange crystals the current increased much more strongly than allowed by Ohm's law.

What Ferdinand Braun had discovered was the first semiconductor device ever: a rectifying diode, which could transform alternating current into direct current, and many years later actually was used in the form of point contact diodes in radio receivers or radar units. Unfortunately, the reason for this strange behavior of his crystals remained entirely mysterious to Ferdinand Braun. The scientific explanation of this phenomenon was given much later in 1939 by Walter Schottky, who at that time developed a theory for the electronic properties of semiconductor/metal interfaces. Honoring this work, rectifying metal/semiconductor contacts now are known as Schottky diodes. As for Ferdinand Braun, he received the Nobel Prize for Physics in 1909 and today mainly is remembered for the invention of the cathode ray tube. As a matter of fact, due to the development of vacuum tubes, solid-state electronics did not gain any practical importance until the first transistor was built by Bardeen, Brattain, and Shockley in 1947. Since then, however, solid-state electronics based on semiconductor devices has revolutionized our world, which no longer can be imagined without this.

So what actually are semiconductors, what makes them so different from metallic conductors, and why are they so interesting for many applications? The scientific answer to these questions and the development of new semiconductor materials and devices is what semiconductor physics and technology is all about. It takes many years of intensive studies to master the basics of this important part of solid-state physics. Here, we will try to convey the most important concepts concerning the unique properties of semiconductors with the help of a simple analogy.

In a regular periodic crystal lattice, electrons as the carriers of electric current are not allowed to move around freely. Instead, they have to obey certain rules enforced by quantum mechanics. As a consequence, electrons have to occupy so-called energy bands which are separated from each other by small or large "bandgaps." As schematically depicted in Fig. 1.1, this situation can be compared to a two-storey building consisting of a ground floor and a first floor. In the language of solid-state physics, these two storeys are called "valence band" and "conduction band," respectively. Both floors are covered by a well-ordered array of quadratic tiles, representing the periodic lattice of atoms in a semiconductor crystal. The movement of electrons in a crystal is then analogous to the movement of inhabitants in our building, whose most important purpose is to transport "charge" from one end of the building to

Fig. 1.1. Two-floor-building analogue of a semiconductor crystal. A semiconductor may be regarded as a two-storey building consisting of a ground floor and a first floor; in the language of solid-state physics, these two storeys are called "valence band" and "conduction band," respectively. Here, both floors are covered by a well-ordered array of quadratic tiles, representing the periodic lattice of atoms in a semiconductor crystal. The movement of electrons in a crystal is then analogous to the movement of inhabitants in our building, whose most important purpose is to transport charge from one end of the building to the other end [1]

the other end. The inhabitants of our "semiconductor house" have to obey one additional important rule: at no time more than one inhabitant is allowed to occupy the space of a given tile! In the same way, electrons in a solid crystal have to obey the quantum-mechanical "exclusion principle" formulated by the famous physicist Wolfgang Pauli.

Now that the blueprints of our semiconductor building and the basic rules for its inhabitants have been defined, let us start to occupy this building with people. At first, all inhabitants can be accommodated on the ground floor, where they can move around more or less freely and transport their cargo across the building. This leads to a steady increase of the amount of cargo transported through the building, until the occupancy of the ground floor has increased so much that the inhabitants start to hinder one another on their way. Eventually, the stream of cargo will come to a complete stop, once all tiles in the ground floor are occupied by an inhabitant, so that nobody is able to move any more. Further inhabitants can only occupy the first floor, where they again have sufficient room to move about. As a consequence, the overall cargo stream through the building will again start to increase, reach a maximum, and eventually come to an end when also the first floor is fully occupied.

The fate of electrons in solids is quite similar to what happens to the inhabitants of our semiconductor building. In particular, it is easy to understand why both, electric conductors and insulators, exist in nature. Crystals in which energy bands are only partially filled will belong to the group of electric conductors, since their electrons can move more or less freely through the crystal lattice. If, on the other hand, all energy bands are fully occupied or completely empty, no electric current can pass through the crystal at all and we are dealing with an electric insulator. Which situation will be encountered for a given solid crystal depends on how many electrons per atom are available to occupy the energy bands of the crystal. For example, almost all metals are very good electric conductors due to a half-filled conduction band, whereas metal oxides very often are good insulators with a completely filled valence band and an empty conduction band.

Now, how do semiconductors fit into this picture? As already suggested by their name, semiconductors are solids which are able to pass an electric current much better than insulators, but at the same time not as efficiently as an electric conductor. Obviously, semiconductors are solids in which for one reason or the other a few of the many tiles on the ground floor remain empty or a few of the conduction band tiles are occupied, or both. This particular constellation can be achieved via three routes, all of which are of fundamental importance in semiconductor physics and, thus, will be treated in more detail in the following.

The starting point of our discussion will be the situation of an electric insulator, where all tiles in the ground floor are occupied by exactly one inhabitant and all tiles in the first floor are empty. Thus, no charge transport can occur. To change this state of affairs, which is very unfavorable for device applications, semiconductor physicists have developed the concept of "doping." Contrary to the very negative image in sports, doping in semiconductors can be used to exactly pre-determine the electric conductivity of a given device by addition of a precisely calculated amount of impurity atoms. In the analogue of our semiconductor building, doping can be achieved by adding special tiles with the following properties. As a first example, so-called "acceptor tiles" can be added to the ground floor. These acceptor tiles have the unpleasant property of swallowing exactly one inhabitant of the fully occupied ground floor, thus creating a "hole" in the overall occupancy. This allows the other inhabitants of the ground floor to move again. The hole created by the acceptor tiles also will move at the same time, however, in the opposite direction as compared to the inhabitants. In the same way, acceptor atoms incorporated into a semiconductor crystal will create a hole in the occupancy of the valence band, which will act as a "missing electron" and, thus, as a positively charged particle in electronic transport. Therefore, doping of a semiconductor crystal with acceptor atoms is referred to as "p-type" doping ("p" as in positive); such process is schematically depicted in Fig. 1.2. The second possibility to induce controlled electric conduction in an insulator is the doping with donor impurities. In our semiconductor building, such "donor tiles" bring along one additional inhabitant, who has to occupy a free tile in the first floor, since all tiles of the ground floor are already occupied. Accordingly, donor atoms added

Fig. 1.2. Schematics of a p-type semiconductor in terms of the two-floor-building analogue of Fig. 1.1. Here, so-called "acceptor tiles" can be added to the ground floor, thus swallowing exactly one inhabitant of the fully occupied ground floor and creating a "hole" in the overall occupancy; this allows the other inhabitants of the ground floor to move again. The hole created by the acceptor tiles will move as well, however, in the opposite direction as compared to the electrons (inhabitants) [1]

to a semiconductor crystal will provide additional electrons in the conduction band, which contribute to electronic charge transport in the expected way ("n-type" doping by additional negatively charged electrons); such process is schematically depicted in Fig. 1.3.

In summary, it is indeed the possibility of doping with additional donor or acceptor atoms which distinguishes semiconductors from insulators or conductors as a third class of materials. Conductors will always pass electric current with little resistance and independent of chemical details, no matter what. In the same way, insulators will always block electric current. In contrast, semiconductors will either behave more like an insulator or more like a metal, depending on the level of doping. This provides semiconductors with the unique property to rectify, switch, or amplify electric signals and, thus, to manipulate electric current as it passes through the semiconductor crystal.

There is yet another way to produce additional holes in the valence band or electrons in the conduction band of a semiconductor without doping, namely by providing external energy in the form of heat or light. We all know from our own experience that it takes energy to walk up the stairs from the ground floor to the first floor. The same holds for the electrons in a semiconductor: electrons in the conduction band (first floor) have a higher energy than electrons in the valence band (ground floor). This difference in energy is determined by the bandgap of the semiconductor, as already mentioned above. Since electrons are lazy, they prefer to stay on the ground floor. In order to move up to the first floor, they have to be stimulated by an external influence. One possibility

Fig. 1.3. Schematics of a n-type semiconductor in terms of the two-floor-building analogue of Fig. 1.1. Here, each additional "donor tile" brings along one additional inhabitant, who has to occupy a free tile in the first floor, since all tiles of the ground floor are already occupied; accordingly, donor atoms added to a semiconductor crystal will provide additional electrons in the conduction band, thus contributing to electronic charge transport in the usual way [1]

is provided by the thermal movement of the atoms. At low temperatures, atoms are frozen at their lattice sites, but at higher temperatures they start to wiggle more and more and to push the electrons around. In the analogue of our semiconductor building, the thermal motion of the atoms can be visualized by a staircase leading from the ground floor to the first floor. The thermal motion of the atoms will push the electrons upwards step by step. The larger the bandgap of the semiconductor, the longer the staircase and the smaller the number of electrons which actually make it all the way up to the first floor; such process is schematically depicted in Fig. 1.4. However, in every well-planned building, there is also another possibility to reach the upper floors more easily: an elevator. In semiconductors, the job of the elevator is done by the elementary particles of light, the photons. If such a flash of light hits a semiconductor, it can directly elevate an electron from the valence band up to the conduction band. The stronger the light beam that falls onto the semiconductor, the more often the photon elevator will make the trip between the two floors, each time taking an electron with it. But also the other direction of electron transport is possible: electrons in the conduction band can return to the valence band, if there is a hole to accommodate the returning electron. This process is called "recombination." To do this, the electrons can either take the staircase down, giving their energy back to the atoms, or they can take the photon elevator. Then, each time the elevator doors open in the valence band and an electron recombines with a hole, an elementary flash of light is emitted by the semiconductor. The energy of the emitted photon is

Fig. 1.4. Schematics of an undoped semiconductor with thermally excited electrons and holes in terms of the two-floor-building analogue of Fig. 1.1. Here, the thermal-activation mechanism can be visualized by a staircase leading from the ground floor (valence band) to the first floor (conduction band); the thermal motion of the atoms will push the electrons upwards step by step. The larger the bandgap of the semiconductor, the longer the staircase and the smaller the number of electrons which actually make it all the way up to the first floor [1]

the same as the bandgap of the semiconductor. Semiconductors with a small bandgap emit red photons, whereas semiconductors with a large bandgap emit blue photons.

This interaction between electrons and photons in semiconductors provides the basis of optoelectronics, another very important application area. The fundamental optoelectronic devices are solar cells and light-emitting diodes (LEDs). In a solar cell, light enters the semiconductor from the outside and, via the photon elevator, lifts electrons from the valence into the conduction band. The excited electrons leave the semiconductor as an electric current. In an LED, on the other hand, electrons are injected into the conduction band of the semiconductor through one contact and extracted from the valence band through a second contact, leaving holes behind. When the injected electrons recombine with these holes, they emit light, as discussed above.

This short introduction into the basic properties and electronic processes of semiconductors should show why semiconductors are so important for electronics. They differ from metals and insulators through the fact that they can be doped, thus enabling complete control over their electric conductivity necessary for tailor-made diodes and transistors. In addition we have seen how the existence of a bandgap between the valence and the conduction band can be used for the absorption and emission of light. Unfortunately, we also have to pay a price for this flexibility. Since small amounts of foreign impurity atoms can considerably alter the electric properties of semiconductors, the preparation and deposition of semiconductors requires ultra-pure environments and utmost care. And to emit or absorb light efficiently at different wavelengths, new semiconducting compounds with complex chemical compositions have to be synthesized, investigated, and optimized.

But semiconductors are worth such an effort! Be it in computer sciences, the control of industrial processes, energy technology, information technology, consumer electronics, medical diagnostics, illumination technology or in airplanes, cars, appliances, phones, or watches: semiconductor devices are omnipresent and indispensable. Today, the direct annual market volume of semiconductor devices is about 300 Billion Euro, and the financial impact of semiconductors as a key enabling technology in many other markets is by far larger, still. Also in the future, semiconductors will be a part of our daily life, hopefully with a positive impact. A very visible example in the true sense of the words is the rapid progress in solid-state lighting. Here, highly efficient LEDs are beginning to replace the old vacuum tube technology, much in the same way as 50 years ago solid-state transistors have replaced vacuum tubes in analog and digital signal processing. We all witness personally the current progress in computer processors, where Moore's law[1] is still alive and well. And last but not least semiconductors will help to provide the world population with clean and sustainable energy. The current prediction is that in the year 2010 alone, solar cells with an accumulated electrical power of 20 GW will be installed, the equivalent of about 20 nuclear power plants.

1.2 Bulk Materials and Nanostructures

In what follows, we shall introduce basic concepts related to the physical properties of semiconductor bulk and nanostructured materials aimed at understanding their electro-optical properties.

1.2.1 Ground State and Excitation Spectra of a Semiconductor Crystal

As for any other crystalline solid [2–18], a semiconductor crystal [19–26] is formed by a partially ordered system of ionized atoms, i.e., the crystal lattice, surrounded by a corresponding "cloud" of valence electrons. Exactly as for the case of atoms and molecules, it is such electronic cloud that, according to a fully quantum-mechanical treatment of the problem, gives rise to an ordered solid-state aggregation of these constituents.

By denoting with $\boldsymbol{r}_i$ and $\boldsymbol{R}_j$ valence electron and ionic coordinates, respectively, and with m and M_j the corresponding masses, the global crystal Hamiltonian may be written as

[1] Moore's law describes a long-term trend in the history of computing hardware, in which the number of transistors that can be placed inexpensively on an integrated circuit has doubled approximately every 2 years.

$$\hat{H} = \sum_i -\frac{\hbar^2\nabla^2_{\boldsymbol{r}_i}}{2m} + \frac{1}{2}\sum_{ii'}{}' \frac{e^2}{|\boldsymbol{r}_i - \boldsymbol{r}_{i'}|}$$
$$+\sum_j -\frac{\hbar^2\nabla^2_{\boldsymbol{R}_j}}{2M_j} + \frac{1}{2}\sum_{jj'}{}' \frac{Z_j Z_{j'} e^2}{|\boldsymbol{R}_j - \boldsymbol{R}_{j'}|}$$
$$-\sum_{ij} \frac{Z_j e^2}{|\boldsymbol{r}_i - \boldsymbol{R}_j|} \,. \tag{1.1}$$

The latter may be regarded as the sum of three contributions: the first line describes the interacting valence-electron gas, the second one the interacting ionic system, while the third one describes the electron–ion coupling term. Here, Z_j denotes the net charge of the jth ion in units of the elementary charge e.

In view of the significant difference between electronic and ionic masses ($\frac{m}{M_j} \sim 10^{-3}$), the lattice dynamics comes out to be adiabatically decoupled from the electronic one; as a result, in such decoupling limit the valence electrons feel the ions as they were at rest, while the ions feel only a temporal average of the electronic motion. This adiabatic decoupling scheme – known as Born–Oppenheimer approximation and originally introduced for the analysis of molecular systems – allows one to rewrite the total Hamiltonian (1.1) as

$$\hat{H} = \hat{H}_\mathrm{e}(\{\boldsymbol{r}_i\}, \{\boldsymbol{R}^\circ_j\}) + \hat{H}_\mathrm{i}(\{\boldsymbol{R}_j\}) + \hat{H}_\mathrm{e\text{–}i}(\{\boldsymbol{r}_i\}, \{\boldsymbol{U}_j\}) \,, \tag{1.2}$$

where

$$\hat{H}_\mathrm{e}(\{\boldsymbol{r}_i\}, \{\boldsymbol{R}^\circ_j\}) = \sum_i -\frac{\hbar^2\nabla^2_{\boldsymbol{r}_i}}{2m} + \frac{1}{2}\sum_{ii'}{}' \frac{e^2}{|\boldsymbol{r}_i - \boldsymbol{r}_{i'}|} - \sum_{ij} \frac{Z_j e^2}{|\boldsymbol{r}_i - \boldsymbol{R}^\circ_j|} \tag{1.3}$$

describes the valence-electron quantum-mechanical problem for a given ionic configuration $\{\boldsymbol{R}^\circ_j\}$, and

$$\hat{H}_\mathrm{i}(\{\boldsymbol{R}_j\}) = \sum_j -\frac{\hbar^2\nabla^2_{\boldsymbol{R}_j}}{2M_j} + V_\mathrm{i}(\{\boldsymbol{R}_j\}) \tag{1.4}$$

describes the lattice dynamics in terms of an effective ionic potential, resulting from the indirect quantum-mechanical action of the valence-electron cloud.

Within such adiabatic-decoupling scheme – valid only for an infinite ion-electron mass ratio – the last term in (1.2) describes the residual (and adiabatically small) interaction of the valence electrons with the lattice vibrations $\boldsymbol{U}_j = \boldsymbol{R}_j - \boldsymbol{R}^\circ_j$. As we shall see, this contribution corresponds to the well-known electron–phonon interaction.

Electronic Band Structure

According to the Born–Oppenheimer approximation scheme previously introduced, the many-electron energy spectrum $\mathcal{E}_\mathrm{e}$ (as well as the corresponding set of wavefunctions Φ_e) is the solution of the following time-independent Schrödinger equation

$$\hat{H}_e \Phi_e(\{\boldsymbol{r}_i\}) = \mathcal{E}_e \Phi_e(\{\boldsymbol{r}_i\}) \ , \tag{1.5}$$

where the explicit form of the valence-electron Hamiltonian is given in (1.3).

As for the case of many-electron atoms and molecules, this Schrödinger problem cannot be treated exactly, and a corresponding mean-field (i.e., Hartree–Fock) approximation scheme is imperative. The key idea is that the effect of all other electrons plus the ionic lattice on a given electron "of interest" may be expressed in terms of an average single-particle potential $V(\boldsymbol{r})$; the corresponding one-electron (1e) Schrödinger equation is of the form

$$\hat{H}_{1e}\phi_\alpha(\boldsymbol{r}) = \left[-\frac{\hbar^2\nabla_{\boldsymbol{r}}^2}{2m} + V(\boldsymbol{r})\right]\phi_\alpha(\boldsymbol{r}) = \epsilon_\alpha \phi_\alpha(\boldsymbol{r}) \ , \tag{1.6}$$

where ϵ_α and $\phi_\alpha(\boldsymbol{r})$ denote single-particle electron energies and wavefunctions, respectively. Here, α is a compact notation for the generic set of electronic quantum numbers, whose nature is determined by the particular symmetry properties of the single-particle potential $V(\boldsymbol{r})$. Indeed, dealing with a semiconductor crystal – or with a low-dimensional version of it – it is possible to show that for each real-space direction characterized by lattice periodicity, we have a sort of generalized Bloch theorem; more specifically, denoting with $\boldsymbol{r}_\parallel$ the crystal periodicity subspace (and with $\boldsymbol{r}_\perp$ its complementary one), we have

$$\phi_\alpha(\boldsymbol{r}) = \mathrm{e}^{\mathrm{i}\boldsymbol{k}_\parallel \cdot \boldsymbol{r}_\parallel} u_{\boldsymbol{k}_\parallel \nu}(\boldsymbol{r}_\parallel, \boldsymbol{r}_\perp) \ , \tag{1.7}$$

where $\boldsymbol{k}_\parallel$ denotes the so-called electron wavevector defined over the Brillouin zone of the periodicity subspace and u is a periodic function of $\boldsymbol{r}_\parallel$. In this case, the generic set of quantum numbers reduces to the periodicity wavevector plus an additional index ν: $\alpha \equiv \boldsymbol{k}_\parallel \nu$. It follows that the electronic energy spectrum $\epsilon_\alpha = \epsilon_{\boldsymbol{k}_\parallel \nu}$ exhibits the well-known band-structure feature, i.e., continuous energy spectra interrupted/separated by so-called energy gaps.

For the particular case of a two-dimensional crystal, e.g., a semiconductor surface/interface, the electronic states in (1.7) correspond to surface states, while for a conventional (i.e., three-dimensional) crystal, the standard version of the Bloch theorem is recovered (see, e.g., [2]):

$$\phi_{\boldsymbol{k}\nu}(\boldsymbol{r}) = \mathrm{e}^{\mathrm{i}\boldsymbol{k}\cdot\boldsymbol{r}} u_{\boldsymbol{k}\nu}(\boldsymbol{r}) \ . \tag{1.8}$$

In this case the index ν reduces to the conventional band index of crystalline solids, and the fully three-dimensional vector $\boldsymbol{k}$ is the so-called electron quasimomentum in units of $\hbar$. More precisely, for each band ν, a unique value of the electron energy $\epsilon_\nu(\boldsymbol{k}) \equiv \epsilon_{\boldsymbol{k}\nu}$ is associated to the electron quasimomentum $\boldsymbol{k}$. The set of functions $\epsilon_\nu(\boldsymbol{k})$ describes what is known as the electronic band structure of the bulk material.

The knowledge of the band structure of a solid is the starting point for the study of its electro-optical properties, and its theoretical determination is still one of the major open problems of solid-state physics (see, e.g., [27]). Figures 1.5 and 1.6 show typical band-structure diagrams for the case of bulk silicon and GaAs, respectively. While for all cubic semiconductor crystals the valence band displays very similar features (see, e.g., [28]), the conduction band of so-called elemental semiconductors (like silicon) is qualitatively different from that of compound materials (like GaAs). More specifically, for silicon (see Fig. 1.5) the conduction band is characterized by six equivalent ellipsoidal minima along the $< 1\,0\,0 >$ directions at about 0.85% of the Brillouin zone; in contrast, for GaAs (see Fig. 1.6) the absolute minimum of the conduction band lies at the center of the Brillouin zone (referred to as Γ point). As we shall see, such differences have a strong impact on their optical properties (see below).

Starting from the electronic band structure, it is also useful to define another key quantity, the so-called density of states (DOS); for any given band ν the corresponding DOS is defined as

$$g(\epsilon) = \sum_{\boldsymbol{k}} \delta(\epsilon - \epsilon_{\boldsymbol{k}\nu}) \; . \tag{1.9}$$

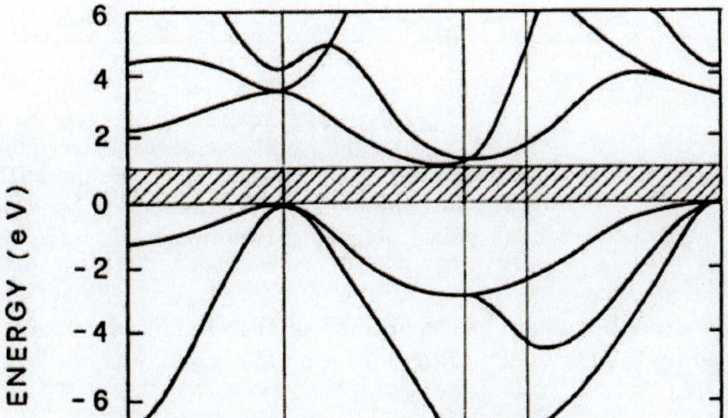

Fig. 1.5. Typical band-structure diagram for bulk silicon. The conduction band is characterized by six equivalent ellipsoidal minima along the $< 1\,0\,0 >$ directions at about 0.85% of the Brillouin zone. Reprinted with permission from [28]

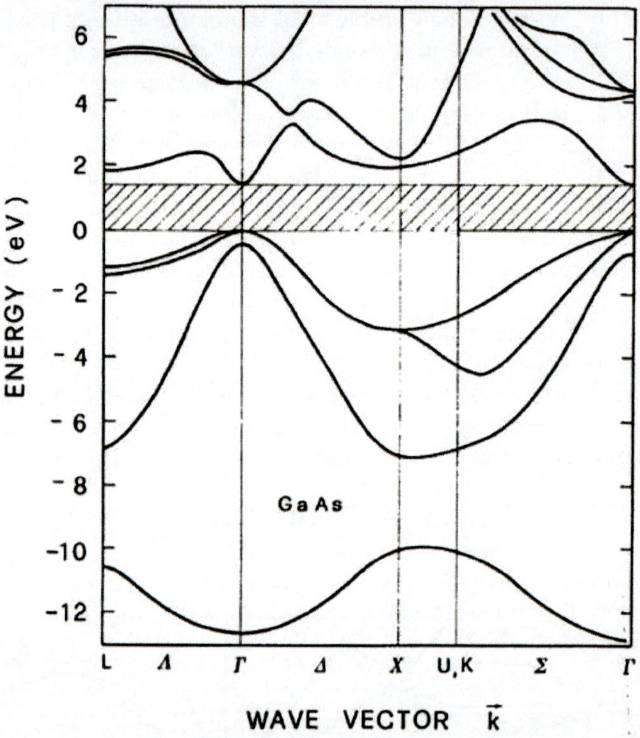

Fig. 1.6. Typical band-structure diagram for bulk GaAs. As for most III–V semiconductor compounds, the absolute minimum of the conduction band lies at the center of the Brillouin zone (Γ point). Reprinted with permission from [28]

The latter describes a sort of "counting" of the electronic states in band ν (performed by the Dirac delta function) around a given energy ϵ.[2] Indeed, by recalling that

$$\delta(x) = \frac{\mathrm{d}}{\mathrm{d}x}\theta(x) \tag{1.10}$$

($\theta(x)$ denoting the usual Heaviside step function) the DOS in (1.9) may also be written as

$$g(\epsilon) = \frac{\mathrm{d}}{\mathrm{d}\epsilon}N(\epsilon) , \qquad N(\epsilon) = \sum_{\boldsymbol{k}} \theta(\epsilon - \epsilon_{\boldsymbol{k}\nu}) , \tag{1.11}$$

[2] In spite of the fact that for an infinitely extended crystal the electronic wavevector is a continuous quantity, we shall always denote any integration over $\boldsymbol{k}$ via a sum over a discrete set of wavevectors corresponding to a large but finite crystal volume Ω. Indeed, in the so-called thermodynamic limit ($\Omega \to \infty$) such a discrete summation may be replaced by a corresponding integral via the usual relationship $\sum_{\boldsymbol{k}} = \frac{\Omega}{(2\pi)^3} \int \mathrm{d}^3k$.

i.e., it may be regarded as the energetic derivative of the total number of electronic states $N(\epsilon)$ in band ν with energy smaller than ϵ. As we shall see in Sect. 1.2.2, the DOS is the key quantity for the basic characterization of low-dimensional semiconductor nanostructures.

As anticipated, semiconductor materials are characterized by their gap of forbidden energies (of width ϵ_g) between the last band whose states are occupied by electrons at zero temperature (valence band) and the first empty band (conduction band). The energetic region of the band structure which is of interest for the investigation of electro-optical phenomena is typically centered around the semiconductor energy gap and extends some ϵ_g's above the minimum of the conduction band and below the top of the valence band. Electrons in the conduction band are simply referred to as electrons, while missing electrons in the valence band are referred to as "holes." At zero temperature, the valence band of a semiconductor is entirely occupied by electrons, and the conduction band is entirely empty. Thus, at $T = 0$ a semiconductor is actually an insulator. In contrast, at finite temperatures some electrons will leave the valence band and occupy some states of the conduction band. In such a case the transport properties of the semiconductor are due to an equal number of electrons and holes. If, however, some impurities create localized states with occupied energy levels at $T = 0$ within the energy gap, for $T > 0$ these electrons will tend to occupy the conduction band. Such impurities are called "donors." On the other hand, if the states created by the impurities are empty at $T = 0$, for $T > 0$ they will tend to accept electrons from the valence band and create holes. Such impurities are called "acceptors."

The density of electrons and holes at equilibrium is controlled by the Fermi–Dirac statistics. More specifically, the average occupation of the generic electronic state $\boldsymbol{k}\nu$ is described by the well-known Fermi–Dirac distribution

$$f^{\circ}_{\boldsymbol{k}\nu} = \frac{1}{e^{\frac{\epsilon_{\boldsymbol{k}\nu} - \mu}{k_B T}} + 1} , \tag{1.12}$$

where μ is the chemical potential, whose value depends on temperature as well as on donor versus acceptor concentrations. If μ is located close to the conduction band, the electric conduction is mainly due to electrons and the semiconductor is called extrinsic of n-type; if μ is close to the valence band, the conduction is mainly due to holes and the semiconductor is said to be extrinsic of p-type. If, finally, the number of electrons and holes due to the presence of impurities is negligible with respect to the number of thermally excited electron–hole pairs (from the valence to the conduction band), then μ is typically located around the center of the bandgap and the semiconductor is said to be intrinsic.

As previously mentioned, for many of the electro-optical phenomena governing the semiconductor quantum devices discussed in this book, the energetic region of interest is typically centered around the semiconductor energy gap ϵ_g, corresponding to high-symmetry points of the crystal Brillouin zone, i.e., characterized by extrema (minima or maxima) of the electronic band structure $\epsilon_{\boldsymbol{k}\nu}$; this general feature suggests to approximate the band structure via a second-order Taylor expansion around each band minimum or maximum. More specifically, by denoting with $\boldsymbol{k}_\circ$ the generic high-symmetry point, we have

$$\epsilon_{\boldsymbol{k}\nu} = \epsilon_{\boldsymbol{k}_\circ+\Delta\boldsymbol{k},\nu} \approx \epsilon_{\boldsymbol{k}_\circ\nu} + \frac{1}{2}\sum_{ll'} \gamma^\nu_{ll'} \Delta k_l \Delta k_{l'} \ , \tag{1.13}$$

where $\gamma^\nu_{ll'}$ is in general a non-diagonal tensor, whose matrix elements may be obtained via the well-known $\boldsymbol{k}\cdot\boldsymbol{p}$ method (see, e.g., [24]). In particular, for zinc-blend structures like III–V semiconductor compounds, e.g., GaAs and InP, both conduction and valence bands exhibit, respectively, an isotropic minimum and maximum at the center of the Brillouin zone ($\boldsymbol{k}_\circ = \boldsymbol{0}$); in this case, the tensor of the second-order Taylor expansion (1.13) is simply a scalar ($\gamma^\nu_{ll'} = \gamma_\nu \delta_{ll'}$) and (1.13) reduces to the well-known "effective-mass approximation":

$$\epsilon_{\boldsymbol{k}\nu} \approx \epsilon_{\boldsymbol{0}\nu} + \frac{\hbar^2 k^2}{2m^*_\nu} \ . \tag{1.14}$$

Here, $m^*_\nu = \frac{\hbar^2}{\gamma_\nu}$ is called "effective mass"; it describes the curvature inverse radius of band ν around $\boldsymbol{k} = \boldsymbol{0}$. Within such approximation scheme, we deal with spherical and parabolic bands, and the only physical parameters describing the electronic properties around the bandgap are the effective masses for electrons and holes. As we shall see, the latter – together with the semiconductor bandgap ϵ_g – are often the few key parameters used in the simulation of bulk as well as nanostructured materials and devices.

Phononic Band Structure

Based again on the Born–Oppenheimer approximation scheme, the vibrational energy spectrum of the crystal is described by the following Schrödinger equation

$$\hat{H}_\mathrm{i}\Phi_\mathrm{i}(\{\boldsymbol{R}_j\}) = \mathcal{E}_\mathrm{i}\Phi_\mathrm{i}(\{\boldsymbol{R}_j\}) \ , \tag{1.15}$$

where the explicit form of the ionic Hamiltonian is given in (1.4). Since the lattice dynamics is typically taking place around the minimum of the effective ionic potential V_i (corresponding to the lattice configuration $\{\boldsymbol{R}^\circ_j\}$), the latter may be safely replaced by its second-order Taylor expansion around such energy minimum, i.e.,

$$V_\mathrm{i}(\{\boldsymbol{R}_j\}) = V_\mathrm{i}(\{\boldsymbol{R}^\circ_j + \boldsymbol{U}_j\}) \approx V_\mathrm{i}(\{\boldsymbol{R}^\circ_j\}) + \frac{1}{2}\sum_{jj'} \boldsymbol{U}_j \cdot \boldsymbol{\Upsilon}_{jj'} \cdot \boldsymbol{U}_{j'} \ , \tag{1.16}$$

where

$$\boldsymbol{\Upsilon}_{jj'} = \left. \frac{\partial^2 V_\mathrm{i}}{\partial \boldsymbol{U}_j \partial \boldsymbol{U}_{j'}} \right|_{\{\boldsymbol{R}^\circ_j\}} \tag{1.17}$$

is the so-called force-constant matrix. Within this second-order approximation – also known as "harmonic limit" – the ionic Hamiltonian in (1.4) reduces to

$$\hat{H}_{\mathrm{i}} = \sum_j -\frac{\hbar^2 \nabla^2_{\boldsymbol{U}_j}}{2M_j} + \frac{1}{2}\sum_{jj'} \boldsymbol{U}_j \cdot \Upsilon_{jj'} \cdot \boldsymbol{U}_{j'} \ . \tag{1.18}$$

Here, $\boldsymbol{U}_j = \boldsymbol{R}_j - \boldsymbol{R}_j^\circ$ denotes again the ionic displacement.

The above ionic Hamiltonian – describing a set of coupled oscillators – may be transformed into a sum of independent (i.e., noninteracting) harmonic oscillators, via the introduction of the so-called normal or collective coordinates (see, e.g., [6]):

$$\hat{H}_{\mathrm{i}} = \sum_\beta \epsilon_\beta \left(\hat{b}^\dagger_\beta \hat{b}_\beta + \frac{1}{2} \right) , \tag{1.19}$$

where $\hat{b}^\dagger_\beta$ and $\hat{b}_\beta$ are operators describing, respectively, the creation and destruction of a vibrational energy quantum $\epsilon_\beta = \hbar\omega_\beta$ corresponding to the normal mode β, while the collective (or resonance) frequencies ω_β are given by the square roots of the eigenvalues of the dynamical matrix

$$D_{jj'} = \frac{\Upsilon_{jj'}}{\sqrt{M_j M_{j'}}} \ . \tag{1.20}$$

Indeed the energy spectrum of the Hamiltonian in (1.19) is simply given by the sum of elementary harmonic-oscillator spectra corresponding to the different resonance frequencies ω_β, i.e.,

$$\mathcal{E}_{\mathrm{i}} = \sum_\beta \epsilon_\beta \left(n_\beta + \frac{1}{2} \right) . \tag{1.21}$$

Apart from the zero-point energy $\frac{1}{2}\epsilon_\beta$, the energy contents of each vibrational mode β are simply expressed by the number n_β of energy quanta ϵ_β.

Exactly as for the case of the electronic properties, the presence of a total or partial crystal symmetry translates once again into a strong simplification of the problem. Indeed, in the presence of a lattice periodicity, the index/quantum number β is replaced by a wavevector $\boldsymbol{q}_\parallel$ (defined once again over the Brillouin zone of the symmetry subspace) plus a residual index η: $\beta \equiv \boldsymbol{q}_\parallel \eta$. Therefore, exactly as for the electronic case ($\alpha \equiv \boldsymbol{k}_\parallel \nu$), in the presence of crystal symmetry the vibrational energy $\epsilon_\beta = \epsilon_{\boldsymbol{q}_\parallel \eta}$ exhibits once again a band structure defined on the very same Brillouin zone.

The previous analysis suggests to describe the vibrational properties of a semiconductor bulk crystal in terms of elementary excitations or quasiparticles called phonons, characterized by a three-dimensional quasimomentum $\hbar\boldsymbol{q}$ and a corresponding energy dispersion $\epsilon_{\boldsymbol{q}\eta}$. Figures 1.7 and 1.8 show typical phononic-band diagrams for the case of bulk silicon and GaAs, respectively. As we can see, such phononic bands are given by so-called acoustic phonons, i.e., vibrational modes whose energy goes to zero for $q \to 0$, as well as by

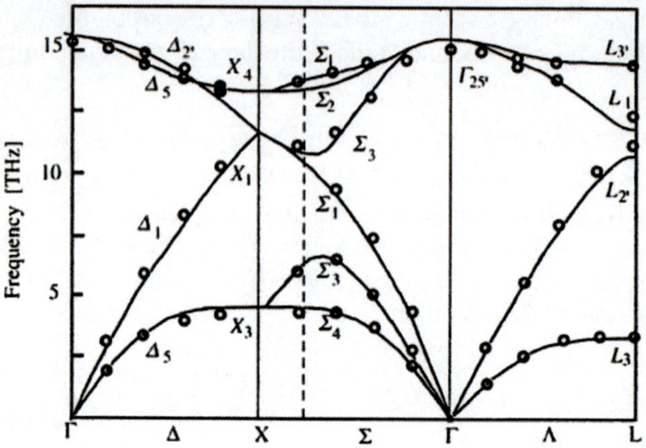

Fig. 1.7. Typical phononic-band diagram for bulk silicon along high-symmetry axes. Here, the *circles* are data points from [29] while the continuous *curves* are calculated with the adiabatic bound charge model [30]. Reprinted with permission from [22]

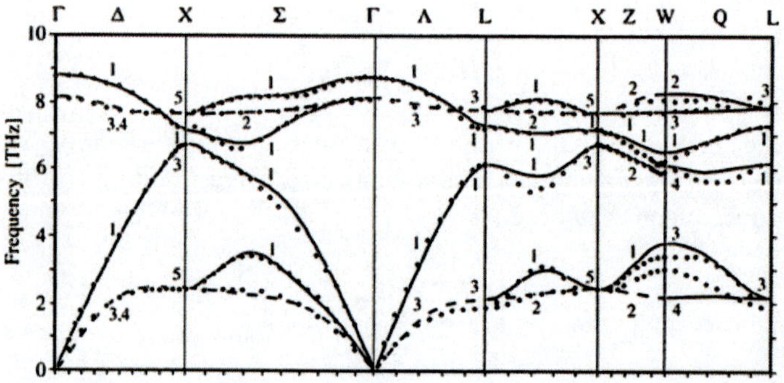

Fig. 1.8. Typical phononic-band diagram for bulk GaAs along high-symmetry axes. Here, the experimental data points were measured at a temperature of 12 K while the *continuous lines* were calculated with a 11-parameter rigid-ion model (the numbers next to the phonon branches label the corresponding irreducible representations) [31]. Reprinted with permission from [22]

optical phonons, i.e., vibrational modes whose energy for $q \to 0$ remains finite. As discussed extensively in [28], these two types of vibrational modes – acoustic versus optical phonons – play a different role in determining phonon-induced electron dissipation and decoherence phenomena in bulk as well as in nanostructured materials (see also Chaps. 8 and 10).

Electro-optical Excitations

Generally speaking, the presence of an external electro-optical excitation may induce significant modifications to the electronic spectrum of our semiconductor crystal. For many of the phenomena discussed in this book, such an external electromagnetic excitation may be treated within a classical framework, by generalizing the single-electron Schrödinger equation (1.6) as follows:

$$\left[-\frac{\left(-i\hbar\boldsymbol{\nabla}_{\boldsymbol{r}}+\frac{e}{c}\boldsymbol{A}(\boldsymbol{r},t)\right)^2}{2m}+V(\boldsymbol{r})-e\varphi(\boldsymbol{r},t)\right]\phi_\alpha(\boldsymbol{r})=\epsilon_\alpha\phi_\alpha(\boldsymbol{r}). \tag{1.22}$$

Here, $\boldsymbol{A}(\boldsymbol{r},t)$ and $\varphi(\boldsymbol{r},t)$ denote, respectively, the vector and scalar potentials corresponding to our external electromagnetic field:

$$\boldsymbol{E}(\boldsymbol{r},t)=-\frac{1}{c}\frac{\partial}{\partial t}\boldsymbol{A}(\boldsymbol{r},t)-\boldsymbol{\nabla}_{\boldsymbol{r}}\varphi(\boldsymbol{r},t)\ , \qquad \boldsymbol{B}(\boldsymbol{r},t)=\boldsymbol{\nabla}_{\boldsymbol{r}}\times\boldsymbol{A}(\boldsymbol{r},t)\ . \tag{1.23}$$

Equation (1.23) reflects the well-known gauge freedom: there is an infinite number of possible combinations of $\boldsymbol{A}$ and φ which correspond to the very same electromagnetic field $\{\boldsymbol{E},\boldsymbol{B}\}$.

The single-particle Schrödinger equation (1.22) may describe the simultaneous presence of both quasi-static electric and/or magnetic fields as well as of optical (i.e., high-frequency) excitations. As discussed previously, due to the crystal symmetry we have $\alpha\equiv\boldsymbol{k}_{\|}\nu$. However, in the presence of nonhomogeneous electromagnetic potentials, such symmetry is lost, and the nature of the set of quantum numbers α depends on the particular problem under investigation. For instance, in the presence of a homogeneous electric field within a scalar-potential gauge we deal with the well-known Wannier–Stark states (see Sect. 3.4), while for the case of a homogeneous magnetic field we deal with Landau-like states. An additional important remark is that, also for the case of static (i.e., time independent) fields, the corresponding electromagnetic potentials may be time dependent; it follows that in this case the single-electron energy spectrum ϵ_α in (1.22) – as well as the corresponding set of states ϕ_α – is time dependent as well. This feature will play an important role in the treatment of high-field transport phenomena in semiconductor nanostructures (see, e.g., [32]).

In the limit of weak electromagnetic fields – often referred to as linear-response regime – quadratic terms in the vector potential $\boldsymbol{A}$ may be safely neglected. More specifically, in the so-called Coulomb gauge ($\nabla\cdot\boldsymbol{A}=0$) the Schrödinger equation (1.22) reduces to

$$\left(\hat{H}_{1\mathrm{e}}+\hat{H}'\right)\phi_\alpha(\boldsymbol{r})=\epsilon_\alpha\phi_\alpha(\boldsymbol{r}) \tag{1.24}$$

with

$$\hat{H}'=-\frac{i\hbar e}{mc}\boldsymbol{A}(\boldsymbol{r},t)\cdot\boldsymbol{\nabla}_{\boldsymbol{r}}-e\varphi(\boldsymbol{r},t)\ . \tag{1.25}$$

The result is that within such linear-response regime the effect of an external electro-optical excitation may be fully described via the above interaction Hamiltonian $\hat{H}'$.

According to the particular problem under investigation, the latter may be treated exactly (e.g., for the investigation of high-field transport and/or ultrafast coherent-optics experiments) or within conventional perturbation-theory schemes (e.g., for the study of incoherent optical processes), where the phenomenon is typically described in terms of transitions between field-free electronic states, whose probability amplitude is proportional to the matrix elements of the perturbation Hamiltonian (1.25). In particular, for slowly varying optical excitations (compared to the lattice space-scale) fully described via a vector-potential gauge ($\varphi(\boldsymbol{r},t)=0$), we have

$$H'_{\boldsymbol{k}'\nu',\boldsymbol{k}\nu} \propto \int \mathrm{d}^3r \phi^*_{\boldsymbol{k}'\nu'}(\boldsymbol{r}) \left[-\mathrm{i}\hbar\nabla_{\boldsymbol{A}}\phi_{\boldsymbol{k}\nu}(\boldsymbol{r})\right] = p_{\boldsymbol{k},\nu'\nu}\delta(\boldsymbol{k}-\boldsymbol{k}') \tag{1.26}$$

with

$$p_{\boldsymbol{k},\nu'\nu} = \int u^*_{\boldsymbol{k}\nu'}(\boldsymbol{r}) \left[-\mathrm{i}\hbar\nabla_{\boldsymbol{A}} u_{\boldsymbol{k}\nu}(\boldsymbol{r})\right] \mathrm{d}^3r \ , \tag{1.27}$$

i.e., the matrix elements of $\hat{H}'$ within the Bloch-state basis (1.8) are simply proportional to the matrix elements of $-\mathrm{i}\hbar\nabla_{\boldsymbol{A}}$ (describing the component of the quasimomentum operator along the direction of the vector potential $\boldsymbol{A}$), which in turn, thanks to the Dirac delta function $\delta(\boldsymbol{k}-\boldsymbol{k}')$, are different from zero only for $\boldsymbol{k}=\boldsymbol{k}'$. It follows that within such approximation scheme – known as "dipole approximation" – an optical excitation may induce electronic transitions from any occupied band ν to any empty band ν', conserving the value of the electron quasimomentum $\hbar\boldsymbol{k}$; such microscopic processes are usually referred to as "vertical transitions," and the quantity $p_{\boldsymbol{k},\nu'\nu}$ in (1.27) is called "dipole matrix element."

In addition to the momentum selection rule $\boldsymbol{k}=\boldsymbol{k}'$, optical-absorption processes in semiconductors are also governed by energy conservation. More specifically, given a monochromatic radiation of frequency ω (and corresponding photon energy $\hbar\omega$), the allowed vertical transitions $\boldsymbol{k}\nu \to \boldsymbol{k}\nu'$ are selected by the following energy-conservation condition: $\epsilon_{\boldsymbol{k}\nu'}-\epsilon_{\boldsymbol{k}\nu}=\hbar\omega$.

As a result of the scenario discussed so far, the macroscopic absorption coefficient will be proportional to the total number of absorbed photons per time unit; the latter depends on the microscopic transition probabilities previously discussed (see (1.26) and (1.27)) as well as on the total number of possible (i.e., energy-conserving) vertical transitions. More specifically, within the standard perturbation-theory scenario we have

$$\mathrm{A}(\hbar\omega) \propto \sum_{\boldsymbol{k}} p^2_{\boldsymbol{k},\nu'\nu}\delta\left(\epsilon_{\boldsymbol{k}\nu'}-\epsilon_{\boldsymbol{k}\nu}-\hbar\omega\right) \ . \tag{1.28}$$

For relatively small variations of the photon energy $\hbar\omega$, the dipole matrix element $p_{\boldsymbol{k},\nu'\nu}$ is usually a slowly varying function of $\boldsymbol{k}$, and therefore the latter can be taken out of the integral and evaluated at some high-symmetry point $\boldsymbol{k}_\circ$ of the Brillouin zone ($\overline{p}_{\nu'\nu} \equiv p_{\boldsymbol{k}_\circ,\nu'\nu}$):

$$\mathrm{A}(\hbar\omega) \propto \overline{p}_{\nu'\nu} g^j(\hbar\omega) \propto g^j(\hbar\omega) \tag{1.29}$$

with

$$g^j(\hbar\omega) = \sum_{\boldsymbol{k}} \delta\left(\epsilon_{\boldsymbol{k}\nu'} - \epsilon_{\boldsymbol{k}\nu} - \hbar\omega\right) \ . \tag{1.30}$$

The quantity $g^j(\hbar\omega)$ is called "joint density of states (J-DOS)": similar to the conventional DOS in (1.9), the latter describes a sort of "counting" of all vertical transitions corresponding to an energy difference $\hbar\omega$.

Equation (1.29) tells us that, for a slowly varying dipole matrix element p, the absorption spectrum $\mathrm{A}(\hbar\omega)$ is simply proportional to the J-DOS. The latter, in turn, exhibits large values for the so-called direct semiconductors (like, e.g., III–V compounds), characterized by a vertical alignment of conduction- and valence-band extrema (see Fig. 1.6). In contrast, the J-DOS is orders of magnitude smaller for the so-called indirect semiconductors (like, e.g., IV group materials), characterized by a misalignment between conduction-band minima and valence-band maxima (see Fig. 1.5).

In the presence of electron–hole Coulomb correlation (see, e.g., [33]) (not included in the previous treatment), the approximation in (1.29) is not valid anymore, and this is particularly true for the case of low-dimensional nanostructures (see Sect. 1.2.2). This suggests to rewrite (1.28) as

$$\mathrm{A}(\hbar\omega) = s(\hbar\omega) g^j(\hbar\omega) \ , \tag{1.31}$$

where the new quantity $s(\hbar\omega)$ is called "oscillator strength". The physical interpretation of the above equation is the following: for each photon energy $\hbar\omega$, the optical absorption is given by the oscillator strength (proportional to the microscopic transition probability) times the J-DOS (corresponding to the total number of allowed vertical transitions). As we shall see, the general scheme in (1.31) applies to semiconductor nanostructures as well.

Additional Interaction Mechanisms

The analysis presented so far – based on the Born–Oppenheimer decoupling scheme previously introduced – has allowed to identify noninteracting electronic and phononic states, described by corresponding energy-band diagrams (see, e.g., Figs. 1.5 and 1.7). However, as already pointed out, such adiabatic-decoupling limit does not account for the so-called electron–phonon interaction, i.e., the interaction of our Bloch-state electrons (see (1.8)) with the lattice vibrations $\boldsymbol{U}_j$. As discussed extensively in [6], such electron–phonon coupling – described by the interaction Hamiltonian $\hat{H}_{\text{e-i}}$ in (1.2) – may have two physically different origins. On the one hand, our Bloch electrons are disturbed by local lattice distortions (i.e., lattice dilatation versus compression) due to ionic vibrations; such effect is known as "deformation-potential contribution." On the other hand, in polar semiconductor crystals the electron dynamics is also disturbed by the dipolar field produced by the crystal vibrations; the latter effect is known as "electrostatic contribution."

Since for most semiconductor materials this electron–phonon coupling is small, the latter is again described within a so-called scattering picture, i.e., similar to the case of optical absorption previously discussed, the effect of such mechanism is accounted for in terms of stochastic transitions between noninteracting electron times phonon basis states according to quantum-mechanical transition probabilities obtained, e.g., via the well-known "Fermi's golden rule."

In addition to the electron–phonon coupling, a number of crystal imperfections – not considered so far – may give rise to other interaction processes; for instance, the presence of ionized impurities may result in a significant electron scattering due to static Coulombic centers.

Let us finally stress that the single-particle band diagram $\epsilon_\alpha = \epsilon_{\boldsymbol{k}_\parallel \nu}$ – based on a proper solution of the mean-field single-particle Schrödinger equation (1.6) – describes the electronic excitation spectrum corresponding to the ground-state configuration (full valence bands and empty conduction ones). However, as previously mentioned, due to thermal and/or optical excitations, a significant fraction of valence electrons may "jump" into empty conduction-band states, leaving charge holes in the initial valence-band ones. Within the well-known electron–hole picture, such excitation processes will result in the generation/recombination of so-called electron–hole pairs.

It is then clear that in such a scenario, carrier–carrier (i.e., electron–electron, hole–hole, and electron–hole) Coulomb interaction should be considered as well. The level of description of such interaction processes will depend on the particular problem under examination. More specifically, for a wide class of semiconductor bulk and nanostructures (including quantum wells and wires), such two-body interactions may be treated again within a mean-field framework. In this case, it is possible and convenient to split such interaction into a short-range contribution described by a screened Coulomb potential, and a long-range one resulting in bosonic-like collective excitations of our carrier gas, called "plasmons." In particular, in the so-called low-density limit, i.e., in the limit of a low-spatial density of electron–hole pairs, the only relevant interaction mechanism is the attractive Coulombic interaction between electrons and holes; indeed, in this regime electrons and holes attract each other, forming bound two-particle complexes called "excitons," characterized by a partially discrete energy spectrum.

In contrast, for genuine zero-dimensional systems (referred to as semiconductor macroatoms or quantum dots) one deals typically with a finite number of carriers (i.e., electrons and/or holes); therefore, in this case a mean-field picture is definitely inadequate, and an exact treatment of the interacting few-carrier system is imperative. A detailed discussion of such few-electron/exciton phenomena and of its potential application to new-generation quantum devices is given in Chap. 9.

Within the scattering picture previously mentioned, our Bloch electrons will undergo stochastic transitions $\boldsymbol{k}\nu \to \boldsymbol{k}'\nu'$ induced by the various disturbance sources previously mentioned. More specifically, for a generic intraband scattering mechanism – e.g., carrier–phonon, carrier–impurity, carrier–carrier – the corresponding inverse life-time (i.e., the probability per time unit that the electron will leave its current state $\boldsymbol{k}$) is always of the form

$$\Gamma(\boldsymbol{k}) = \sum_{\boldsymbol{k}'} P_{\boldsymbol{k}',\boldsymbol{k}} = \sum_{\boldsymbol{k}'} F_{\boldsymbol{k}',\boldsymbol{k}} \delta\left(\epsilon_{\boldsymbol{k}'} - \epsilon_{\boldsymbol{k}} - \Delta\epsilon\right) . \tag{1.32}$$

Since for the relevant phase-space region (selected by the energy-conserving delta function) the quantity F is a slowly varying function of the final wavevector $\boldsymbol{k}'$, we have

$$\Gamma(\epsilon_{\boldsymbol{k}}) \propto g(\epsilon_{\boldsymbol{k}} + \Delta\epsilon) , \tag{1.33}$$

i.e., the inverse life-time is simply proportional to the DOS of the final state (see (1.9)).

It is important to stress the strong similarity between (1.33) and (1.29): both the inverse life-time Γ and the absorption spectrum A are strongly influenced by the DOS (or J-DOS) of the material under examination; it follows that both electronic-transport and optical properties may be significantly improved by a proper tailoring of the material DOS; indeed, this is the primary goal of semiconductor nanoscience.

1.2.2 Low-Dimensional Heterostructures

As anticipated, modern semiconductor technology has opened the way to the massive fabrication of so-called semiconductor heterostructures, obtained by properly combining different semiconductor materials within the same growth process. The typical space-scale of the constituent materials may vary from the micrometric down to the nanometric scale, in which case such heterostructures – also referred to as "nanostructures" [34–40] – may display genuine quantum-mechanical effects due to the spatial carrier confinement induced by band discontinuities at the various semiconductor interfaces. Indeed, as we shall see, it is exactly such quantum-confinement regime which constitutes the basic ingredient for many of the semiconductor quantum devices discussed in this book.

The specific geometry of the nanostructure under investigation, i.e., number and spatial orientation of the various material interfaces, gives rise to a corresponding carrier-confinement potential profile, thus determining what we shall refer to as the "dimensionality" of our nanostructure. More specifically, the latter is defined as the number D of interface-free directions ($0 \leq D \leq 3$), which in turn corresponds to the dimensionality of the periodicity subspace $\boldsymbol{r}_{\parallel}$ introduced in Sect. 1.2.1.

The Envelope-Function Picture

Due to the reduced crystal symmetry, the investigation of the electronic properties of these low-dimensional nanostructures is, in general, much more demanding, compared to conventional (i.e., bulk) band-structure calculations.

However, for low-dimensional structures characterized by a confinement space-scale much larger than the lattice period $a_\circ$, the problem may be greatly simplified by employing the well-known "envelope-function approximation," a general spatial-decoupling scheme introduced originally in the 1960s to study impurity as well as excitonic states in solids (see, e.g., [41]). A general and detailed derivation of the envelope-function approximation is given in Appendix A. In the following, we shall just recall the basic idea and main results of such spatial-decoupling scheme, describing its application to semiconductor nanostructures.

As discussed extensively in Appendix A, the general formulation of the envelope-function theory starts by considering a total electronic Hamiltonian given by the sum of the single-electron Hamiltonian $\hat{H}_{1e}$ in (1.6) plus a non-periodic perturbation potential $\overline{V}(\boldsymbol{r})$. In general, the corresponding energy spectrum is obtained by solving the following eigenvalue equation:

$$\left[\hat{H}_{1e} + \overline{V}(\boldsymbol{r})\right] \psi(\boldsymbol{r}) = \epsilon \psi(\boldsymbol{r}) \ . \tag{1.34}$$

However, due to the lack of symmetry (introduced by the non-periodic potential $\overline{V}$), this task is extremely demanding.

For perturbations $\overline{V}$ which are slowly varying on the atomic scale, the problem in (1.34) may be greatly simplified by adopting the envelope-function approximation previously mentioned. As shown in Appendix A, within this spatial-decoupling scheme the unknown wavefunction ψ in (1.34) – parameterized by the perfect-crystal band index ν and wavevector $\boldsymbol{k}_\circ$ – may be written as

$$\psi_{\boldsymbol{k}_\circ \nu}(\boldsymbol{r}) \approx \sqrt{\Omega}\overline{\psi}_{\boldsymbol{k}_\circ \nu}(\boldsymbol{r}) \phi_{\boldsymbol{k}_\circ \nu}(\boldsymbol{r}) \ , \tag{1.35}$$

where the effective wavefunction $\overline{\psi}$ – called "envelope function" – is obtained by solving the following effective Schrödinger equation:

$$\epsilon_\nu(\boldsymbol{k}_\circ - \mathrm{i}\boldsymbol{\nabla}_{\boldsymbol{r}})\overline{\psi}_{\boldsymbol{k}_\circ \nu}(\boldsymbol{r}) + \overline{V}(\boldsymbol{r})\overline{\psi}_{\boldsymbol{k}_\circ \nu}(\boldsymbol{r}) = \epsilon \overline{\psi}_{\boldsymbol{k}_\circ \nu}(\boldsymbol{r}) \ . \tag{1.36}$$

As discussed extensively in Appendix A, here $\epsilon_\nu(\boldsymbol{k}_\circ - \mathrm{i}\boldsymbol{\nabla}_{\boldsymbol{r}})$ denotes the operatorial version ($\Delta\boldsymbol{k} \rightarrow -\mathrm{i}\boldsymbol{\nabla}_{\boldsymbol{r}}$) of the crystal band structure $\epsilon_{\boldsymbol{k}\nu} \equiv \epsilon_\nu(\boldsymbol{k}_\circ + \Delta\boldsymbol{k})$.

Compared to the original Schrödinger equation (1.34), the effective equation in (1.36) – also known as "envelope-function equation" – involves the potential $\overline{V}$ only. Indeed, the effect of the crystalline periodic potential V in (1.6) is fully expressed via the band-structure operator $\epsilon_\nu(\boldsymbol{k}_\circ - \mathrm{i}\boldsymbol{\nabla}_{\boldsymbol{r}})$. As a result, it is possible to solve the original problem in (1.34) starting directly from the knowledge of the perfect-crystal band structure – or from an approximated version of it (see below) – and solving an extremely simplified Schrödinger-like equation involving the perturbation potential $\overline{V}$ only.

In spite of its generality and formal elegance, the envelope-function equation (1.36) can never be solved directly. Indeed, also for analytical bulk band profiles $\epsilon_\nu(\boldsymbol{k}_\circ + \Delta\boldsymbol{k})$ (like non-parabolic models), the corresponding operatorial function $\epsilon_\nu(\boldsymbol{k}_\circ - \mathrm{i}\boldsymbol{\nabla}_{\boldsymbol{r}})$ involves all powers of the quasimomentum operator

$-\mathrm{i}\boldsymbol{\nabla}_{\boldsymbol{r}}$, giving rise to an infinite-order differential equation. However, as discussed in Sect. 1.2.1, for many of the electro-optical phenomena discussed in this book, the energetic region of interest is typically centered around the semiconductor gap ϵ_{g}, which corresponds to high-symmetry points of the Brillouin zone given by extrema, i.e., minima or maxima, of the electronic band structure. This suggests a crucial simplification of the envelope-function equation (1.36) based again on the second-order Taylor expansion (1.13). Indeed, by taking as reference wavevector $\boldsymbol{k}_\circ$ of our envelope function (see (1.35)) a given high-symmetry point, and by approximating our energy-band operator (around such point) as

$$\epsilon_\nu(\boldsymbol{k}_\circ - \mathrm{i}\boldsymbol{\nabla}_{\boldsymbol{r}}) \approx \epsilon_{\boldsymbol{k}_\circ\nu} - \frac{1}{2}\sum_{ll'} \nabla_l \gamma^\nu_{ll'} \nabla_{l'} \ , \tag{1.37}$$

the original envelope-function equation (1.36) reduces to the following second-order differential equation:

$$-\frac{1}{2}\sum_{ll'} \nabla_l \gamma^\nu_{ll'} \nabla_{l'} \overline{\psi}_{\boldsymbol{k}_\circ\nu}(\boldsymbol{r}) + \overline{V}(\boldsymbol{r})\overline{\psi}_{\boldsymbol{k}_\circ\nu}(\boldsymbol{r}) = (\epsilon - \epsilon_{\boldsymbol{k}_\circ\nu})\overline{\psi}_{\boldsymbol{k}_\circ\nu}(\boldsymbol{r}) \ . \tag{1.38}$$

It is imperative to stress the generality of the above result: for any given high-symmetry point (like conduction-band ellipsoidal minima in silicon), one is able to describe the effect of a generic slowly varying potential $\overline{V}$, no matter if the latter is due to a ionized impurity, a band discontinuity, a two-body Coulomb interaction, or to any other disturbance mechanism.

For the particular case of isotropic band extrema located at the center of the Brillouin zone ($\boldsymbol{k} = \boldsymbol{0}$) – a typical feature of all III–V compound materials used in optoelectronic applications – it is possible to employ the operatorial version of the effective-mass approximation (1.14), i.e.,

$$\epsilon_\nu(-\mathrm{i}\boldsymbol{\nabla}_{\boldsymbol{r}}) \approx \epsilon_{0\nu} - \frac{\hbar^2\nabla^2_{\boldsymbol{r}}}{2m^*_\nu} \ , \tag{1.39}$$

and the general envelope-function equation (1.36) reduces to

$$-\frac{\hbar^2\nabla^2_{\boldsymbol{r}}}{2m^*_\nu}\overline{\psi}_{0\nu}(\boldsymbol{r}) + \overline{V}(\boldsymbol{r})\overline{\psi}_{0\nu}(\boldsymbol{r}) = (\epsilon - \epsilon_{0\nu})\overline{\psi}_{0\nu}(\boldsymbol{r}) \ . \tag{1.40}$$

It is important to underline the extremely simple structure of the above envelope-function equation: the latter describes a particle of mass m^*_ν subjected to the perturbation potential $\overline{V}(\boldsymbol{r})$ only; indeed, the presence of the semiconductor crystal is totally expressed by the value of the effective mass m^*_ν.

Let us now come to the application of the above envelope-function theory to semiconductor nanostructures (see, e.g., [34]), whose general and rigorous derivation is a formidable task. Here, we shall present a simplified treatment of the problem, which is, however, fully adequate to treat all the nanostructure quantum devices discussed in this book.

As anticipated, the key requirement in the fabrication of high-quality nanostructures is the choice of the two (or more) constituent materials (see, e.g., [37–40]), since this will crucially influence the physical properties of the various semiconductor interfaces and thus the electro-optical response of the resulting low-dimensional nanomaterial. To this aim, primary goal is to employ semiconductor materials with the same crystalline structure and very similar lattice constants, and at the same time with significantly different values of their electronic energy gaps. In this way, thanks to properly designed epitaxial-growth protocols (see below), the resulting semiconductor interface – often referred to as heterojunction – is characterized by a very similar and regular crystalline structure (across the interface) and shows an abrupt, i.e., discontinuous, electronic-band profile at the semiconductor interface, also referred to as "band offset."

For such ideal scenario – also referred to as "ideal-heterojunction limit" – the application of the previous envelope-function approximation is straightforward. Indeed, by neglecting small differences in the bulk Bloch states of the constituent materials, the electronic properties of such an ideal interface may be well described by the Schrödinger equation (1.34), where the non-periodic potential $\overline{V}$ is simply given by a step-like potential describing the energy-band offset at the interface. The above approximation scheme may be easily extended to a many-interface nanostructure with arbitrary geometry, like semiconductor quantum wells, wires and dots (see, e.g., [42]). In this case, the effective tree-dimensional potential $\overline{V}(\boldsymbol{r})$ is piece-wise constant (over the different materials).

More specifically, adopting the conventional electron–hole picture and taking as energy reference the corresponding band minimum, the nanostructure envelope function $\overline{\psi}$ (for both electrons and holes) will be described by a simplified version of (1.40):

$$\hat{H}^{\mathrm{env}}\overline{\psi}_{\overline{\alpha}}(\boldsymbol{r}) \equiv \left[-\frac{\hbar^2\nabla^2_{\boldsymbol{r}}}{2m^*} + \overline{V}(\boldsymbol{r})\right]\overline{\psi}_{\overline{\alpha}}(\boldsymbol{r}) = \epsilon_{\overline{\alpha}}\overline{\psi}_{\overline{\alpha}}(\boldsymbol{r}) \,. \tag{1.41}$$

Here, $\overline{\alpha}$ denotes a generic set of quantum numbers, whose nature (discrete versus continuous) will depend on the particular shape and dimensionality D of the nanostructure under examination. As anticipated, this envelope-function equation may describe conduction electrons as well as holes close to their band minima/maxima via the corresponding effective mass m^* as well as their nanostructure potential profile $\overline{V}$; the latter, in view of its piece-wise constant profile, may give rise to the so-called spatial quantum confinement (see, e.g., [34]).

The Concept of Spatial Quantum Confinement

To enlighten this fundamental concept, let us start by considering the simplest example of low-dimensional nanostructure, called semiconductor quantum well [34, 42, 43]. The latter corresponds to a nanometric slice of a small-gap material A (e.g., GaAs) surrounded on both sides by a large-gap material B (e.g., AlAs). Due to the identical band-offset value at the two interfaces, on a macroscopic scale both conduction electrons and holes will experience a square well potential profile, called "quantum well"; therefore, the carrier motion – still free along the interface plane – will be spatially confined within the nanometric slice of material A. Since in this example electrons are characterized by a bidimensional free-motion subspace, such quantum-well nanostructures are also called two-dimensional systems ($D = 2$).

Thanks to the impressive and continuous progress in epitaxial-growth nanotechnology (see, e.g., [37–40]), in addition to quantum wells ($D = 2$), over the last two decades it has been possible to fabricate state-of-the-art nanostructures of lower dimensionality, namely quantum wires ($D = 1$) and quantum dots ($D = 0$) [42, 44–49].

Generally speaking, semiconductor quantum wires are nanostructures where electrons and/or holes are subjected to a bidimensional confinement profile and therefore are allowed to travel freely within a wire-like nanometric object only. Among various experimental realizations of quantum-wire systems, it is worth mentioning the so-called T-shaped wires depicted in Fig. 1.9. The T concept – originally proposed by Leona Esaki and co-workers [52] – has been experimentally realized via the so-called cleaved-edge overgrowth method, a molecular beam epitaxy (MBE) technique that uses high-quality regrowth on the cleaved edge of a multilayer sample (see, e.g., [53]).

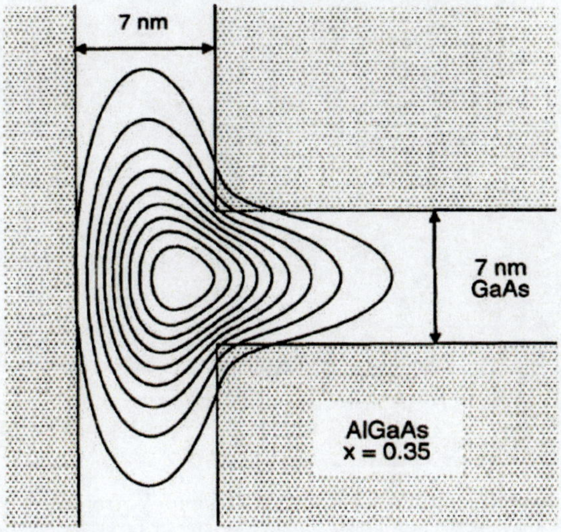

Fig. 1.9. T-shaped intersection of two quantum wells in cross section. The contours are lines of constant charge density corresponding to electrons spatially confined within the quantum wire (see text). Reprinted with permission from [50]

The electronic bound state created by two intersecting quantum wells is depicted in Fig. 1.9; a carrier in such a bound state is free to move along the line defined by the intersecting planes of the two quantum wells. As can be seen from the cross-sectional transmission electron micrograph (TEM) of the T-shaped quantum wire depicted in Fig. 1.10, the high degree of structural perfection of these one-dimensional structures attainable via cleaved-edge overgrowth is manifested by the planarity and abruptness of the interfaces along both growth directions. This gives rise to a relatively strong carrier confinement as well as good optical properties, key prerequisites for optoelectronic applications.

As discussed extensively in Chap. 9, opposite to quantum wells and wires, in semiconductor quantum dots (see Figs. 1.11 and 1.12) – often referred to as

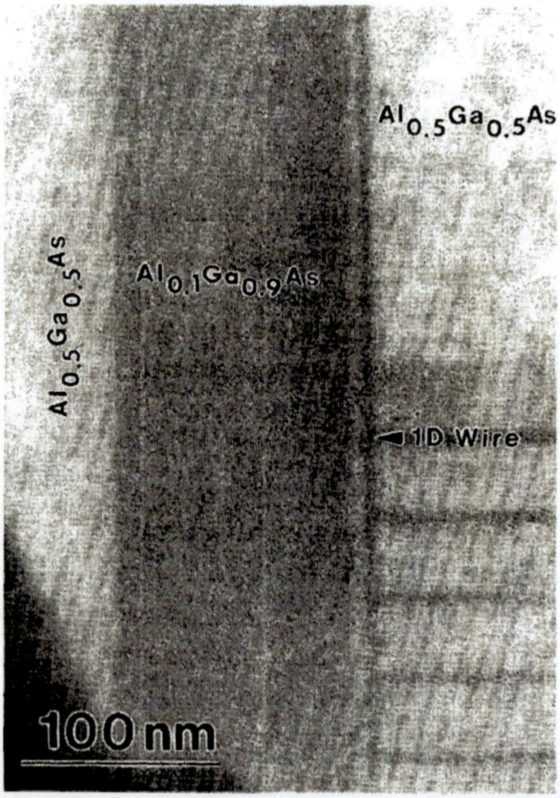

Fig. 1.10. Cross-sectional bright-field transmission electron micrograph (TEM) of a T-shaped quantum-wire structure taken in the [1 1 0] zone axis, i.e., with the electron beam aligned along the quantum wire. *Dark areas* correspond to GaAs or AlAs rich regions. The location of one T-shaped quantum-well intersection is marked by an *arrow* (see text). Reprinted with permission from [50]

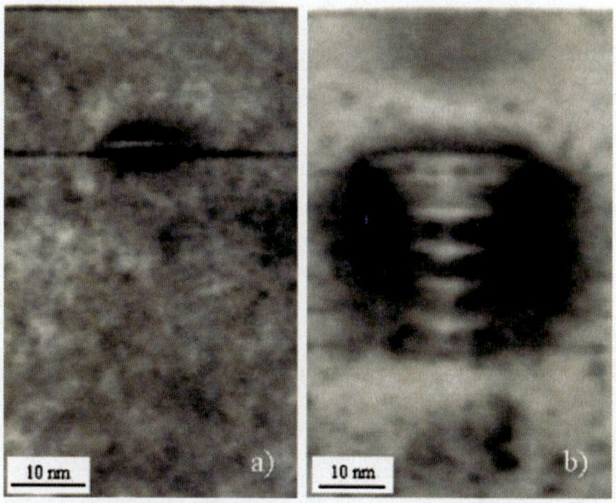

Fig. 1.11. Cross-sectional low-magnification transmission electron micrograph (TEM) images of a prototypical molecular-chemical-vapor-deposition (MOCVD) grown Ga(In)As/GaAs single-dot sample (**a**) and of a non-uniform vertically stacked dot heterostructure (**b**) (see text). Reprinted with permission from [51]

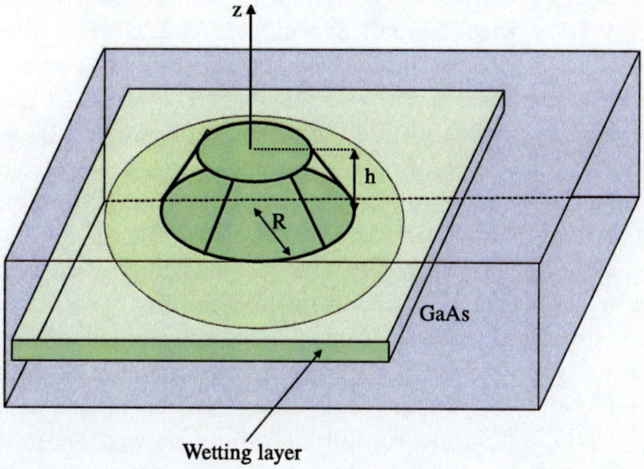

Fig. 1.12. Schematic representation of a single Ga(In)As/GaAs single quantum dot (see text). Reprinted with permission from [51]

semiconductor macroatoms (see, e.g., [47]) – electrons and holes are subjected to a three-dimensional confinement potential, and thus there are absolutely no free-motion directions ($D = 0$); as we shall see, this key feature gives rise to partially discrete single-electron energy spectra $\epsilon_{\overline{\alpha}}$, never present for systems of higher dimensionality ($D > 0$), like quantum wells and wires.

The fabrication of semiconductor-based quantum-dot structures may be performed employing different materials via a number of properly designed experimental protocols. Among the most successful fabrication strategies, self-assembled quantum dots seem to be the ideal choice for the realization of new-generation optoelectronic devices, mainly due to their optical efficiency combined with their relatively high degree of carrier confinement.[3] As described extensively in [46], by suitably controlling the growth parameters (substrate temperature, elemental fluxes of Ga, In, As), islands of InAs or Ga(In)As are formed on the GaAs substrate. The islands are subsequently covered by GaAs (see Fig. 1.11a); then, one may iterate the process and grow a second layer of quantum dots, which may form vertical stacks owing to strain-induced alignment (see Fig. 1.11b).

An interesting property of this fabrication strategy is that one can, in principle, vary the separation between the dot planes at will, from zero to several tens of nanometers, thereby spanning a wide range of interdot electronic coupling between vertically aligned dots, from very strong coupling to no vertical coupling at all. In addition, control of the vertical and lateral position of the quantum dots is pursued by growth on patterned substrates with different types of modulations or by locally modulating the potential by nano-stressors (see, e.g., [51]). Self-assembled (or self-organized) Ga(In)As/GaAs quantum dots are often lens shaped or look like truncated pyramids with well-defined facets. In a first approximation, lens-shaped dots are modeled by truncated cones (see Fig. 1.12) with a small aspect ratio ($\frac{h}{R} \leq 0.3$, where h and R denote, respectively, the quantum-dot height and radius).

It is important to point out the existence of the so-called Ga(In)As wetting layer below the quantum dots (see Figs. 1.11 and 1.12). Such wetting layer is an important source of carrier decoherence: the electrons moving in the wetting layers are relatively close to the dots and have intermediate energy between bound electronic states of the dots and the three-dimensional barrier-like states. The wetting-layer states therefore play an important part in electronically connecting the dot to its surrounding.

Let us now discuss in general terms the solution of the envelope-function equation (1.41). By denoting with $\boldsymbol{r}_\perp$ the quantum-confinement subspace (of dimension $3-D$) and with $\boldsymbol{r}_\parallel$ the remaining free-motion subspace (of dimension D), the envelope-function Hamiltonian may be written as the sum of a perpendicular ($\perp$) plus a parallel ($\parallel$) part:

[3] Such spontaneous formation of Ga(In)As islands on GaAs was discovered as a failed attempt to grow by molecular beam epitaxy (MBE) sufficiently thick InAs quantum wells embedded in a GaAs barrier (see, e.g., [54]). Indeed, the crystal-lattice mismatch between InAs and GaAs is about 7%; as a result, the coherent two-dimensional growth of strained InAs films on a GaAs substrate accumulates elastic energy. The two-dimensional growth proceeds until this energy exceeds the one required to form a dislocation in a relaxed InAs layer. In the InAs/GaAs system this happens when the InAs thickness exceeds about 1.8 monolayer (roughly 0.5 nm).

$$\hat{H}^{\text{env}} = \hat{H}_\perp + \hat{H}_\parallel = \left(-\frac{\hbar^2\nabla_\perp^2}{2m^*} + \overline{V}(\boldsymbol{r}_\perp)\right) + \left(-\frac{\hbar^2\nabla_\parallel^2}{2m^*}\right) . \tag{1.42}$$

The above additivity property allows one to factorize the envelope function $\overline{\psi}_{\overline{\alpha}}$ as

$$\overline{\psi}_{\overline{\alpha}}(\boldsymbol{r}) = \overline{\psi}_n^\perp(\boldsymbol{r}_\perp) \cdot \frac{\mathrm{e}^{i\boldsymbol{k}_\parallel \cdot \boldsymbol{r}_\parallel}}{\sqrt{\Omega_\parallel}} , \tag{1.43}$$

where the new envelope function $\overline{\psi}_n^\perp$ – defined over the confinement subspace only – is obtained by solving the following Schrödinger equation:

$$\hat{H}_\perp \overline{\psi}_n^\perp(\boldsymbol{r}_\perp) = \left[-\frac{\hbar^2\nabla_\perp^2}{2m^*} + \overline{V}(\boldsymbol{r}_\perp)\right] \overline{\psi}_n^\perp(\boldsymbol{r}_\perp) = \epsilon_n^\perp \overline{\psi}_n^\perp(\boldsymbol{r}_\perp) . \tag{1.44}$$

The corresponding energy spectrum is additive as well; the latter is given by the sum of a confinement plus a free-particle spectrum:

$$\epsilon_{\overline{\alpha}} = \epsilon_n^\perp + \epsilon^\parallel(k_\parallel) = \epsilon_n^\perp + \frac{\hbar^2 k_\parallel^2}{2m^*} . \tag{1.45}$$

For conventional semiconductor nanostructures, the confinement potential $\overline{V}(\boldsymbol{r}_\perp)$ in (1.44) is characterized by a minimum, and therefore the lowest part of the confinement energy spectrum $\epsilon^\perp$ is always discrete, i.e., the index n identifies discrete energy levels; as a result, the low-energy part of the electronic spectrum $\epsilon_{\overline{\alpha}}$ in (1.45) exhibits the well-known "subband structure" (see, e.g., [34]): attached to each discrete energy level $\epsilon_n^\perp$ there is a D-dimensional spherical and parabolic band, called "subband." In the absence of any quantum confinement ($D = 3$), the conventional three-dimensional parabolic band of the effective-mass approximation is recovered; in the opposite limit of a three-dimensional confinement potential ($D = 0$) there are no subbands, and the lowest part of the spectrum in (1.45) is indeed discrete (quantum dots) (see, e.g., [47]). For both intermediate cases – quantum wells ($D = 2$) and wires ($D = 1$) – the global single-electron spectrum $\epsilon_{\overline{\alpha}}$ is always continuous.

Based on the analysis presented so far, the single-electron spectrum of a semiconductor nanostructure is obtained by solving the envelope-function equation (1.44); the latter is defined over the $(3-D)$-dimensional confinement subspace $\boldsymbol{r}_\perp$ only, and its solution is not necessarily straightforward. Indeed, while for $D = 2$ the corresponding one-dimensional equation may be solved analytically employing standard methods of elementary quantum mechanics (i.e., by imposing the continuity of the envelope function $\overline{\psi}$ and of its derivative at the various interfaces), for $D < 2$ (quantum wires and dots) the problem may not be treated analytically, and a numerical solution based on a multi-dimensional plane-wave expansion is typically performed (see Sect. 5.1.2).

Let us now discuss the DOS of the generic subband structure in (1.45). In this case, the corresponding function $N(\epsilon)$ in (1.11) – defined as the total number of electronic states with energy smaller than ϵ – may be written as

$$N(\epsilon) = \sum_n \theta\left(\epsilon - \epsilon_n^\perp\right) N_D^\parallel\left(\epsilon - \epsilon_n^\perp\right) . \tag{1.46}$$

Indeed, the latter is obtained by summing over all relevant subbands n, i.e., the only subbands n involved are those whose minimum $\epsilon_n^\perp$ is smaller than ϵ (this is accomplished via the θ step function). The individual contribution for each subband is given by the function

$$N_D^\parallel(\epsilon^\parallel) = \sum_{\boldsymbol{k}'_\parallel} \theta\left(\epsilon^\parallel - \frac{\hbar^2 {k'_\parallel}^2}{2m^*}\right) = \frac{\Omega}{(2\pi)^D} V_D(k_\parallel) , \tag{1.47}$$

corresponding to the total number of states of a D-dimensional parabolic subband with energy smaller than the parallel energy $\epsilon^\parallel = \epsilon - \epsilon_n^\perp$. Here, $V_D(k_\parallel)$ is simply the volume of a D-dimensional sphere of radius $k_\parallel = \frac{\sqrt{2m^*\epsilon^\parallel}}{\hbar}$, and therefore we have

$$V_D \propto \left(k_\parallel\right)^D \propto \left(\epsilon^\parallel\right)^{\frac{D}{2}} . \tag{1.48}$$

Combining (1.46), (1.47), and (1.48) we obtain

$$N(\epsilon) \propto \sum_n \theta\left(\epsilon - \epsilon_n^\perp\right)\left(\epsilon - \epsilon_n^\perp\right)^{\frac{D}{2}} . \tag{1.49}$$

Inserting this last result into (1.11) and employing again the property of the step function in (1.10), we finally get

$$g(\epsilon) \propto \sum_n \left[\delta\left(\epsilon - \epsilon_n^\perp\right)\left(\epsilon - \epsilon_n^\perp\right)^{\frac{D}{2}} + \frac{D}{2}\theta\left(\epsilon - \epsilon_n^\perp\right)\left(\epsilon - \epsilon_n^\perp\right)^{\frac{D}{2}-1}\right] . \tag{1.50}$$

The above DOS is the sum of two terms, but they never contribute simultaneously; indeed, the first one is different from zero for $D = 0$ only (quantum dots), while the second one is different from zero only for $D > 0$. In particular, in the absence of confinement ($D = 3$) we deal with a three-dimensional parabolic band, and the corresponding DOS is simply proportional to $\sqrt{\epsilon^\parallel}$. For $D = 2$ (quantum wells) the DOS exhibits a step-like profile. Indeed, since in this case the exponent $\frac{D}{2} - 1$ is equal to zero, for each new subband n involved, we have a corresponding step of constant height. Contrary to the two previous cases, for $D = 1$ (quantum wires), the DOS in (1.50) displays a diverging behavior like $\frac{1}{\sqrt{\epsilon^\parallel}}$ at each subband edge $\epsilon_n^\perp$; as we shall see, such divergences – known as Van Hove singularities – have been the main technological motivation for the realization of quantum-wire optoelectronic devices.

As anticipated, the remaining $D = 0$ case – corresponding to a fully three-dimensional quantum confinement – is radically different from all the high-dimensional cases ($D > 0$) discussed so far. Indeed, the corresponding DOS, given now by the first term in (1.50), describes a partially discrete single-electron spectrum with energy levels $\epsilon_n^\perp$ (mathematically described via the conventional Dirac delta function). As discussed extensively in Chaps. 9 and 10, such discrete-spectrum feature – corresponding to a significant suppression of some scattering mechanisms – has motivated most of the research activity on semiconductor quantum dots.

To exemplify and better illustrate the nanostructure classification presented so far, in Figs. 1.13 and 1.14 we show the DOS corresponding to an ideal family of GaAs/AlAs nanostructures (bulk (3D), quantum well (2D), quantum wire (1D), and quantum dot (0D)) for electrons and holes, respectively; here, a relatively small spatial confinement length $l = 15\,\mathrm{nm}$ in all directions – corresponding to the GaAs width – has been considered (corresponding to the strong-confinement regime discussed below), which allows us to better identify all the dimensionality-dependent features previously intro-

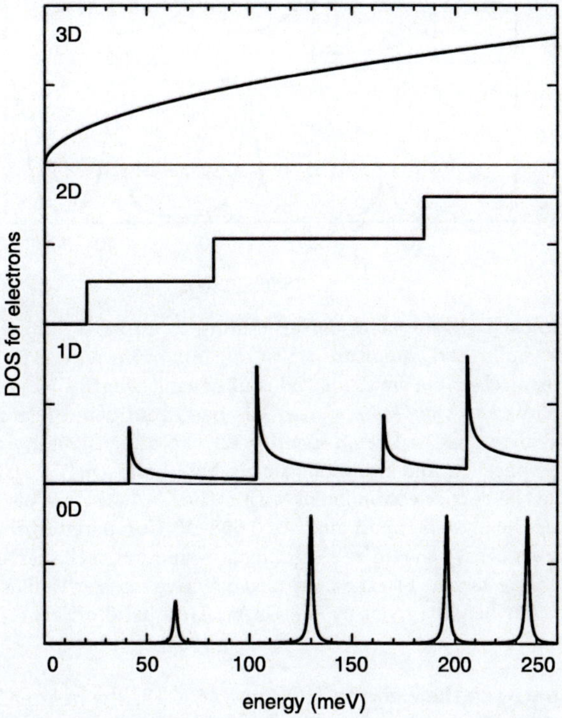

Fig. 1.13. Typical DOS for electrons corresponding to an ideal family of GaAs/AlAs nanostructures (bulk (3D), quantum well (2D), quantum wire (1D), and quantum dot (0D)). Here, a relatively small spatial confinement length $l = 15\,\mathrm{nm}$ in all directions (corresponding to the GaAs width) has been considered (strong-confinement regime), which allows us to better identify all the dimensionality-dependent features. A single spherical- and parabolic-band model ($m_e^* = 0.067\,m$) has been employed, while the carrier confinement within the GaAs region has been described via a GaAs/AlAs electron band offset of 1.04 eV

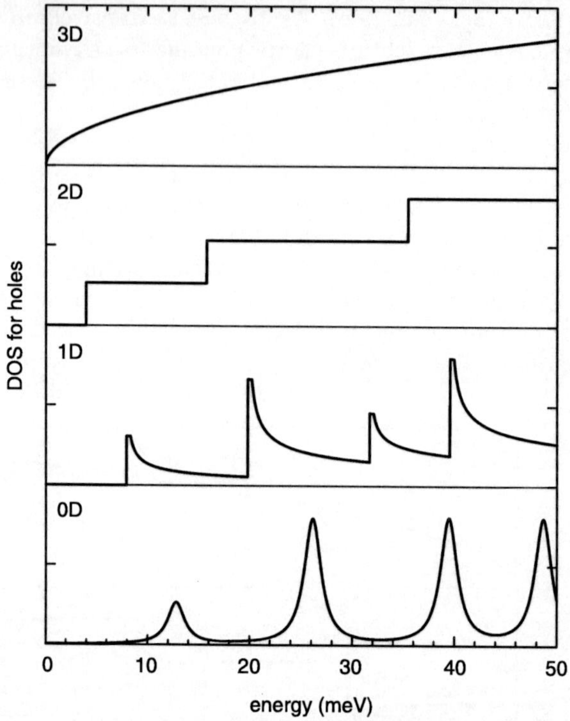

Fig. 1.14. Typical DOS for holes corresponding to an ideal family of GaAs/AlAs nanostructures (bulk (3D), quantum well (2D), quantum wire (1D), and quantum dot (0D)). Here, a relatively small spatial confinement length $l = 15\,\text{nm}$ in all directions (corresponding to the GaAs width) has been considered (strong-confinement regime), which allows us to better identify all the dimensionality-dependent features. A single spherical and parabolic heavy-hole band ($m_h^* = 0.38\,m$) has been employed, while the carrier confinement within the GaAs region has been described via a GaAs/AlAs heavy-hole band offset of 0.558 eV. Compared to the electron DOS in Fig. 1.13, here it is easy to realize a significant reduction of the carrier-confinement energy scale, mainly ascribed to the increased effective mass as well as to the reduced value of the barrier height (given by the GaAs/AlAs band offset)

duced.[4] Compared to the electron DOS in Fig. 1.13, the hole DOS in Fig. 1.14 shows a significant reduction of the carrier-confinement energy scale, mainly ascribed to the increased effective mass as well as to the reduced value of the barrier height (given by the GaAs/AlAs band offset).

[4] To better illustrate the quantum-dot (0D) discrete energy spectrum together with the degeneracy of the individual peaks (due to the high-symmetry character of our nanostructures induced by the same GaAs width l in all confinement directions), the latter have been depicted via a Lorentzian linewidth of 1 meV.

As already pointed out, only for quantum dots ($D = 0$) we deal with discrete energy levels corresponding to fully three-dimensional confined states; however, for energies higher than the confinement potential (not shown here), such quantum-dot (or macroatom) spectra become always continuous, exactly as for the case of real atoms.

The scenario reported in Figs. 1.13 and 1.14 confirms that the dimensionality D of the nanostructure under examination may be unambiguously determined from the corresponding DOS. However, to this end, the quantum confinement should play a role energetically comparable to the free-motion contribution (see (1.45)). To enlighten this crucial point, in Fig. 1.15 we show the DOS for electrons corresponding again to GaAs/AlAs nanostructures of different dimensionality ($3 \geq D \geq 1$) but now characterized by a much larger confinement length ($l = 100\,\mathrm{nm}$). As we can see, all quantum-confinement effects are strongly reduced, and in the low-energy region the overall shape of the various DOS is almost the same for all values of D. Indeed, in the limit $l \to \infty$, the partially-discrete confinement energies $\epsilon_n^{\perp}$ "collapse" into a

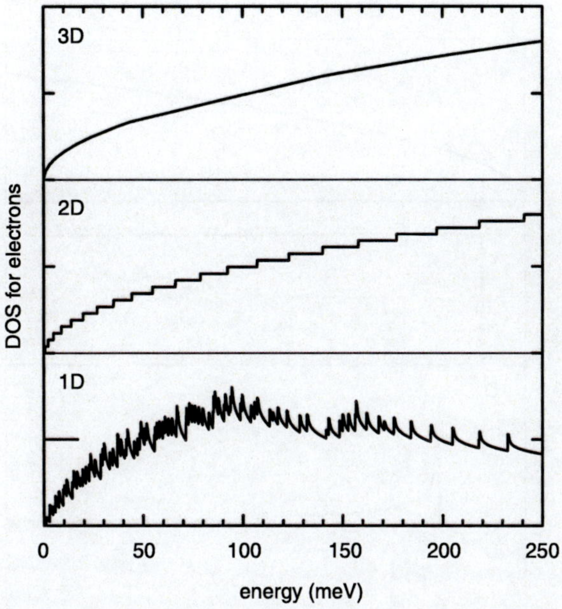

Fig. 1.15. Typical DOS for electrons corresponding to an ideal family of GaAs/AlAs nanostructures (bulk (3D), quantum well (2D), quantum wire (1D), and quantum dot (0D)) in the weak-confinement regime. Opposite to the results of Fig. 1.13, here a much larger spatial-confinement length $l = 100\,\mathrm{nm}$ in all directions (corresponding to the GaAs width) has been considered. Again, a single spherical- and parabolic-band model ($m_e^* = 0.067\,m$) has been employed and the carrier confinement within the GaAs region has been described via a GaAs/AlAs electron band offset of 1.04 eV

continuous spectrum, and for any value of D the corresponding DOS tends to the bulk one.

At this point an important comment is in order. Looking into the vast literature on the theoretical modeling of semiconductor nanostructures, it is frequent to find simplified treatments based on a low-dimensional description of the system in real space, i.e., by simply neglecting all real-space confinement coordinates $\boldsymbol{r}_\perp$. Such a highly artificial picture corresponds to the simultaneous limits of an infinite-height confinement barrier and of a vanishing confinement length ($l \to 0$); for such extreme and non-physical conditions, the confinement energies $\epsilon_n^\perp$ are much larger than the corresponding free-motion contributions, and therefore the many-subband structure in (1.45) may be safely neglected by considering the lowest subband only. By applying such approximation – often referred to as single-subband model – to the realistic (many-subband) scenario of Fig. 1.13, one gets the artificial DOS profiles reported in Fig. 1.16. Comparing Figs. 1.13 and 1.16, the first obvious consideration is that such single-subband description may be employed only for the investigation of phenomena taking place close to the first-subband minimum.

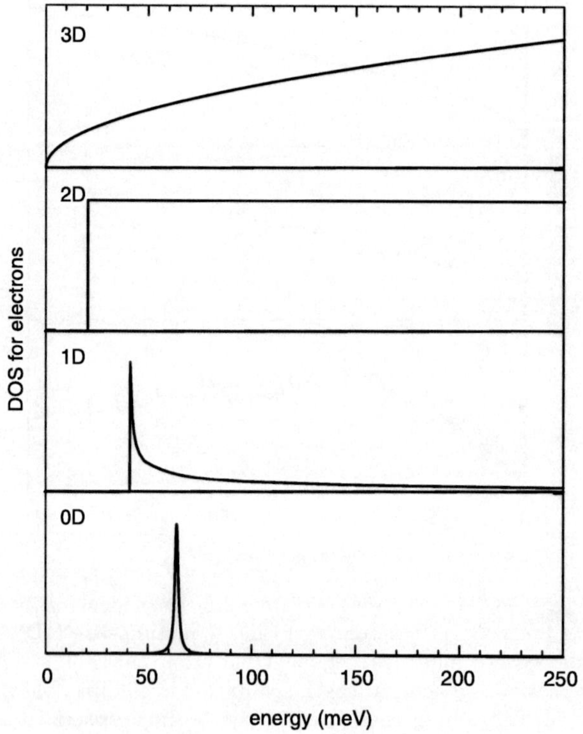

Fig. 1.16. Same as in Fig. 1.13 but in the single-subband approximation (see text)

Moreover, such single-subband picture does not allow to properly describe the nanostructure versus bulk transition reported in Fig. 1.15.

Generally speaking, a physical system "living" in a real space with less than three dimensions is intrinsically meaningless, i.e., it may describe only a few limiting situations, but such a simplified description is far from that of a real semiconductor nanostructure. Indeed, it is imperative to stress that the latter is always a three-dimensional system, and its dimensionality D refers to the free-motion reciprocal subspace only. A peculiar situation showing the intrinsic inadequacy of real-space low-dimensional models is the description of semiconductor quantum wires via a single electronic coordinate z only. Indeed, as discussed below, in the low-density limit, the previous single-particle DOS scenario is significantly modified by electron–hole Coulomb correlation (see, e.g., [55]). However, within such a purely one-dimensional description all Coulomb contributions are known to diverge. In contrast, by employing a fully three-dimensional real-space description, all Coulomb contributions are well defined and in good agreement with experiments.

Let us finally discuss the case of periodic nanostructures, i.e., nanostructures obtained as artificial periodic repetition of a given elementary unit. The simplest example is given by a periodic repetition of the single quantum-well structure previously considered. Indeed, the resulting nanostructure – known as "semiconductor superlattice" (see, e.g., [34]) – is obtained choosing, as elementary unit, a single interface between two slices of materials A and B with total length $d = l_A + l_B$. The resulting one-dimensional confinement potential $\overline{V}$ is therefore periodic along the so-called growth direction (i.e., the direction normal to the interface planes); the latter will act on the electronic motion as a sort of macroscopic lattice (of period $d = l_A + l_B$), called "superlattice."

For any periodic nanostructure – i.e., a nanostructure characterized by a periodic confinement potential $\overline{V}$ – the corresponding envelope-function equation (1.44) is formally identical to the single-electron Schrödinger equation (1.6). It follows that the corresponding envelope functions will be Bloch-like, i.e.,

$$\overline{\psi}_n^{\perp}(\boldsymbol{r}_\perp) \equiv \overline{\psi}_{\boldsymbol{k}_\perp \overline{\nu}}^{\perp}(\boldsymbol{r}_\perp) = \mathrm{e}^{\mathrm{i}\boldsymbol{k}_\perp \cdot \boldsymbol{r}_\perp} \overline{u}_{\boldsymbol{k}_\perp \overline{\nu}}^{\perp}(\boldsymbol{r}_\perp) \ , \tag{1.51}$$

and the corresponding confinement spectrum will display again a band-structure character: $\epsilon_n^{\perp} \equiv \epsilon_{\boldsymbol{k}_\perp \overline{\nu}}$.

At this point, it is imperative to stress the different origin/nature of the nanostructure confinement potential $\overline{V}$ compared to the atomic potential V in (1.6). Indeed, within the present envelope-function picture, the latter is expressed via the effective mass m^* only, and the confinement potential $\overline{V}$ is characterized by an artificial periodicity space-scale – also called "superperiodicity" – much larger than the natural crystalline period $a_\circ$. Due to such space-scale difference, the reciprocal-space Brillouin zone corresponding to this superperiodicity is much smaller than the conventional one; for this reason, the latter is usually referred to as "minizone," and the corresponding band structure $\epsilon_n^{\perp} \equiv \epsilon_{\boldsymbol{k}_\perp \overline{\nu}}$ as "miniband structure" (see, e.g., [34]).

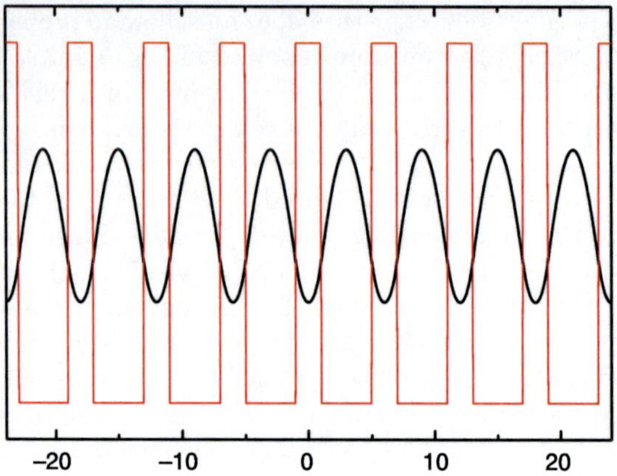

Fig. 1.17. Electronic confinement-potential profile and charge distribution for a GaAs/AlGaAs superlattice with a band offset of 0.28 eV and well (GaAs) and barrier (AlGaAs) widths of 4 and 2 nm, respectively. Here, the spatial charge distribution corresponds to the squared moduli of the carrier envelope wavefunction evaluated at the minizone center for the miniband of Fig. 1.18. As a result of the quantum confinement, the charge distribution displays an electron localization within the superlattice GaAs wells accompanied by a significant interwell tunneling across the thin AlGaAs barriers

Figure 1.17 shows the electronic confinement-potential profile together with the corresponding charge distribution for a typical GaAs/AlGaAs superlattice characterized by well (GaAs) and barrier (AlGaAs) widths of 4 and 2 nm, respectively, while Fig. 1.18 displays the corresponding electronic miniband. As anticipated, in Fig. 1.18 we notice the reduced size of the minizone and of the electronic-band width, compared to the GaAs bulk band structure of Fig. 1.6. Moreover, as a result of the quantum confinement, the spatial charge distribution in Fig. 1.17 – proportional to the squared moduli of the carrier envelope wavefunctions evaluated at the minizone center (see Fig. 1.18) – displays an electron localization within the superlattice GaAs wells accompanied by a significant interwell tunneling across the thin AlGaAs barriers.

The analysis presented so far has clearly shown that the presence of a strong quantum confinement leads to significant modifications of the DOS (see Figs. 1.13 and 1.14), which in turn may improve the global electro-optical response of our nanomaterial; this is particularly true for the case of semiconductor quantum dots. Moreover, a strong reduction of the DOS (i.e., of the final scattering states available) may translate in extremely long coherence life-times (see (1.33)), a crucial prerequisite for the realization of genuine quantum devices.

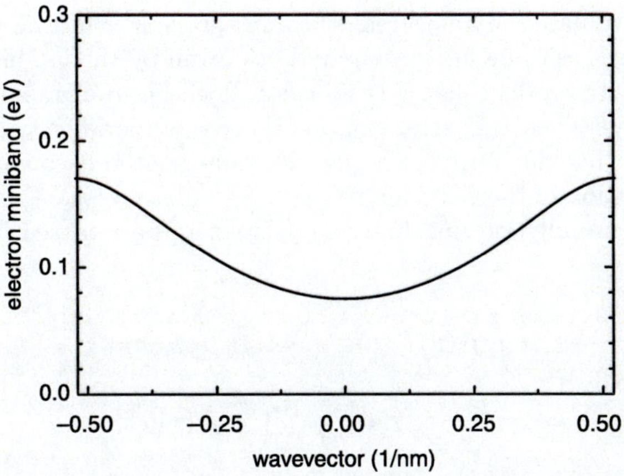

Fig. 1.18. Electronic miniband corresponding to the GaAs/AlGaAs superlattice structure of Fig. 1.17. Here, we notice the reduced size of the minizone (*horizontal axis*) and of the electronic-band width (*vertical axis*), compared to the GaAs bulk band structure of Fig. 1.6

Free-Carrier Versus Excitonic Absorption

Let us now come to the optical properties of semiconductor nanostructures (see, e.g., [34]). To this aim, we shall start from the perturbation approach introduced in Sect. 1.2.1 for the case of bulk semiconductors; according to (1.26), in the presence of a slowly varying optical excitation, the absorption probability amplitudes are simply proportional to the matrix elements of the quasimomentum operator between field-free electronic states, given in the bulk case by conventional Bloch functions. Moving from bulk to nanostructures, this general property is still valid, provided to replace in (1.26) the Bloch states ϕ with the non-periodic states ψ in (1.35) corresponding to our nanostructure problem in $\boldsymbol{k} = \mathbf{0}$, i.e.,

$$\psi_{\nu\overline{\alpha}}(\boldsymbol{r}) = \sqrt{\Omega}\overline{\psi}_{\nu\overline{\alpha}}(\boldsymbol{r})\phi_\nu(\boldsymbol{r}) \ , \tag{1.52}$$

where, to simplify the notation, the label $\boldsymbol{k} = \mathbf{0}$ has been omitted. More specifically, we have

$$\begin{aligned} H'_{\nu'\overline{\alpha}',\nu\overline{\alpha}} &\propto \int \mathrm{d}^3r \psi^*_{\nu'\overline{\alpha}'}(\boldsymbol{r}) \left[-\mathrm{i}\hbar\nabla_{\boldsymbol{A}}\psi_{\nu\overline{\alpha}}(\boldsymbol{r})\right] \\ &= \Omega \int \mathrm{d}^3r \overline{\psi}^*_{\nu'\overline{\alpha}'}(\boldsymbol{r})\phi^*_{\nu'}(\boldsymbol{r}) \left[-\mathrm{i}\hbar\nabla_{\boldsymbol{A}} \left(\overline{\psi}_{\nu\overline{\alpha}}(\boldsymbol{r})\phi_\nu(\boldsymbol{r})\right)\right] \\ &= \Omega \int \mathrm{d}^3r \left\{\overline{\psi}^*_{\nu'\overline{\alpha}'}(\boldsymbol{r}) \left[-\mathrm{i}\hbar\nabla_{\boldsymbol{A}}\overline{\psi}_{\nu\overline{\alpha}}(\boldsymbol{r})\right]\right\} \left\{\phi^*_{\nu'}(\boldsymbol{r})\phi_\nu(\boldsymbol{r})\right\} \\ &+ \Omega \int \mathrm{d}^3r \left\{\overline{\psi}^*_{\nu'\overline{\alpha}'}(\boldsymbol{r})\overline{\psi}_{\nu\overline{\alpha}}(\boldsymbol{r})\right\} \left\{\phi^*_{\nu'}(\boldsymbol{r}) \left[-\mathrm{i}\hbar\nabla_{\boldsymbol{A}}\phi_\nu(\boldsymbol{r})\right]\right\} \ . \end{aligned} \tag{1.53}$$

In order to evaluate the above real-space integrals, it is crucial to notice that the two quantities to be integrated are both given by the product of an envelope (i.e., slowly varying) part $f_1(\boldsymbol{r})$ times a Bloch-like (i.e., rapidly oscillating) part $f_2(\boldsymbol{r})$; moreover, the latter exhibits the crystal periodicity of the original Bloch states. Invoking once again the very same spatial-decoupling limit used for the derivation of the envelope-function formulation (see Appendix A), the periodic and rapidly varying function $f_2(\boldsymbol{r})$ may be replaced by its average value $\overline{f}_2$, i.e.,

$$\Omega \int f_1(\boldsymbol{r}) f_2(\boldsymbol{r}) \mathrm{d}^3 r = \Omega \overline{f}_2 \int f_1(\boldsymbol{r}) \mathrm{d}^3 r$$
$$= \left(\int f_1(\boldsymbol{r}) \mathrm{d}^3 r \right) \cdot \left(\int f_2(\boldsymbol{r}) \mathrm{d}^3 r \right) . \tag{1.54}$$

By applying this spatial-decoupling scheme to (1.53), we finally get

$$H'_{\nu'\overline{\alpha}',\nu\overline{\alpha}} \propto p^{\mathrm{env}}_{\nu,\overline{\alpha}'\overline{\alpha}} \delta_{\nu'\nu} + f^{\mathrm{env}}_{\nu'\overline{\alpha}',\nu\overline{\alpha}} p^{\mathrm{bulk}}_{\nu'\nu} , \tag{1.55}$$

where

$$p^{\mathrm{env}}_{\nu,\overline{\alpha}'\overline{\alpha}} = \int \overline{\psi}^*_{\nu\overline{\alpha}'}(\boldsymbol{r}) \left[-\mathrm{i}\hbar \nabla_{\boldsymbol{A}} \overline{\psi}_{\nu\overline{\alpha}}(\boldsymbol{r}) \right] \mathrm{d}^3 r \tag{1.56}$$

are dipole matrix elements between our envelope functions,

$$f^{\mathrm{env}}_{\nu'\overline{\alpha}',\nu\overline{\alpha}} = \int \overline{\psi}^*_{\nu'\overline{\alpha}'}(\boldsymbol{r}) \overline{\psi}_{\nu\overline{\alpha}}(\boldsymbol{r}) \mathrm{d}^3 r \tag{1.57}$$

are corresponding overlap integrals, and $p^{\mathrm{bulk}}_{\nu'\nu} \equiv p_{\boldsymbol{0},\nu'\nu}$ denote the bulk dipole matrix elements (1.27) evaluated in $\boldsymbol{k} = \boldsymbol{0}$. As we can see, we deal with two different types of optical processes: the first term in (1.55) describes processes within the same band ν, called intraband transitions, while the second term describes processes connecting different bands ($\nu \neq \nu'$), called interband transitions. For the peculiar case of III–V semiconductor compounds, the only relevant bands are the valence and conduction ones. Since – in addition to the vertical character of these transitions – the latter are also dictated/restricted by energy conservation, the first term in (1.55) describes absorption/emission processes in the infrared spectral region between different subbands ($\overline{\alpha} \rightarrow \overline{\alpha'}$) of the same band ν, and for this reason they are also referred to as intersubband transitions. In contrast, due to the symmetry properties of the Bloch states, the intraband bulk dipole matrix elements $p^{\mathrm{bulk}}_{\nu\nu}$ are always equal to zero (see (1.27)); it follows that for the second term in (1.55) the only contributions correspond to absorption/emission processes around the visible spectral region, connecting valence- to conduction-band states via interband transitions.

As for the case of bulk materials, in addition to the microscopic probability amplitudes (1.55), the key ingredient for the evaluation of intraband as well as interband absorption spectra is the subband structure in (1.45).

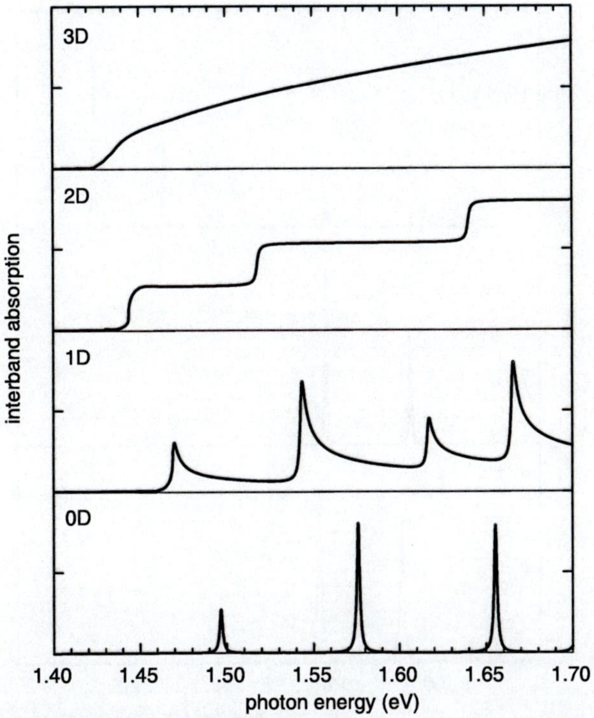

Fig. 1.19. Interband absorption spectra corresponding to the low-dimensional electron and hole DOS reported in Figs. 1.13 and 1.14. Similar to the bulk case (see (1.29)), the absorption spectrum is again a fingerprint of the interband J-DOS; in the presence of electron–hole Coulomb correlation; however, this feature is not valid anymore (see Fig. 1.21)

Figure 1.19 shows the interband absorption spectra corresponding to the low-dimensional electron and hole DOS of Figs. 1.13 and 1.14. Similar to the bulk case (see (1.29)), the interband absorption spectrum is indeed a fingerprint of the interband J-DOS. As we shall see, in the presence of electron–hole Coulomb correlation this feature is not valid anymore (see, e.g., [55]). Finally, Fig. 1.20 shows the conduction intraband (or intersubband) absorption spectra corresponding again to the electron DOS of Fig. 1.13 for all three relevant cases: quantum well (2D), quantum wire (1D), and quantum dot (0D).[5] Since all vertical transitions connecting two given subbands $\overline{\alpha}$ and $\overline{\alpha'}$ correspond to the same photon energy, contrary to the interband case in Fig. 1.19, the

[5] To better illustrate the main features of interband versus intraband absorption, in both Figs. 1.19 and 1.20 a Lorentzian energy broadening of 1 meV has been considered; as discussed in Chap. 2, the latter is a fingerprint of energy-dissipation and decoherence processes induced by the host material (see also Fig. 2.5).

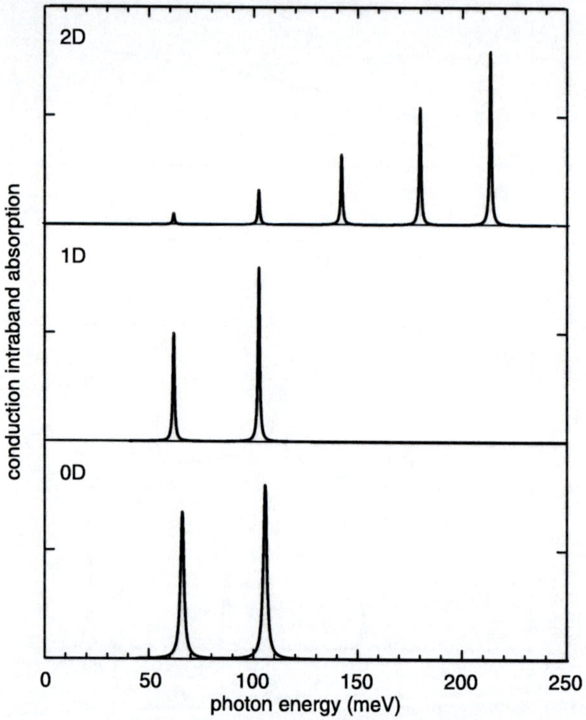

Fig. 1.20. Conduction intraband (or intersubband) absorption spectra corresponding to the low-dimensional electron DOS of Fig. 1.13 for all three relevant cases: quantum well (2D), quantum wire (1D), and quantum dot (0D). Since all vertical transitions connecting two given subbands $\overline{\alpha}$ and $\overline{\alpha'}$ correspond to the same photon energy, contrary to the interband case in Fig. 1.19, the intraband spectrum exhibits a discrete structure also for the case of quantum wells and wires

intraband spectrum exhibits a discrete structure also for the case of quantum wells and wires.

The above spectral classification, in terms of intra- versus interband processes, corresponds to two different families of quantum optoelectronic devices (see, e.g., [56]): on the one hand, we shall investigate intraband (or intersubband) optoelectronic devices, like infrared quantum-cascade lasers and photodetectors, whose principle of operation involves conduction-band electrons only [57–66]; on the other hand, we shall discuss a number of interband quantum devices operating in the visible spectral region; the latter include conventional interband lasers and detectors (see, e.g., [67, 68]) as well as new-generation quantum nanodevices based on the generation/recombination of Coulomb-correlated electron–hole pairs (see, e.g., [51, 69]).

As anticipated, the optical-response analysis presented so far does not include Coulomb-correlation effects. In particular, in the low-density limit, i.e., in the limit of a low-spatial density of electron–hole pairs, the only relevant interaction mechanism is the attractive Coulombic interaction between electrons and holes; indeed, in this regime electrons and holes attract each other, forming bound two-particle complexes called "excitons" (see, e.g., [33]), characterized by a partially discrete energy spectrum. The qualitative idea is that any electron, promoted from the valence to the conduction band via an absorption process, is attracted by the corresponding hole left in the valence band; this tells us that, in order to incorporate this effect in our envelope-function picture, one needs a reformulation of the treatment in Appendix A (not discussed in this book) via a two-particle description. More specifically, for all semiconductor materials characterized by bounded electron–hole pairs with average distance much larger than the crystal space-scale – called Mott–Wannier excitons (see, e.g., [22]) – it is possible to write the global electron–hole wavefunction as

$$\Psi_\xi(\boldsymbol{r}_\mathrm{e}, \boldsymbol{r}_\mathrm{h}) \approx \Omega \overline{\Psi}_\xi(\boldsymbol{r}_\mathrm{e}, \boldsymbol{r}_\mathrm{h}) \phi_\mathrm{e}(\boldsymbol{r}_\mathrm{e}) \phi^*_\mathrm{h}(\boldsymbol{r}_\mathrm{h}) \ , \tag{1.58}$$

where the excitonic envelope function $\overline{\Psi}$ is the solution of the following two-body Schrödinger equation:

$$\left[\hat{H}^\mathrm{env}_\mathrm{e} + \hat{H}^\mathrm{env}_\mathrm{h} - \frac{e^2}{\varepsilon |\boldsymbol{r}_\mathrm{e} - \boldsymbol{r}_\mathrm{h}|}\right] \overline{\Psi}_\xi(\boldsymbol{r}_\mathrm{e}, \boldsymbol{r}_\mathrm{h}) = \mathcal{E}_\xi \overline{\Psi}_\xi(\boldsymbol{r}_\mathrm{e}, \boldsymbol{r}_\mathrm{h}) \ . \tag{1.59}$$

Here, ε is the static dielectric constant and $\hat{H}^\mathrm{env}_\mathrm{e/h}$ denotes the single-particle envelope Hamiltonian in (1.41), describing our electron/hole within the corresponding confinement potential.

To better illustrate the role and properties of such excitonic envelope function $\overline{\Psi}$, let us start by considering the limiting case of a vanishing Coulomb attraction between electrons and holes; in this limit, the problem is totally factorized and the excitonic envelope function is the product of the Coulomb-free envelope functions for electrons and holes. In contrast, in the presence of Coulomb correlation, the problem is not factorized anymore; however, it is always possible to expand our excitonic wavefunction as a linear combination of products of Coulomb-free states:

$$\overline{\Psi}_\xi(\boldsymbol{r}_\mathrm{e}, \boldsymbol{r}_\mathrm{h}) = \sum_{\overline{\alpha'}\overline{\alpha}} U^\xi_{\overline{\alpha'}\overline{\alpha}} \overline{\psi}_{\mathrm{e}\overline{\alpha'}}(\boldsymbol{r}_\mathrm{e}) \overline{\psi}^*_{\mathrm{h}\overline{\alpha}}(\boldsymbol{r}_\mathrm{h}) \ . \tag{1.60}$$

The coefficients $U^\xi_{\overline{\alpha'}\overline{\alpha}}$ can be regarded as the matrix elements of a unitary transformation $\hat{U}$ connecting the Coulomb-free two-body (i.e., electron–hole) states $\overline{\psi}_{\mathrm{e}\overline{\alpha'}}\overline{\psi}^*_{\mathrm{h}\overline{\alpha}}$ to the Coulomb-correlated (i.e., excitonic) states $\overline{\Psi}_\xi$. As we shall see in Chap. 9, the above expansion in terms of Coulomb-free states constitutes the starting point for a numerical solution of the two-body Schrödinger equation (1.59); indeed, by substituting the linear combination (1.60) into (1.59), one gets an eigenvalue problem, with eigenvectors $U^\xi_{\overline{\alpha'}\overline{\alpha}}$ and eigenvalues $\mathcal{E}_\xi$ (see Sect. 9.2).

In the presence of electron–hole correlation, the single-particle result in (1.55) is not valid anymore: while the latter has been obtained by evaluating the matrix elements of the light-matter Hamiltonian $\hat{H}'$ between Coulomb-free states, in order to fully incorporate excitonic effects one needs to consider absorption processes connecting states corresponding to the absence and presence of one correlated electron–hole pair. The so-called excitonic absorption is then obtained by applying to the interband part ($\nu = h, \nu' = e$) of (1.55) the exciton versus free-carrier transformation (1.60). More specifically, we have

$$H'_\xi = \sum_{\overline{\alpha'\alpha}} U^{\xi *}_{\overline{\alpha'\alpha}} H'_{\mathrm{e}\overline{\alpha'},\mathrm{h}\overline{\alpha}} \propto p^{\mathrm{bulk}}_{\mathrm{eh}} \sum_{\overline{\alpha'\alpha}} U^{\xi *}_{\overline{\alpha'\alpha}} f^{\mathrm{env}}_{\mathrm{e}\overline{\alpha'},\mathrm{h}\overline{\alpha}} \, . \tag{1.61}$$

Combining (1.57) and (1.60), we finally get

$$H'_\xi \propto p^{\mathrm{bulk}}_{\mathrm{eh}} \left(\int \overline{\Psi}_\xi(\boldsymbol{r}, \boldsymbol{r}) \mathrm{d}^3 \boldsymbol{r} \right)^* \, . \tag{1.62}$$

As we can see, in order to incorporate excitonic effects in the absorption spectrum, the first step is to replace the single-particle overlap integral (1.57) with the above integral of the excitonic wavefunction $\overline{\Psi}$ evaluated in $\boldsymbol{r} = \boldsymbol{r}_\mathrm{e} = \boldsymbol{r}_\mathrm{h}$. The second step is to replace the joint (electron–hole) subband structure with the excitonic spectrum $\mathcal{E}_\xi$ in (1.59).

Figure 1.21 shows a comparison between free-carrier and excitonic absorption spectra as a function of the so-called excess energy[6] corresponding to the low-dimensional electron and hole DOS reported in Figs. 1.13 and 1.14 in the single-subband approximation (see Fig. 1.16). As we can see, electron–hole Coulomb correlation (solid curves) induces dramatic modifications to the free-carrier spectra (dashed curves). Already for the bulk case, while the free-carrier result goes to zero at the band edge, the excitonic absorption exhibits a discrete spectrum within the gap and, more important, the absorption in the continuum does not go to zero at the band edge. This effect is known as "Coulomb enhancement" (see, e.g., [33]). For quantum wells, we have again the appearance of a discrete absorption spectrum within the gap, but the continuum part is not dramatically changed. The most significant excitonic effects are present for the case of quantum wires, where, in addition to the discrete spectrum, we have a suppression of the band-edge singularity [70, 71]. As discussed extensively in [71], this is due to a significant modification of the oscillator strength close to the band edge. Finally, for the quantum-dot case, electron–hole correlation simply translates into a significant red shift of the free-carrier peak (see, e.g., [47]). As anticipated, in addition to these modifications to the continuum part of the spectrum, the presence of Coulomb

[6] The excess energy is defined as the difference between the photon energy and the lowest free-carrier interband transition: $\epsilon_{\mathrm{excess}} = \hbar\omega - (\epsilon_\mathrm{g} + \epsilon^\perp_{n_\mathrm{e}=1} + \epsilon^\perp_{n_\mathrm{h}=1})$. It is crucial to notice that the position of the ground-state exciton expressed in terms of the excess energy corresponds to the so-called exciton binding energy.

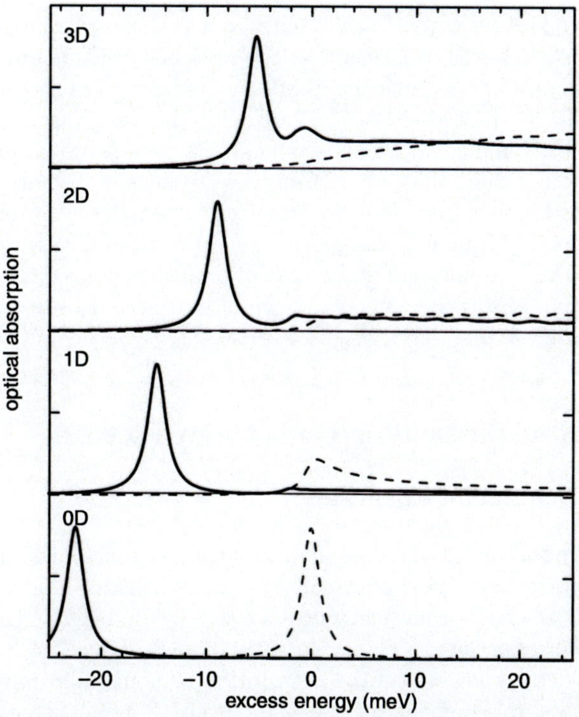

Fig. 1.21. Comparison between free-carrier (*dashed curves*) and excitonic absorption spectra (*solid curves*) corresponding to the low-dimensional electron and hole DOS reported in Figs. 1.13 and 1.14 in the single-subband approximation (see Fig. 1.16). Here, electron–hole Coulomb correlation induces dramatic modifications to the free-carrier spectra (see text)

correlation gives rise to bound electron–hole states, whose binding energy – called exciton binding energy – increases with the nanostructure dimensionality, ranging from a bulk (3D) value of about 5 meV to the quantum-dot (0D) value of about 22 meV. We notice that, contrary to quantum wells and wires, in the quantum-dot case the concept of exciton binding energy, i.e., the energy difference between the lowest excitonic state and the continuum, is ill-defined; therefore, it is difficult to distinguish experimentally between single-particle confinement and excitonic contributions.[7]

At this point an important remark is in order. While in the free-carrier case the absorption response of our nanostructure shows relevant differences going

[7] All single-particle properties of the GaAs/AlAs nanostructures previously considered (see Figs. 1.13, 1.14, 1.15, 1.16, 1.17, 1.18, 1.19 and 1.20) as well as the corresponding Coulomb-correlated results (see Fig. 1.21) have been obtained via realistic fully three-dimensional calculations based on a plane-wave expansion (see Sect. 5.1.2).

from bulk to quantum wires, in the excitonic case such differences are strongly reduced. In other words, it seems that the presence of Coulomb correlation – which is always a three-dimensional effect – tends to reduce the role of the dimensionality D; indeed, both the bulk and the wire case exhibit a finite spectrum at the band edge, which is typical of a bidimensional system. One is then forced to conclude that, apart from the presence of discrete energy levels within the gap, going from bulk to quantum wires, the excitonic absorption does not show the dramatic changes of the Coulomb-free case. The situation is totally different, however, for the case of quantum dots; the latter, in view of their discrete absorption spectrum, are often referred to as semiconductor macroatoms [47, 51].

1.3 The Semiclassical or Boltzmann Picture

1.3.1 The Boltzmann Equation

The starting point of the so-called semiclassical picture recalled in this section is the well-known Boltzmann equation [72–74], an integro-differential equation proposed at the end of the nineteenth century by Ludwig Boltzmann [75] in his kinetic theory of gases [76].

Goal of such equation was the description of nonequilibrium diffusion as well as transport phenomena for a generic gas of N point-like classical particles, whose phase-space distribution is described via a single-particle picture in terms of the so-called distribution function $f(\boldsymbol{r}, \boldsymbol{p})$, proportional to the probability of finding a particle with position $\boldsymbol{r}$ and momentum $\boldsymbol{p}$. Each particle will evolve according to the corresponding single-particle Hamilton equations, but it will also undergo stochastic collision processes with other particles. The Boltzmann equation is the equation of motion for the distribution function: by denoting with $\boldsymbol{F}$ a classical force (corresponding to any external single-particle potential) and employing the classical Hamilton equations

$$\dot{\boldsymbol{r}} = \frac{\boldsymbol{p}}{m} , \qquad \dot{\boldsymbol{p}} = \boldsymbol{F} , \tag{1.63}$$

we have

$$\left.\frac{\mathrm{d}f}{\mathrm{d}t}\right|_{\mathrm{det}} = \left(\frac{\partial}{\partial t} + \frac{\boldsymbol{p}}{m} \cdot \boldsymbol{\nabla}_{\boldsymbol{r}} + \boldsymbol{F} \cdot \boldsymbol{\nabla}_{\boldsymbol{p}}\right) f(\boldsymbol{r}, \boldsymbol{p}) = \left.\frac{\partial f}{\partial t}\right|_{\mathrm{coll}} \tag{1.64}$$

with

$$\left.\frac{\partial f}{\partial t}\right|_{\mathrm{coll}} = \int \mathrm{d}^3 p' \left[W(\boldsymbol{p}, \boldsymbol{p}') f(\boldsymbol{r}, \boldsymbol{p}') - W(\boldsymbol{p}', \boldsymbol{p}) f(\boldsymbol{r}, \boldsymbol{p})\right] . \tag{1.65}$$

The Boltzmann equation (1.64) simply states that the total variation of the distribution function due to the deterministic single-particle motion is equal to the corresponding variation due to particle–particle collisions; the latter are described via the so-called collision term in (1.65), whose physical interpretation is straightforward: denoting with $W(\boldsymbol{p}', \boldsymbol{p})$ the collision or scattering probability per time unit from $\boldsymbol{p}$ to $\boldsymbol{p}'$, the net variation of f in the current point $\boldsymbol{r}, \boldsymbol{p}$ is given by all particles scattered from any $\boldsymbol{p}'$ into the current value $\boldsymbol{p}$ ("in-scattering" contribution) minus all particles scattered out of $\boldsymbol{p}$ to any value $\boldsymbol{p}'$ ("out-scattering" contribution). Within the Boltzmann kinetic theory, such collision processes are assumed to be instantaneous and point-like (see, e.g., [72]), i.e., $\boldsymbol{r}, \boldsymbol{p} \rightarrow \boldsymbol{r}, \boldsymbol{p}'$.

A closer inspection of the Boltzmann equation (1.64) allows us to identify a few crucial features: in the absence of collisions ($W(\boldsymbol{p}', \boldsymbol{p}) = 0$) it reduces to the well-known Liouville equation, i.e., the total derivative of the phase-space distribution function is always equal to zero; the latter, in turn, contains a so-called diffusion term (involving the real-space gradient of f) as well as a drift contribution (involving the gradient of f in momentum space). As we shall see, under suitable physical conditions, such purely classical transport theory may be successfully employed for the description of non-classical systems; this constitutes the essence of the so-called semiclassical picture.

1.3.2 Application to Electron Dynamics in Semiconductors

In spite of its unambiguous quantum-mechanical nature, the behavior of a Bloch electron $\boldsymbol{k}\nu$ subjected to a slowly varying electromagnetic force may be described via a classical particle of momentum $\boldsymbol{p} = \hbar \boldsymbol{k}$. More precisely, the crucial idea is to describe the electronic state in band ν via a wavepacket characterized by a central (or reference) position $\boldsymbol{r}$ and wavevector (momentum in units of $\hbar$) $\boldsymbol{k} = \frac{\boldsymbol{p}}{\hbar}$; in the presence of a slowly varying applied force $\boldsymbol{F}$, it is possible to show that the time evolution of the wavepacket reference coordinates $\boldsymbol{r}, \boldsymbol{k}$ is dictated by the following semiclassical Hamilton equations:

$$\dot{\boldsymbol{r}}_\nu = \boldsymbol{v}_{\boldsymbol{k}\nu} \equiv \frac{1}{\hbar} \boldsymbol{\nabla}_{\boldsymbol{k}} \epsilon_{\boldsymbol{k}\nu} , \qquad \dot{\boldsymbol{k}} = \frac{\boldsymbol{F}}{\hbar} . \tag{1.66}$$

Compared to the classical Hamilton equations (1.63), here the particle velocity $\frac{\boldsymbol{p}}{m}$ is replaced by the so-called group velocity $\boldsymbol{v}_{\boldsymbol{k}\nu}$, proportional to the gradient of the electronic energy band $\epsilon_{\boldsymbol{k}\nu}$.

As discussed in [2], a rigorous derivation of the above semiclassical equations is a formidable task. Here, we shall just underline two important issues: (i) Such a particle-like description may be employed provided to fulfill the uncertainty principle $\Delta r \Delta k \geq \frac{1}{2}$; this means that such semiclassical picture becomes inadequate on very short space-scales (see Chap. 2). (ii) In the absence of external forces and of additional interaction mechanisms, the time evolution of a localized wavepacket in a crystal will tend unavoidably to a Bloch state, thus loosing any information on its spatial coordinate; however, such diffusion effect – not predicted by the semiclassical equations (1.66) – is partially contrasted by the presence of scattering events, which tend to destroy such phase-interference phenomenon; it follows that the semiclassical equations (1.66) are better justified in the presence of a significant electron-scattering dynamics.

In analogy with the kinetic theory of gases previously recalled, we shall introduce a corresponding distribution function $f_{\boldsymbol{k}\nu}(\boldsymbol{r})$ [28, 77, 78], describing the probability of finding an electron in the Bloch state $\boldsymbol{k}\nu$ with position $\boldsymbol{r}$. By employing the semiclassical result in (1.66), it is easy to derive the following electronic-transport

$$\left.\frac{\mathrm{d}f_{\boldsymbol{k}\nu}}{\mathrm{d}t}\right|_{\mathrm{det}} = \left(\frac{\partial}{\partial t} + \boldsymbol{v}_{\boldsymbol{k}\nu}\cdot\boldsymbol{\nabla}_{\boldsymbol{r}} + \frac{\boldsymbol{F}}{\hbar}\cdot\boldsymbol{\nabla}_{\boldsymbol{k}}\right) f_{\boldsymbol{k}\nu} = \left.\frac{\partial f_{\boldsymbol{k}\nu}}{\partial t}\right|_{\mathrm{coll}} \tag{1.67}$$

with a semiclassical collision term of the form

$$\left.\frac{\partial f_{\boldsymbol{k}\nu}}{\partial t}\right|_{\mathrm{coll}} = \sum_{\boldsymbol{k}'\nu'} \left[P_{\boldsymbol{k}\nu,\boldsymbol{k}'\nu'} f_{\boldsymbol{k}'\nu'} - P_{\boldsymbol{k}'\nu',\boldsymbol{k}\nu} f_{\boldsymbol{k}\nu}\right] . \tag{1.68}$$

Here, the effective collision rates

$$P_{\boldsymbol{k}\nu,\boldsymbol{k}'\nu'} = (1 - f_{\boldsymbol{k}\nu}) P^0_{\boldsymbol{k}\nu,\boldsymbol{k}'\nu'} \tag{1.69}$$

are written as the product of the so-called Pauli factor $(1 - f_{\boldsymbol{k}\nu})$ (accounting for the average occupation of the final state $\boldsymbol{k}\nu$) times the total transition rate (from the initial state $\boldsymbol{k}'\nu'$ to the empty state $\boldsymbol{k}\nu$)

$$P^0_{\boldsymbol{k}\nu,\boldsymbol{k}'\nu'} = \sum_s P^s_{\boldsymbol{k}\nu,\boldsymbol{k}'\nu'} \tag{1.70}$$

due to all possible interaction mechanisms. More specifically, the scattering rates $P^s_{\boldsymbol{k}\nu,\boldsymbol{k}'\nu'}$ in (1.70) describe again instantaneous and point-like scattering processes $\boldsymbol{k}'\nu' \to \boldsymbol{k}\nu$ induced by the generic interaction mechanism s; compared to the original Boltzmann collision term (1.65), in addition to interparticle (carrier–carrier) scattering, our Bloch electrons will undergo stochastic transitions induced by a number of interaction mechanisms ascribed to the host material, like carrier–phonon, carrier–plasmon, carrier–impurity scattering.

Compared to the purely classical Boltzmann equation (1.64), the electronic-transport equation (1.67) is semiclassical in nature, i.e., it is grounded on the following quantum-mechanical ingredients: (i) the electron group velocity $\boldsymbol{v}_{\boldsymbol{k}\nu}$ (see (1.66)), (ii) the Pauli factor $(1 - f_{\boldsymbol{k}\nu})$ (see (1.69)), and (iii) the microscopic scattering rates $P^s_{\boldsymbol{k}\nu,\boldsymbol{k}'\nu'}$. Indeed, the latter – describing intraband ($\nu = \nu'$) as well as interband ($\nu \neq \nu'$) transitions – are typically derived in quantum-mechanical terms via the well-known Fermi's golden rule (see Chaps. 2 and 3).

In the low-density limit ($f_{\boldsymbol{k}\nu} \ll 1$) and for close-to-equilibrium phenomena, the Boltzmann collision term (1.68) may be replaced by a conventional relaxation-time approximation [28, 77, 78], i.e.,

$$\left.\frac{\partial f_{\boldsymbol{k}\nu}}{\partial t}\right|_{\mathrm{coll}} = -\Gamma^{\mathrm{coll}}_{\boldsymbol{k}\nu}(f_{\boldsymbol{k}\nu} - f^\circ_{\boldsymbol{k}\nu}) , \tag{1.71}$$

where $\frac{1}{\Gamma^{\text{coll}}_{\boldsymbol{k}\nu}}$ denotes the scattering-induced relaxation time for a carrier in state $\boldsymbol{k}\nu$, and $f^{\circ}_{\boldsymbol{k}\nu}$ the corresponding equilibrium distribution in (1.12).

1.3.3 Generalization to Low-Dimensional Nanostructures

Starting from the Boltzmann-like equation (1.67), it is possible to formulate a corresponding transport equation also for the low-dimensional semiconductor nanostructures previously considered. To this aim, based on the envelope-function picture introduced in Sect. 1.2.2, we shall consider a corresponding envelope distribution function $f_{\nu\overline{\alpha}}(\boldsymbol{r}_{\|}) \equiv f_{\nu,\boldsymbol{k}_{\|}n}(\boldsymbol{r}_{\|})$ describing carriers in band ν and subband n. The resulting equation of motion is still Boltzmann like; in particular, neglecting drift as well as diffusion contributions within the confinement subspace $\boldsymbol{r}_{\perp}$, we get

$$\left.\frac{\mathrm{d}f_{\nu,\boldsymbol{k}_{\|}n}}{\mathrm{d}t}\right|_{\text{det}} = \left(\frac{\partial}{\partial t} + \boldsymbol{v}_{\|} \cdot \boldsymbol{\nabla}_{\boldsymbol{r}_{\|}} + \frac{\boldsymbol{F}_{\|}}{\hbar} \cdot \boldsymbol{\nabla}_{\boldsymbol{k}_{\|}}\right) f_{\nu,\boldsymbol{k}_{\|}n} = \left.\frac{\partial f_{\nu,\boldsymbol{k}_{\|}n}}{\partial t}\right|_{\text{coll}} \tag{1.72}$$

with

$$\left.\frac{\partial f_{\alpha}}{\partial t}\right|_{\text{coll}} = \sum_{\alpha'} \left[P_{\alpha\alpha'} f_{\alpha'} - P_{\alpha'\alpha} f_{\alpha}\right] . \tag{1.73}$$

Here, the compact notation $\alpha \equiv \nu, \boldsymbol{k}_{\|}n$ has been employed. In analogy to (1.69) and (1.70), the effective collision probabilities

$$P_{\alpha\alpha'} = (1 - f_{\alpha}) \sum_{s} P^{s}_{\alpha\alpha'} \tag{1.74}$$

describe again instantaneous and point-like scattering processes $\alpha' \equiv \nu', \boldsymbol{k}'_{\|}n' \to \alpha \equiv \nu, \boldsymbol{k}_{\|}n$ weighted by the Pauli factor; the latter include intrasubband $(\nu, n = \nu, n')$ as well as intersubband $(\nu, n \neq \nu, n')$ transitions.

Contrary to the bulk transport equation (1.67), here the semiclassical particle dynamics (drift versus diffusion) applies to the parallel subspace only. In particular, for nanostructures with homogeneous carrier distribution along $\boldsymbol{r}_{\|}$ and for $\boldsymbol{F}_{\|} = 0$, the Boltzmann transport equation (1.72) reduces to

$$\frac{\partial f_{\alpha}}{\partial t} = \sum_{\alpha'} \left[P_{\alpha\alpha'} f_{\alpha'} - P_{\alpha'\alpha} f_{\alpha}\right] . \tag{1.75}$$

Exactly as for the bulk case (see (1.71)), in the low-density limit ($f_{\alpha} \ll 1$) and for close-to-equilibrium phenomena, the Boltzmann collision term in (1.75) may be replaced again by a corresponding relaxation-time term, i.e.,

$$\frac{\partial f_{\alpha}}{\partial t} = -\Gamma^{\text{coll}}_{\alpha}(f_{\alpha} - f^{\circ}_{\alpha}) , \tag{1.76}$$

where $\frac{1}{\Gamma^{\text{coll}}_{\alpha}}$ denotes the scattering-induced relaxation time for a carrier in state α, and f°_{α} the corresponding equilibrium distribution.

The semiclassical Boltzmann-like equation (1.67) – together with its nanostructure version (1.72) – comes out to be the key instrument for the description of nonequilibrium optoelectronic processes in semiconductor bulk materials as well as in low-dimensional systems, provided that the phenomenon under examination is compatible with such a particle-like picture. As we shall see, the most effective strategy for the solution of the semiclassical nonlinear equations (1.67) and (1.72) is the well-known Monte Carlo method [28, 77, 79] presented in Chap. 5.

1.4 From Materials to Devices: "Closed" Versus "Open" Systems

The semiclassical description presented so far applies to the so-called closed systems, i.e., semiconductor materials extending over the whole coordinate space $\boldsymbol{r}$. In contrast, a semiconductor device [79, 81–84] is typically an "open system", i.e., a portion of material characterized by a well-precise volume Ω_d and spatial boundaries $\boldsymbol{r}^b$ acting as electric contacts. It follows that, in order to describe the carrier distribution within the device active region Ω_d, it is crucial to impose proper spatial boundary conditions; to this end, we shall employ the conventional boundary scheme discussed in Appendix B and depicted in Fig. 1.22 for the one-dimensional case.

More specifically, in order to properly impose the desired spatial boundary conditions to the Boltzmann equation (1.67), we shall fix the value of the carrier distribution $f^b_{\boldsymbol{k}\nu}$ entering the device spatial region Ω_d, which in turn will fix the flux of incoming particles. Following the general prescription discussed in Appendix B, it is easy to show that such a first-order boundary condition may be successfully imposed by adding to the original (closed-system) Boltzmann equation (1.67) a boundary source term (see (B.6))

$$S_{\boldsymbol{k}\nu}(\boldsymbol{r}) = \delta(\boldsymbol{r} - \boldsymbol{r}^b) v^{\mathrm{in}}_{\boldsymbol{k}\nu} f^b_{\boldsymbol{k}\nu} \tag{1.77}$$

and a (negative) boundary loss term (see (B.7))

$$-\Lambda_{\boldsymbol{k}\nu}(\boldsymbol{r}) = -\delta(\boldsymbol{r} - \boldsymbol{r}^b) v^{\mathrm{in}}_{\boldsymbol{k}\nu} f_{\boldsymbol{k}\nu}(\boldsymbol{r}) \;, \tag{1.78}$$

where v^{in} denotes the incoming part of the carrier group velocity $\boldsymbol{v}_{\boldsymbol{k}\nu}$ normal to the boundary surface.

The resulting open-system Boltzmann equation is then given by

$$\left.\frac{\mathrm{d}f_{\boldsymbol{k}\nu}}{\mathrm{d}t}\right|_{\mathrm{det}} = \left.\frac{\partial f_{\boldsymbol{k}\nu}}{\partial t}\right|_{\mathrm{coll}} + \left.\frac{\partial f_{\boldsymbol{k}\nu}}{\partial t}\right|_{b} \tag{1.79}$$

with

$$\left.\frac{\partial f_{\boldsymbol{k}\nu}}{\partial t}\right|_{b} = S_{\boldsymbol{k}\nu} - \Lambda_{\boldsymbol{k}\nu} \;. \tag{1.80}$$

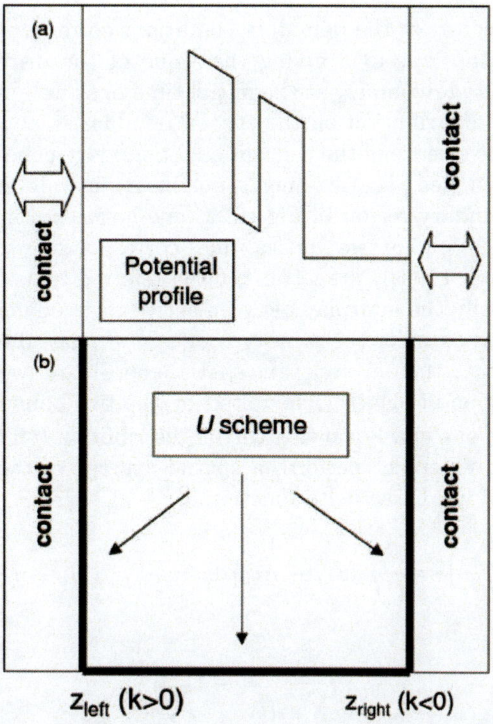

Fig. 1.22. Schematic representation of a typical device active region showing **(a)** the potential profile sandwiched between its electric contacts, and **(b)** the usual boundary-condition scheme for a one-dimensional problem. The latter, discussed in Appendix B and often referred to as U scheme, implies, in particular, the knowledge of the incoming carrier distribution function $f(z^b, k)$, i.e., $f(z_{\mathrm{left}}, k > 0)$ and $f(z_{\mathrm{right}}, k < 0)$ (see text). Reprinted from [80]

As we can see, the proper inclusion of spatial boundary conditions gives rise to the additional contribution in (1.80); the latter, in turn, may be regarded as a source or injection term plus a sort of loss contribution, i.e., proportional to the current value of the carrier distribution function. In view of the local character of the additional boundary term (1.80), the latter does not alter the form of the original Boltzmann equation inside the simulated region Ω_{d}, while on the boundary region ($\boldsymbol{r} = \boldsymbol{r}^b$) it imposes to the incoming carrier distribution the desired boundary value $f_{\boldsymbol{k}\nu}(\boldsymbol{r}^b) = f^b_{\boldsymbol{k}\nu}$.

A relevant comment is now in order. While the original (closed-system) version of the Boltzmann equation in (1.67) is a homogeneous first-order differential equation defined over the whole real-space domain, the open-system version in (1.79) contains a nonhomogeneous source term. Indeed, the boundary conditions corresponding to the closed-system case are fully specified by assigning the distribution function at the initial time $t = t_0$ over the whole real space $\boldsymbol{r}$; in contrast, the boundary conditions corresponding to the open-system problem are specified giving the value of the distribution function inside the simulated region Ω_d at the initial time t_0 as well as the value of the incoming carrier distribution on the spatial boundaries $\boldsymbol{r}^b$ at any later time $t \geq t_0$. This is the essence of the conventional boundary-condition scheme previously mentioned (see Fig. 1.22 and Appendix B); in view of the geometrical shape of the boundary region in Fig. 1.22 (one horizontal line corresponding to the initial time t_0 plus two vertical lines corresponding to the two spatial boundaries at later times), the latter is often referred to as U scheme.

To better clarify the interplay between active-region dynamics and spatial boundaries, let us consider the prototypical case of a one-dimensional device. By denoting with z the only relevant spatial coordinate, we shall consider a device active region of length l, delimited by the two boundary points $z^b_\pm = \pm\frac{l}{2}$. For the case of a slowly varying carrier distribution (on the device space-scale l) it is convenient to perform a spatial average of the boundary term (1.80); in view of the Dirac delta function $\delta(z - z^b_\pm)$ we get

$$\left.\frac{\partial f_{k\nu}}{\partial t}\right|_b = \frac{1}{l}\int [S_{k\nu} - \Lambda_{k\nu}]\,\mathrm{d}z = -\Gamma^b_{k\nu}\left(f_{k\nu} - f^b_{k\nu}\right) \tag{1.81}$$

with

$$\Gamma^b_{k\nu} = \frac{1}{\tau^\mathrm{t}_{k\nu}} = \frac{|v_{k\nu}|}{l} \,. \tag{1.82}$$

The characteristic time $\tau^\mathrm{t}_{k\nu}$ – also called "transit time" – is the time needed by a carrier in state $k\nu$ to travel ballistically across our one-dimensional device of length l.

Recalling that our boundary-condition scheme amounts to fixing the incoming $f_{k\nu}$ typically via a corresponding quasiequilibrium distribution, it is quite natural to regard the boundary contribution (1.81) as a relaxation-time term. Indeed, by comparing the result in (1.81) with the relaxation-time approximation (1.71), we are forced to conclude that the global effect of our spatial boundaries is equivalent to an additional scattering mechanism, whose typical time-scale is directly related to the transit time τ^t in (1.82). It follows that the one-dimensional result (1.81) may be regarded as a formal derivation of the fact that the presence of spatial boundaries is equivalent to an additional scattering channel (see, e.g., [85]), thus giving rise to additional energy-dissipation and decoherence phenomena. As discussed in Chap. 4, such a scenario applies to fully quantum-mechanical treatments of the problem as well.

Based on the previous one-dimensional example, it is clear that the role played by spatial boundaries is dominant when the device transit time is much shorter than any other scattering time; in this case the carrier will undergo a negligible number of scattering processes traveling across the device, and for this reason this is usually referred to as "ballistic-transport regime." In contrast, for transit times much longer than all relevant scattering times we deal with a so-called diffusive-transport regime, where the carrier dynamics is primarily governed/controlled by carrier scattering within the simulated region.

Let us now come to the nanostructure version of the Boltzmann equation (1.75). As anticipated, in this case the particle-like description applies to the parallel subspace only; it follows that the open-system boundary-condition scheme introduced in (1.80) is rigorously justified and applicable to spatial boundaries within $\boldsymbol{r}_\parallel$. On the other hand, semiconductor quantum devices are often characterized by spatial boundaries along the confinement subspace (see, e.g., [56]), for which the previous open-system treatment is ill-defined. However, inspired by the physical interpretation of the one-dimensional result in (1.81), it is quite natural to incorporate such nanostructure spatial boundaries again via a relaxation-time-like term. More specifically, the closed-system Boltzmann equation (1.75) may be replaced by

$$\frac{\partial f_\alpha}{\partial t} = \sum_{\alpha'} \left[P_{\alpha\alpha'} f_{\alpha'} - P_{\alpha'\alpha} f_\alpha \right] + \left. \frac{\partial f_\alpha}{\partial t} \right|_b \tag{1.83}$$

with

$$\left. \frac{\partial f_\alpha}{\partial t} \right|_b = S_\alpha - \Gamma^b_\alpha f_\alpha = -\Gamma^b_\alpha (f_\alpha - f^\circ_\alpha) \; . \tag{1.84}$$

Such effective spatial-boundary term may be regarded again as a source/injection contribution $S_\alpha = \Gamma^b_\alpha f^\circ_\alpha$ describing a sort of thermal injection rate into state α, minus a loss term $\Gamma^b_\alpha f_\alpha$ describing the escape of carriers from state α according to the escape rate Γ^b_α. It is important to stress that, contrary to the rigorous spatial-boundary scheme in (1.80), the nanostructure model (1.84) is purely phenomenological, i.e., the escape rate Γ^b_α may still be regarded as an effective inverse transit time, but its value will unavoidably enter our device modeling as a phenomenological/fitting parameter.

We finally stress that by describing the scattering dynamics via the relaxation-time approximation, the open-system Boltzmann equation (1.83) has again the form of the closed-system transport equation (1.76), provided to add to the scattering relaxation rate $\Gamma^{\text{coll}}_\alpha$ the escape rate Γ^b_α in (1.84). This confirms once again that the presence of spatial boundaries is fully equivalent to an additional dissipation/decoherence channel; such feature will play a key role in the design and optimization of optoelectronic quantum devices.

2

Ultrashort Space- and Time-Scales: Need for a Quantum Description

In Sect. 1.3 we have introduced the basic concepts of the semiclassical picture. In that formulation several assumptions were more or less explicitly made that require some considerations. As we shall see, the latter are not always justified in new-generation semiconductor nanomaterials and nanodevices, so that fully quantum-mechanical treatments of the problem are imperative.

2.1 Intrinsic Limitations of the Semiclassical Picture

In order to decide if the semiclassical treatment is good enough, it is crucial to decide how good is good enough. To this aim, an acceptable energy uncertainty $\delta\epsilon$ is to be established. Its definition is somewhat arbitrary and it depends upon the particular phenomenon under investigation: It may coincide with a characteristic system quantity, e.g., the thermal energy, or may correspond to the energy resolution of our experimental apparatus.

In order to proceed with our analysis, a few characteristic quantities should be introduced. Their definitions are, for the moment, not rigorous; only their intuitive meaning is needed for the purpose of the present discussion. They are the mean scattering time τ_s (between two successive scattering processes), the corresponding electronic mean free path ℓ, and the collision duration τ_c; furthermore, in ultrafast electro-optical processes two other experimental times are relevant: the observation time, i.e., the time elapsed from the preparation of the initial condition to the time of the optical measurement, and the time resolution of the experimental apparatus.

Let us now reconsider the most relevant assumptions made in the semiclassical treatment.

(i) During the so-called free flight (between two successive scattering processes) electrons are treated as classical particles with well-defined positions and momenta. For this approximation to be acceptable, one must be able to conceive electronic wavepackets with a momentum uncertainty much smaller than their average momentum ($\Delta p \ll p$) and, at the same time, with a position uncertainty much smaller than the electronic mean free path ($\Delta r \ll \ell$); taking into account that $\ell \sim \frac{p}{m^*}\tau_s$, from the uncertainty relation we have

$$\hbar \sim \Delta r \Delta p \ll \ell p \sim \frac{p^2}{m^*}\tau_s \ . \tag{2.1}$$

Since the energy value $\frac{p^2}{m^*}$ is always defined within our energy uncertainty $\delta\epsilon$, in order to fulfill the condition in (2.1) for any value of the electron momentum p, it is necessary to ask that:

$$\tau_s \gg \frac{\hbar}{\delta\epsilon} \ . \tag{2.2}$$

We are then forced to conclude that such semiclassical scenario becomes questionable for values of the mean scattering time τ_s comparable or smaller than the characteristic time $\frac{\hbar}{\delta\epsilon}$; as we shall see, this is equivalent to state that the mean scattering time τ_s should be much larger than the so-called collision duration.

(ii) In the traditional semiclassical picture, collisions are assumed to be point-like and instantaneous. Since the interaction process between the electron and a given scattering agent has always a finite duration, this last assumption is not correct in general, not even within a purely classical framework; however, in the usual weak-coupling limit, when scattering events are sufficiently rare, the collision duration τ_c may be safely neglected (with respect to the mean scattering time τ_s) and the assumption of instantaneous scattering processes is reasonable. In contrast, in the presence of a strong scattering dynamics, the collision duration τ_c becomes comparable with the mean scattering time τ_s, and for such regimes a detailed analysis of the collision duration is imperative; however, the estimate of τ_c is somewhat arbitrary because the concept of collision duration itself is ill-defined. In very general terms, from basic perturbation theory (see, e.g., [86]) we know that the final state of a scattering process is defined within an energy range $\delta\epsilon$ of the order of $\frac{\hbar}{\tau_c}$, where τ_c is the time elapsed after the initial condition/preparation of the scattering process. The collision duration can then be estimated, in general, with reference to the desired precision $\delta\epsilon$ of the electron energy according to

$$\tau_c \sim \frac{\hbar}{\delta\epsilon} \ . \tag{2.3}$$

As anticipated, the result in (2.2) may also be written as

$$\tau_s \gg \tau_c \ , \tag{2.4}$$

i.e., the mean scattering time should be much longer than the collision duration. When such a condition is not fulfilled, we deal with the so-called multiple-scattering phenomenon, i.e., a scattering process starts before the completion of the previous ones; the neglect of such multiple scattering effects is known as "completed-collision limit."

For the study of quasiequilibrium electron-transport phenomena at room temperature, the energy uncertainty $\delta\epsilon$ is of the order of $k_B T \sim 25\,\text{meV}$, and therefore, we have

$$\tau_c \sim \frac{\hbar}{k_B T} \sim 10^{-14}\,\text{s} . \tag{2.5}$$

We stress, however, that in low-temperature optical experiments the energy uncertainty $\delta\epsilon$ may be strongly influenced/limited also by the resolution of our experimental apparatus, and therefore, the corresponding collision duration τ_c may strongly depend on the particular phenomenon under investigation.

(iii) During the collision time τ_c, the presence of an intense electro-optical excitation may induce a significant drift of both the initial and final electron wavevectors $\boldsymbol{k}$ and $\boldsymbol{k}'$; however, since the energy dispersion is not a linear function of $\boldsymbol{k}$, the energy difference between initial and final states $(\epsilon_{\boldsymbol{k}'\nu} - \epsilon_{\boldsymbol{k}\nu})$ varies with time, and the quantum-mechanical interference process that generates the transition may be significantly modified by the presence of the field. Such a process, called "intracollisional field effect" (ICFE) (see, e.g., [87, 88] and references therein), may introduce significant modifications to the standard scattering rates obtained via the conventional second-order perturbation theory within a field-free basis. In order to estimate the relevance of this effect, one may compare the energy imparted to the electron by the external field during the collision, with the average electron energy. A well-established result is that for static fields larger than $10^5\,\text{V/cm}$ the ICFE starts to play a significant role. A rigorous microscopic treatment of this phenomenon is given in Sect. 3.4.

To the above discussion we should add that modern technology provides conventional semiconductor devices much smaller than $0.1\,\mu\text{m}$ (see, e.g., [89]), characterized by local fields as high as $10^6\,\text{V/cm}$; furthermore, ultrafast spectroscopy has now reached a time resolution of a few femtoseconds [90–92], i.e., shorter than $10^{-14}\,\text{s}$. A few basic conclusions are then in order.

(a) In conventional semiconductor devices, electric fields are attainable, for which the ICFE has to be taken into account.
(b) At such fields electrons reach energies of the order of or larger than $1\,\text{eV}$, for which the mean scattering time τ_s between collisions may reach the limiting value of $10^{-14}\,\text{s}$, and therefore many of the semiclassical approximations fail.
(c) The time resolution of modern ultrafast electro-optical technology may be comparable or shorter than the collision duration τ_c itself.

As a general conclusion, it seems clear that a quantum approach to the problem [93–105] – essential for the design and optimization of genuine quantum devices – is also imperative for the simulation of conventional semiconductor devices, i.e., devices based on classical operation principles.

2.2 Semiclassical Versus Quantum Treatments

In order to clarify the concept of quantum-mechanical phase coherence – and thus the fundamental link between semiclassical and quantum treatments – let us consider the simplest physical model for the description of light–matter interaction, i.e., an optically driven two-level system (see, e.g., [106]). Within a two-level picture, a ground state a with energy ϵ_a and an excited state b with energy ϵ_b are mutually coupled by a driving force, e.g., due to an external electromagnetic field. The total system Hamiltonian written in the noninteracting-level basis $\{|a\rangle, |b\rangle\}$ is given by

$$\hat{H} = \hat{H}_\circ + \hat{H}' = \begin{pmatrix} \epsilon_a & 0 \\ 0 & \epsilon_b \end{pmatrix} + \begin{pmatrix} 0 & U_{ab} \\ U_{ba} & 0 \end{pmatrix} . \qquad (2.6)$$

This is the sum of a diagonal free-level contribution and of a non-diagonal coupling term; the latter will induce transitions from state $|a\rangle$ to state $|b\rangle$ and vice versa according to the coupling constant $U_{ba} = U_{ab}^*$.

In the absence of interlevel coupling ($U_{ab} = 0$), one deals with two noninteracting stationary states

$$|\psi_a(t)\rangle = \mathrm{e}^{-\frac{i\epsilon_a t}{\hbar}}|a\rangle , \qquad |\psi_b(t)\rangle = \mathrm{e}^{-\frac{i\epsilon_b t}{\hbar}}|b\rangle , \qquad (2.7)$$

corresponding to our electron in level a or b, respectively.

In contrast, in the presence of interlevel coupling, the quantum-mechanical state of the system will be given by a linear superposition of the noninteracting states[1]:

$$|\psi(t)\rangle = c_a(t)|a\rangle + c_b(t)|b\rangle . \qquad (2.8)$$

By inserting the above linear combination into the time-dependent Schrödinger equation

$$i\hbar\frac{\mathrm{d}|\psi\rangle}{\mathrm{d}t} = \hat{H}|\psi\rangle , \qquad (2.9)$$

it is easy to get the following set of Schrödinger-like equations for the two coefficients:

[1] As we shall discuss in Chap. 10, the present quantum-mechanical two-level system is the physical realization of the so-called quantum bit (qubit), i.e., the quantum generalization of the concept of classical-information bit. In this context, the so-called computational state $|\psi\rangle$ is expressed as a generic superposition of the two basis states $|a\rangle$ and $|b\rangle$; the latter are typically referred to as "computational basis," and are denoted by $|0\rangle$ and $|1\rangle$, respectively.

$$\begin{aligned} i\hbar\frac{dc_a}{dt} &= \epsilon_a c_a + U_{ab}c_b \\ i\hbar\frac{dc_b}{dt} &= \epsilon_b c_b + U_{ba}c_a \ . \end{aligned} \tag{2.10}$$

Again, we see that in the absence of interlevel coupling ($U = 0$) the time variation of the coefficients is simply given by their energy rotations in (2.7), i.e., if the system is prepared in state a or state b it will remain in such eigenstate. In contrast, the presence of the interlevel coupling induces a nontrivial time variation of the two coefficients.

As anticipated, the above two-level model provides the simplest description of light–matter interaction in atomic and molecular systems (see, e.g., [106]) as well as in solids (see, e.g., [90]). Let us now start our analysis by considering two limiting cases: the ultrashort- and continuous-excitation regimes.

An ultrashort optical excitation can be described in terms of a delta-like light pulse: $U_{ab}(t) = \eta\delta(t)$. In this case, the equations of motion (2.10) can be solved analytically: Due to such excitation, the two-level system at time $t = 0$ will undergo an instantaneous transition from its ground state $\{c_a = 1, c_b = 0\}$ to the excited state $\{c_a = \cos\alpha, c_b = -i\sin\alpha\}$, with $\alpha = \frac{\eta}{\hbar}$. Therefore, after the pulse the two-level system will be in the excited state

$$|\psi(t)\rangle = \left(\cos\alpha e^{-\frac{i\epsilon_a t}{\hbar}}\right)|a\rangle + \left(-i\sin\alpha e^{-\frac{i\epsilon_b t}{\hbar}}\right)|b\rangle \ , \tag{2.11}$$

which is a coherent quantum-mechanical superposition of the two noninteracting states; indeed, in addition to a finite occupation probability $|c_b|^2 = \sin^2\alpha$ of the excited state, there exists a well-defined phase coherence between the ground- and the excited-state contributions, i.e., apart from their amplitudes, the coefficients c_a and c_b exhibit a time-dependent phase difference given by

$$c_b(t)c_a^*(t) \propto e^{-i\omega_\circ t} \ , \tag{2.12}$$

where $\omega_\circ = \frac{\epsilon_b - \epsilon_a}{\hbar}$ is the interlevel energy splitting in units of $\hbar$. As discussed in [107], this phenomenon is what is generally meant by optically induced phase coherence.

As second limiting case, let us now consider a continuous optical excitation resonant with our two-level system: $U_{ab}(t) = U_\circ e^{i\omega_\circ t}$. Also for this case, the set of equations (2.10) can be solved analytically. In particular, taking again as initial condition the system ground state, in this case we get

$$|\psi(t)\rangle = \left(\cos\left(\frac{1}{2}\omega_R t\right)e^{-\frac{i\epsilon_a t}{\hbar}}\right)|a\rangle + \left(-i\sin\left(\frac{1}{2}\omega_R t\right)e^{-\frac{i\epsilon_b t}{\hbar}}\right)|b\rangle \ , \tag{2.13}$$

where $\omega_R = \frac{2U_\circ}{\hbar}$ is the so-called Rabi frequency. Compared to the previous case in (2.11), the continuous excitation gives rise to a periodic population and depopulation of the excited state according to $|c_b|^2 = \sin^2(\frac{1}{2}\omega_R t)$. This purely coherent phenomenon is known as “Rabi-oscillation regime.” Moreover, exactly as for the previous case, the excited state in (2.13) exhibits again the quantum-mechanical phase relation in (2.12). This clearly shows that the time-dependent quantum state (2.13) is characterized by two distinct time-scales, corresponding to the Rabi frequency ω_R and to the interlevel frequency $\omega_\circ$: the first one – proportional to the amplitude $U_\circ$ of the applied field – may be detected by measuring semiclassical quantities only, namely the ground- and excited-state populations $|c_a|^2 = \cos^2\left(\frac{1}{2}\omega_R t\right)$ and $|c_b|^2 = \sin^2\left(\frac{1}{2}\omega_R t\right)$; the second one, in contrast, does not influence the state populations and can only be detected via phase-sensitive optical experiments (see Sect. 2.5), i.e., measurements able to detect the quantum-mechanical phase coherence in (2.12).

For an electromagnetic excitation of arbitrary temporal shape $U_{ab}(t)$ no analytical solution is available, and the set in (2.10) needs to be solved numerically. More specifically, the intermediate regime – between the ultrafast-and the continuous-excitation ones introduced so far – will correspond to an excitation pulse of finite duration, typically written as

$$U_{ab}(t) = U_p(t)e^{i\omega_p t} , \tag{2.14}$$

where the pulse shape $U_p(t)$ is characterized by a time-scale τ_p. When the latter is much longer than the interlevel time-scale (i.e., $\tau_p\omega_\circ \gg 1$) and the excitation amplitude is small (i.e., $|U_{ab}(t)| \ll \hbar\omega_\circ$), the excitation process may be considered adiabatically slow, and the previous fully quantum-mechanical description may be safely replaced by an effective semiclassical description based on the well-known Fermi's golden rule.

To this aim, let us go back to the set of equations (2.10), whose formal solution can always be written as

$$\begin{aligned} c_a(t) &= e^{-\frac{i\epsilon_a t}{\hbar}} c_a(0) + \int_0^t dt' e^{-\frac{i\epsilon_a(t-t')}{\hbar}} \frac{U_{ab}(t')}{i\hbar} c_b(t') \\ c_b(t) &= e^{-\frac{i\epsilon_b t}{\hbar}} c_b(0) + \int_0^t dt' e^{-\frac{i\epsilon_b(t-t')}{\hbar}} \frac{U_{ba}(t')}{i\hbar} c_a(t') . \end{aligned} \tag{2.15}$$

Following a standard perturbation scheme discussed in general terms in Chap. 3, by iteratively substituting this formal solution into itself and by limiting the resulting expansion to first-order contributions in the electromagnetic field U_{ab}, we get

$$\begin{aligned} c_a(t) &= e^{-\frac{i\epsilon_a t}{\hbar}} c_a(0) + \frac{2\pi U_{ab}(t) D(\omega_\circ - \omega_p)}{i\hbar} c_b(t) \\ c_b(t) &= e^{-\frac{i\epsilon_b t}{\hbar}} c_b(0) + \frac{2\pi U_{ba}(t) D^*(\omega_\circ - \omega_p)}{i\hbar} c_a(t) \end{aligned} \tag{2.16}$$

with

$$D(\omega, t) = \frac{1}{2\pi} \int_0^t e^{i\omega\tau} d\tau . \tag{2.17}$$

In deriving the result in (2.16), we have employed the slowly varying nature of the pulse shape U_p in (2.14), namely $U_{ab}(t') = U_{ab}(t)\mathrm{e}^{-\mathrm{i}\omega_\mathrm{p}(t-t')}$.

As anticipated, the goal of the desired simplified treatment is the derivation of a set of semiclassical equations of motion for the level populations only. To this end, by employing the original set of equations (2.10) and by denoting the two level populations with f_a and f_b, it is easy to get the following result:

$$\frac{\mathrm{d}f_a}{\mathrm{d}t} = \frac{\mathrm{d}|c_a|^2}{\mathrm{d}t} = 2\Re\left(\frac{U_{ab}(t)c_b(t)c_a^*(t)}{\mathrm{i}\hbar}\right) = -\frac{\mathrm{d}|c_b|^2}{\mathrm{d}t} = -\frac{\mathrm{d}f_b}{\mathrm{d}t} \ . \tag{2.18}$$

By assuming as initial condition the system ground state ($c_a(0) = 1, c_b(0)=0$), replacing c_a and c_b with their first-order expansions (2.16), and considering terms up to second order in U_{ab} only, we finally obtain[2]

$$\begin{aligned} \frac{\mathrm{d}f_a}{\mathrm{d}t} &= P_{ab}f_b - P_{ba}f_a \\ \frac{\mathrm{d}f_b}{\mathrm{d}t} &= P_{ba}f_a - P_{ab}f_b \end{aligned} \tag{2.19}$$

with

$$P_{ab}(t) = P_{ba}(t) = \frac{2\pi}{\hbar^2}\left|U_{ab}\right|^2 F(\omega_\circ - \omega_\mathrm{p}, t) \tag{2.20}$$

and

$$F(\omega, t) = D(\omega, t) + D^*(\omega, t) = \frac{1}{\pi}\int_0^t \cos(\omega\tau)\mathrm{d}\tau = \frac{1}{\pi}\frac{\sin(\omega t)}{\omega} \ . \tag{2.21}$$

A few important comments are now in order. In spite of its Boltzmann-like structure (see (1.68)), the semiclassical result (2.19) does not constitute a genuine set of rate equations: The quantities in (2.20) are not positive-definite, and therefore they cannot be regarded as transition rates. Indeed, the function $F(\omega)$ in (2.21) is the result of a fully quantum-mechanical interference process, and – for any finite value of t – the latter will exhibit negative-value regions. However, for increasing values of time, such rapidly oscillating interference pattern will tend to a Dirac delta function, i.e.,

$$\lim_{t\to\infty} \frac{1}{\pi}\frac{\sin(\omega t)}{\omega} = \delta(\omega) \ , \tag{2.22}$$

and the quantity P_{ab} in (2.20) reduces to the well-known Fermi's golden rule

$$P_{ab} = \frac{2\pi}{\hbar^2}\left|U_{ab}\right|^2 \delta(\omega_\circ - \omega_\mathrm{p}) = \frac{2\pi}{\hbar}\left|U_{ab}\right|^2 \delta(\hbar\omega_\circ - \hbar\omega_\mathrm{p}) \ : \tag{2.23}$$

[2] Within the general framework of a semiclassical treatment, here various first-order terms involving the interlevel phase coherence in (2.12) have been disregarded; the latter, in contrast, will play a central role for the description of nonlinear ultrafast optical processes (see below).

in the limit $t \to \infty$ – i.e., far from the initial condition – we recover a genuine transition probability per time unit proportional to the squared modulus of the interaction matrix element U_{ab} as well as to the energy-conserving Dirac delta function; it follows that in this longtime regime, typically referred to as completed-collision limit (see below), the semiclassical description in (2.19) is now fully equivalent to the semiclassical Boltzmann picture introduced in Sect. 1.3.

Figure 2.1 shows the interference pattern in (2.21) as a function of the so-called detuning energy $\hbar\omega$ at three different times: $t = 150\,\mathrm{fs}$, $t = 300\,\mathrm{fs}$, and $t = 600\,\mathrm{fs}$. Recalling that in our case the detuning energy ($\hbar\omega = \hbar(\omega_\circ - \omega_\mathrm{p})$) is just the difference between the interlevel energy splitting $\hbar\omega_\circ = \epsilon_b - \epsilon_a$ and the photon energy $\hbar\omega_\mathrm{p}$, the scenario depicted in Fig. 2.1 clearly shows that at short times (after the state preparation) one deals with the so-called energy-nonconserving transitions (see, e.g., [108]), i.e., transitions characterized by a non-zero detuning energy $\hbar\omega$[3]: Indeed, immediately after the state

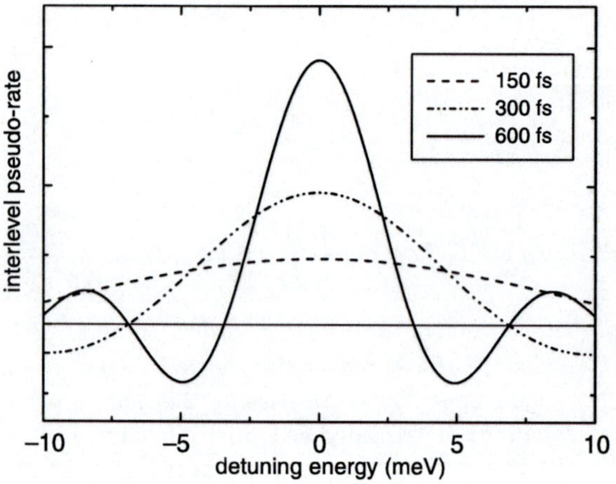

Fig. 2.1. Interference pattern (2.21) as a function of the detuning energy $\hbar\omega$ at three different times: $t = 150\,\mathrm{fs}$, $t = 300\,\mathrm{fs}$, and $t = 600\,\mathrm{fs}$. This scenario clearly shows that at short times (after the state preparation) one deals with the so-called energy-nonconserving transitions, i.e., transitions characterized by a non-zero detuning energy $\hbar\omega$: Indeed, immediately after the state preparation ($t = 150\,\mathrm{fs}$) we deal with an extremely broad profile; for increasing values of time ($t = 300\,\mathrm{fs}$) we assist to the progressive formation of an energy-conserving resonance peak ($\hbar\omega = 0$) accompanied by negative tails; at longer times ($t = 600\,\mathrm{fs}$) we are left with a much sharper central peak surrounded by an oscillatory behavior, which can be regarded as a precursor of the Dirac delta function obtained in the limit $t \to \infty$

[3] The presence of electron–phonon energy-nonconserving transitions in photoexcited semiconductors, originally pointed out in 1989 by Carlo Jacoboni and co-workers [108], has been experimentally demonstrated by Alfred Leitenstorfer and co-workers in 1997 [109].

preparation ($t = 150\,\text{fs}$) we deal with an extremely broad profile; for increasing values of time ($t = 300\,\text{fs}$) we assist to the progressive formation of an energy-conserving resonance peak ($\hbar\omega = 0$) accompanied by negative tails; at longer times ($t = 600\,\text{fs}$) we are left with a much sharper central peak surrounded by an oscillatory behavior, which can be regarded as a precursor of the Dirac delta function obtained in the limit $t \to \infty$. From our analysis it is easy to recognize the typical features of the time–energy uncertainty relation governing this quantum-mechanical interaction process: At very short times we deal with a significant uncertainty of our transition energy, which implies that immediately after the state preparation the electron may undergo transitions from state a to state b induced by non-resonant photons; for increasing values of time such energy-nonconserving transitions become less and less important, and in the longtime limit we deal with energy-conserving processes only. This leads to the natural conclusion that for the present case the concept of collision-duration time introduced in the previous section coincides with the measurement time t for which energy-nonconserving transitions exceeding a given energy uncertainty $\delta\epsilon$ may be safely neglected.

A second crucial point needs to be addressed now. According to the analysis presented so far, on the one hand, in the longtime limit Fermi's golden rule (2.23) will induce energy-conserving processes only; on the other hand, at short times (immediately after the state preparation) the semiclassical equations (2.19) may allow for energy-nonconserving transitions. It is thus crucial to understand if such short-time energy-nonconserving processes will also influence the state populations at longer times. To this end, let us evaluate the probability of finding our electron in the excited state b at time t; by neglecting higher-order contributions (in P_{ab}), such occupation probability is simply given by the time integral of P_{ab}:

$$f_b(t) = \int_0^t P_{ab}(t')\,\mathrm{d}t' \,. \tag{2.24}$$

Recalling the explicit form of P_{ab} in (2.20) and the slowly varying character of our excitation, the final result is

$$f_b(t) \propto \int_0^t F(\omega, t')\mathrm{d}t' = \frac{2}{\pi}\frac{\sin^2\left(\frac{1}{2}\omega t\right)}{\omega^2} \,. \tag{2.25}$$

Figure 2.2 shows the excited-level occupation f_b in (2.25) as a function of the detuning energy $\hbar\omega$ at the same three times considered in Fig. 2.1: $t = 150\,\text{fs}$, $t = 300\,\text{fs}$, and $t = 600\,\text{fs}$. As expected, at very short times we have a significant excited-level occupation also for non-zero detuning energies, a clear fingerprint of energy-nonconserving processes; however, for increasing times the excited-level occupation displays the progressive formation of an energy-conserving resonance peak accompanied again by oscillatory tails; at longer times we are left with a much sharper central peak, which can be regarded again as a precursor of a Dirac delta-like behavior obtained in the limit

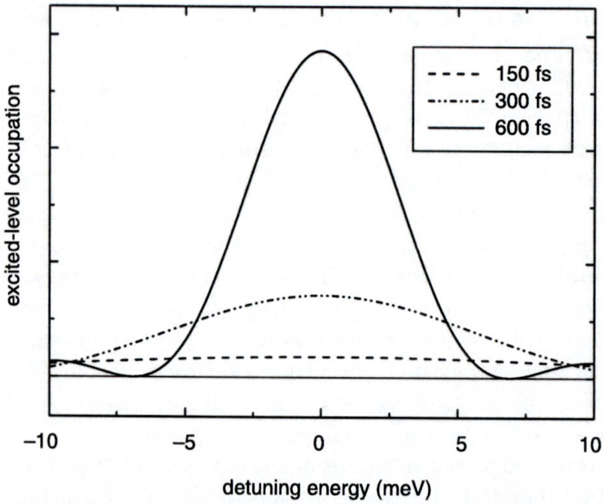

Fig. 2.2. Excited-level occupation f_b in (2.25) as a function of the detuning energy $\hbar\omega$ at the same three times considered in Fig. 2.1: t = 150 fs, t = 300 fs, and t = 600 fs. At very short times we have a significant excited-level occupation also for non-zero detuning energies, a clear fingerprint of energy-nonconserving processes; however, for increasing times the excited-level occupation displays the progressive formation of an energy-conserving resonance peak accompanied again by oscillatory tails; at longer times we are left with a much sharper central peak, which can be regarded again as a precursor of a Dirac delta-like behavior

$t \to \infty$. The scenario depicted in Fig. 2.2 tells us that – in spite of energy-nonconserving processes at short times – the longtime level occupation does not exhibit significant signatures of such ultrafast dynamics; indeed, as discussed in [110], it is possible to show that these ultrafast non-resonant electronic excitations are subsequently canceled by corresponding non-resonant stimulated-emission processes (see negative tails in Fig. 2.1), so that the longtime result does not show any energy-conservation violation. As we shall see, this is true provided that our quantum-mechanical interference process is not disturbed by other ultrafast interaction mechanisms.

In summary, we have seen that for weak as well as slowly varying optical excitations, both ultrafast and longtime level occupations are properly described by the semiclassical Boltzmann-like equations (2.19). More specifically, while in the ultrafast regime it is crucial to include energy-nonconserving processes via the transition pseudo-rate (2.20), for longtime investigations the latter may be safely replaced by the conventional transition rates given by Fermi's golden rule in (2.23). It follows that in spite of the purely quantum-mechanical character of the original set of equations (2.10), for slowly varying excitations and long times the latter may be replaced by a fully semiclassical set of Boltzmann-like equations, describing the light–matter interaction via purely stochastic scattering processes (see, e.g., [28]).

Let us finally consider the case of an ultrafast optical excitation characterized by a spectral function $\tilde{U}(\omega)$, i.e.,

$$U_{ab}(t) = \int \tilde{U}(\omega) \mathrm{e}^{\mathrm{i}\omega t} \, \mathrm{d}\omega \, . \tag{2.26}$$

Assuming again to work within the weak-excitation regime ($|U_{ab}(t)| \ll \hbar\omega_\circ$), the set of equations (2.16), corresponding to a slowly varying excitation, in this case should be replaced by

$$\begin{aligned} c_a(t) &= \mathrm{e}^{-\frac{\mathrm{i}\epsilon_a t}{\hbar}} c_a(0) + \frac{2\pi\overline{U}(\omega_\circ, t)}{\mathrm{i}\hbar} \mathrm{e}^{\mathrm{i}\omega_\circ t} c_b(t) \\ c_b(t) &= \mathrm{e}^{-\frac{\mathrm{i}\epsilon_b t}{\hbar}} c_b(0) + \frac{2\pi\overline{U}^*(\omega_\circ, t)}{\mathrm{i}\hbar} \mathrm{e}^{-\mathrm{i}\omega_\circ t} c_a(t) \end{aligned} \tag{2.27}$$

with

$$\overline{U}(\omega_\circ, t) = \frac{1}{2\pi} \int_0^t U_{ab}(t') \mathrm{e}^{-\mathrm{i}\omega_\circ t'} \mathrm{d}t' \, . \tag{2.28}$$

By inserting these new solutions into (2.18) and considering again terms up to second order in U_{ab} only, it is possible to obtain again the set of Boltzmann-like equations (2.19) with new pseudo-rates given by

$$P_{ab}(t) = P_{ba}(t) = \frac{4\pi}{\hbar^2} \Re \left(U_{ba}(t) \mathrm{e}^{\mathrm{i}\omega_\circ t} \overline{U}(\omega_\circ, t) \right) \, . \tag{2.29}$$

Also in this case, it is possible to show that – similar to the slowly varying-excitation result in (2.20) – at short times the system will exhibit energy-nonconserving transitions. The crucial question is now if such an ultrafast excitation will lead to energy-conservation violations also in the longtime limit. To this end, let us consider again the probability of finding our electron in the excited state b at time t; by inserting the pseudo-rate (2.29) into (2.24), the final result is

$$f_b(t) \propto \int_0^t \mathrm{d}t' 2\Re \left(U_{ba}(t') \mathrm{e}^{\mathrm{i}\omega_\circ t'} \overline{U}(\omega_\circ, t') \right) \propto \left| \overline{U}(\omega_\circ, t) \right|^2 \, , \tag{2.30}$$

i.e., the probability of finding our electron in the excited state at time t is proportional to the squared modulus of the function $\overline{U}$ in (2.28); the latter, in turn, depends crucially on the duration as well as on the central time of the optical pulse in (2.26). In particular, for ultrafast excitations fully contained within our time interval ($0 < t' < t$), the time integral in (2.28) may be extended from $-\infty$ to $+\infty$, and thus in this case the function $\overline{U}$ coincides with the pulse spectral function $\tilde{U}$ in (2.26), i.e., with the Fourier transform of the original ultrafast excitation. It follows that when the pulse is fully contained in our time interval, we are in the completed-collision limit previously mentioned, and the final electron occupation is determined by the pulse Fourier transform only; in contrast, for observation times t comparable to or shorter than the pulse duration, the electron population will display additional energy broadening due to short-time energy-nonconserving transitions.

In order to better elucidate the scenario previously discussed, let us consider, as prototypical example of ultrafast excitation, a Gaussian optical pulse

$$U_{ab}(t) = U_{\circ} e^{-\left(\frac{t-t_0}{2\tau_p}\right)^2} e^{i\omega_p t} \tag{2.31}$$

with central photon energy $\hbar\omega_p$, central time t_0, and pulse duration τ_p. Figure 2.3 shows the excited-level occupation (2.30) as a function of the detuning energy $\hbar\omega = \hbar\omega_p - \hbar\omega_{\circ}$ at different observation times for a Gaussian pulse of duration $\tau_p = 100\,\mathrm{fs}$ and central time $t_0 = 200\,\mathrm{fs}$. As expected, at early stages ($t = 100\,\mathrm{fs}$) we observe an extremely broad electron distribution due to energy-nonconserving processes; for increasing times ($t = 200\,\mathrm{fs}$ and $t = 300\,\mathrm{fs}$) such energy broadening is progressively reduced, and close to the end of our excitation ($t = 400\,\mathrm{fs}$) the latter approaches the square of the excitation Fourier transform, which for our Gaussian pulse is Gaussian as well

$$|\tilde{U}(\omega)|^2 \propto e^{-\frac{(\hbar\omega)^2}{2\Delta\epsilon^2}} , \tag{2.32}$$

with $\Delta\epsilon = \frac{\hbar}{2\tau_p} \sim 3\,\mathrm{meV}$.

We are then led to the conclusion that opposite to the case of a slowly varying excitation, the application of an ultrafast optical pulse gives rise to a longtime electronic energy broadening, which is just a fingerprint of the spectral broadening of our time-dependent excitation. Indeed, we deal with

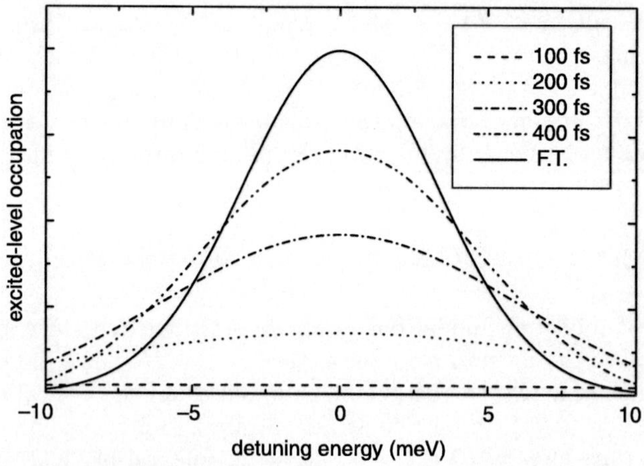

Fig. 2.3. Excited-level occupation (2.30) as a function of the detuning energy $\hbar\omega = \hbar\omega_p - \hbar\omega_{\circ}$ at different observation times for a Gaussian pulse of duration $\tau_p = 100\,\mathrm{fs}$ and central time $t_0 = 200\,\mathrm{fs}$. At early stages ($t = 100\,\mathrm{fs}$) we observe an extremely broad electron distribution due to energy-nonconserving processes; for increasing times ($t = 200\,\mathrm{fs}$ and $t = 300\,\mathrm{fs}$) such energy broadening is progressively reduced, and close to the end of our excitation ($t = 400\,\mathrm{fs}$) the latter approaches the square of the excitation Fourier transform

a sort of superposition principle: The ultrafast pulse (2.26) may always be regarded as a coherent superposition of monochromatic components with different photon energies; each of these components at long times will result in electron–photon energy-conserving processes, then producing an overall energy broadening described by $|\tilde{U}(\omega)|^2$. As we shall discuss in more detail in Chaps. 9 and 10, the above time–energy uncertainty relation between pulse duration and electronic energy broadening, i.e.,

$$\Delta\epsilon\tau_{\rm p} \geq \frac{\hbar}{2} , \tag{2.33}$$

constitutes one of the most severe limitations in the design and optimization of ultrafast electro-optical quantum devices.

So far, we have limited our attention to the case of weak optical excitations; within our two-level picture this means that the excited-state occupation in (2.25) as well as (2.30) is always very small ($f_b(t) \ll 1$). In order to describe light–matter interaction induced by strong optical pulses, the perturbative scheme previously considered – as well as the resulting semiclassical equations (2.19) – is definitely inadequate. Moreover, the interlevel quantum-mechanical phase coherence is not affecting the level populations $f_a = |c_a|^2$ and $f_b = |c_b|^2$ (see (2.11) and (2.13)); in contrast, the latter is well described via the quantity $p = c_b c_a^*$ in (2.12) also called interlevel polarization (see below). It is then clear that in order to describe phase-sensitive ultrafast experiments induced by strong optical excitations, we need to replace the semiclassical equations (2.19) with a new set of equations involving the interlevel polarization as well. More specifically, combining the original equations of motion (2.10) with the definitions of the level occupations f_a and f_b as well as of the interlevel polarization p, we get

$$\begin{aligned} \frac{\mathrm{d}f_a}{\mathrm{d}t} &= 2\Re\left(\frac{U_{ab}p}{\mathrm{i}\hbar}\right) = -\frac{\mathrm{d}f_b}{\mathrm{d}t} \\ \frac{\mathrm{d}p}{\mathrm{d}t} &= \frac{\epsilon_b - \epsilon_a}{\mathrm{i}\hbar}p + \frac{U_{ba}}{\mathrm{i}\hbar}(f_a - f_b) . \end{aligned} \tag{2.34}$$

The latter are known as optical Bloch equations (see, e.g., [33, 90]) in analogy with the equations first derived by Felix Bloch in 1946 [111] for spin systems. They constitute the simplest application of the well-known density-matrix formalism (see, e.g., [112, 113]) presented in Chap. 3: Our new variables may be regarded as the elements of the following two-by-two matrix:

$$\hat{\rho} = \begin{pmatrix} \rho_{aa} & \rho_{ab} \\ \rho_{ba} & \rho_{bb} \end{pmatrix} = \begin{pmatrix} f_a & p^* \\ p & f_b \end{pmatrix} = \begin{pmatrix} c_a c_a^* & c_a c_b^* \\ c_b c_a^* & c_b c_b^* \end{pmatrix} . \tag{2.35}$$

This is the so-called density matrix for our two-level system; indeed, recalling that $c_a = \langle a|\psi\rangle$ and $c_b = \langle b|\psi\rangle$, the density matrix (2.35) may also be regarded as the operator

$$\hat{\rho} = |\psi\rangle\langle\psi| \tag{2.36}$$

written in our two-level basis. The operator in (2.36) – called density-matrix operator – in this particular case (corresponding to a so-called pure state [112, 113]) is simply given by the projector of the system state vector $|\psi\rangle$ (see (3.7) in Chap. 3).

Let us consider again the case of a continuous optical excitation resonant with our two-level system: $U_{ab}(t) = U_\circ \mathrm{e}^{\mathrm{i}\omega_\circ t}$. If we choose as initial condition at time $t = 0$ the system ground state ($\{f_a = 1, f_b = 0, p = 0\}$), the solution of the optical Bloch equations (2.34) is given by

$$\begin{aligned} f_a(t) &= \cos^2\left(\frac{1}{2}\omega_\mathrm{R} t\right) \\ f_b(t) &= \sin^2\left(\frac{1}{2}\omega_\mathrm{R} t\right) \\ p(t) &= \frac{1}{4}\left[\mathrm{e}^{-\mathrm{i}\omega^+ t} - \mathrm{e}^{-\mathrm{i}\omega^- t}\right] , \end{aligned} \tag{2.37}$$

where $\omega^\pm = \omega_\circ \pm \omega_\mathrm{R}$. As anticipated, the above solution describes a Rabi oscillation regime. In particular, the interlevel density-matrix element p originates from the superposition of the two frequency components ω^+ and ω^-. They differ from $\omega_\circ$ by the Rabi frequency ω_R. Such modification of the two-level frequency $\omega_\circ$ due to its coupling with the external field is known as "Rabi splitting" (see, e.g., [114]). If we now rewrite the interlevel density-matrix element p in (2.37) as

$$p(t) = -\frac{\mathrm{i}}{2}\mathrm{e}^{-\mathrm{i}\omega_\circ t}\sin(\omega_\mathrm{R} t) , \tag{2.38}$$

we see that apart from the quantum-mechanical phase factor corresponding to the interlevel energy separation $\hbar\omega_\circ$, its amplitude exhibits Rabi oscillations according to $\sin(\omega_\mathrm{R} t)$. More specifically, we have

$$|p|^2 \propto \sin^2(\omega_\mathrm{R} t) \propto \cos^2\left(\frac{1}{2}\omega_\mathrm{R} t\right)\sin^2\left(\frac{1}{2}\omega_\mathrm{R} t\right) \propto f_a(t) f_b(t) , \tag{2.39}$$

i.e., the quantity $|p|^2$ is proportional to the product of the two occupation numbers, thus reflecting the total (or macroscopic) dipole moment of our two-level system at time t. This elucidates the link between optically induced phase coherence and polarization: A coherent optical excitation gives rise to a coherent quantum-mechanical superposition of the two states which results in a macroscopic polarization of the system. Such polarization field is fully described by the non-diagonal matrix element p in (2.38).[4]

[4] It is important to stress that opposite to the Boltzmann-like treatment previously considered (see (2.19)), the optical Bloch equations (2.34) may describe light-induced population inversion ($f_b > f_a$) and therefore stimulated-emission coherent processes (see also Chap. 8).

The optical Bloch equations (2.34) and the Schrödinger equations (2.10) are totally equivalent; however, as we shall see, the density-matrix description allows, in addition to the study of coherent phenomena, for the analysis of incoherent phenomena, not possible within a simple Schrödinger-equation formalism.

As a matter of fact, the above simplified treatment of light–matter interaction – based on a single isolated two-level system – neglects any kind of incoherent – i.e., energy-dissipation as well as phase-breaking – phenomena induced by the host material; indeed, dealing with optically excited solid-state electrons, such incoherent processes will lead to a decay of the excited-state population as well as of the interlevel polarization, thus destroying the optically induced phase coherence previously discussed (see (2.12)). As anticipated, such incoherent phenomena may be easily incorporated within our simplified two-level system description via the optical-Bloch-equation scheme previously introduced; more specifically, the set of equations (2.34) may be generalized as

$$\frac{\mathrm{d}f_a}{\mathrm{d}t} = 2\Re\left(\frac{U_{ab}p}{\mathrm{i}\hbar}\right) - \frac{f_a - f_a^\circ}{T_1}$$
$$\frac{\mathrm{d}f_b}{\mathrm{d}t} = -2\Re\left(\frac{U_{ab}p}{\mathrm{i}\hbar}\right) - \frac{f_b - f_b^\circ}{T_1}$$
$$\frac{\mathrm{d}p}{\mathrm{d}t} = \frac{\epsilon_b - \epsilon_a}{\mathrm{i}\hbar}p + \frac{U_{ba}}{\mathrm{i}\hbar}(f_a - f_b) - \frac{p}{T_2} \,. \tag{2.40}$$

Here, the incoherent phenomena previously mentioned are accounted for via simple relaxation-time approximations: On the longtime limit, the level populations f_a and f_b tend to their thermal-equilibrium values f_a° and f_b° via a relaxation time T_1, while the interlevel polarization p will decay to 0 according to a second relaxation time T_2. The theoretical scheme in (2.40) is known as T_1/T_2 model (see, e.g., [33, 90]).

Let us consider again the ultrafast-excitation regime previously discussed (see (2.11)) and, in particular, the case of a $\frac{\pi}{4}$ pulse ($\alpha = \frac{\pi}{4}$) able to generate an after-pulse equipopulated coherent state: $f_a = f_b = \frac{1}{2}, p = -\frac{\mathrm{i}}{2}$. In the limit $T_1 \gg T_2$ the population relaxation terms in (2.40) may be safely neglected, and the resulting interlevel polarization is simply given by

$$p(t) = p(0)\mathrm{e}^{-\mathrm{i}\omega_\circ t}\mathrm{e}^{-\frac{t}{T_2}} \propto \mathrm{i}\mathrm{e}^{-\mathrm{i}\omega_\circ t}\mathrm{e}^{-\frac{t}{T_2}} \,. \tag{2.41}$$

In addition to the interlevel quantum-mechanical phase factor in (2.12), the presence of incoherent scattering sources manifests itself in a decay of the interlevel polarization described by the relaxation time T_2. It follows that the overall polarization dynamics is the result of a non-trivial interplay between coherent and incoherent phenomena. More specifically, in the limit $\omega_\circ T_2 \gg 1$ – also referred to as coherent regime – the polarization decay is negligible on the interlevel time-scale $\omega_\circ^{-1}$; in contrast, in the limit $\omega_\circ T_2 \ll 1$ – also referred to

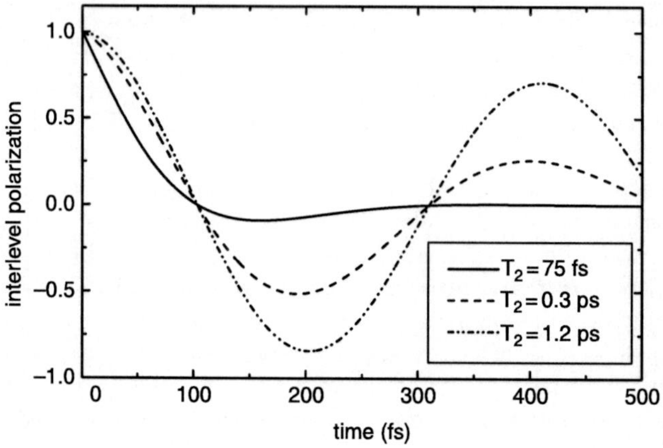

Fig. 2.4. Imaginary part of the interlevel polarization (2.41) corresponding to an interlevel energy splitting of 10 meV for three different values of the relaxation time T_2: for $T_2 = 1.2$ ps, we have a nearly periodic oscillatory behavior confirming a coherent-like regime where the incoherent polarization decay plays a very minor role; for $T_2 = 75$ fs, we deal with an incoherent-like regime, characterized by a nearly exponential decay and by a corresponding suppression of the coherent oscillatory behavior; for $T_2 = 300$ fs, we deal with an intermediate regime characterized by a strong interplay between coherent oscillations and incoherent decay

as incoherent regime – coherent oscillations are totally dominated/suppressed by a polarization decay much faster than the interlevel time-scale.

To better illustrate the scenario discussed so far, in Fig. 2.4 we show the imaginary part of the interlevel polarization (2.41) corresponding to an interlevel energy splitting of 10 meV for three different values of the relaxation time T_2: for $T_2 = 1.2$ ps, we have a nearly periodic oscillatory behavior confirming a coherent-like regime where the incoherent polarization decay plays a very minor role; in contrast, for $T_2 = 75$ fs, we deal with an incoherent-like regime, characterized by a nearly exponential decay and by a corresponding suppression of the coherent oscillatory behavior; finally, for $T_2 = 300$ fs, we deal with an intermediate regime characterized by a strong interplay between coherent oscillations and incoherent decay.

In this context, it is important to recall that the magnitude of the polarization decay in (2.41) may also be detected looking to the imaginary part of its Fourier transform, which in turn is proportional to the optical-absorption spectrum, i.e.,

$$A(\hbar\omega) \propto \Im\left(\int_0^\infty p(t')\mathrm{e}^{\mathrm{i}\omega t'}\mathrm{d}t'\right) \propto \frac{\Gamma}{(\omega-\omega_\circ)^2+\Gamma^2} . \tag{2.42}$$

Here, $\Gamma = \frac{\hbar}{T_2}$ is the energy broadening of the resulting Lorentzian spectral line, usually referred to as “homogeneous broadening.”

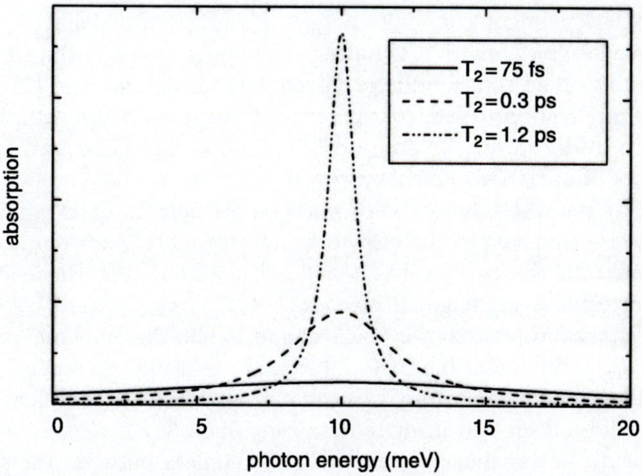

Fig. 2.5. Optical-absorption spectra corresponding to the three results of Fig. 2.4. For decreasing values of the decay time T_2 we observe an increase of the absorption line-width Γ (see (2.42))

Figure 2.5 shows the optical-absorption spectra corresponding to the three results of Fig. 2.4: As expected, for decreasing values of the decay time T_2 we observe an increase in the absorption line-width Γ.

As we shall discuss in the following section, while for the coherent as well as for the incoherent regime previously identified (see Fig. 2.4) the quantitative description of semiconductor devices may still be grounded on simplified – i.e., Schrödinger-like or semiclassical – models, any realistic simulation of the intermediate regime requires a fully microscopic approach able to treat on the same footing both coherent and incoherent phenomena.

Let us now discuss the extension of the simplified two-level description presented so far to the realistic case of a semiconductor crystal. As recalled in Sect. 1.2, a slowly varying optical excitation induces vertical transitions from a given occupied band $\nu = v$ (called valence band) to a given empty band $\nu = c$ (called conduction band); each of these transitions may be regarded as a two-level system formed by the Bloch states $\boldsymbol{k}v$ and $\boldsymbol{k}c$. It follows that, qualitatively speaking, our semiconductor crystal may be regarded as a collection of $\boldsymbol{k}$-dependent two-level systems driven by the same applied excitation. Indeed, by performing an heuristic and straightforward generalization of the optical Bloch equations (2.40), one gets

$$\begin{aligned}
\frac{\mathrm{d}f_{\boldsymbol{k}v}}{\mathrm{d}t} &= 2\Re\left(\frac{U_{\boldsymbol{k}}p_{\boldsymbol{k}}}{\mathrm{i}\hbar}\right) - \frac{f_{\boldsymbol{k}v} - f^\circ_{\boldsymbol{k}v}}{T_1} \\
\frac{\mathrm{d}f_{\boldsymbol{k}c}}{\mathrm{d}t} &= -2\Re\left(\frac{U_{\boldsymbol{k}}p_{\boldsymbol{k}}}{\mathrm{i}\hbar}\right) - \frac{f_{\boldsymbol{k}c} - f^\circ_{\boldsymbol{k}c}}{T_1} \\
\frac{\mathrm{d}p_{\boldsymbol{k}}}{\mathrm{d}t} &= \frac{\epsilon_{\boldsymbol{k}c} - \epsilon_{\boldsymbol{k}v}}{\mathrm{i}\hbar}p_{\boldsymbol{k}} + \frac{U^*_{\boldsymbol{k}}}{\mathrm{i}\hbar}(f_{\boldsymbol{k}v} - f_{\boldsymbol{k}c}) - \frac{p_{\boldsymbol{k}}}{T_2}\,, \qquad (2.43)
\end{aligned}$$

where the $\boldsymbol{k}$-dependent coupling energies $U_{\boldsymbol{k}}$ are proportional to the dipole matrix elements $p_{\boldsymbol{k},cv}$ in (1.27). Indeed, the above result is the simplest version of the so-called semiconductor Bloch equations (see, e.g., [33, 90, 115]), i.e., the generalization to semiconductors of the optical Bloch equations (2.40) originally introduced for the study of atomic and molecular systems. Again, the new set of variables, given by the Bloch-state occupations $f_{\boldsymbol{k}v}$ and $f_{\boldsymbol{k}c}$ as well as by the interband ($v \rightarrow c$) microscopic polarizations $p_{\boldsymbol{k}}$, correspond to well precise elements of the electronic density matrix written in the Bloch basis: Similar to the two-level-system case in (2.35), the Bloch-state occupations correspond to diagonal elements ($f_{\boldsymbol{k}v} = \rho_{\boldsymbol{k}v,\boldsymbol{k}v}$ and $f_{\boldsymbol{k}c} = \rho_{\boldsymbol{k}c,\boldsymbol{k}c}$), while the interband polarizations correspond to non-diagonal matrix elements ($p_{\boldsymbol{k}} = \rho_{\boldsymbol{k}c,\boldsymbol{k}v}$), thus describing the interband quantum-mechanical phase coherence. The meaning and role of diagonal versus non-diagonal density-matrix elements will be discussed in more detail in Chap. 3.

In order to better illustrate the main differences between the single two-level system and the semiconductor crystal, let us consider again the ultrafast-optical-excitation regime in (2.41), whose semiconductor version is given by

$$p_{\boldsymbol{k}}(t) = p_{\boldsymbol{k}}(0)\mathrm{e}^{-\mathrm{i}\omega_{\boldsymbol{k}}t}\mathrm{e}^{-\frac{t}{T_2}} , \tag{2.44}$$

where $\hbar\omega_{\boldsymbol{k}} = \epsilon_{\boldsymbol{k}c} - \epsilon_{\boldsymbol{k}v}$ denotes the interband energy difference. This tells us that within such model, for any $\boldsymbol{k}$ value, we deal with an interband microscopic polarization characterized by a constant decay time T_2 and by a $\boldsymbol{k}$-dependent rotation frequency $\omega_{\boldsymbol{k}}$. As discussed in [110], it is possible to show that conventional optical experiments are not able to detect such microscopic polarizations individually; in contrast, the measured quantity is the total polarization p^{tot} induced by our optical excitation, i.e.,

$$p^{\mathrm{tot}}(t) = \sum_{\boldsymbol{k}} U_{\boldsymbol{k}} p_{\boldsymbol{k}}(t) . \tag{2.45}$$

By neglecting the $\boldsymbol{k}$-dependence of the dipole matrix element, the above total polarization will be of the form.

$$p^{\mathrm{tot}}(t) \propto \mathrm{i}\sum_{\boldsymbol{k}} \mathrm{e}^{-\mathrm{i}\omega_{\boldsymbol{k}}t}\mathrm{e}^{-\frac{t}{T_2}} . \tag{2.46}$$

To elucidate the main features of the total-polarization dynamics, Fig. 2.6 shows the modulus of the total polarization (2.46) as a function of time for the same three T_2 values considered in Fig. 2.4; here, a simplified parabolic-band GaAs bulk model has been employed. As we can see, in spite of the relatively different T_2 values, the ultrafast polarization dynamics (i) is nearly T_2 independent and (ii) is decaying on a much faster time-scale. In order to understand this peculiar feature, it is useful to rewrite the total polarization (2.46) as

$$p^{\mathrm{tot}}(t) \propto \mathrm{i}\mathrm{e}^{-\frac{t}{T_2}} f_p(t) \tag{2.47}$$

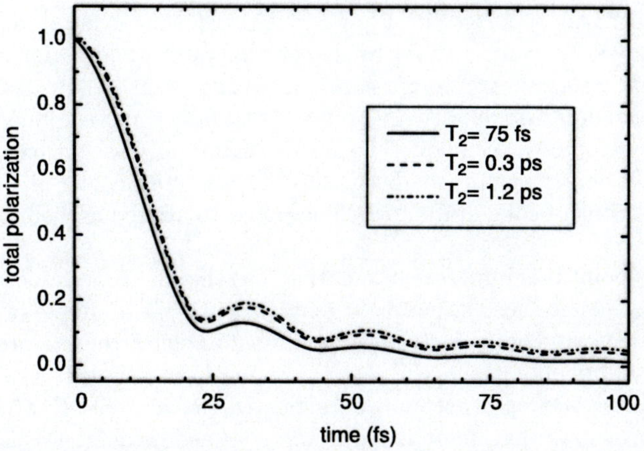

Fig. 2.6. Modulus of the total polarization (2.46) as a function of time for the same three T_2 values considered in Fig. 2.4; here, a simplified parabolic-band GaAs bulk model has been employed. In spite of the relatively different T_2 values, the ultrafast polarization dynamics is nearly T_2 independent and is decaying on a much faster time-scale (see text)

with

$$f_{\boldsymbol{p}}(t) = \sum_{\boldsymbol{k}} \mathrm{e}^{-\mathrm{i}\omega_{\boldsymbol{k}}t} \ . \tag{2.48}$$

While at the initial time ($t = 0$) all the quantum-mechanical phase factors in (2.48) sum up constructively, for increasing times the latter will start to rotate with different frequencies, leading to a destructive-interference dynamics on a time-scale much shorter than the individual microscopic-polarization decay T_2; such phenomenon is usually referred to as polarization dephasing.

Such basic difference between individual and total-polarization decay manifests itself also via the corresponding absorption response; indeed, by evaluating the imaginary part of the Fourier transform of the total polarization in (2.46), one gets

$$A^{\mathrm{tot}}(\hbar\omega) \propto \Im\left(\int_0^\infty p^{\mathrm{tot}}(t')\mathrm{e}^{\mathrm{i}\omega t'}\,\mathrm{d}t'\right) \propto \sum_{\boldsymbol{k}} \frac{\Gamma}{(\omega-\omega_{\boldsymbol{k}})^2+\Gamma^2} \ . \tag{2.49}$$

As we can see, the total absorption $A^{\mathrm{tot}}(\hbar\omega)$ is the sum of several Lorentzian peaks with different central frequencies $\omega_{\boldsymbol{k}}$. Recalling that in the limit $\Gamma \to 0$ the latter behave as Dirac delta functions, the total absorption (2.49) is qualitatively proportional to the J-DOS (1.30), which is fully consistent with the general result in (1.29). This shows that opposite to the previous single two-level case, the energetic broadening displayed by the total absorption – commonly referred to as "inhomogeneous broadening" – is basically a fingerprint of the system J-DOS, which is typically much larger than the individual Lorentzian line-width Γ.

The scenario reported in Fig. 2.6 tells us that conventional optical-absorption experiments – able to detect the total polarization in (2.45) – are not able to access the microscopic polarizations in (2.44); to this end, as discussed in detail in [90], it is necessary to adopt more refined ultrafast-optical-spectroscopy techniques based on many-pulse sequences, like, e.g., photon-echo experiments (see Sect. 2.5). This is indeed one of the most severe limitations of solid-state optics compared to atomic as well as molecular systems.

At this point it is imperative to notice that the semiconductor Bloch equations (2.43) – introduced as heuristic generalization of the simple two-level case in (2.40) – describe both electron–environment and electron–electron coupling via the T_1/T_2 phenomenological model previously mentioned. As discussed in detail in [110], a proper account of carrier–carrier as well as carrier–phonon interactions shows that, in addition to more refined incoherent-dynamics contributions, one deals with coherent quantum-mechanical corrections to the basic light–matter dynamics in (2.43), leading, e.g., to Hartree–Fock as well as polaronic renormalization effects. It follows that a rigorous (i.e., non-phenomenological) treatment of ultrafast optical excitations in semiconductors requires fully microscopic treatments far beyond the T_1/T_2 model in (2.43).

As a very final step, let us briefly discuss the extension of the phenomenological semiconductor Bloch equations (2.43) to the case of a semiconductor nanostructure. As shown in Chap. 3, a microscopic treatment of electro-optical processes in semiconductor nanomaterials requires a fully quantum-mechanical many-body approach; the latter is commonly based on an effective set of equations for the relevant single-particle density-matrix elements. More specifically, by denoting with α the generic electronic state of our semiconductor nanostructure, it is possible to show that the nanostructure generalization of the semiconductor Bloch equations (2.43) is always of the form

$$\frac{d\rho_{\alpha\alpha'}}{dt} = \left.\frac{d\rho_{\alpha\alpha'}}{dt}\right|_{1e} + \left.\frac{d\rho_{\alpha\alpha'}}{dt}\right|_{mb}, \tag{2.50}$$

i.e., the global time evolution of the density-matrix element $\rho_{\alpha\alpha'}$ is given by a one-electron (1e) contribution, describing the electronic dynamics in the presence of our electro-optical excitation, plus a many-body (mb) contribution, describing electron–electron as well as various electron–quasiparticle couplings (e.g., carrier–phonon, carrier–plasmon). While the first term may be treated exactly, the second one is treated within some approximation scheme.

By recalling that for the case of a semiconductor nanostructure, an optical excitation – according to its spectral range – may induce transitions between different electronic bands (interband transitions) or within the same band (intraband transitions), it follows that the relevant elements of the electronic density matrix involved depend crucially on the excitation spectral range. More specifically, by employing the envelope-function picture introduced in Sect. 1.2.2 we have $\alpha \equiv \nu, \overline{\alpha} \equiv \nu, \boldsymbol{k}_\parallel n$, i.e., the generic envelope-function state α is specified by its band index ν plus its envelope-function index $\overline{\alpha}$

(see (1.41)); the latter depends on the nanostructure dimensionality D and is given by the parallel-subspace wavevector $\boldsymbol{k}_\parallel$ and by the corresponding subband index n (see (1.43)). It follows that in addition to the state populations corresponding to the diagonal density-matrix elements ($f_\alpha = \rho_{\alpha\alpha}$), for intraband ($\nu = \nu'$) excitations the only relevant quantities are given by intraband polarizations $p^{\rm intra}_{\nu\boldsymbol{k}_\parallel,nn'}$ corresponding to the non-diagonal density-matrix elements $\rho_{\nu\boldsymbol{k}_\parallel n,\nu\boldsymbol{k}_\parallel n'}$; the latter are also referred to as intersubband ($n \to n'$) polarizations. In contrast, for interband ($\nu \neq \nu'$) excitations the only relevant quantities are given by interband polarizations $p^{\rm inter}_{\boldsymbol{k}_\parallel,\nu n,\nu' n'}$ corresponding to the non-diagonal density-matrix elements $\rho_{\nu\boldsymbol{k}_\parallel n,\nu'\boldsymbol{k}_\parallel n'}$; the latter describe interband optical transitions connecting the generic subband n in band ν with the generic subband n' in band ν'. Compared to the original semiconductor Bloch equations (2.43), moving from bulk to nanostructures the three-dimensional wavevector $\boldsymbol{k}$ is replaced by the parallel-subspace wavevector $\boldsymbol{k}_\parallel$, while – in addition to the band index – the subband index n should be considered as well.

As we shall discuss extensively in the following chapters, the approximation level employed for the description of the many-body contribution in (2.50) depends crucially on the specific problem under investigation; generally speaking, we shall adopt specific perturbation expansions combined with corresponding adiabatic-decoupling schemes. In particular, for slowly varying electro-optical excitations in the longtime regime, it is possible to show that all non-diagonal density-matrix elements may be adiabatically eliminated, leaving as relevant variables the electronic-state populations f_α of the conventional semiclassical picture introduced in Sect. 1.3. It follows that in this limit the nanostructure version of the semiconductor Bloch equations (2.50) reduces to the Boltzmann-like equation (1.73), where the light–matter interaction may be treated stochastically via corresponding intra- as well as interband transition probabilities $P_{\alpha\alpha'}$. This suggests that the semiclassical picture is just a limiting case of the density-matrix formalism presented in Chap. 3.

As a final remark, let us discuss the concept of phase coherence in connection with the choice of representation. On the basis of the density-matrix picture – whose general formulation will be given in the following chapter – the quantum-mechanical phase coherence is described by the non-diagonal density-matrix elements $\rho_{\alpha\alpha'}$ in (2.50). However, this separation in diagonal versus non-diagonal terms is clearly representation dependent: what is diagonal in a given basis is in general non-diagonal in a different basis and vice versa. Indeed, if one considers, as basis set, the eigenstates of the global Hamiltonian (describing system plus electro-optical excitation), the density matrix written in this new basis is always diagonal, i.e., no phase coherence. Thus, in order to speak of optically induced phase coherence, we need to regard the global Hamiltonian as the sum of the system Hamiltonian plus an external driving-force term (see (2.6)). Provided such separation between system-of-interest and driving force, the non-diagonal density-matrix elements within the representation given by the system eigenstates will describe the degree of quantum-mechanical phase coherence induced on the system by the driving force.

As anticipated, in a semiconductor material we deal with a variety of many-body effects, not described/included in the usual Bloch state picture α; it follows that the many-body term in (2.50) induces, in general, a sort of phase coherence $\rho_{\alpha\alpha'}$ between the single-particle Bloch states α and α'. Such a coherent phenomenon is obviously fictitious and is simply ascribed to the fact that the Bloch state basis differs from the rigorous eigenstates of the full many-body semiconductor Hamiltonian. Two typical examples of such fictitious phase coherence may help in clarifying this point.

The first case, discussed in Chap. 9, is that of Coulomb-induced phase coherence: Within a Coulomb-free representation, even at the simplest Hartree–Fock level, Coulomb interaction induces phase coherence between Coulomb-free states. This is clearly not the case within an excitonic basis (see (1.58)), where the density matrix is always diagonal.

The second and more intriguing case, discussed in Sect. 3.4, is that of a semiconductor superlattice in the presence of a constant and homogeneous electric field (see, e.g., [32]). As we shall see, in this case we deal with two physically equivalent representations (simply connected by a gauge transformation) leading to different descriptions of transport phenomena: Within the so-called Wannier–Stark representation the Bloch-oscillation dynamics is the result of phase coherence between the various Wannier–Stark states; in contrast, within the accelerated Bloch-state representation the same phenomenon is purely described on a semiclassical level in terms of carrier populations, i.e., no off-diagonal density-matrix elements.

2.3 Space-Dependent Phenomena

The discussion presented so far has been intentionally restricted to the case of slowly varying electro-optical excitations. In this case, the neglect of the photon momentum – also referred to as dipole approximation – leads to the vertical-transition picture previously introduced; indeed, according to the semiconductor Bloch equation (2.43), a slowly varying optical excitation creates a quantum-mechanical phase coherence between valence and conduction Bloch states characterized by the same wavevector $\boldsymbol{k}$ – expressed by the interband polarization $p_{\boldsymbol{k}}$ – which, in turn, leads to after-pulse populations $f_{\boldsymbol{k}v}$ and $f_{\boldsymbol{k}c}$. As anticipated, such kinetic variables correspond to interband and intraband density-matrix elements, always diagonal in $\boldsymbol{k}$: $f_{\boldsymbol{k}\,v/c} = \rho_{\boldsymbol{k}\,v/c,\boldsymbol{k}\,v/c}$ and $p_{\boldsymbol{k}} = \rho_{\boldsymbol{k}c,\boldsymbol{k}v}$. As discussed in more detail in [110], given a generic electronic-state basis α, the electron spatial density may be easily expressed in terms of the density matrix $\rho_{\alpha\alpha'}$ previously introduced according to

$$n(\boldsymbol{r}) = \sum_{\alpha\alpha'} \rho_{\alpha\alpha'} \psi_\alpha(\boldsymbol{r}) \psi^*_{\alpha'}(\boldsymbol{r}) \ . \qquad (2.51)$$

For the particular case of the two-band model of the semiconductor Bloch equations (2.43), we get

$$n(\boldsymbol{r}) = \sum_{\boldsymbol{k}} \left[f_{\boldsymbol{k}v} |\phi_{\boldsymbol{k}v}(\boldsymbol{r})|^2 + f_{\boldsymbol{k}c} |\phi_{\boldsymbol{k}c}(\boldsymbol{r})|^2 + 2\Re \left(p_{\boldsymbol{k}} \phi_{\boldsymbol{k}c}(\boldsymbol{r}) \phi^*_{\boldsymbol{k}v}(\boldsymbol{r}) \right) \right] . \quad (2.52)$$

In view of the periodicity properties of the conventional Bloch states $\phi_{\boldsymbol{k}\nu}$ in (1.8), the electronic spatial density (2.52) is always periodic, thus extending over the whole semiconductor crystal; this is indeed consistent with the assumption of a slowly varying, i.e., nearly constant, electromagnetic excitation.

Let us now consider the problem of a spatially localized excitation. In this case, opposite to the conventional semiconductor Bloch equations (2.43), it is possible to show that such an optical excitation induces diagonal ($\boldsymbol{k} = \boldsymbol{k}'$) as well as non-diagonal ($\boldsymbol{k} \neq \boldsymbol{k}'$) interband polarizations $p_{\boldsymbol{k}\boldsymbol{k}'} = \rho_{\boldsymbol{k}c,\boldsymbol{k}'v}$, giving rise to diagonal as well as non-diagonal valence and conduction density-matrix elements $f^v_{\boldsymbol{k}\boldsymbol{k}'} = \rho_{\boldsymbol{k}v,\boldsymbol{k}'v}$ and $f^c_{\boldsymbol{k}\boldsymbol{k}'} = \rho_{\boldsymbol{k}c,\boldsymbol{k}'c}$. Therefore, for the case of a spatially localized excitation the electron density (2.52) should be replaced by

$$n(\boldsymbol{r}) = \sum_{\boldsymbol{k}\boldsymbol{k}'} \left[f^v_{\boldsymbol{k}\boldsymbol{k}'} \phi_{\boldsymbol{k}v}(\boldsymbol{r}) \phi^*_{\boldsymbol{k}'v}(\boldsymbol{r}) + f^c_{\boldsymbol{k}\boldsymbol{k}'} \phi_{\boldsymbol{k}c}(\boldsymbol{r}) \phi^*_{\boldsymbol{k}'c}(\boldsymbol{r}) + 2\Re \left(p_{\boldsymbol{k}\boldsymbol{k}'} \phi_{\boldsymbol{k}c}(\boldsymbol{r}) \phi^*_{\boldsymbol{k}'v}(\boldsymbol{r}) \right) \right] . \quad (2.53)$$

Contrary to the result in (2.52), due to non-vertical ($\boldsymbol{k} \neq \boldsymbol{k}'$) contributions, the electron spatial density (2.53) is not periodic anymore; in particular, it is possible to show that the latter exhibits the same spatial localization of the applied optical excitation, a quite intuitive result. It follows that, also within a periodic crystal, a spatially localized excitation may produce a strongly nonhomogeneous charge distribution, whose after-pulse dynamics requires a non-vertical density-matrix formulation.

In a similar way, a slowly varying intraband excitation applied to a semiconductor nanostructure will induce intersubband polarizations, resulting in nonhomogeneous valence and conduction carrier densities. Indeed, as discussed previously, for the case of semiconductor nanostructures the time evolution of the electronic density matrix $\rho_{\alpha\alpha'}$ is governed by the semiconductor Bloch equations (2.50) which, in turn, allows us to evaluate the time evolution of the electronic spatial density $n(\boldsymbol{r})$ in (2.51).

According to the semiclassical picture introduced in Sect. 1.3, for space-scales compatible with the position–momentum uncertainty principle, such a space-dependent carrier dynamics is well described via the conventional Boltzmann transport theory; in contrast, on very short space-scales the semiclassical picture fails, and fully quantum-mechanical approaches are imperative. For a quantitative study of space-dependent phenomena it is then crucial to establish a direct link between the classical picture and the quantum-mechanical density-matrix formulation; this is realized by the well-known Wigner-function formalism (see, e.g., [85] and references therein) presented in Chap. 4.

The key ingredient of the Wigner formulation of quantum mechanics is the so-called Wigner function, i.e., a function – originally introduced by Eugene

Wigner in 1932 [116] – defined over the ordinary phase-space $\boldsymbol{r}, \boldsymbol{p}$ as the following Weyl–Wigner transform of the single-particle density matrix:

$$f_W(\boldsymbol{r}, \boldsymbol{p}) = \sum_{\alpha\alpha'} \mathcal{W}_{\alpha\alpha'}(\boldsymbol{r}, \boldsymbol{p}) \rho_{\alpha\alpha'} \tag{2.54}$$

with

$$\mathcal{W}_{\alpha\alpha'}(\boldsymbol{r}, \boldsymbol{p}) = \int d^3 r' \psi_\alpha \left(\boldsymbol{r} + \frac{\boldsymbol{r}'}{2}\right) \frac{e^{-\frac{i\boldsymbol{p}\cdot\boldsymbol{r}'}{\hbar}}}{(2\pi\hbar)^{\frac{3}{2}}} \psi^*_{\alpha'} \left(\boldsymbol{r} - \frac{\boldsymbol{r}'}{2}\right) . \tag{2.55}$$

Combining the above definition with the general expression for the electron spatial density (2.51), it is easy to verify that

$$n(\boldsymbol{r}) \propto \int f_W(\boldsymbol{r}, \boldsymbol{p}) d^3 p , \tag{2.56}$$

i.e., the spatial density is proportional to the integral over $\boldsymbol{p}$ of the Wigner function (2.54). In spite of the strong similarities with the classical phase-space picture, it is important to stress that, in general, the Wigner function (2.54) is not positive-definite, and therefore the latter does not allow for a probabilistic interpretation. However, for slowly varying (in space) optical excitations and confinement-potential profiles, the Wigner function (2.54) is always positive and is well described via effective semiclassical transport models. In contrast, as already pointed out, in the presence of ultrashort spatial confinement and/or strongly inhomogeneous optical excitations, the fully non-diagonal density-matrix formalism of the semiconductor Bloch equations (2.50) is imperative, giving rise to highly non-classical Wigner-function profiles.

As a final remark, it is important to stress the strong similarities between ultrashort time and space phenomena. On the one hand, for the case of ultrafast phenomena the semiclassical picture comes out to be highly inadequate, since in addition to state populations it is crucial to consider intraband as well as interband polarizations; the latter – describing quantum-mechanical interference phenomena among different energy states – give rise to the well-known time–energy uncertainty relation. However, in the presence of slowly varying excitations, the polarizations may be adiabatically eliminated (see, e.g., [110]) and – in spite of the quantum-mechanical nature of the problem – the electronic dynamics may be well described via effective semiclassical equations. On the other hand, in the presence of strongly localized excitations, in addition to electron populations, it is crucial to consider non-vertical ($\boldsymbol{k} \neq \boldsymbol{k}'$) density-matrix elements. The latter – describing quantum-mechanical interference among different momentum values – reflect the well-known position-momentum uncertainty relation. Again, for slowly varying perturbations (in space) all non-diagonal density-matrix elements may be adiabatically eliminated (see, e.g., [117]), and our space-dependent problem may be safely treated via effective Boltzmann equations.

2.4 Quantum Systems with Spatial Boundaries

The quantum-mechanical description presented so far applies to the so-called closed systems, i.e., semiconductor materials extending over the whole coordinate space $\boldsymbol{r}$. In contrast, as already pointed out in Sect. 1.4, a semiconductor device [79, 81–84] is typically an "open system," i.e., a portion of material characterized by a well precise volume Ω_d and spatial boundaries $\boldsymbol{r}^b$ acting as electric contacts. It follows that, generally speaking, the presence of spatial boundaries will induce modifications to the semiconductor-nanostructure Bloch equations (2.50), whose open-system version may be schematically written as

$$\frac{d\rho_{\alpha\alpha'}}{dt} = \left.\frac{d\rho_{\alpha\alpha'}}{dt}\right|_{1e} + \left.\frac{d\rho_{\alpha\alpha'}}{dt}\right|_{mb} + \left.\frac{d\rho_{\alpha\alpha'}}{dt}\right|_{b} . \tag{2.57}$$

Here, the last term describes the effect on the time evolution of the density matrix $\rho_{\alpha\alpha'}$ induced by our spatial boundaries.

As discussed in great detail in Chap. 4, the incorporation of spatial boundary conditions within a quantum-mechanical framework is a highly non-trivial task. In view of the intrinsic space-dependent character of the problem under examination, the most natural strategy seems to move from the density-matrix formalism to the Wigner picture via the Weyl–Wigner transform (2.55), and then to apply to the Wigner function (2.54) the standard U boundary-condition scheme discussed in Appendix B. However, in spite of the formal similarity between semiclassical and Wigner pictures, it comes out that such conventional boundary-condition scheme is definitely not compatible with the non-local character of quantum mechanics, leading to highly non-physical results like, e.g., negative electronic spatial densities (see Sect. 4.2).

To overcome this severe limitation, two alternative strategies may be adopted: (i) properly designed projection techniques (see Sect. 4.3) allowing, still within an open-system density-matrix framework, to incorporate spatial boundary effects via generalized injection/loss contributions and (ii) simplified closed-system approaches (see Sect. 4.4) describing the device-environment coupling dynamics via properly designed particle-conserving scattering superoperators.

2.5 Experimental Techniques

The investigation of nonequilibrium carrier dynamics in optically excited semiconductors started in the late 1960s with the analysis of the energy-relaxation process (see, e.g., [118]) by using continuous-wave excitation. The measurement of the carrier temperature as a function of the laser intensity – obtained from the luminescence spectrum – gave insight into the power loss from the carriers to the lattice, i.e., about carrier–phonon scattering processes. In subsequent years, these investigations were extended to different materials and excitation conditions. While band-to-band luminescence spectra gave only a combination of electron and hole temperatures (see, e.g., [90]), direct information on the electronic distribution function was obtained by studying band-to-acceptor luminescence spectra of doped semiconductors (see, e.g., [119]).

However, as anticipated, only by using a pulsed excitation, dynamical processes can be directly investigated. Here, the pulse duration limits the temporal resolution and therefore restricts the phenomena which can be studied. The typical time-scales for many of the processes discussed in this book are in the range between a few femtoseconds and a few picoseconds. Therefore, the application of time-resolved nonlinear optical spectroscopy to the study of dynamical processes in semiconductors has been strongly linked to the ability of producing laser pulses on these time-scales. Such laser sources became available for semiconductor studies in the late 1970s (see, e.g., [120]). Since then a great number of phenomena have been studied, first mainly focusing on incoherent dynamics, i.e., the nonequilibrium dynamics of distribution functions, and subsequently analyzing more and more coherent phenomena, i.e., the dynamics of optically created interband and intraband polarizations.

A typical scenario for the dynamics of carrier distribution functions is plotted schematically in Fig. 2.7: The laser pulse with a given photon energy and a certain spectral width determined by its duration creates electron–hole pairs in a more or less localized region in $\boldsymbol{k}$-space. This initial distribution then relaxes due to the presence of scattering processes. In polar semiconductors on ultrafast time-scales there are typically two mechanisms of particular importance: Due to the polar coupling to longitudinal optical (LO) phonons

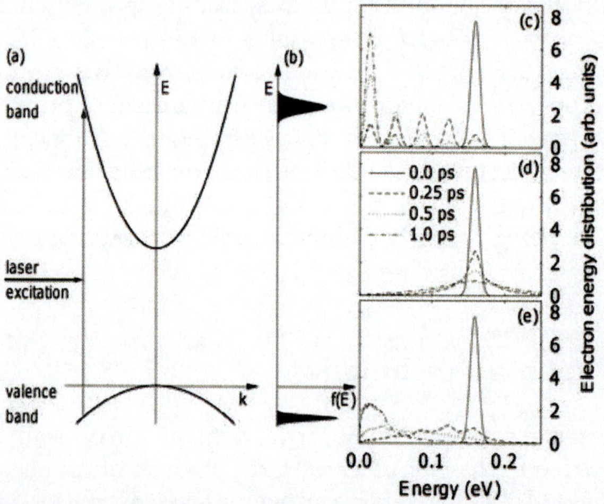

Fig. 2.7. Schematic plot of (**a**) the excitation process by a short laser pulse with a certain excess energy above the bandgap and (**b**) the resulting carrier distribution of electrons and holes as well as its subsequent energy relaxation due to (**c**) electron–phonon scattering, (**d**) electron–electron scattering, and (**e**) both types of processes (see text). Reprinted from [110]

the carriers may transfer a significant fraction of their initial kinetic energy to the lattice. Since optical phonons in the relevant region close to the center of the Brillouin zone have a negligible dispersion, this leads to the build-up of the so-called phonon-replica of the initial distribution shifted downward by multiples of the phonon energy (see Fig. 2.7c). The scattering among the electrons themselves due to the Coulomb interaction, on the other hand, conserves the total kinetic energy; however it leads to a spreading in $\boldsymbol{k}$-space (see Fig. 2.7d) and eventually, at later times, to a Fermi–Dirac distribution where the temperature is determined by the initial excess energy. If, as is always the case in a real semiconductor, both mechanisms are present, both energy relaxation and thermalization toward a Fermi–Dirac distribution occur simultaneously (see Fig. 2.7e), the respective time-scales, however, being strongly dependent on the excitation conditions, in particular on the carrier density.

These thermalization and relaxation processes have been studied in great detail in the past three decades both experimentally and theoretically in bulk semiconductor materials as well as in a variety of nanostructures. The most commonly used experimental techniques for these studies have been luminescence, where the photons created by the radiative recombination of electrons and holes are detected, and pump-probe (PP) measurements, where the change in the absorption (or reflection) of a probe beam caused by the prior excitation of electron–hole pairs by the pump beam is observed. While it is nearly impossible to quote all the work, we just mention some of the phenomena which have turned out to be important under certain excitation conditions. In the case of sufficiently high excitation densities it has been found that the distribution function of LO-phonons is driven substantially out-of-equilibrium and that this so-called hot-phonon effect, described in Sect. 6.3, may drastically reduce the cooling process (see, e.g., [121]). The dynamics of the nonequilibrium phonons has been studied directly by Raman measurements (see, e.g., [122]). If the excitation energy is above the threshold for transitions to satellite valleys in the conduction band, intervalley scattering due to carrier–phonon interaction is a very effective process mainly because of the high density of states in these valleys (see, e.g., [123]).

Starting from the mid-1980s the field of coherent excitations in semiconductors became an increasingly active research area. Even if in the measurements mentioned above essentially carrier relaxation processes are monitored, it turned out that also in these signals features related to coherence are present; in particular, under certain conditions coherent aspects may be dominant. In the case of PP spectra this holds most prominently if the pump pulse is non-resonant with optical transitions, i.e., if it is tuned into the bandgap region where it does not create real populations. Here, it gives rise to shifts and splittings of the exciton line which is known as the optical Stark effect. Starting from the mid-1980s this effect has been extensively investigated (see, e.g., [124]).

Besides PP and luminescence measurements, however, there are other techniques which rely completely on the phase coherence in the carrier system, thus providing a direct information on the dynamics of coherent interband and intraband polarizations. The most prominent among these techniques are four-wave-mixing (FWM) experiments and the detection of coherently emitted radiation in the terahertz spectral region.

As anticipated, many solid-state systems exhibit an inhomogeneous broadening (see Fig. 2.6). In the case of a semiconductor in the excitonic region this is typically due to some structural disorder in the sample while in the band-to-band continuum region the $\boldsymbol{k}$-dependence of transition energies may also be interpreted as such a broadening. Then, as anticipated, the total polarization rapidly decays due to destructive interference of the different frequency components and, thus, it is difficult to extract information on the true loss of phase coherence due to decoherence processes. In the case of magnetic resonance the technique of spin echoes has been introduced already in 1950 by Erwin L. Hahn [125]; the latter is able to eliminate the decay due to inhomogeneous broadening (see (2.48) in Sect. 2.2), and decoherence times – the so-called T_2 times – can be measured. In the 1960s, due to the availability of laser sources, echo experiments have been brought into the optical spectral range; photon echoes have been first observed in ruby [126]. Since in semiconductor materials decoherence is much faster, very short pulses are required here for such techniques. In 1985 photon echoes from delocalized excitons in semiconductors have been observed by using 7 ps laser pulses [127], and a few years later, photon echoes from band-to-band transitions have been measured with 6 fs pulses [128]. These photon echoes are typically studied by means of degenerate FWM measurements which will be described below. With such experiments an excitonic phase-coherence time-scale of about 7 ps and a well precise density dependence have been established.

But such experiments do not only provide information on the decay of carrier phase coherence; instead, many other coherent phenomena have been studied in the past, like quantum beats due to quantum-mechanical superpositions of electronic states (see, e.g., [129–132]). If such superpositions are excited between states with different spatial localizations, they are the source of an electromagnetic radiation with a frequency given by the energy splitting (between these states). This frequency is typically in the terahertz spectral range, and in this regime the electric-field strength may be directly measured, in contrast to the optical regime where typically only intensities can be measured. Such a terahertz emission has first been observed from asymmetric double quantum-well structures [133], thus opening the way to the field of terahertz spectroscopy in semiconductors (see, e.g., [134] and references therein).

Another direct approach to coherent phenomena is the technique of coherent control by two (temporally separated) phase-locked optical pulses (see, e.g., [135]). If the optical polarization created by the first pulse is still present in the sample, this polarization can constructively or destructively interfere with the second pulse, leading to a quantum-mechanical carrier dynamics which strongly depends on the relative phase of the two pulses. Such non-trivial dynamics can then be probed by the reflection or transmission change via, e.g., PP measurements, or by monitoring the FWM signal induced by a third pulse. It should be noted that there is a second type of coherent-control experiments where a superposition of a one-photon and a two- or three-photon excitation by two simultaneous pulses is used to control the final state in the case of degeneracy (see, e.g., [136] and references therein).

Generally speaking, coherent control makes use of the full time-dependence of the electric-field vector of the light pulse including intensity, phase, and polarization. An overview of different applications of coherent-control techniques can be found in [137].

Coherent phenomena do not only take place in the electronic subsystem of the semiconductor. Indeed, in spatially inhomogeneous systems or in systems with sufficiently low symmetry the optical excitation may also give rise to the generation of coherent phonons, i.e., phonons with a non-vanishing expectation value of the lattice displacement, in contrast to incoherent phonons where only the mean square displacement is non-zero. The excitation of coherent phonons in semiconductors has first been observed by optical excitation in the surface field of n-doped GaAs [138]. Here, the differential reflectivity change exhibited clear modulations with the phonon frequency. If the generated phonons are infrared-active they will also directly emit an electromagnetic radiation with the corresponding frequency in the terahertz range which again can be detected in the way described above.

As mentioned previously, for the study of carrier energy-relaxation processes essentially two different classes of experiments have been employed: luminescence and pump-probe measurements. In both cases a pump pulse is used to generate electron–hole pairs, thus bringing the semiconductor material in a state far from thermal equilibrium. In a luminescence experiment the radiation emitted in a direction different from that of the incident pulse due to recombination processes is analyzed spectrally and/or temporally. This is shown schematically in Fig. 2.8a. Depending on the temporal resolution, different techniques have to be employed: A temporal resolution in the range of 10 ps can be achieved by direct techniques, either by using fast photodiodes or a streak camera which provides spectral and temporal information simultaneously. Higher time resolution is obtained by gating the luminescence signal with a second delayed laser pulse. Both signal and delayed pulse are focused on a nonlinear crystal, which creates a sum-frequency signal only in the presence of such gating pulse. In this upconversion technique the temporal resolution is limited by the laser-pulse duration only; thus, resolutions in the 10 fs range are possible.

In an intrinsic semiconductor the luminescence is due to the recombination of an electron in the conduction band with a hole in the valence band. In a fully incoherent picture, according to Fermi's golden rule, the signal is essentially proportional to the product of the distribution functions of electrons and holes. This complicates the interpretation of experimental results. An alternative approach is the use of doped semiconductors where, e.g., in p-doped samples,

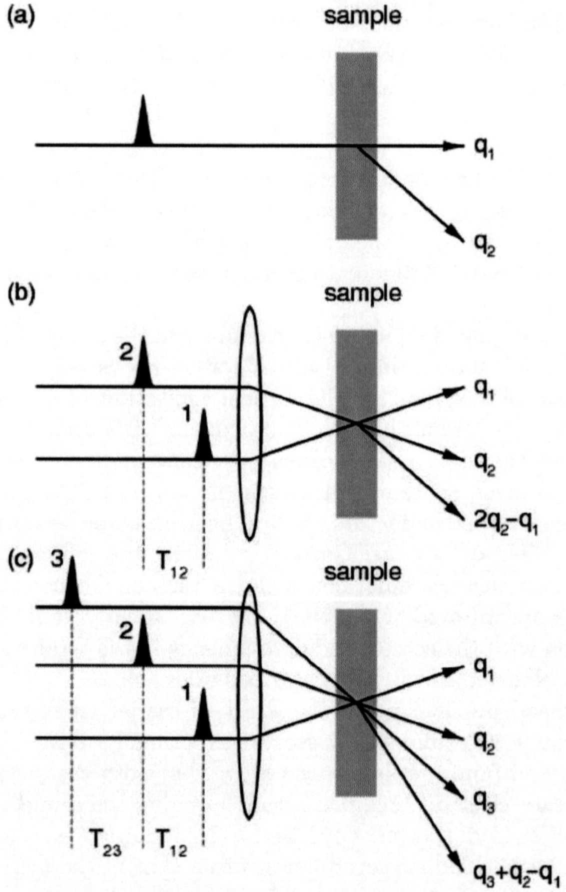

Fig. 2.8. Schematics of typical experimental setups for the study of ultrafast phenomena in semiconductor materials. In (**a**) the sample is excited by a single pulse and the secondary emission (resonant Rayleigh scattering or luminescence) is detected in a direction different from the incident one; (**b**) refers to pump-probe experiments or four-wave-mixing in the two-pulse self-diffraction geometry; and (**c**) refers to general three-pulse four-wave-mixing experiments (see text). Reprinted from [110]

the band-to-acceptor luminescence monitors directly the distribution function of electrons.

In a pump-probe experiment the semiconductor is excited by a pump pulse traveling in a direction $\boldsymbol{q}_1$ (see pulse 1 in Fig. 2.8b) and the dynamics of the carriers induced by this excitation is studied by looking at some property related to a delayed probe pulse in a direction $\boldsymbol{q}_2$. The most commonly used technique is transmission or reflection spectroscopy, where the change in the transmission or reflection of the probe pulse – induced by the pump – is measured as a function of the time delay between the two pulses.[5] By using a broadband probe pulse, differential transmission/reflection spectra are therefore obtained. In a purely incoherent free-carrier picture the absorption is changed due to phase-space filling, and these signals provide information on the sum of the electron and hole populations in the optically coupled states. Again, the interpretation of the results is facilitated if the spectra are determined by a single distribution function. This can be achieved by exploiting optical transitions for the probe in a different spectral range, e.g., by pumping heavy and light holes to conduction-band transitions and probing the split-off to conduction-band transition. A variation of the PP technique previously introduced is the electro-optic sampling, where the difference between two polarization components of the transmitted/reflected signal is analyzed; the latter provides information, e.g., on a birefringence induced by the optically excited dynamics. Instead of measuring the change in the transmitted/reflected signal, the change in the Raman-scattering signal generated by the probe pulse can also be measured. By using this technique the dynamics of photoexcited phonons as well as electronic excitations can be investigated.

It is clear that the interpretation of luminescence and PP experiments in terms of a fully incoherent free-carrier picture is valid only under limited conditions. In intrinsic semiconductors at sufficiently low carrier densities, absorption and luminescence spectra in the region close to the bandgap are strongly dominated by excitonic effects (see Fig. 1.21). Even high up in the band, pump-induced changes in the Coulomb enhancement (see Sect. 1.2.2) may significantly influence PP spectra. Furthermore, on time-scales comparable to or shorter than the characteristic decoherence times, the signals may be considerably modified by coherence effects. Therefore, a detailed analysis of luminescence and PP spectra in the ultrafast regime also provides information on coherent phenomena in the semiconductor material under examination.

The most popular technique which provides direct information on the carrier coherence in the semiconductor is four-wave-mixing spectroscopy. It can be performed both in a two-pulse and in a three-pulse configuration, as shown schematically in Fig. 2.8b,c.[6] For reasons of clarity, let us start with the three-pulse configuration by assuming that the time delay T_{12} between the two laser pulses 1 and 2 with wavevectors $\boldsymbol{q}_1$ and $\boldsymbol{q}_2$ is equal to 0. In this case these pulses create an interference pattern with wavevector $\pm(\boldsymbol{q}_2 - \boldsymbol{q}_1)$ on the sample which translates into a density grating when the light is absorbed. Such density grating results in a refractive index grating which may diffract a third pulse with incident wavevector $\boldsymbol{q}_3$ into various diffraction orders $\boldsymbol{q}_3 + n(\boldsymbol{q}_2 - \boldsymbol{q}_1)$, n being an integer number. In FWM measurements the first diffracted order $n = 1$ is measured; in this case, three incident waves interact giving rise to a fourth emitted wave which explains the name of this

[5] In Fig. 2.8 only the transmission case is plotted.

[6] Four-wave-mixing experiments can be performed in a reflection geometry as well.

technique. Access to the coherent polarization in the sample is now obtained if the pulses 1 and 2 are temporally separated. In this case there is no more direct interference pattern on the sample. However, as long as the microscopic interband polarization created by pulse 1 is at least partly still present when pulse 2 arrives, the interaction of a pulse with wavevector $\boldsymbol{q}_2$ with the interband polarization in the direction $\boldsymbol{q}_1$ again results in a transient grating which can diffract pulse 3. Thus, by varying T_{12} information is obtained on the dynamics and life-time of the polarization, i.e., on the decoherence time. It should be noted that a microscopic interband polarization may still be present even if, in the case of a continuous spectrum due to destructive interference of different microscopic components, there is no more macroscopic polarization in the sample (see Fig. 2.6). This is exactly the reason why FWM spectroscopy may discriminate between homogeneous and inhomogeneous broadening.

In the most frequently used two-pulse setup, pulse 2 simultaneously creates the grating and is diffracted by this grating; therefore it is also called the "self-diffraction geometry." Besides analyzing the signal in a time-integrated way it can also be spectrally dispersed in a monochromator or temporally resolved by means of an upconversion technique as discussed previously in the case of luminescence, which provides additional information on the dynamics of the interband polarization.

Both PP and FWM experiments can be employed to study coherent-control phenomena. In this case the pulse 1 is replaced by a pair of phase-locked pulses with variable delay traveling in the same direction $\boldsymbol{q}_1$. The carrier dynamics induced by these pulses then depends on the relative phase between these two pulses, and it can be analyzed by the second pulse traveling along the direction $\boldsymbol{q}_2$. Measuring the transmitted signal along $\boldsymbol{q}_2$ (or the corresponding reflected direction) and the FWM signal along $2\boldsymbol{q}_2 - \boldsymbol{q}_1$ yields in general complementary information on the dynamics of carrier distribution functions and polarizations.

If the optically excited interband polarizations couple electronic states with different spatial localizations, the dynamics is associated with a time-dependent dipole moment which, according to classical electrodynamics, acts as a source of electromagnetic radiation. Often this radiation is in the terahertz spectral range. Indeed, in a typical terahertz-emission experiment a short laser pulse excites the system in a superposition of states, thus creating an intraband polarization; The latter oscillates according to the energy splitting of the corresponding states and emits a pulse of electromagnetic radiation with the respective frequency. This radiation is then collimated and transmitted to an optically gated photoconductive antenna which measures the electric field of the radiation. Thus, the general setup is again according to Fig. 2.8a; the emitted signal in direction $\boldsymbol{q}_2$, however, being in the terahertz spectral region.

For a more detailed discussion of the experimental techniques previously recalled, we refer the reader to the book by Jagdeep Shah [90].

2.6 The Wide Family of Quantum Devices

The wide family of the so-called quantum devices [51, 56–69] can be divided into two main classes: a first one grouping semiconductor devices characterized by a genuine quantum-mechanical behavior of their electronic subsystem and a second one which comprises low-dimensional nanostructures whose optoelectronic response in steady-state conditions may be safely treated within the semiclassical picture previously introduced.

Devices within the first class – characterized by a relatively weak coupling of the electronic subsystem to the host material – are natural candidates for the implementation of quantum-information/computation processing [139–152]. These include, in particular, semiconductor quantum-dot structures [42, 44–49], for which all-optical implementations have been recently proposed (see, e.g., [51, 69] and references therein). In this case, the pure quantum-mechanical carrier dynamics is only weakly disturbed by dissipation and decoherence processes; therefore, the latter are usually described in terms of relatively simplified models.

Conversely, quantum devices in the second class – in spite of their partially discrete energy spectrum due to spatial quantum confinement – exhibit a carrier dynamics which for normal operation conditions can still be described via a semiclassical scattering picture. Such optoelectronic nanostructured devices include multiple-quantum-well and superlattice structures, like quantum-well infrared photodetectors (see, e.g., [64] and references therein) and quantum-cascade lasers (see, e.g., [66] and references therein). These systems are characterized by a strong interplay between coherent dynamics and energy-relaxation/decoherence processes; it follows that for a quantitative description of such non-trivial coherence/dissipation coupling the latter needs to be treated via fully microscopic models (see, e.g., [110]).

Based on the above subdivision, it is quite natural to identify two distinct regimes, determined both by the peculiar features of the nanomaterial involved and by the particular operation conditions. More specifically, as schematically summarized in Fig. 2.9, we deal with two different regimes, the semiclassical and the quantum-mechanical one. For both regimes it is possible to adopt a phenomenological description or a microscopic treatment of the problem. In particular, according to the classification scheme of Fig. 2.9, semiconductor devices operating within the semiclassical regime may be described either phenomenologically via simplified rate-equation models or microscopically via

	Semiclassical Regime	Quantum-mechanical Regime
Phenomenological Treatments	*Simplified Rate-Equation Models*	*Simplified Bloch-Equation Models*
Microscopic Treatments	*Realistic Boltzmann-like Approaches*	*Realistic Density-Matrix Approaches*

Fig. 2.9. General classification scheme of the various approaches employed for the study of semiconductor quantum devices (see text)

realistic Boltzmann-like treatments, while for devices operating within the quantum-mechanical regime we may adopt either a phenomenological description based on simplified Bloch-equation models or a microscopic description based, e.g., on realistic density-matrix or Green's function treatments (see Chap. 3).

A common feature of all phenomenological treatments is the description of electron–electron as well as electron–environment interaction mechanisms via relaxation-time models (see, e.g., (1.76), (2.40), and (2.43)); conversely, all microscopic treatments are based on a detailed knowledge of the various interaction Hamiltonians, and therefore they do not require any phenomenological parameter.

As anticipated, the primary goal of this book is to provide a cohesive description of many diverse quantum nanodevices, via a unified and fully microscopic density-matrix treatment of coherent versus incoherent electro-optical processes in semiconductor nanostructures.

Part I

Microscopic Description and Simulation Techniques

3

The Density-Matrix Approach

In this chapter we shall recall the fundamentals of the well-known density-matrix formalism (see, e.g., [112, 113]) applied to the investigation of the electro-optical properties of semiconductor nanomaterials and nanodevices. To this end, the reader is assumed to have some acquaintance with the general formulation of quantum mechanics (see, e.g., [86, 153]) including the fundamentals of the second-quantization formalism (see, e.g., [154]), as well as with basic statistical physics (see, e.g., [155, 156]).

It is worth mentioning that an alternative approach, equivalent to the density-matrix formalism considered in this book, is given by the nonequilibrium Green's function technique; the latter can be regarded as an extension of the well-known equilibrium or zero-temperature Green's function theory to nonequilibrium regimes, introduced in the 1960s by Leo P. Kadanoff and Gordon Baym [157] as well as by Leonid V. Keldysh [158]. An introduction to the theory of nonequilibrium Green's functions with applications to many problems in transport and optics of semiconductors can be found in the book by Hartmut Haug and Antti-Pekka Jauho [102]; by employing – and further developing and extending – such nonequilibrium Green's function formalism, a number of groups have proposed efficient quantum-transport treatments for the study of various meso- and nanoscale structures as well as of corresponding micro- and optoelectronic devices (see, e.g., [101] and references therein).

Within the general density-matrix formalism, two different strategies are commonly employed: (i) quantum-kinetic treatments (see, e.g., [110]) and (ii) global descriptions based on the Liouville–von Neumann equation (see, e.g., [108]).

The primary goal of a quantum-kinetic theory is the study of the temporal evolution of a reduced set of single- or few-particle quantities directly related to the electro-optical phenomenon under examination, the so-called kinetic variables of the system. However, due to the many-body nature of the problem, an exact solution is in general not possible; it follows that, for a detailed understanding, realistic semiconductor models have to be considered, which can only be treated approximately. Within the kinetic theory approach one starts directly with the equations of motion for the single-particle density matrix. Due to the many-body nature of the problem, the resulting set of equations of motion is not closed; instead, it constitutes the starting point of an infinite hierarchy of higher order density matrices. Besides differences related to the quantum statistics of the quasiparticles involved, the latter is equivalent to the well-known Bogoliubov-Born-Green-Kirkwood-Yvon (BBGKY) hierarchy of classical gas dynamics (see, e.g., [159]). The central approximation in this formalism is the truncation of such hierarchy. This can be based on different physical pictures. A common approach (see, e.g., [110]) is to use the argument that correlations involving an increasing number of particles will become less and less important; an alternative quantum-kinetic scheme – based on an expansion in powers of the exciting laser field – has been introduced by Vollrath M. Axt and Arne Stahl, the so-called dynamics-controlled truncation (DCT) (see, e.g., [160] and references therein).

Within the treatment based on the Liouville–von Neumann equation, the starting point is the equation of motion for the global density-matrix operator, describing many electrons plus various quasiparticle excitations. The physical quantities of interest for the electronic subsystem are then typically derived via a suitable "reduction procedure," aimed at tracing out non-relevant degrees of freedom. Contrary to the quantum-kinetic theory, such global approach – originally proposed by Carlo Jacoboni and co-workers [108] – has allowed for a fully quantum-mechanical treatment of high-field transient transport phenomena in semiconductor bulk and nanostructures [161–167], thus overcoming some of the basic limitations of conventional kinetic treatments, e.g., the completed-collision limit and the Markov approximation.

Primary goal of the present chapter is to discuss in very general terms the physical properties and validity limits of the so-called adiabatic or Markov approximation. Within the traditional semiclassical or Boltzmann theory previously introduced (see Sect. 1.3), the latter is typically introduced together with the so-called diagonal approximation, i.e., the neglect of non-diagonal density-matrix elements. However, as described in [110], the Markov limit can also be performed within a fully non-diagonal density-matrix treatment of the problem; this leads to the introduction of generalized in- and out-scattering superoperators, whose general properties and physical interpretation are not straightforward. In particular, it is imperative to understand if – and under which conditions – the adiabatic or Markov approximation preserves the positive-definite character of our reduced density matrix; indeed, this distinguished property is generally lost within the quantum-kinetic approaches previously mentioned.

To this aim, starting from the Liouville–von Neumann equation approach, in this chapter we shall provide a very general treatment of the Markov approximation. More specifically, contrary to the conventional single-particle correlation expansion of the kinetic theory, we shall investigate separately the effects of the Markov limit and of the reduction procedure. Our fully operatorial approach – originally introduced in [168] – will allow us to better identify the general properties of the scattering superoperators entering our effective quantum-transport theory at various description levels, e.g., N electrons-plus-quasiparticles, N electrons only, and single-particle picture.

3.1 Physical System and Liouville–von Neumann Equation

In order to provide a general and fully quantum-mechanical description of the electro-optical properties of semiconductor nanostructures, let us consider an electron gas within a semiconductor crystal in the presence of electromagnetic fields. The corresponding Hamiltonian can be schematically written as

$$\hat{H} = \hat{H}^{\mathrm{c}} + \hat{H}^{\mathrm{qp}} + \hat{H}^{\mathrm{cc}} + \hat{H}^{\mathrm{c-qp}} + \hat{H}^{\mathrm{qp-qp}} \ . \tag{3.1}$$

Here, the first two terms are the free-carrier and free-quasiparticle Hamiltonians, while the last three contributions describe, respectively, carrier–carrier, carrier–quasiparticle (e.g., carrier–phonon, carrier–photon, carrier–plasmon), and quasiparticle–quasiparticle interactions (e.g., phonon–phonon, photon–photon, plasmon–plasmon).

The global semiconductor Hamiltonian in (3.1) can never be treated exactly; in contrast, within the spirit of the usual perturbation theory recalled in Sect. 3.2, the latter may be written as the sum of a so-called unperturbed contribution $\hat{H}_\circ$ which can be treated exactly, plus a perturbation term $\hat{H}'$ which is typically treated within some approximation scheme, i.e.,

$$\hat{H} = \hat{H}_\circ + \hat{H}' \ . \tag{3.2}$$

As we shall see, the separation of the global Hamiltonian (3.1) into unperturbed contribution and perturbation term is definitely not unique and depends crucially on the physical phenomenon under investigation. For example, while for many-electron systems – like bulk materials as well as quantum-well/wire nanostructures – carrier–carrier interaction ($\hat{H}^{\mathrm{cc}}$) may be safely treated as a perturbation (and thus included into $\hat{H}'$), in order to account for few-carrier effects in zero-dimensional systems (see Chap. 9) the latter needs to be treated exactly (and thus included in $\hat{H}$); in a similar way, in order to describe incoherent optical properties of semiconductor materials/devices the carrier–photon interaction ($\hat{H}^{\mathrm{c-qp}}$) may be treated as a perturbation, while for the understanding of ultrafast coherent phenomena the latter needs to be treated exactly.

Regardless of the explicit form of the separation in (3.2), the unperturbed basis states $|\lambda\rangle$ are defined as the eigenstates of $\hat{H}_\circ$, i.e.,

$$\hat{H}_\circ|\lambda\rangle = \epsilon_\lambda|\lambda\rangle \ , \tag{3.3}$$

where ϵ_λ denote the corresponding (unperturbed) energy levels. According to the general prescription of the perturbation theory (see, e.g., [153]), the interaction Hamiltonian $\hat{H}'$ in (3.2) will induce transitions between the unperturbed (or noninteracting) states $\{|\lambda\rangle\}$ previously introduced; as we shall see, in this context the basic ingredients are the matrix elements of $\hat{H}'$ within our noninteracting basis:

$$H'_{\lambda\lambda'} = \langle\lambda|\hat{H}'|\lambda'\rangle \ . \tag{3.4}$$

In view of the huge number of degrees of freedom involved in the microscopic treatment of any solid-state system (of the order of 10^{22} per cubic centimeter), a statistical description of the problem is imperative. As we shall see, this will result in a suitable statistical average over "non-relevant" degrees of freedom.

Given a physical quantity a – described by the operator $\hat{a}$ – its quantum plus statistical average value is given by

$$a = \overline{\langle\Psi|\hat{a}|\Psi\rangle} \ , \tag{3.5}$$

the overbar denoting the ensemble average previously mentioned. Equation (3.5) is usually written as

$$a = \mathrm{tr}\left\{\hat{a}\hat{\rho}\right\} \ , \tag{3.6}$$

where

$$\hat{\rho} = \overline{|\Psi\rangle\langle\Psi|} \tag{3.7}$$

is the so-called density-matrix operator. The latter is defined as a suitable statistical average of the projection operator corresponding to the generic state vector $|\Psi\rangle$ of the system. The density-matrix operator $\hat{\rho}$ in (3.7) can then be regarded as the statistical generalization of the concept of state vector.

Starting from the global Schrödinger equation describing our interacting carrier-plus-quasiparticle many-body system

$$\mathrm{i}\hbar\frac{\mathrm{d}|\Psi\rangle}{\mathrm{d}t} = \hat{H}|\Psi\rangle \ , \tag{3.8}$$

the following equation of motion for the density-matrix operator can be readily obtained:

$$\frac{\mathrm{d}\hat{\rho}}{\mathrm{d}t} = \mathcal{L}\left(\hat{\rho}\right) = \frac{1}{\mathrm{i}\hbar}\left[\hat{H},\hat{\rho}\right] \ , \tag{3.9}$$

where $\mathcal{L}$ is usually referred to as Liouville superoperator. The latter, called Liouville–von Neumann equation, can be regarded as the statistical generalization of the Schrödinger equation (3.8). Its exact (formal) solution is given by

$$\hat{\rho}(t) = \mathrm{e}^{\mathcal{L}(t-t_0)}\hat{\rho}(t_0) = \hat{U}(t-t_0)\hat{\rho}(t_0)\hat{U}^{\dagger}(t-t_0) \ , \tag{3.10}$$

where

$$\hat{U}(t-t_0) = \mathrm{e}^{\frac{\hat{H}(t-t_0)}{\mathrm{i}\hbar}} \tag{3.11}$$

denotes the evolution operator corresponding to the total Hamiltonian $\hat{H}$ in (3.1). Such exact solution corresponds to a fully quantum-mechanical unitary evolution of the whole many-body system, i.e., no energy dissipation/decoherence. Indeed, it is easy to verify that the total quantum entropy (called von Neumann entropy)

$$S = -k_{\mathrm{B}}\mathrm{tr}\left\{\hat{\rho}\log\hat{\rho}\right\} \tag{3.12}$$

is not affected by the unitary transformation (3.11).[1]

As anticipated, the total many-body Hamiltonian (3.1) cannot be treated exactly. Aim of a quantum-transport theory is to derive effective equations describing the carrier subsystem of interest within some approximation scheme; this is typically realized via the following two basic steps: first an adiabatic decoupling between the different time-scales induced by $\hat{H}_\circ$ and $\hat{H}'$ – called Markov limit – and then a projection of the global system dynamics over a subsystem of interest via the introduction of a so-called reduced density-matrix operator.

3.2 The Interaction Picture

Starting from the separation $\hat{H} = \hat{H}_\circ + \hat{H}'$ in (3.2), the Liouville–von Neumann equation (3.9) may be written as

$$\frac{\mathrm{d}\hat{\rho}}{\mathrm{d}t} = \left.\frac{\mathrm{d}\hat{\rho}}{\mathrm{d}t}\right|_{\hat{H}_\circ} + \left.\frac{\mathrm{d}\hat{\rho}}{\mathrm{d}t}\right|_{\hat{H}'} , \tag{3.13}$$

where

$$\left.\frac{\mathrm{d}\hat{\rho}}{\mathrm{d}t}\right|_{\hat{H}_\circ} = \frac{1}{\mathrm{i}\hbar}\left[\hat{H}_\circ, \hat{\rho}\right] \tag{3.14}$$

and

$$\left.\frac{\mathrm{d}\hat{\rho}}{\mathrm{d}t}\right|_{\hat{H}'} = \frac{1}{\mathrm{i}\hbar}\left[\hat{H}', \hat{\rho}\right] \tag{3.15}$$

describe, respectively, the time evolution induced by the noninteracting Hamiltonian $\hat{H}_\circ$ and by the interaction term $\hat{H}'$.

As usual, the first contribution can be treated exactly within the standard interaction picture (see, e.g., [86, 153]). More specifically, the form of a generic operator $\hat{a}$ within the interaction picture is given by

$$\hat{a}^i(t) = \hat{U}_\circ^\dagger(t-t_0)\hat{a}\hat{U}_\circ(t-t_0) , \tag{3.16}$$

where

$$\hat{U}_\circ(t-t_0) = \mathrm{e}^{\frac{\hat{H}_\circ(t-t_0)}{\mathrm{i}\hbar}} \tag{3.17}$$

is the evolution operator defined as the solution of the Schrödinger-like equation

[1] The quantum entropy (3.12) is time independent since the trace is invariant under the unitary transformation in (3.10).

$$i\hbar \frac{d\hat{U}_\circ}{dt} = \hat{H}_\circ \hat{U}_\circ \tag{3.18}$$

corresponding to the noninteracting Hamiltonian $\hat{H}_\circ$ only.[2] It is easy to show that the time evolution of the density-matrix operator in the interaction picture,

$$\hat{\rho}^i = \hat{U}_\circ^\dagger(t - t_0)\hat{\rho}\hat{U}_\circ(t - t_0) , \tag{3.19}$$

is simply given by

$$\frac{d\hat{\rho}^i}{dt} = -i\left[\hat{\mathcal{H}}^i, \hat{\rho}^i\right] , \tag{3.20}$$

where

$$\hat{\mathcal{H}}^i(t) = \hat{U}_\circ^\dagger(t - t_0)\frac{\hat{H}'}{\hbar}\hat{U}_\circ(t - t_0) \tag{3.21}$$

denotes the Hamiltonian $\hat{H}'$ in units of $\hbar$ written in the interaction picture.

3.3 Three-Key Approximation Levels

In this section we shall introduce and discuss the three-key approximations usually employed in the theoretical description of electro-optical processes in semiconductor nanostructures, namely the Markov limit, the electronic or reduced description, and the single-particle picture.

3.3.1 The Adiabatic or Markov Limit

The key idea beyond any perturbation approach is that the effect of the interaction Hamiltonian $\hat{H}'$ (see (3.15)) is "small" compared to the free evolution dictated by the noninteracting Hamiltonian $\hat{H}_\circ$ (see (3.14)). More precisely, this amounts to saying that the interaction matrix elements (3.4) are small compared to the typical energy difference $\epsilon_\lambda - \epsilon_{\lambda'}$ (see (3.3)).

Following this spirit, by formally integrating the result in (3.20) from t_0 to the current time t we get

$$\hat{\rho}^i(t) = \hat{\rho}^i(t_0) - i\int_{t_0}^{t} dt' \left[\hat{\mathcal{H}}^i(t'), \hat{\rho}^i(t')\right] . \tag{3.22}$$

By inserting the above formal solution for $\hat{\rho}^i$ on the right-hand side of (3.20) we obtain an integro-differential equation of the form

$$\frac{d}{dt}\hat{\rho}^i(t) = -i\left[\hat{\mathcal{H}}^i(t), \hat{\rho}^i(t_0)\right] - \int_{t_0}^{t} dt' \left[\hat{\mathcal{H}}^i(t), \left[\hat{\mathcal{H}}^i(t'), \hat{\rho}^i(t')\right]\right] . \tag{3.23}$$

[2] We stress that the explicit form of the noninteracting evolution operator (3.17) applies to a time-independent Hamiltonian only; the case of a time-dependent Hamiltonian – due, e.g., to a time-dependent external perturbation or to a particular gauge choice – will be treated in Sect. 3.4.

We stress that so far no approximation has been introduced: Equations (3.20), (3.22), and (3.23) are all fully equivalent, we have just isolated the first-order contribution from the full time evolution in (3.20). It is then clear that, by iteratively substituting the formal solution (3.22) into itself, the above procedure can be extended to any perturbation order. This leads to the well-known Neumann series,

$$\hat{\rho}^i(t) = \sum_{n=0}^{\infty}(-\mathrm{i})^n \int_{t_0}^{t} \mathrm{d}t_n \ldots \int_{t_0}^{t_2} \mathrm{d}t_1 [\hat{\mathcal{H}}^i(t_n), \ldots [\hat{\mathcal{H}}^i(t_1), \hat{\rho}^i(t_0)] \ldots] \,, \tag{3.24}$$

which constitutes the starting point of the so-called Quantum Monte Carlo (QMC) method[3] [108] for the study of high-field quantum-transport phenomena in semiconductors [161–167].

In order to introduce the so-called adiabatic or Markov approximation, let us now focus on the time integral in (3.23). Here, the two quantities to be integrated over t' are the interaction Hamiltonian $\hat{\mathcal{H}}^i$ and the density-matrix operator $\hat{\rho}^i$. In the spirit of the perturbation approach previously recalled, the time variation of $\hat{\rho}^i$ can be considered adiabatically slow compared to that of the Hamiltonian $\hat{\mathcal{H}}^i$ within the interaction picture; indeed, the latter will exhibit rapid oscillations due to the noninteracting unitary transformation $\hat{U}_\circ$ (see (3.17) and (3.21)). As a result, the density-matrix operator $\hat{\rho}^i$ can be taken out of the time integral and evaluated at the current time t.

Within such adiabatic limit we get the following effective Liouville–von Neumann equation:

$$\frac{\mathrm{d}}{\mathrm{d}t}\hat{\rho}^i(t) = -\mathrm{i}\left[\hat{\mathcal{H}}^i(t), \hat{\rho}^i(t_0)\right] - \left[\hat{\mathcal{H}}^i(t), \left[\hat{\mathcal{K}}^i(t), \hat{\rho}^i(t)\right]\right] \tag{3.25}$$

with

$$\hat{\mathcal{K}}^i(t) = \int_{t_0}^{t} \hat{\mathcal{H}}^i(t')\mathrm{d}t' \,. \tag{3.26}$$

The above equation is still characterized by the double-commutator structure in (3.23) but it is now local in time.

Going back to the original Schrödinger picture, we finally get

$$\frac{\mathrm{d}\hat{\rho}}{\mathrm{d}t} = \left.\frac{\mathrm{d}\hat{\rho}}{\mathrm{d}t}\right|_{\hat{H}_\circ} + \hat{C} - \left[\hat{\mathcal{H}}, \left[\hat{\mathcal{K}}, \hat{\rho}\right]\right] \tag{3.27}$$

with

$$\hat{C}(t) = -\mathrm{i}\left[\hat{\mathcal{H}}, \hat{U}_\circ(t-t_0)\hat{\rho}(t_0)\hat{U}_\circ^\dagger(t-t_0)\right] \tag{3.28}$$

and

$$\hat{\mathcal{K}} = \hat{U}_\circ(t-t_0)\hat{\mathcal{K}}^i(t)\hat{U}_\circ^\dagger(t-t_0) \,. \tag{3.29}$$

Combining the results in (3.21), (3.26), and (3.29), we get

[3] A brief account of the QMC simulation technique is given in Sect. 5.2.

$$\hat{\mathcal{K}} = \int_{t_0}^{t} dt' \hat{U}_\circ(t-t')\hat{\mathcal{H}}\hat{U}_\circ^\dagger(t-t') \, . \tag{3.30}$$

The time-dependent operator $\hat{C}$ in (3.28) acts as a source term and describes how the quantum-mechanical correlations at the initial time t_0 propagate to the current time t; more specifically,the initial correlations propagate from t_0 to t via the interaction-free dynamics described by the evolution operator $\hat{U}_\circ$ in (3.17). As discussed extensively in [110], the quantum-correlation operator (3.28) is responsible for a number of purely quantum-mechanical phenomena, like Hartree–Fock single-particle renormalizations and coherent-phonon effects.

As a result of the adiabatic or Markov limit previously introduced, the original Liouville–von Neumann equation (3.9) is then replaced by the following effective equation of motion:

$$\frac{d\hat{\rho}}{dt} = \frac{1}{i\hbar}\left[\hat{H}_\circ, \hat{\rho}\right] + \hat{C} + \Gamma\left(\hat{\rho}\right) \, . \tag{3.31}$$

Here, the first term describes again the free evolution dictated by the noninteracting Hamiltonian $\hat{H}_\circ$, the second one the initial-correlation source term (3.28), while

$$\Gamma\left(\hat{\rho}\right) = -\left[\hat{\mathcal{H}}, \left[\hat{\mathcal{K}}, \hat{\rho}\right]\right] \tag{3.32}$$

can be regarded as a generalized scattering superoperator describing the effect of the interaction Hamiltonian $\hat{H}'$ within the Markov limit previously introduced. Equation (3.31) may also be written in a more compact form as

$$\frac{d\hat{\rho}}{dt} = \overline{\mathcal{L}}\left(\hat{\rho}\right) + \hat{C} \, , \tag{3.33}$$

where

$$\overline{\mathcal{L}}\left(\hat{\rho}\right) = \frac{1}{i\hbar}\left[\hat{H}_\circ, \hat{\rho}\right] - \left[\hat{\mathcal{H}}, \left[\hat{\mathcal{K}}, \hat{\rho}\right]\right] \tag{3.34}$$

is the effective Liouville superoperator within our approximation scheme. The general solution of the effective equation (3.33) is of the form

$$\hat{\rho}(t) = T\left[e^{\int_{t_0}^{t}\overline{\mathcal{L}}(t')dt'}\right]\hat{\rho}(t_0) + \int_{t_0}^{t} T\left[e^{\int_{t'}^{t}\overline{\mathcal{L}}(t'')dt''}\right]\hat{C}(t')dt' \, , \tag{3.35}$$

where $T[\ldots]$ denotes the usual time- or chronological-ordering operator (see, e.g., [102, 157]).

At this point a few comments are in order. So far, the only approximation introduced in our theoretical description is the adiabatic decoupling between free-carrier evolution and various many-body interactions; this leads to a significant modification of the system dynamics: while the exact quantum-mechanical evolution (3.10) corresponds to a fully reversible and isoentropic unitary transformation, the instantaneous double-commutator structure in (3.34) describes, in general, a non-reversible (i.e., non-unitary) dynamics (see

(3.35)) characterized by energy dissipation and decoherence; it follows that the system quantum entropy (3.12) is no more a constant. For the present global level of description this behavior comes out to be totally unphysical, clearly showing the potential failure and intrinsic limitations of the Markov approximation. However, as discussed below (see Sect. 3.3.2), the Markov limit previously introduced is usually employed together with a reduced description of the system, for which such irreversible dynamics is physically justified.

Let us finally focus on the nature of the effective Liouville superoperator (3.34). As stressed before, this is the sum of a single-commutator term plus a double-commutator contribution. In the absence of carrier–carrier as well as carrier–quasiparticle interactions, i.e., $\hat{H}' = 0$, the second term vanishes and the system undergoes a reversible unitary transformation induced by the single-commutator term, which preserves the trace and the positive character of our density-matrix operator $\hat{\rho}$. In contrast, the perturbation Hamiltonian $\hat{H}'$ within the Markov limit previously introduced will induce, in general, a non-unitary evolution. Since any effective Liouville superoperator should describe correctly the time evolution of $\hat{\rho}$ and since the latter, by definition, needs to be trace invariant and positive-definite at any time, it is important to determine if – and under which conditions – the superoperator $\overline{\mathcal{L}}$ fulfills these two basic requirements.

As far as the first issue is concerned, recalling that the trace of a commutator is always equal to zero, and taking the trace of the quantity in (3.33), it is easy to verify that the time derivative of the trace of $\hat{\rho}$ is equal to zero, i.e., that our effective dynamics is trace preserving.

Let us now come to the positivity issue. As discussed in [168], the effective scattering superoperator Γ in (3.32) – derived via the conventional Markov limit previously recalled – does not ensure that for any initial condition the density-matrix operator $\hat{\rho}$ will remain positive-definite at later times. Indeed, as originally shown by Göran Lindblad in 1976 [169], this fulfillment is accomplished by employing any effective superoperator of the form

$$\left.\frac{d\hat{\rho}}{dt}\right|_{\text{Lindblad}} = -\sum_s \left(\frac{1}{2}\left\{\hat{A}_s^\dagger \hat{A}_s, \hat{\rho}\right\} - \hat{A}_s \hat{\rho} \hat{A}_s^\dagger\right) , \tag{3.36}$$

where $\{\hat{A}_s\}$ is a generic set of operators.[4] The latter, known as Lindblad superoperator, for the particular case of Hermitian operators (i.e., $\hat{A}_s = \hat{A}_s^\dagger$)

[4] For purely atomic and/or photonic systems, dissipation and decoherence phenomena are successfully described via adiabatic-decoupling procedures (see, e.g., [170]) in terms of extremely simplified models via phenomenological parameters; within such effective treatments, the main goal/requirement is to identify a suitable form of the Liouville superoperator, able to ensure/maintain the positive-definite character of the corresponding density-matrix operator (see, e.g., [171]). This is usually accomplished by identifying proper Lindblad-like decoherence superoperators, expressed in terms of a few crucial system–environment coupling parameters. In contrast, as discussed in Chap. 2, solid-state devices are often characterized by a complex many-electron quantum evolution, resulting in a non-trivial interplay between coherent dynamics and energy-dissipation/decoherence processes; it fol-

reduces to

$$\left.\frac{d\hat{\rho}}{dt}\right|_{\text{Lindblad}} = -\frac{1}{2}\sum_s \left[\hat{A}_s, \left[\hat{A}_s, \hat{\rho}\right]\right] . \tag{3.37}$$

We stress that the Markov superoperator (3.32) exhibits the same double-commutator structure in (3.37); however, due to its strongly asymmetric form, the latter is not Lindblad like. In order to overcome such severe limitation, in Sect. 3.5 we shall discuss an alternative adiabatic procedure [172–174] which (i) is physically justified under the same validity restrictions of the conventional Markov limit previously recalled, (ii) in the semiclassical limit reduces to the standard Fermi's golden rule, and (iii) describes a genuine Lindblad evolution, thus providing a reliable/robust treatment of energy dissipation and decoherence in state-of-the-art quantum devices.

Since one of our goals is the investigation of stationary quantum-transport phenomena, let us finally focus on the steady-state solution of the equation of motion (3.33). In the absence of initial quantum-mechanical correlations ($\hat{C} = 0$) it is easy to verify that the identity operator, properly normalized ($\hat{\rho}(t) \propto \hat{\mathcal{I}}$), is the stationary solution we are looking for. As anticipated, this clearly shows that within such approximation scheme our effective dynamics describes a sort of decoherence, since all non-diagonal terms of the density matrix will vanish on the long timescale. This is again an artifact of the Markov limit.

Let us now discuss the physical meaning of the steady-state solution $\hat{\rho}(t \rightarrow \infty) \propto \hat{\mathcal{I}}$. Within our noninteracting basis λ (see (3.3)) we have

$$\rho_{\lambda_1\lambda_2}(t \rightarrow \infty) \propto \delta_{\lambda_1\lambda_2} . \tag{3.38}$$

This tells us that, physically speaking, the steady-state solution of our transport equation corresponds to an equally probable population of all the microscopic states λ without any interstate quantum coherence ($\rho_{\lambda_1 \neq \lambda_2} = 0$). This scenario is typical of the present global (carrier + quasiparticle) description; in contrast, within a reduced description (involving, e.g., the carrier subsystem only) the steady-state solution differs from the identity operator, since in this case the trace over non-relevant degrees of freedom will translate into a thermal distribution over our electronic states (see Sect. 3.3.2).

Generalized Scattering Rates

As a starting point, let us consider the effective Liouville–von Neumann equation (3.31) written in the noninteracting-states basis $\{|\lambda\rangle\}$ previously introduced (see Sect. (3.3)):

$$\frac{d\rho_{\lambda_1\lambda_2}}{dt} = \frac{\epsilon_{\lambda_1} - \epsilon_{\lambda_2}}{i\hbar}\rho_{\lambda_1\lambda_2} + C_{\lambda_1\lambda_2} + \Gamma\left(\hat{\rho}\right)_{\lambda_1\lambda_2} . \tag{3.39}$$

---

lows that for a quantitative description of such coherence/dissipation coupling the latter needs to be treated via fully microscopic approaches (like the one presented in this chapter), whose Lindblad-like character is not guaranteed.

As we can see, the first term describes the exact free rotation induced by the noninteracting Hamiltonian $\hat{H}_\circ$, the second one the initial many-body quantum correlations, and the last one the interaction dynamics within the Markov limit.

Let us now analyze the explicit form of the matrix elements $\lambda_1\lambda_2$ of the scattering superoperator Γ; as shown in Appendix C, by expanding the double commutator in (3.32) one gets

$$\Gamma(\hat{\rho})_{\lambda_1\lambda_2} = \frac{1}{2}\sum_{\lambda_1'\lambda_2'}\left[\mathcal{P}_{\lambda_1\lambda_2,\lambda_1'\lambda_2'}\rho_{\lambda_1'\lambda_2'} - \mathcal{P}^*_{\lambda_1'\lambda_1',\lambda_1\lambda_2'}\rho_{\lambda_2'\lambda_2}\right] + \text{H.c.} \qquad (3.40)$$

with generalized scattering rates given by

$$\mathcal{P}_{\lambda_1\lambda_2,\lambda_1'\lambda_2'} = 2\mathcal{H}_{\lambda_1\lambda_1'}\mathcal{K}^*_{\lambda_2\lambda_2'} , \qquad (3.41)$$

where "H.c." denotes the Hermitian conjugate.

For the case of a time-independent perturbation $\hat{H}' = \hbar\hat{\mathcal{H}}$ the operator $\hat{\mathcal{K}}$ can be rewritten as

$$\hat{\mathcal{K}} = \int_0^{t-t_0} \mathrm{d}\tau \hat{U}_\circ(\tau)\hat{\mathcal{H}}\hat{U}_\circ^\dagger(\tau) . \qquad (3.42)$$

Taking into account that within the λ-representation the noninteracting evolution operator $\hat{U}_\circ$ is simply given by

$$U_\circ^{\lambda\lambda'}(\tau) = \mathrm{e}^{\frac{\epsilon_\lambda\tau}{i\hbar}}\delta_{\lambda\lambda'} , \qquad (3.43)$$

the matrix elements of the operator $\hat{\mathcal{K}}$ in (3.42) come out to be

$$\mathcal{K}_{\lambda\lambda'} = 2\pi\mathcal{H}_{\lambda\lambda'}D_{\lambda\lambda'} \qquad (3.44)$$

with

$$D_{\lambda\lambda'} = \frac{1}{2\pi}\int_0^{t-t_0} \mathrm{e}^{\frac{(\epsilon_\lambda-\epsilon_{\lambda'})\tau}{i\hbar}}\mathrm{d}\tau = D^*_{\lambda'\lambda} . \qquad (3.45)$$

By inserting the above result into (3.41) and recalling that $\mathcal{H}_{\lambda\lambda'} = \frac{H'_{\lambda\lambda'}}{\hbar}$, we finally get

$$\mathcal{P}_{\lambda_1\lambda_2,\lambda_1'\lambda_2'} = \frac{4\pi}{\hbar^2}H'_{\lambda_1\lambda_1'}H'^*_{\lambda_2\lambda_2'}D^*_{\lambda_2\lambda_2'} . \qquad (3.46)$$

We finally stress that, in general, the scattering superoperator $\mathcal{P}$ is a function of time; however, in the limit $t_0 \to -\infty$ – i.e., the completed-collision limit introduced in Chap. 2 – the function D in (3.45) becomes time independent:

$$D^{-\infty}_{\lambda\lambda'} = \frac{1}{2\pi}\int_0^{\infty} \mathrm{e}^{\frac{(\epsilon_\lambda-\epsilon_{\lambda'})\tau}{i\hbar}}\mathrm{d}\tau . \qquad (3.47)$$

It follows that in such limit the operator $\mathcal{K}$ as well as the superoperators Γ and $\overline{\mathcal{L}}$ are time independent as well. In this case there is no need for the time-ordering operator T in (3.35). Moreover, the real part of the function $D^{-\infty}$ in (3.47) gives the well-known energy-conserving Dirac delta function, i.e.,

$$D_{\lambda\lambda'}^{-\infty} = \frac{\hbar}{2}\delta(\epsilon_\lambda - \epsilon_{\lambda'}) + \mathrm{i}\mathcal{R}_{\lambda\lambda'} \,, \tag{3.48}$$

while its imaginary part – denoted by $\mathcal{R}_{\lambda\lambda'}$ – describes, in general, energy-renormalization effects. Within the validity limits of the present Markov treatment, such renormalization effects can be safely neglected: if, as requested, the perturbation Hamiltonian is small compared to the noninteracting one, then the resulting energy-level renormalization is small compared to the noninteracting energy levels ϵ_λ; in this case, the scattering superoperator (3.46) simply reduces to

$$\mathcal{P}_{\lambda_1\lambda_2,\lambda_1'\lambda_2'} = \frac{2\pi}{\hbar} H'_{\lambda_1\lambda_1'} H'^{*}_{\lambda_2\lambda_2'} \delta(\epsilon_{\lambda_2} - \epsilon_{\lambda_2'}) \,. \tag{3.49}$$

The Semiclassical Limit

The semiclassical or Boltzmann picture introduced in Sect. 1.3 can be easily derived from the quantum-transport formulation presented so far by introducing the so-called diagonal or semiclassical limit: the latter corresponds to neglecting all non-diagonal density-matrix elements, i.e.,

$$\rho_{\lambda_1\lambda_2} = f_{\lambda_1}\delta_{\lambda_1\lambda_2} \,, \tag{3.50}$$

where the diagonal elements f_λ describe the semiclassical distribution function over our noninteracting basis states λ. As anticipated in Chap. 2, from a physical point of view this corresponds to the neglect of any quantum-mechanical phase coherence between the generic states λ_1 and λ_2.

By introducing the above diagonal or semiclassical density matrix into the effective Liouville–von Neumann equation (3.39), for the diagonal elements ($\lambda = \lambda_1 = \lambda_2$) we get the following Boltzmann-like equation:

$$\frac{\mathrm{d}f_\lambda}{\mathrm{d}t} = \sum_{\lambda'} \left[P_{\lambda\lambda'} f_{\lambda'} - P_{\lambda'\lambda} f_\lambda\right] \tag{3.51}$$

with

$$P_{\lambda\lambda'} = \Re\left(\mathcal{P}_{\lambda\lambda,\lambda'\lambda'}\right) = P_{\lambda'\lambda} \,. \tag{3.52}$$

Indeed, within the semiclassical limit the free-rotation term in (3.39) vanishes; the same applies to the initial quantum-correlation source term: by rewriting the diagonal matrix elements of the correlation operator (3.28) within the semiclassical limit we get

$$C_{\lambda\lambda} = -\mathrm{i}\left(\mathcal{H}_{\lambda\lambda} f_\lambda(t_0) - f_\lambda(t_0)\mathcal{H}_{\lambda\lambda}\right) = 0 \,. \tag{3.53}$$

By inserting into (3.52) the explicit form of the generalized scattering rates (3.49) – obtained in the completed-collision limit ($t_0 \to -\infty$) by neglecting renormalization contributions – the semiclassical result (3.51) reduces to the usual Boltzmann transport equation written in our basis states λ, where the quantities

$$P_{\lambda\lambda'} = \frac{2\pi}{\hbar}|H'_{\lambda\lambda'}|^2\delta\left(\epsilon_\lambda - \epsilon_{\lambda'}\right) = P_{\lambda'\lambda} \tag{3.54}$$

are genuine scattering rates given by the well-known Fermi's golden rule: in addition to the square of the interaction matrix element $H'_{\lambda\lambda'}$, they contain the energy-conserving Dirac delta function.

At this point an important remark is in order. As anticipated, at short times, i.e., far from the completed-collision limit, the generalized scattering rates $\mathcal{P}$ are time dependent, which implies that the semiclassical rates P in (3.51) are time dependent as well. More specifically, by inserting into (3.52) the time-dependent result (3.46), we get

$$P_{\lambda\lambda'} = \frac{2\pi}{\hbar^2}|H'_{\lambda\lambda'}|^2 F_{\lambda\lambda'} \tag{3.55}$$

with

$$F_{\lambda\lambda'} = D_{\lambda\lambda'} + D^*_{\lambda\lambda'} = \frac{1}{\pi}\int_0^{t-t_0} \cos(\omega_{\lambda\lambda'}\tau)\mathrm{d}\tau \ , \quad \hbar\omega_{\lambda\lambda'} = \epsilon_\lambda - \epsilon_{\lambda'} \ . \tag{3.56}$$

This result may be regarded as the generalization (to a many-level system) of the two-level analysis presented in Sect. 2.2 . Indeed, at short times we deal again with pseudo-rates, i.e., quantities which are not positive-definite; the function $F_{\lambda\lambda'}$ is the generalization of the detuning interference pattern $F(\omega, t)$ in (2.21) (see Fig. 2.1) and describes, in general, the so-called energy-nonconserving processes discussed in Chap. 2.

The analysis presented so far clearly shows that the quantum-transport equation (3.39) can be regarded as the quantum-mechanical generalization of the Boltzmann equation (3.51). Indeed, the generalized scattering rates (3.46) are the quantum-mechanical generalizations of the standard in- and out-scattering rates entering the Boltzmann collision operator in (3.51).

As a confirmation of the fact that the Markov approximation alone leads to a totally unphysical non-reversible (i.e., non-unitary) system evolution, it is possible to show that the semiclassical version of the von Neumann entropy (3.12), i.e.,

$$S = -k_\mathrm{B} \sum_\lambda f_\lambda \log f_\lambda \ , \tag{3.57}$$

is a non-decreasing function of time:

$$\frac{\mathrm{d}S}{\mathrm{d}t} \geq 0 \ . \tag{3.58}$$

It is imperative to stress that, contrary to the usual semiclassical transport theory recalled in Sect. 1.3, the Boltzmann-like equation (3.51) describes a scattering dynamics within the whole (i.e., carrier-plus-quasiparticle) λ space; in other words, so far no reduction procedure to a subsystem of interest – e.g., the carrier gas – has been performed. This explains why, contrary to the usual Boltzmann theory, the scattering probabilities in (3.54) are symmetric, i.e., invariant under time reversal: $P_{\lambda\lambda'} = P_{\lambda'\lambda}$; moreover, the Dirac delta function in (3.54) leads to the conservation of the total energy of the system.

A second crucial remark is that, contrary to the non-diagonal density-matrix description previously introduced, the Markov limit combined with the semiclassical or diagonal approximation (3.50) ensures that at any time t our semiclassical distribution function f_λ is always positive-definite.

Let us finally discuss the steady-state solution of the Boltzmann transport (3.51). From the detailed-balance principle (see, e.g., [72]), i.e.,

$$P_{\lambda\lambda'} f_{\lambda'} = P_{\lambda'\lambda} f_\lambda \,, \tag{3.59}$$

and considering that the semiclassical scattering rates (3.54) are symmetric ($P_{\lambda\lambda'} = P_{\lambda'\lambda}$), we get

$$\frac{f_{\lambda'}}{f_\lambda} = \frac{P_{\lambda'\lambda}}{P_{\lambda\lambda'}} = 1 \,. \tag{3.60}$$

Exactly as for the quantum-mechanical case, the steady-state solution corresponds to a uniform distribution over the noninteracting states, which coincides with the identity-operator solution (3.38). This is not the case when our description is reduced to a given subsystem of interest (see below).

3.3.2 The Reduced or Electronic Description

As discussed previously, the average value of any given physical quantity a can be easily expressed in terms of the density-matrix operator $\hat{\rho}$ according to (3.6); the former, written in our λ-representation, is given by

$$a = \sum_{\lambda\lambda'} a_{\lambda\lambda'} \rho_{\lambda'\lambda} \,. \tag{3.61}$$

In order to study electro-optical phenomena in semiconductor nanostructures, many of the physical quantities of interest depend on the electronic-subsystem coordinates only (e.g., carrier drift velocity, total electronic energy, carrier–carrier correlation functions); this suggests to reduce the global description presented so far to a given subsystem of interest.

Generally speaking, it is convenient to subdivide the degrees of freedom of the global problem in (3.1) into a relevant set ξ plus a non-relevant set Ξ; this amounts to assuming a noninteracting Hamiltonian of the form

$$\hat{H}_\circ = \hat{H}_\xi + \hat{H}_\Xi \tag{3.62}$$

and to treat any relevant versus non-relevant coordinate coupling via the perturbation Hamiltonian $\hat{H}'$. The above subdivision in terms of relevant and non-relevant degrees of freedom depends crucially on the particular phenomenon under examination. As anticipated, the relevant coordinates ξ describe, in general, a given carrier subsystem of interest, while the non-relevant ones may describe quasiparticle coordinates, e.g., phonons, and/or additional carrier coordinates.

By inserting into (3.3) the unperturbed Hamiltonian (3.62), the generic noninteracting state $|\lambda\rangle$ is given by the tensor product of relevant times non-relevant coordinate states:

$$|\lambda\rangle = |\xi\rangle \otimes |\Xi\rangle \, . \tag{3.63}$$

In turn, the noninteracting energy spectrum is simply the sum of the energetic spectra corresponding to relevant and non-relevant degrees of freedom:

$$\epsilon_\lambda = \epsilon_\xi + \epsilon_\Xi \, . \tag{3.64}$$

For any physical quantity depending on the relevant-coordinate set ξ only, we have

$$a_{\lambda\lambda'} = a_{\xi\Xi,\xi'\Xi'} = a_{\xi\xi'}\delta_{\Xi\Xi'} \, . \tag{3.65}$$

In this case it is convenient to rewrite (3.61) as

$$a = \sum_{\xi\xi'} a_{\xi\xi'}\rho^{\xi}_{\xi'\xi} \, , \tag{3.66}$$

where

$$\rho^{\xi}_{\xi\xi'} = \sum_{\Xi} \rho_{\xi\Xi,\xi'\Xi} \tag{3.67}$$

is the so-called reduced or electronic density matrix. Equation (3.67) may also be written in an operatorial form as

$$\hat\rho^{\xi} = \mathrm{tr}\,\{\hat\rho\}_\Xi \, , \tag{3.68}$$

which shows that the reduced density-matrix operator $\hat\rho^{\xi}$ is obtained by performing a trace operation over the non-relevant coordinates Ξ. Since $\hat\rho^{\xi}$ is the only quantity entering the evaluation of the average value (3.66), it is desirable to derive a corresponding equation of motion for the reduced density-matrix operator. Combining the effective Liouville–von Neumann equation (3.33) with the definition of the reduced density-matrix operator (3.68) we get

$$\frac{\mathrm{d}\hat\rho^{r}}{\mathrm{d}t} = \mathrm{tr}\,\{\overline{\mathcal{L}}\,(\hat\rho)\}_\Xi + \mathrm{tr}\left\{\hat C\right\}_\Xi \, . \tag{3.69}$$

In general, the trace over the non-relevant coordinates Ξ does not commute with the effective Liouville superoperator $\overline{\mathcal{L}}$ in (3.34), which does not allow to obtain a closed equation of motion for the reduced density-matrix operator $\hat\rho^{\xi}$.

Carrier–Carrier Interaction

As a starting point, let us consider as relevant coordinates ξ those of the carrier subsystem, and as coupling mechanism carrier–carrier interaction only, i.e.,

$$\hat H_\circ = \hat H^{\mathrm{c}} + \hat H^{\mathrm{qp}} \, , \qquad \hat H' = \hat H^{\mathrm{cc}} \, . \tag{3.70}$$

In this case the perturbation Hamiltonian $\hat{\mathcal{H}}^{cc} = \frac{\hat{H}^{cc}}{\hbar}$ does not depend on non-relevant coordinates Ξ (as, e.g., phonons), i.e.,

$$\mathcal{H}_{\xi\Xi,\xi'\Xi'} = \mathcal{H}^{cc}_{\xi\xi'} \delta_{\Xi\Xi'} \ . \tag{3.71}$$

It follows that the trace operation in (3.69) commutes with the Liouville superoperator $\overline{\mathcal{L}}$, which allows us to write a closed effective Liouville equation for the reduced or electronic density-matrix operator; denoting with $\hat{\rho}^{\xi} = \hat{\rho}^{c}$ the reduced density-matrix operator corresponding to the carrier subsystem, we get

$$\frac{d\hat{\rho}^c}{dt} = \overline{\mathcal{L}}^c \left(\hat{\rho}^c\right) + \hat{C}^c \tag{3.72}$$

with

$$\overline{\mathcal{L}}^c \left(\hat{\rho}^c\right) = \frac{1}{i\hbar}\left[\hat{H}^c, \hat{\rho}^c\right] - \left[\hat{\mathcal{H}}^{cc}, \left[\hat{\mathcal{K}}^{cc}, \hat{\rho}^c\right]\right] \tag{3.73}$$

and

$$\hat{C}^c = -i\left[\hat{\mathcal{H}}^{cc}, \hat{U}^c(t-t_\circ)\hat{\rho}^c(t_\circ)\hat{U}^{c\dagger}(t-t_\circ)\right] \ , \tag{3.74}$$

$\hat{U}^c$ denoting the evolution operator corresponding to the noninteracting-carrier Hamiltonian $\hat{H}^c$ in (3.1).

Carrier–Quasiparticle Interaction

Contrary to the carrier–carrier interaction previously considered, in the presence of carrier–quasiparticle coupling we may adopt the following perturbation scheme[5]:

$$\hat{H}_\circ = \hat{H}^c + \hat{H}^{cc} + \hat{H}^{qp} \ , \qquad \hat{H}' = \hat{H}^{c-qp} \ ; \tag{3.75}$$

it follows that the trace operation over non-relevant degrees of freedom (given in this case by quasiparticle coordinates) does not commute with the Liouville superoperator (3.34). Thus, in order to get a closed equation of motion for the reduced density-matrix operator $\hat{\rho}^{\xi} = \hat{\rho}^{c}$ additional approximations are needed.

In very general terms, by adopting once again the separation between relevant and non-relevant degrees of freedom, the typical assumption is to consider the Ξ-subsystem as characterized by a huge number of degrees of freedom (compared to the subsystem ξ of interest). In other words this amounts to saying that the Ξ subsystem has an infinitely high heat capacity, i.e., it behaves as a thermal bath; this allows one to consider the Ξ-subsystem always in thermal equilibrium, i.e., not significantly perturbed by the subsystem ξ. Within such approximation scheme, the global ($\xi+\Xi$) density-matrix operator

[5] We stress that, opposite to the perturbative treatment of carrier–carrier coupling previously considered, here the carrier–carrier interaction Hamiltonian $\hat{H}^{cc}$ is part of the noninteracting Hamiltonian $\hat{H}_\circ$; as we shall see, this feature plays a crucial role for the investigation of few-carrier phenomena discussed in Chap. 9.

$\hat{\rho}$ can be written as the product of the reduced density-matrix operator times the equilibrium density-matrix operator for the Ξ subsystem:

$$\hat{\rho} = \hat{\rho}^{\xi} \hat{\rho}^{\circ} \,, \qquad \hat{\rho}^{\circ} = \frac{\mathrm{e}^{-\frac{\hat{H}_{\Xi}}{k_B T}}}{\mathrm{tr}\left\{ \mathrm{e}^{-\frac{\hat{H}_{\Xi}}{k_B T}} \right\}} \,. \tag{3.76}$$

The corresponding matrix elements within our basis states $\lambda = \xi, \Xi$ are then given by

$$\rho_{\lambda\lambda'} = \rho^{\xi}_{\xi\xi'} f^{\circ}_{\Xi} \delta_{\Xi\Xi'} \tag{3.77}$$

with

$$f^{\circ}_{\Xi} = \frac{\mathrm{e}^{-\frac{\epsilon_{\Xi}}{k_B T}}}{\sum_{\Xi} \mathrm{e}^{-\frac{\epsilon_{\Xi}}{k_B T}}} \,. \tag{3.78}$$

Inserting the general factorization scheme (3.76) into the effective Liouville–von Neumann equation (3.33) and identifying once again the relevant coordinates ξ with those of the carrier subsystem ($\hat{\rho}^{\xi} = \hat{\rho}^{c}$), by performing the trace over the non-relevant (quasiparticle) coordinates it is easy to get again the effective electronic equation (3.33) with

$$\overline{\mathcal{L}}^{c}\left(\hat{\rho}^{c}\right) = \frac{1}{\mathrm{i}\hbar}\left[\hat{H}^{c+cc}, \hat{\rho}^{c}\right] + \Gamma^{c}\left[\hat{\rho}^{c}\right] \,, \tag{3.79}$$

$$\Gamma^{c}\left[\hat{\rho}^{c}\right] = \mathrm{tr}\left\{\Gamma\left[\hat{\rho}^{c}\hat{\rho}^{\circ}\right]\right\}_{\Xi} \,, \tag{3.80}$$

and

$$\hat{C}^{c} = \mathrm{tr}\left\{\hat{C}\right\}_{\Xi} \,. \tag{3.81}$$

Here

$$\hat{H}^{c+cc} = \hat{H}^{c} + \hat{H}^{cc} \tag{3.82}$$

is the Hamiltonian describing the interacting-carrier system.

This shows that, by assuming the factorized solution (3.76), also in the presence of carrier–quasiparticle interaction – e.g., carrier–phonon, carrier–photon, carrier–plasmon – one is able to derive a closed equation of motion for the reduced density-matrix operator $\hat{\rho}^{c}$ formally identical to (3.72). However, contrary to the case of carrier–carrier interaction (see (3.73)), the new effective Liouville superoperator (3.79) does not exhibit the double-commutator structure previously discussed. This point will be extensively addressed later in this section.

For a wide class of quasiparticle excitations – including phonons, photons, plasmons – the corresponding Hamiltonian $\hat{H}^{qp}$ in (3.1) may be conveniently expressed as

$$\hat{H}^{qp} = \sum_{\boldsymbol{q}} \epsilon_{\boldsymbol{q}} \hat{b}^{\dagger}_{\boldsymbol{q}} \hat{b}_{\boldsymbol{q}} \,, \tag{3.83}$$

where the bosonic operators $\hat{b}_{\boldsymbol{q}}^\dagger$ ($\hat{b}_{\boldsymbol{q}}$) denote creation (destruction) of a generic quasiparticle excitation with wavevector $\boldsymbol{q}$ and energy $\epsilon_{\boldsymbol{q}}$. It follows that the quasiparticle eigenstates

$$|\Xi\rangle = |\{n_{\boldsymbol{q}}\}\rangle \,, \tag{3.84}$$

and thus the non-relevant coordinates, correspond to the quasiparticle occupation numbers ($\Xi = \{n_{\boldsymbol{q}}\}$), and the quasiparticle energy spectrum is given by

$$\epsilon_\Xi = \sum_{\boldsymbol{q}} \epsilon_{\boldsymbol{q}} n_{\boldsymbol{q}} \,. \tag{3.85}$$

For all relevant carrier–quasiparticle interaction mechanisms in semiconductor nanostructures – e.g., carrier–phonon, carrier–photon, carrier–plasmon – the perturbation Hamiltonian $\hat{H}'$ can be written as

$$\hat{H}' = \hbar\hat{\mathcal{H}} = \hbar\sum_{\boldsymbol{q}} \left(\hat{\mathcal{H}}_{\boldsymbol{q}}^- \hat{b}_{\boldsymbol{q}} + \hat{\mathcal{H}}_{\boldsymbol{q}}^+ \hat{b}_{\boldsymbol{q}}^\dagger\right) = \hat{H}^{\mathrm{ab}} + \hat{H}^{\mathrm{em}} \,. \tag{3.86}$$

Here, $\hat{\mathcal{H}}_{\boldsymbol{q}}^- = \hat{\mathcal{H}}_{\boldsymbol{q}}^{+\dagger}$ are electronic operators (parameterized by the quasiparticle wavevector $\boldsymbol{q}$) acting on the ξ coordinates only. The two terms in (3.86) – corresponding to quasiparticle destruction and creation – describe quasiparticle absorption (ab) and emission (em) processes.

Let us consider again the definition of the effective carrier–quasiparticle scattering superoperator Γ^{c}. By introducing into (3.80) the explicit double-commutator form in (3.32) and setting $\Xi = \{n_{\boldsymbol{q}}\}$, we get

$$\Gamma^{\mathrm{c}}\left(\hat{\rho}^{\mathrm{c}}\right) = -\mathrm{tr}\left\{\left[\hat{\mathcal{H}}, \left[\hat{\mathcal{K}}, \hat{\rho}^{\mathrm{c}}\hat{\rho}^{\circ}\right]\right]\right\}_{\{n_{\boldsymbol{q}}\}} \,. \tag{3.87}$$

By inserting into (3.42) and (3.87) the explicit form of the carrier–quasiparticle Hamiltonian (3.86) and employing the bosonic commutation relations ($[\hat{b}_{\boldsymbol{q}}, \hat{b}_{\boldsymbol{q}'}^\dagger] = \delta_{\boldsymbol{q}\boldsymbol{q}'}$), it is possible to obtain an explicit form of the effective carrier–quasiparticle scattering superoperator (3.80). More specifically we get

$$\Gamma^{\mathrm{c}}\left(\hat{\rho}^{\mathrm{c}}\right) = -\sum_{\boldsymbol{q}\pm}\left(N_{\boldsymbol{q}}^{\circ} + \frac{1}{2} \pm \frac{1}{2}\right)\left(\hat{\mathcal{H}}_{\boldsymbol{q}}^{\pm\dagger}\hat{\mathcal{K}}_{\boldsymbol{q}}^{\pm}\hat{\rho}^{\mathrm{c}} - \hat{\mathcal{H}}_{\boldsymbol{q}}^{\pm}\hat{\rho}^{\mathrm{c}}\hat{\mathcal{K}}_{\boldsymbol{q}}^{\pm\dagger}\right) + \mathrm{H.c.} \tag{3.88}$$

with

$$\hat{\mathcal{K}}_{\boldsymbol{q}}^{\pm} = \int_0^{t-t_0} \mathrm{d}\tau \mathrm{e}^{\frac{\hat{H}^{\mathrm{c}+\mathrm{cc}}\tau}{\mathrm{i}\hbar}} \hat{\mathcal{H}}_{\boldsymbol{q}}^{\pm} \mathrm{e}^{-\frac{\hat{H}^{\mathrm{c}+\mathrm{cc}}\tau}{\mathrm{i}\hbar}} \mathrm{e}^{\pm\frac{\epsilon_{\boldsymbol{q}}\tau}{\mathrm{i}\hbar}} = \hat{\mathcal{K}}_{\boldsymbol{q}}^{\mp\dagger} \,. \tag{3.89}$$

Here

$$N_{\boldsymbol{q}}^{\circ} = \mathrm{tr}\left\{\hat{b}_{\boldsymbol{q}}^\dagger\hat{b}_{\boldsymbol{q}}\hat{\rho}^{\circ}\right\} = \frac{1}{\mathrm{e}^{\frac{\epsilon_{\boldsymbol{q}}}{k_B T}} - 1} \tag{3.90}$$

denotes the equilibrium average occupation number for the quasiparticle state $\boldsymbol{q}$, the so-called Bose occupation number. As we can see, for each quasiparticle state $\boldsymbol{q}$ we have two contributions ($\pm$) describing quasiparticle emission and absorption. The detailed derivation of the effective scattering superoperator (3.88) is given in Appendix C.

It is important to stress that the effective scattering superoperator Γ^{c} in (3.88) does not exhibit the double-commutator structure typical of the global description (see (3.32)). Indeed, denoting with $\hat{\mathcal{H}}_{\boldsymbol{q}}^{\pm\mathrm{h/a}}$ and $\hat{\mathcal{K}}_{\boldsymbol{q}}^{\pm\mathrm{h/a}}$ the Hermitian/anti-Hermitian (h/a) parts of the operators $\hat{\mathcal{H}}_{\boldsymbol{q}}^{\pm}$ and $\hat{\mathcal{K}}_{\boldsymbol{q}}^{\pm}$, respectively, the superoperator Γ^{c} in (3.88) may also be expressed as

$$\begin{aligned}
\Gamma^{\mathrm{c}}\left(\hat{\rho}^{\mathrm{c}}\right) = &-\sum_{\boldsymbol{q}\pm}\left(N_{\boldsymbol{q}}^{\circ}+\frac{1}{2}\pm\frac{1}{2}\right)\left[\hat{\mathcal{H}}_{\boldsymbol{q}}^{\pm\mathrm{h}},\left[\hat{\mathcal{K}}_{\boldsymbol{q}}^{\pm\mathrm{h}},\hat{\rho}^{\mathrm{c}}\right]\right] \\
&-\sum_{\boldsymbol{q}\pm}\left(N_{\boldsymbol{q}}^{\circ}+\frac{1}{2}\pm\frac{1}{2}\right)\left[\hat{\mathcal{H}}_{\boldsymbol{q}}^{\pm\mathrm{h}},\left\{\hat{\mathcal{K}}_{\boldsymbol{q}}^{\pm\mathrm{a}},\hat{\rho}^{\mathrm{c}}\right\}\right] \\
&+\sum_{\boldsymbol{q}\pm}\left(N_{\boldsymbol{q}}^{\circ}+\frac{1}{2}\pm\frac{1}{2}\right)\left[\hat{\mathcal{H}}_{\boldsymbol{q}}^{\pm\mathrm{a}},\left\{\hat{\mathcal{K}}_{\boldsymbol{q}}^{\pm\mathrm{h}},\hat{\rho}^{\mathrm{c}}\right\}\right] \\
&+\sum_{\boldsymbol{q}\pm}\left(N_{\boldsymbol{q}}^{\circ}+\frac{1}{2}\pm\frac{1}{2}\right)\left[\hat{\mathcal{H}}_{\boldsymbol{q}}^{\pm\mathrm{a}},\left[\hat{\mathcal{K}}_{\boldsymbol{q}}^{\pm\mathrm{a}},\hat{\rho}^{\mathrm{c}}\right]\right] . \qquad (3.91)
\end{aligned}$$

As anticipated, the scattering superoperator involves again double-commutator terms, but we have also commutator–anticommutator contributions.

To better underline the physical role played by commutator versus anticommutator contributions, let us recall a simplified model usually invoked to qualitatively describe the quantum-mechanical evolution of open systems, i.e., subsystems interacting with their environment. Within the Schrödinger picture, the latter are typically treated by adding to the system Hamiltonian $\hat{H}^{\mathrm{s}}$ an anti-Hermitian part (imaginary potential) $\hat{V}^{\mathrm{env}}$ describing the system–environment coupling:

$$\mathrm{i}\hbar\frac{\mathrm{d}|\Psi^{\mathrm{s}}\rangle}{\mathrm{d}t} = \left(\hat{H}^{\mathrm{s}}+\hat{V}^{\mathrm{env}}\right)|\Psi^{\mathrm{s}}\rangle . \qquad (3.92)$$

Starting from the above-modified Schrödinger equation, it is easy to obtain a corresponding version of the Liouville–von Neumann equation (3.9):

$$\frac{\mathrm{d}\hat{\rho}^{\mathrm{s}}}{\mathrm{d}t} = \frac{1}{\mathrm{i}\hbar}\left[\hat{H}^{\mathrm{s}},\hat{\rho}^{\mathrm{s}}\right]+\frac{1}{\mathrm{i}\hbar}\left\{\hat{V}^{\mathrm{env}},\hat{\rho}^{\mathrm{s}}\right\} . \qquad (3.93)$$

In addition to the commutator-like dynamics typical of a closed system, we deal with an anticommutator term, describing dissipation induced by the system–environment coupling. Indeed, contrary to the closed dynamics in (3.9), the latter leads to a non-reversible dynamics. Such a simplified model is known to be highly unphysical, since it does not preserve the trace of the density-matrix operator; however, it clearly shows how the commutator structure is intimately related to a closed evolution, while anticommutator-like terms always describe dissipation processes.

Contrary to the above simplified model, the scattering superoperator (3.88) is trace preserving. However, the presence of the various commutator/anticommutator contributions is a clear fingerprint of carrier–quasiparticle dissipation phenomena leading to genuine energy-dissipation/decoherence processes.

Let us now come to the correlation term (3.81):

$$\hat{C}^{\mathrm{c}} = -\mathrm{i}\,\mathrm{tr}\left\{\left[\hat{\mathcal{H}}, \hat{U}_\circ(t-t_0)\hat{\rho}^{\mathrm{c}}(t_0)\hat{\rho}^{\mathrm{qp}}\hat{U}_\circ^\dagger(t-t_0)\right]\right\}_{\{n_{\boldsymbol{q}}\}} . \tag{3.94}$$

By inserting into the above commutator the explicit form of the carrier–quasiparticle interaction Hamiltonian (3.86) and considering that

$$\mathrm{tr}\left\{\hat{b}^\dagger_{\boldsymbol{q}}\hat{\rho}^\circ\right\} = \mathrm{tr}\left\{\hat{b}_{\boldsymbol{q}}\hat{\rho}^\circ\right\} = 0 , \tag{3.95}$$

it is easy to verify that for the linear-coupling carrier–quasiparticle Hamiltonian (3.86) the initial-correlation term (3.94) vanishes.

In summary, within the approximation scheme considered so far we get the following effective equation of motion for the reduced density-matrix operator $\hat{\rho}^{\mathrm{c}}$:

$$\frac{\mathrm{d}\hat{\rho}^{\mathrm{c}}}{\mathrm{d}t} = \overline{\mathcal{L}}^{\mathrm{c}}\left(\hat{\rho}^{\mathrm{c}}\right) . \tag{3.96}$$

In the completed-collision limit ($t_0 \to -\infty$) previously considered, the effective Liouville superoperator (3.79) becomes time independent, and the general solution of the homogeneous equation (3.96) is of the form

$$\hat{\rho}^{\mathrm{c}}(t) = \mathrm{e}^{\overline{\mathcal{L}}^{\mathrm{c}}(t-t_0)}\hat{\rho}^{\mathrm{c}}(t_0) . \tag{3.97}$$

Again, contrary to the isoentropic and fully-reversible unitary evolution (3.10), the instantaneous commutator-plus-anticommutator structures in (3.91) describe a non-reversible (i.e., non-unitary) dynamics characterized by energy dissipation and decoherence induced by the carrier–quasiparticle coupling (3.86).

Let us now focus on the steady-state solution of the quantum-transport equation (3.96). Contrary to the global (carrier + quasiparticle) equation (3.33), the identity operator $\hat{\mathcal{I}}$ is no more a solution. Indeed, the latter fulfills the double commutator but not the additional anticommutator-like structures in (3.91). As discussed extensively in [172], by neglecting energy-renormalization effects, it is possible to show that the steady-state solution is simply given by the equilibrium density-matrix operator corresponding to our electronic subsystem:

$$\hat{\rho}^{\mathrm{c}} = \frac{\mathrm{e}^{-\frac{\hat{H}^{\mathrm{c+cc}}}{k_B T}}}{\mathrm{tr}\left\{\mathrm{e}^{-\frac{\hat{H}^{\mathrm{c+cc}}}{k_B T}}\right\}} . \tag{3.98}$$

The latter is again diagonal within the basis given by the eigenstates $|\xi\rangle$ of the interacting-carrier Hamiltonian $\hat{H}^{\mathrm{c+cc}}$ in (3.82):

$$\hat{H}^{\mathrm{c+cc}}|\xi\rangle = \epsilon_\xi|\xi\rangle \ , \tag{3.99}$$

where ϵ_ξ denote the energy levels corresponding to the interacting-carrier states ξ. This diagonal form corresponds to the absence of any interstate polarization. In contrast, in the presence of renormalization contributions, the steady-state solution differs from the thermal one, showing non-diagonal density-matrix elements; the latter may be regarded as a fingerprint of scattering-induced quantum-mechanical phase coherence (see, e.g., [168]).

Let us now come back to the positivity issue. As discussed in [168], the electronic scattering superoperator (3.88) does not ensure that for any initial condition the density-matrix operator $\hat{\rho}^{\mathrm{c}}$ will remain positive-definite at later times; moreover, due to anomalous spectral properties of the scattering superoperator (see, e.g., [174]), for some initial conditions the system will never reach any steady-state regime. As anticipated, in order to overcome these severe limitations, in Sect. 3.5 we shall discuss an alternative formulation of the Markov limit [172–174].

As a final step, let us discuss the explicit form of the electronic scattering superoperator (3.88) within the interacting-carrier basis ξ. To this end, by expanding the various terms in (3.88) it is easy to obtain the following equation of motion for our electronic density matrix:

$$\Gamma^{\mathrm{c}}\left(\hat{\rho}^{\mathrm{c}}\right)_{\xi_1\xi_2} = \frac{1}{2}\sum_{\xi'_1\xi'_2}\left[\mathcal{P}_{\xi_1\xi_2,\xi'_1\xi'_2}\rho^{\mathrm{c}}_{\xi'_1\xi'_2} - \mathcal{P}^*_{\xi'_1\xi'_1,\xi_1\xi'_2}\rho^{\mathrm{c}}_{\xi'_2\xi_2}\right] + \mathrm{H.c.} \tag{3.100}$$

with electronic generalized scattering rates given by

$$\mathcal{P}_{\xi_1\xi_2,\xi'_1\xi'_2} = 2\sum_{q\pm}\left(N^\circ_{\boldsymbol{q}} + \frac{1}{2} \pm \frac{1}{2}\right)\mathcal{H}^{q\pm}_{\xi_1\xi'_1}\mathcal{K}^{q\pm *}_{\xi_2\xi'_2} \ . \tag{3.101}$$

As we can see, one gets exactly the generalized in-minus out-scattering structure in (3.40); however, while the global scattering operator (3.41) is invariant under time reversal ($\mathcal{P}_{\lambda_1\lambda_2,\lambda'_1\lambda'_2} = \mathcal{P}^*_{\lambda'_1\lambda'_2,\lambda_1\lambda_2}$), the same does not apply to the reduced or electronic rates (3.101).

As for the global (carrier + quasiparticle) description (see (3.44)), the matrix elements $\mathcal{K}^{q\pm}_{\xi\xi'}$ of the operator $\hat{\mathcal{K}}^{\pm}_{\boldsymbol{q}}$ in (3.89) may be expressed in terms of the matrix elements $\mathcal{H}^{q\pm}_{\xi\xi'}$ of the operator $\hat{\mathcal{H}}^{\pm}_{\boldsymbol{q}}$; more specifically we get

$$\mathcal{K}^{q\pm}_{\xi\xi'} = 2\pi\mathcal{H}^{q\pm}_{\xi\xi'}D^{q\pm}_{\xi\xi'} \tag{3.102}$$

with

$$D^{q\pm}_{\xi\xi'} = \frac{1}{2\pi}\int_0^{t-t_0} \mathrm{e}^{\frac{(\epsilon_\xi - \epsilon_{\xi'} \pm \epsilon_{\boldsymbol{q}})\tau}{\mathrm{i}\hbar}}\,\mathrm{d}\tau \ . \tag{3.103}$$

Again, in the completed-collision limit ($t_0 \to -\infty$), the real part of D provides the energy-conserving Dirac delta function, while its imaginary part describes carrier–quasiparticle energy-renormalization effects.

By inserting the result (3.102) into the generalized scattering rate (3.101), the latter comes out to be

$$\mathcal{P}_{\xi_1\xi_2,\xi_1'\xi_2'} = 4\pi \sum_{q\pm} \left(N_q^\circ + \frac{1}{2} \pm \frac{1}{2} \right) \mathcal{H}_{\xi_1\xi_1'}^{q\pm} \mathcal{H}_{\xi_2\xi_2'}^{q\pm *} D_{\xi_2\xi_2'}^{q\pm *} \ . \tag{3.104}$$

It is imperative to stress that the reduced or electronic Markov superoperator discussed so far is linear, i.e., ρ-independent. As we shall see, this feature – typical of the present many-electron basis ξ – will be lost in the single-particle picture discussed below (see Sect. 3.3.3).

Also for the present reduced description we can consider the semiclassical or Boltzmann limit; this corresponds to neglecting the non-diagonal matrix elements of the reduced density matrix, i.e.,

$$\rho_{\xi_1\xi_2}^{\mathrm{c}} = f_{\xi_1}^{\mathrm{c}} \delta_{\xi_1\xi_2} \ . \tag{3.105}$$

Within such approximation, the effective Markov superoperator (3.100) reduces to the following Boltzmann equation for the many-electron subsystem:

$$\frac{\mathrm{d}f_\xi^{\mathrm{c}}}{\mathrm{d}t} = \sum_{\xi'} \left(P_{\xi\xi'} f_{\xi'}^{\mathrm{c}} - P_{\xi'\xi} f_\xi^{\mathrm{c}} \right) \ , \tag{3.106}$$

where in the completed-collision limit

$$P_{\xi\xi'} = \Re \left(\mathcal{P}_{\xi\xi,\xi'\xi'} \right) = \sum_{q\pm} \frac{2\pi}{\hbar} \left(N_q^\circ + \frac{1}{2} \pm \frac{1}{2} \right) |H_{\xi\xi'}^{q\pm}|^2 \delta \left(\epsilon_\xi - \epsilon_{\xi'} \pm \epsilon_q \right) \tag{3.107}$$

are the usual carrier–quasiparticle semiclassical scattering rates given by the well-known Fermi's golden rule. Here, $H_{\xi\xi'}^{q\pm} = \hbar \mathcal{H}_{\xi\xi'}^{q\pm}$.

We stress once again that, contrary to the global (carrier + quasiparticle) description, the scattering rates (3.107) are not symmetric: $P_{\xi\xi'} \neq P_{\xi'\xi}$; this is a direct fingerprint of the irreversible nature of the transport problem induced by energy-dissipation and decoherence processes.

3.3.3 The Single-Particle Picture

As pointed out in Sect. 3.1, the average value of any given physical quantity a can be easily expressed in terms of the global (carrier + quasiparticle) density-matrix operator $\hat{\rho}$ in (3.7) according to

$$a = \mathrm{tr} \left\{ \hat{a} \hat{\rho} \right\} \ . \tag{3.108}$$

In the study of electro-optical processes in semiconductors, however, many of the physical quantities of interest depend on the electronic-subsystem coordinates only; this suggests the introduction of the reduced description previously analyzed via the electronic density-matrix operator:

$$\hat{\rho}^{\rm c} = {\rm tr}\left\{\hat{\rho}\right\}_{\{n_q\}} \ , \tag{3.109}$$

defined as the trace over the (non-relevant) quasiparticle coordinates ($\Xi = \{n_q\}$) of the global density-matrix operator (3.7). Moreover, many of the electronic properties of interest in the analysis of semiconductor nanostructures are single-particle quantities, i.e., physical quantities ascribed to the generic particle (i.e., carrier) in our electronic subsystem, like carrier drift velocity, mean kinetic energy. This suggests to adopt also for the electronic subsystem the second-quantization picture previously introduced for the description of quasiparticles (see (3.83)). More specifically, the noninteracting-carrier Hamiltonian $\hat{H}^{\rm c}$ in (3.1) may be written as

$$\hat{H}^{\rm c} = \sum_\alpha \epsilon_\alpha \hat{c}^\dagger_\alpha \hat{c}_\alpha \ , \tag{3.110}$$

where the fermionic operators $\hat{c}^\dagger_\alpha$ ($\hat{c}_\alpha$) denote creation (destruction) of an electron in the single-particle state α with energy ϵ_α. In the absence of carrier–carrier interaction ($\hat{H}^{\rm cc} = 0$), the many-electron states in (3.63),

$$|\xi\rangle = |\{n_\alpha\}\rangle \ , \tag{3.111}$$

and thus the relevant-coordinate label, correspond to the single-particle occupation numbers ($\xi = \{n_\alpha\}$), and the noninteracting-carrier energy spectrum is simply given by

$$\epsilon_\xi = \sum_\alpha \epsilon_\alpha n_\alpha \ . \tag{3.112}$$

In contrast, in the presence of carrier–carrier interaction, the single-particle occupation numbers $\{n_\alpha\}$ are not "good" quantum numbers anymore (see Chap. 9).

Adopting such second-quantization picture, the generic operator $\hat{a}$ describing a single-particle quantity is of the form

$$\hat{a} = \sum_{\alpha\alpha'} a^{\rm sp}_{\alpha\alpha'} \hat{c}^\dagger_\alpha \hat{c}_{\alpha'} \ . \tag{3.113}$$

Inserting the above single-particle operator into (3.108) we get

$$a = \sum_{\alpha\alpha'} a^{\rm sp}_{\alpha\alpha'} {\rm tr}\left\{\hat{c}^\dagger_\alpha \hat{c}_{\alpha'} \hat{\rho}\right\} \ . \tag{3.114}$$

This suggests to rewrite our average value as

$$a = \sum_{\alpha\alpha'} a^{\rm sp}_{\alpha\alpha'} \rho^{\rm sp}_{\alpha'\alpha} \ , \tag{3.115}$$

where

$$\rho^{\rm sp}_{\alpha_1\alpha_2} = {\rm tr}\left\{\hat{c}^\dagger_{\alpha_2} \hat{c}_{\alpha_1} \hat{\rho}\right\} \tag{3.116}$$

is the so-called single-particle density matrix (see, e.g., [110] and references therein). As we can see, this is defined as the average of the product of creation and destruction operators; its diagonal elements ($\alpha_1 = \alpha_2$) correspond to the single-particle carrier distribution of the semiclassical Boltzmann theory (see Sect. 1.3) while the non-diagonal contributions ($\alpha_1 \neq \alpha_2$), also called interstate polarizations (see Sect. 2.2), describe quantum-mechanical phase coherence between the single-particle states α_1 and α_2.

Equation (3.115) can also be formally written as the original average-value in (3.108):

$$a = \mathrm{tr}\left\{\hat{a}^{\mathrm{sp}}\hat{\rho}^{\mathrm{sp}}\right\}_{\alpha} . \tag{3.117}$$

Here, however, the global operator $\hat{\rho}$ has been replaced by the single-particle density-matrix operator

$$\hat{\rho}^{\mathrm{sp}} = \sum_{\alpha\alpha'} |\alpha\rangle \rho^{\mathrm{sp}}_{\alpha\alpha'} \langle\alpha'| \tag{3.118}$$

and the trace acts on the single-particle Hilbert space α only. We stress that, while the reduced density-matrix operator $\hat{\rho}^{\mathrm{c}}$ describes the whole many-electron system, the single-particle operator $\hat{\rho}^{\mathrm{sp}}$ provides an average or mean-field treatment of the carrier subsystem; indeed, as discussed extensively in Chap. 9, the latter fails in describing many-particle correlations, like Coulomb-correlation effects in quasi zero-dimensional systems.

Since $\rho^{\mathrm{sp}}_{\alpha_1\alpha_2}$ is the only quantity entering the evaluation of the average value (3.115), it is desirable to derive a corresponding equation of motion for the single-particle density matrix (3.116), i.e.,

$$\frac{d\rho^{\mathrm{sp}}_{\alpha_1\alpha_2}}{dt} = \mathrm{tr}\left\{\hat{c}^{\dagger}_{\alpha_2}\hat{c}_{\alpha_1}\frac{d\hat{\rho}}{dt}\right\} . \tag{3.119}$$

Inserting into the above expression the equation of motion for the global density-matrix operator $\hat{\rho}$ in (3.31) one gets

$$\frac{d\rho^{\mathrm{sp}}_{\alpha_1\alpha_2}}{dt} = \left.\frac{d\rho^{\mathrm{sp}}_{\alpha_1\alpha_2}}{dt}\right|_{\hat{H}_\circ} + \left.\frac{d\rho^{\mathrm{sp}}_{\alpha_1\alpha_2}}{dt}\right|_{\hat{C}} + \left.\frac{d\rho^{\mathrm{sp}}_{\alpha_1\alpha_2}}{dt}\right|_{\Gamma} , \tag{3.120}$$

where

$$\left.\frac{d\rho^{\mathrm{sp}}_{\alpha_1\alpha_2}}{dt}\right|_{\hat{H}_\circ} = \frac{1}{i\hbar}\mathrm{tr}\left\{\hat{c}^{\dagger}_{\alpha_2}\hat{c}_{\alpha_1}\left[\hat{H}_\circ, \hat{\rho}\right]\right\} \tag{3.121}$$

is the time variation induced by the noninteracting Hamiltonian $\hat{H}_\circ$,

$$\left.\frac{d\rho^{\mathrm{sp}}_{\alpha_1\alpha_2}}{dt}\right|_{\hat{C}} = \mathrm{tr}\left\{\hat{c}^{\dagger}_{\alpha_2}\hat{c}_{\alpha_1}\hat{C}\right\} \tag{3.122}$$

is the contribution due to the initial quantum-correlation operator $\hat{C}$ in (3.28), and

$$\left.\frac{\mathrm{d}\rho^{\mathrm{sp}}_{\alpha_1\alpha_2}}{\mathrm{d}t}\right|_{\Gamma} = -\mathrm{tr}\left\{\hat{c}^{\dagger}_{\alpha_2}\hat{c}_{\alpha_1}[\hat{\mathcal{H}},[\hat{\mathcal{K}},\hat{\rho}]]\right\} \tag{3.123}$$

is the time evolution dictated by the scattering superoperator Γ.

For a better evaluation of the various contributions (from (3.121) to (3.123)) it is convenient to expand the commutators entering the trace, regrouping the various terms in a different way. More specifically, by inserting the explicit form of the single and double commutators, and using the cyclic property of the trace in (C.7), we finally get

$$\left.\frac{\mathrm{d}\rho^{\mathrm{sp}}_{\alpha_1\alpha_2}}{\mathrm{d}t}\right|_{\hat{H}_\circ} = \frac{1}{\mathrm{i}\hbar}\mathrm{tr}\left\{\left[\hat{c}^{\dagger}_{\alpha_2}\hat{c}_{\alpha_1},\hat{H}_\circ\right]\hat{\rho}\right\}\,, \tag{3.124}$$

$$\left.\frac{\mathrm{d}\rho^{\mathrm{sp}}_{\alpha_1\alpha_2}}{\mathrm{d}t}\right|_{\hat{C}} = -\mathrm{i}\,\mathrm{tr}\left\{\left[\hat{c}^{\dagger}_{\alpha_2}\hat{c}_{\alpha_1},\hat{\mathcal{H}}\right]\hat{U}_\circ(t-t_0)\hat{\rho}(t_0)\hat{U}^{\dagger}_\circ(t-t_0)\right\}\,, \tag{3.125}$$

and

$$\left.\frac{\mathrm{d}\rho^{\mathrm{sp}}_{\alpha_1\alpha_2}}{\mathrm{d}t}\right|_{\Gamma} = -\mathrm{tr}\left\{\left[\left[\hat{c}^{\dagger}_{\alpha_2}\hat{c}_{\alpha_1},\hat{\mathcal{H}}\right],\hat{\mathcal{K}}\right]\hat{\rho}\right\}\,. \tag{3.126}$$

As we can see, the various contributions to the time evolution of the single-particle density matrix can be written as global average values (see (3.108)) of single as well as double commutators.

When the carrier–carrier interaction Hamiltonian $\hat{H}^{\mathrm{cc}}$ is treated as a perturbation, the term (3.124) – describing the time evolution dictated by the noninteracting Hamiltonian $\hat{H}_\circ = \hat{H}^{\mathrm{c}} + \hat{H}^{\mathrm{qp}}$ – can be evaluated exactly. Indeed, recalling the explicit form of the noninteracting-carrier Hamiltonian (3.110) and using the fermionic anticommutation relations $\left\{\hat{c}_{\alpha},\hat{c}^{\dagger}_{\alpha'}\right\} = \delta_{\alpha\alpha'}$, we obtain

$$\left[\hat{c}^{\dagger}_{\alpha_2}\hat{c}_{\alpha_1},\hat{H}_\circ\right] = (\epsilon_{\alpha_1}-\epsilon_{\alpha_2})\,\hat{c}^{\dagger}_{\alpha_2}\hat{c}_{\alpha_1}\,. \tag{3.127}$$

By inserting this result into (3.124), we get a closed equation for the single-particle density matrix:

$$\left.\frac{\mathrm{d}\rho^{\mathrm{sp}}_{\alpha_1\alpha_2}}{\mathrm{d}t}\right|_{\hat{H}_\circ} = \frac{\epsilon_{\alpha_1}-\epsilon_{\alpha_2}}{\mathrm{i}\hbar}\rho^{\mathrm{sp}}_{\alpha_1\alpha_2}\,. \tag{3.128}$$

In contrast, for the first- and second-order interaction contributions in (3.125) and (3.126) it is not possible to obtain closed equations of motion for the single-particle density matrix ρ^{sp}: indeed such contributions involve higher order correlations, e.g., two-body and/or quasiparticle-assisted density matrices (see, e.g., [110]). In order to get a closed equation for $\rho^{\mathrm{sp}}_{\alpha_1\alpha_2}$, an additional approximation is needed, the so-called mean-field approximation. The latter consists in a factorization of higher order correlation functions into products of single-particle density matrices ρ^{sp} and/or quasiparticle populations $N^\circ_{\boldsymbol{q}}$. The required mean-field procedure and the explicit form of the resulting closed equation of motion depends on the particular form of the interaction Hamiltonian considered, e.g., carrier–carrier, carrier–quasiparticle. However, the free-evolution term (3.124) together with the first-order contribution

(3.125) describes, in general, coherent phenomena – including Hartree–Fock renormalizations and coherent phonons – while the second-order term (3.126) describes energy-dissipation and decoherence processes within the Markov approximation previously introduced.

At this point a few comments are in order. The single-particle description discussed so far is based on the Schrödinger picture: the equation of motion for $\rho^{\rm sp}_{\alpha_1\alpha_2}$ (see (3.119)) is derived by treating the operators $\hat{c}^\dagger$ and $\hat{c}$ in (3.116) as time independent, while the time variation is fully ascribed to the density-matrix operator $\hat{\rho}$. Actually, the most popular and commonly used approach (see, e.g., [110]) to derive the equations of motion governing the time evolution of the single-particle density matrix is based on the Heisenberg picture: the density-matrix operator $\hat{\rho}$ entering (3.116) in the Heisenberg scheme is time independent, while the time evolution is fully ascribed to the fermionic operators via their corresponding Heisenberg equations of motion:

$$\frac{d\hat{c}_\alpha}{dt} = \frac{1}{i\hbar}[\hat{c}_\alpha, \hat{H}] \, . \tag{3.129}$$

More precisely, within the Heisenberg picture the result in (3.119) is replaced by

$$\frac{d\rho^{\rm sp}_{\alpha_1\alpha_2}}{dt} = {\rm tr}\left\{\frac{d}{dt}\left(\hat{c}^\dagger_{\alpha_2}\hat{c}_{\alpha_1}\right)\hat{\rho}\right\} = \frac{1}{i\hbar}{\rm tr}\left\{[\hat{c}^\dagger_{\alpha_2}\hat{c}_{\alpha_1}, \hat{H}]\hat{\rho}\right\} \, . \tag{3.130}$$

Contrary to the theoretical approach reviewed in this chapter, the usual Heisenberg treatment is based on a correlation expansion of the trace in (3.130): starting again from the Hamiltonian separation (3.2), a hierarchy of kinetic equations involving higher order as well as quasiparticle-assisted density matrices is established; the different contributions are classified in terms of their perturbation orders. Such infinite hierarchy is truncated/closed via the mean-field approximation previously recalled, and only at this level the Markov limit is usually introduced. Spirit of the present chapter, in contrast, was to analyze the Markov limit from a more general point of view; it is for this reason that the latter has been introduced in very general terms in Sect. 3.3.1 before addressing any reduced-description procedure.

Carrier–Carrier Interaction

As first interaction mechanism we shall consider two–body Coulomb coupling. The corresponding interaction Hamiltonian can be written as

$$\hat{\mathcal{H}}^{\rm cc} = \frac{1}{2\hbar} \sum_{\alpha_1\alpha_2,\alpha'_1\alpha'_2} V^{\rm cc}_{\alpha_1\alpha_2,\alpha'_1\alpha'_2} \hat{c}^\dagger_{\alpha_1}\hat{c}^\dagger_{\alpha_2}\hat{c}_{\alpha'_2}\hat{c}_{\alpha'_1} \, , \tag{3.131}$$

where $V^{\rm cc}_{\alpha_1\alpha_2,\alpha'_1\alpha'_2}$ is the Coulomb matrix element for the generic two-body transition $\alpha'_1\alpha'_2 \to \alpha_1\alpha_2$.

A detailed derivation of the second-order contribution to the single-particle dynamics in (3.126) is out of the scope of the present book. As discussed in

[168], to this end two key quantities need to be evaluated: the inner commutator $\left[\hat{c}^\dagger_{\alpha_2}\hat{c}_{\alpha_1}, \hat{\mathcal{H}}\right]$ and the explicit form of the operator $\hat{\mathcal{K}}$. Given these two quantities, it is possible to show that for the two-body interaction in (3.131) their commutator – key ingredient in (3.126) – involves products of six fermionic operators.

As anticipated, in order to get a closed equation of motion for the single-particle density matrix $\rho^{\rm sp}$ we are forced to employ the mean-field approximation; the latter allows in this case to write the average values of six fermionic operators as products of three single-particle density-matrix elements. By applying such mean-field factorization procedure, the final result can be cast into the general form

$$\left.\frac{\mathrm{d}\rho^{\rm sp}_{\alpha_1\alpha_2}}{\mathrm{d}t}\right|_{\rm cc} = \mathcal{F}^{\rm cc}_{\rm in}\left(\rho^{\rm sp}\right)_{\alpha_1\alpha_2} - \mathcal{F}^{\rm cc}_{\rm out}\left(\rho^{\rm sp}\right)_{\alpha_1\alpha_2} . \tag{3.132}$$

As for the case of the reduced description (see Sect. 3.3.2), the time variation of the single-particle density matrix is the sum of a positive-like – in-scattering – and a negative-like – out-scattering – contribution. However, contrary to the global and reduced descriptions previously considered, now the in- and out-scattering contributions – not reported here – are nonlinear functions of the single-particle density matrix $\rho^{\rm sp}$. In particular, for the present case of two–body interaction between a main (M) and a partner (P) carrier, the superoperators $\mathcal{F}^{\rm cc}_{\rm in}$ and $\mathcal{F}^{\rm cc}_{\rm out}$ both involve a product structure of the form $\hat{\rho}^{\rm sp}_{\rm M}\hat{\rho}^{\rm sp}_{\rm P}(\hat{\mathcal{I}} - \hat{\rho}^{\rm sp}_{\rm M})(\hat{\mathcal{I}} - \hat{\rho}^{\rm sp}_{\rm P})$.

Also for the present single-particle description it is possible to consider the semiclassical or Boltzmann limit introduced in Sect. 3.3.1. This amounts again to neglecting non-diagonal density-matrix elements:

$$\rho^{\rm sp}_{\alpha_1\alpha_2} = f^{\rm sp}_{\alpha_1}\delta_{\alpha_1\alpha_2} . \tag{3.133}$$

By inserting the above diagonal form of $\rho^{\rm sp}$ into the explicit form (not reported here) of the in- and out-scattering superoperators $\mathcal{F}^{\rm cc}_{\rm in}$ and $\mathcal{F}^{\rm cc}_{\rm out}$, the following Boltzmann-like equation for the semiclassical single-particle distribution $f^{\rm sp}_\alpha$ may be derived

$$\left.\frac{\mathrm{d}f^{\rm sp}_\alpha}{\mathrm{d}t}\right|_{\rm cc} = \sum_{\alpha'}\left[(1-f^{\rm sp}_\alpha)P^{\rm cc}_{\alpha\alpha'}f^{\rm sp}_{\alpha'} - (1-f^{\rm sp}_{\alpha'})P^{\rm cc}_{\alpha'\alpha}f^{\rm sp}_\alpha\right] , \tag{3.134}$$

where

$$P^{\rm cc}_{\alpha\alpha'} = \frac{2\pi}{\hbar}\sum_{\tilde{\alpha}\,\tilde{\alpha}'}(1-f^{\rm sp}_{\tilde{\alpha}})\left|V^{\rm cc}_{\alpha\tilde{\alpha},\alpha'\tilde{\alpha}'}\right|^2 f^{\rm sp}_{\tilde{\alpha}'}\delta\left((\epsilon_\alpha+\epsilon_{\tilde{\alpha}}) - (\epsilon_{\alpha'}+\epsilon_{\tilde{\alpha}'})\right) \tag{3.135}$$

are two-body carrier–carrier scattering rates describing the main-carrier transition $\alpha' \to \alpha$ accompanied by the partner-carrier transition $\tilde{\alpha}' \to \tilde{\alpha}$. As we can see, also in the semiclassical limit we deal with a nonlinear transport equation; such nonlinearities are ascribed (i) to the presence of the carrier distribution f^{sp} of the initial partner carrier and (ii) to the two Pauli-blocking factors $(1 - f^{\mathrm{sp}})$ corresponding to the final states of both main and partner carriers. Comparing the semiclassical transport equation (3.134) to its quantum-mechanical generalization in (3.132), we clearly see that the various terms of the form $(\delta_{\alpha\alpha'} - \rho^{\mathrm{sp}}_{\alpha\alpha'})$ previously mentioned are the natural generalization of the Pauli-blocking factors $(1 - f^{\mathrm{sp}}_\alpha)$ of the semiclassical theory.

Carrier–Quasiparticle Interaction

Let us now come to the carrier–quasiparticle coupling mechanism. By adopting as explicit form of the carrier–quasiparticle quantities $\hat{\mathcal{H}}^\pm_{\boldsymbol{q}} = \hat{\mathcal{H}}^{\mp\dagger}_{\boldsymbol{q}}$ single-particle operators of the form

$$\hat{\mathcal{H}}^\pm_{\boldsymbol{q}} = \frac{1}{\hbar}\sum_{\alpha\alpha'} g^\pm_{\alpha\alpha',\boldsymbol{q}} \hat{c}^\dagger_\alpha \hat{c}_{\alpha'} , \tag{3.136}$$

the carrier–quasiparticle interaction Hamiltonian (3.86) is given by

$$\hbar\hat{\mathcal{H}}^{\mathrm{c-qp}} = \sum_{\alpha\alpha',\boldsymbol{q}} \left(g^-_{\alpha\alpha',\boldsymbol{q}} \hat{c}^\dagger_\alpha \hat{b}_{\boldsymbol{q}} \hat{c}_{\alpha'} + g^+_{\alpha\alpha',\boldsymbol{q}} \hat{c}^\dagger_{\alpha'} \hat{b}^\dagger_{\boldsymbol{q}} \hat{c}_\alpha \right) , \tag{3.137}$$

where $g^\pm_{\alpha\alpha',\boldsymbol{q}} = g^{\mp *}_{\alpha'\alpha,\boldsymbol{q}}$ are carrier–quasiparticle matrix elements for the single-particle transition $\alpha' \to \alpha$ induced by the quasiparticle mode $\boldsymbol{q}$, whose explicit form depends on the particular carrier–quasiparticle interaction mechanism considered (e.g., carrier–phonon due to deformation potential or to electrostatic interaction, carrier–photon, carrier–plasmon).

As for the carrier–carrier interaction previously considered, a detailed derivation of the carrier–quasiparticle contribution to the single-particle dynamics in (3.126) is out of the scope of the present book. As discussed in [168], to this end, one has to evaluate again the inner commutator $\left[\hat{c}^\dagger_{\alpha_2}\hat{c}_{\alpha_1}, \hat{\mathcal{H}}\right]$ as well as the explicit form of the operator $\hat{\mathcal{K}}$. Given these two quantities, it is possible to show that their commutator – key ingredient in (3.126) – involves products of two bosonic and two fermionic operators as well as products of four fermionic operators. As for the carrier–carrier interaction previously discussed, in order to get a closed equation of motion for the single-particle density matrix ρ^{sp} we are forced to employ again the mean-field approximation; the latter allows one to write (i) the average value of two fermionic times two bosonic operators as the product of single-particle density-matrix elements times quasiparticle distributions and (ii) the average values of four fermionic operators as products of two single-particle density-matrix elements. By applying such mean-field factorization procedure, the final result can be cast into the same form of the one for carrier–carrier interaction in (3.132), i.e.,

$$\left.\frac{d\rho^{\mathrm{sp}}_{\alpha_1\alpha_2}}{dt}\right|_{\mathrm{c-qp}} = \mathcal{F}^{\mathrm{c-qp}}_{\mathrm{in}}\left(\rho^{\mathrm{sp}}\right)_{\alpha_1\alpha_2} - \mathcal{F}^{\mathrm{c-qp}}_{\mathrm{out}}\left(\rho^{\mathrm{sp}}\right)_{\alpha_1\alpha_2} . \tag{3.138}$$

As for the case of carrier–carrier interaction, the above in- and out-scattering contributions are again nonlinear functions of the single-particle density matrix ρ^{sp}. More specifically, their general structure is of the form $\hat{\rho}^{sp}(\hat{\mathcal{I}} - \hat{\rho}^{sp})$; such nonlinearities – ascribed to Pauli-blocking effects – vanish in the so-called low-density limit $(\hat{\mathcal{I}} - \hat{\rho}^{sp} \to \hat{\mathcal{I}})$.

Let us consider again the semiclassical limit. By inserting the diagonal density-matrix approximation (3.133) into the explicit form (not reported here) of the in- and out-scattering terms in (3.138), one gets a nonlinear Boltzmann equation formally identical to the single-particle transport equation (3.134):

$$\left.\frac{d f_\alpha^{sp}}{dt}\right|_{c-qp} = \sum_{\alpha'} \left[(1 - f_\alpha^{sp}) P_{\alpha\alpha'}^{c-qp} f_{\alpha'}^{sp} - (1 - f_{\alpha'}^{sp}) P_{\alpha'\alpha}^{c-qp} f_\alpha^{sp}\right] , \qquad (3.139)$$

where

$$P_{\alpha\alpha'}^{c-qp} = \frac{2\pi}{\hbar} \sum_{\boldsymbol{q}\pm} \left(N_{\boldsymbol{q}}^\circ + \frac{1}{2} \pm \frac{1}{2}\right) \left|g_{\alpha\alpha',\boldsymbol{q}}^{\pm}\right|^2 \delta(\epsilon_\alpha - \epsilon_{\alpha'} \pm \epsilon_{\boldsymbol{q}}) \qquad (3.140)$$

denote semiclassical carrier–quasiparticle scattering rates for the single-particle transition $\alpha' \to \alpha$ as given by the standard Fermi's golden rule. We deal again with a nonlinear transport equation; in this case, however, such nonlinearities are only ascribed to the Pauli-blocking factor of the final state, and they vanish in the low-density limit.

The Low-Density Limit

Let us finally consider the so-called low-density limit. To this aim, let us recall that within the single-particle description previously introduced the average occupation number for the generic state α is simply given by the diagonal elements of the single-particle density matrix (3.116):

$$f_\alpha^{sp} = \rho_{\alpha\alpha}^{sp} = \mathrm{tr}\left\{\hat{c}_\alpha^\dagger \hat{c}_\alpha \hat{\rho}\right\} . \qquad (3.141)$$

It is then clear that at low carrier concentrations (low densities) the magnitude of all density-matrix elements is much smaller than 1. More precisely, in the so-called low-density limit we have

$$\rho_{\alpha\alpha'}^{sp} \to 0 , \qquad \delta_{\alpha\alpha'} - \rho_{\alpha\alpha'}^{sp} \to \delta_{\alpha\alpha'} . \qquad (3.142)$$

In this limit the carrier–carrier scattering contributions in (3.132) vanish since, as anticipated, they involve the single-particle density matrix of the partner carrier $\hat{\rho}_P^{sp}$.

As a result, in the low-density limit the only non-vanishing contribution to the interaction dynamics is given by the carrier–quasiparticle terms in (3.138). More precisely, by inserting the low-density condition (3.142) in our quantum-transport equation (3.138), it is possible to get the following linear single-particle equation

$$\left.\frac{d\rho^{sp}_{\alpha_1\alpha_2}}{dt}\right|_{c-qp} = \frac{1}{2}\sum_{\alpha'_1\alpha'_2}\left[\mathcal{P}_{\alpha_1\alpha_2,\alpha'_1\alpha'_2}\rho^{sp}_{\alpha'_1\alpha'_2} - \mathcal{P}^*_{\alpha'_1\alpha'_1,\alpha_1\alpha'_2}\rho^{sp}_{\alpha'_2\alpha_2}\right] + \text{H.c.} \quad (3.143)$$

with generalized single-particle scattering rates

$$\mathcal{P}_{\alpha_1\alpha_2,\alpha'_1\alpha'_2} = \frac{4\pi}{\hbar^2}\sum_{\boldsymbol{q}\pm}\left(N^\circ_{\boldsymbol{q}} + \frac{1}{2} \pm \frac{1}{2}\right) g^{\pm}_{\alpha_1\alpha'_1,\boldsymbol{q}} g^{\pm *}_{\alpha_2\alpha'_2,\boldsymbol{q}} D^{\pm *}_{\alpha_2\alpha'_2,\boldsymbol{q}} \quad (3.144)$$

and

$$D^{\pm}_{\alpha\alpha',\boldsymbol{q}} = \frac{1}{2\pi}\int_0^{t-t_0} e^{\frac{(\epsilon_\alpha - \epsilon_{\alpha'} \pm \epsilon_{\boldsymbol{q}})\tau}{i\hbar}} d\tau \, . \quad (3.145)$$

We stress that the single-particle Eq. (3.143) is formally identical to the reduced-description one in (3.100). This can be easily understood considering that the present low-density limit is physically equivalent to consider a system of just one electron interacting with the quasiparticle degrees of freedom, which corresponds to substituting the many-electron configuration label ξ with the state α of the only electron considered; this is confirmed by comparing the many-electron scattering operator (3.104) with the single-particle one in (3.144).

By inserting into the equation of motion (3.120) the results in (3.128) and (3.143) and by neglecting the first-order term (3.125), we finally get the following quantum-transport equation:

$$\frac{d\rho^{sp}_{\alpha_1\alpha_2}}{dt} = \sum_{\alpha'_1\alpha'_2} \mathcal{L}^{sp}_{\alpha_1\alpha_2,\alpha'_1\alpha'_2}\rho^{sp}_{\alpha'_1\alpha'_2} \quad (3.146)$$

with

$$\mathcal{L}^{sp}_{\alpha_1\alpha_2,\alpha'_1\alpha'_2} = \frac{\epsilon_{\alpha_1} - \epsilon_{\alpha_2}}{i\hbar}\delta_{\alpha_1\alpha_2,\alpha'_1\alpha'_2} + \frac{1}{2}\left(\Gamma_{\alpha_1\alpha_2,\alpha'_1\alpha'_2} + \Gamma^*_{\alpha_2\alpha_1,\alpha'_2\alpha'_1}\right) \quad (3.147)$$

and

$$\Gamma_{\alpha_1\alpha_2,\alpha'_1\alpha'_2} = \mathcal{P}_{\alpha_1\alpha_2,\alpha'_1\alpha'_2} - \delta_{\alpha_2\alpha'_2}\sum_{\alpha'}\mathcal{P}^*_{\alpha'\alpha',\alpha_1\alpha'_1} \, . \quad (3.148)$$

The linear superoperator (3.147) can be regarded as an effective single-particle Liouville operator in the low-density limit.

Again, within the semiclassical limit the linear quantum-transport equation (3.143) reduces to the following linear Boltzmann equation:

$$\left.\frac{df^{sp}_\alpha}{dt}\right|_{c-qp} = \sum_{\alpha'}\left[P^{c-qp}_{\alpha\alpha'} f^{sp}_{\alpha'} - P^{c-qp}_{\alpha'\alpha} f^{sp}_{\alpha}\right] \, , \quad (3.149)$$

where the carrier–quasiparticle scattering rates $P^{c-qp}_{\alpha\alpha'}$ are given in (3.140). We stress that the above equation can be obtained directly from the nonlinear Boltzmann equation (3.139) by neglecting Pauli-blocking factors.

3.4 Need for a Gauge-Invariant Formulation of the Problem

Since the early days of quantum mechanics [175] the field-induced coherent dynamics of an electron wavepacket within a crystal, known as "Bloch oscillations," has attracted significant and increasing interest (see, e.g., [32]). Indeed, the problem of properly describing the scattering-free motion of an electron in a solid has led to a three-decade controversy on the existence of Bloch oscillations (see, e.g., [176] and references therein); This originated from the different approaches employed for the description of the applied field, namely the vector-potential or accelerated-Bloch-state picture [177] and the scalar-potential or Wannier–Stark description [178]. As discussed extensively in [179], these two pictures are now recognized to be fully equivalent, since they correspond to different quantum-mechanical representations connected by a gauge transformation.

The presence of scattering as well as tunneling processes strongly modifies such ideal Bloch oscillations scenario (see, e.g., [180, 181]). In particular, nonelastic interaction mechanisms – like carrier–optical phonon scattering – tend to spoil such coherent dynamics, leading to a nearly semiclassical or Boltzmann-like transport behavior. In the presence of strong applied fields, however, the use of the conventional scattering picture – involving transitions between field-free Bloch states via the standard Fermi's golden rule – becomes questionable.

As originally pointed out by Levinson and Yasevichyute [182] as well as by Barker and Ferry [183], the effect of the field during the scattering process, usually referred to as ICFE (see Sect. 2.1), may lead to significant deviations from the semiclassical scenario. On the one hand, the role played by the ICFE has been extensively investigated by means of rigorous quantum-transport approaches (see, e.g., [108, 184, 185]). Their application, however, was often limited to highly simplified physical models and conditions, thus preventing from any quantitative comparison with experiments. On the other hand, strong effort has been devoted to incorporate the ICFE within conventional – and more realistic – Monte Carlo simulations (see, e.g., [28]). In this case, the basic idea is that, due to the field-induced carrier drift, also in the completed-collision limit, energy conservation in the scattering process is relaxed; as a consequence, as discussed below, the Dirac delta function of the Fermi's golden rule is typically replaced by broad spectral functions (see, e.g., [186]). It is imperative to stress that this scenario, intimately related to the vector-potential or accelerated picture, has no counterpart in the scalar-potential one. Indeed, within the Wannier–Stark basis there is no carrier drift, and energy conservation is preserved. It is thus clear that such an effective semiclassical description of the ICFE is not gauge invariant.[6] Following the

[6] The need for a gauge-invariant formulation of the problem was originally pointed out by Rita Bertoncini and Antti-Pekka Jauho in [184].

approach presented in [87, 88], aim of this section is to explain and remove this apparent contradiction by providing a gauge-invariant formulation of the problem.

As recalled in Sect. 1.2.1, in the presence of an electro-optical excitation – described via the corresponding scalar and vector potentials φ and $\boldsymbol{A}$ (see (1.23)) – the single-particle states α are defined by the eigenvalue problem in (1.22), i.e.,

$$\left[-\frac{\left(-i\hbar\boldsymbol{\nabla}_{\boldsymbol{r}} + \frac{e}{c}\boldsymbol{A}(\boldsymbol{r},t)\right)^2}{2m} + V(\boldsymbol{r}) - e\varphi(\boldsymbol{r},t)\right]\phi_\alpha(\boldsymbol{r}) = \epsilon_\alpha\phi_\alpha(\boldsymbol{r})\ . \tag{3.150}$$

For the relevant case of a semiconductor bulk crystal subjected to a constant and homogeneous applied electric field $\boldsymbol{e}$, the corresponding electromagnetic potentials may be written as

$$\boldsymbol{A}(\boldsymbol{r},t;\eta) = -c\eta\boldsymbol{E}t\ , \qquad \varphi(\boldsymbol{r};\eta) = -(1-\eta)\boldsymbol{E}\cdot\boldsymbol{r}\ . \tag{3.151}$$

Here, the gauge freedom previously mentioned is expressed in terms of the transformation parameter η; indeed, the quantum numbers α in (3.150) – as well as the corresponding eigenfunctions $\phi_\alpha(\boldsymbol{r}) = \langle\boldsymbol{r}|\alpha\rangle$ and single-particle energies ϵ_α – are functions of the transformation parameter η, and for $\eta \neq 0$ the latter are also time dependent. In particular, for $\eta = 0$ (scalar-potential gauge) we recover the well-known Wannier–Stark ladder [178]: $\epsilon_\alpha = \epsilon_n = \epsilon_0 + n\Delta\epsilon$ with $\Delta\epsilon = eEd$, d denoting the crystal periodicity along the field direction; In contrast, for $\eta = 1$ (vector-potential gauge) we deal with the Houston or accelerated Bloch states [177]: $\epsilon_\alpha = \epsilon_{\boldsymbol{k}(t)\nu}$, where $\boldsymbol{k}(t) = \boldsymbol{k}_0 + \dot{\boldsymbol{k}}t$ is the instantaneous carrier wavevector in band ν, $\dot{\boldsymbol{k}} = -\frac{e}{\hbar}\boldsymbol{E}$ being its field-induced time variation.

It is then clear that, within the single-particle picture introduced in Sect. 3.3.3, the corresponding density matrix $\rho^{\mathrm{sp}}_{\alpha\alpha'}$ in (3.116) as well as its effective Liouville superoperator $\mathcal{L}^{\mathrm{sp}}$ in (3.146) will be η-dependent as well. In particular, in the presence of a time-dependent basis (e.g., accelerated Bloch states) the function $D^{\pm}_{\alpha\alpha',\boldsymbol{q}}$ in (3.145) is usually replaced by

$$D^{\pm}_{\alpha\alpha',\boldsymbol{q}} = \frac{1}{2\pi}\int_0^{t-t_0} \mathrm{e}^{\frac{1}{i\hbar}\int_0^{\tau}(\epsilon_{\alpha(\tau')}-\epsilon_{\alpha'(\tau')}\pm\epsilon_{\boldsymbol{q}})\mathrm{d}\tau'}\mathrm{d}\tau\ . \tag{3.152}$$

As anticipated, also in the completed-collision limit ($t - t_0 \to \infty$) the delta function of the Fermi's golden rule – typical of a time-independent basis – is replaced by a broad spectral function. This scenario, intimately related to the vector-potential or accelerated picture, has no counterpart in the scalar-potential gauge.

As discussed extensively in Sect. 3.3.3, in order to derive a closed equation of motion for our single-particle density matrix, two alternative strategies are typically employed: on the one hand, within the Schrödinger picture the time evolution is fully ascribed to the global density-matrix operator $\hat{\rho}$, while within the Heisenberg picture the latter is fully attributed to the single-particle creation and destruction operators. In this second case, the starting point of the derivation is their Heisenberg equations of motion: for a time-independent basis α, the latter are given in (3.129); however, for a time-dependent basis we have

$$\frac{\mathrm{d}\hat{c}_\alpha}{\mathrm{d}t} = \frac{1}{\mathrm{i}\hbar}[\hat{c}_\alpha, \hat{H}] + \left.\frac{\mathrm{d}\hat{c}_\alpha}{\mathrm{d}t}\right|_\phi . \tag{3.153}$$

Compared to the standard equations of motion (3.129), the possible time variation of our basis states $\phi_{\alpha(t)}$ gives rise to an additional term; the latter has been usually neglected, giving rise to the apparent discrepancies between scalar- and vector-potential treatments of high-field transport, mentioned in the introductory part of this section.

For a better understanding of the problem, it is crucial to realize that in the presence of external applied fields the one-electron Hamiltonian (1.22) – as well as its second-quantization version in (3.110) – may also be time dependent. Indeed, according to the eigenvalue problem (3.150), also in the presence of a time-independent applied electric field $\boldsymbol{E}$, the corresponding η-dependent electromagnetic potentials (3.151) give rise, in general, to a time-dependent Hamiltonian $\hat{H}_\circ$; the only exception is the scalar-potential gauge ($\eta = 0$) corresponding to the Wannier–Stark basis states previously mentioned.

It is then crucial to discuss how the conventional interaction picture introduced in Sect. 3.2 changes in the presence of a time-dependent Hamiltonian $\hat{H}_\circ$; in this case, the noninteracting evolution operator $\hat{U}_\circ$ in (3.17) needs to be replaced by

$$\hat{U}_\circ(t, t_0) = \sum_{n=0}^{\infty} \left(\frac{1}{\mathrm{i}\hbar}\right)^n \int_{t_0}^{t} \mathrm{d}t_n \ldots \int_{t_0}^{t_2} \mathrm{d}t_1 \hat{H}_\circ(t_n) \ldots \hat{H}_\circ(t_1) . \tag{3.154}$$

The latter may also be written in a compact form via the time-ordering operator T introduced in (3.35):

$$\hat{U}_\circ(t, t_0) = T\left[\mathrm{e}^{\frac{1}{\mathrm{i}\hbar}\int_{t_0}^{t} \hat{H}_\circ(t')\mathrm{d}t'}\right] . \tag{3.155}$$

It is easy to verify that for a time-independent Hamiltonian the various $\hat{H}_\circ$ commute, and the generalized solution (3.154) – which is still a solution of the Schrödinger-like equation (3.18) – reduces to the conventional expression in (3.17).

Starting from the generalized solution (3.154), it is still possible to define our density-matrix operator in the interaction picture according to the general prescription (3.19); combining the definition of $\hat{\rho}^i$ with the Schrödinger-like equation (3.18), it is easy to get the following Liouville–von Neumann equation within the interaction picture:

$$\frac{\mathrm{d}\hat{\rho}^i}{\mathrm{d}t} = -\mathrm{i}\left[\hat{\mathcal{H}}^i, \hat{\rho}^i\right] + \left.\frac{\mathrm{d}\hat{\rho}^i}{\mathrm{d}t}\right|_{\hat{H}_\circ} \tag{3.156}$$

with

$$\left.\frac{\mathrm{d}\hat{\rho}^i}{\mathrm{d}t}\right|_{\hat{H}_\circ} = \frac{1}{\mathrm{i}\hbar}\left[(\hat{H}^i_\circ - \hat{H}_\circ), \hat{\rho}^i\right] . \tag{3.157}$$

Compared to the conventional interaction-picture equation (3.20), the presence of a time-dependent Hamiltonian $\hat{H}_\circ$ gives rise to the additional time variation in (3.157). The latter involves the difference between the noninteracting Hamiltonians written in the Schrödinger and interaction pictures, which is different from zero since, for a time-dependent Hamiltonian, $\hat{H}_\circ$ does not commute with the evolution operator $\hat{U}_\circ$ in (3.154). We stress that the extra-term (3.157) is the Schrödinger-picture counterpart of the additional time variation entering the Heisenberg equation (3.153) induced by the possible time variation of our basis states.[7]

A closer inspection of the new Liouville–von Neumann equation (3.156) shows that – contrary to the conventional interaction-picture scenario (see (3.20)) – in the small-coupling limit ($\hat{\mathcal{H}} \to 0$) the density-matrix operator $\hat{\rho}^i$ is not necessarily slowly varying, and this is due to the presence of the extra-term (3.157). This implies that in the presence of a time-dependent Hamiltonian $\hat{H}_\circ$ the adiabatic-decoupling scheme presented in Sect. 3.3.1 is not applicable. It follows that for the case of a constant and homogeneous applied electric field (see (3.151)), the only gauge for which the adiabatic-decoupling strategy may be employed is the scalar-potential gauge ($\eta = 0$).

This explains the apparent discrepancies between scalar- and vector-potential treatments: the only gauge in which the Markov limit is justified is the scalar-potential one, and therefore the use of the effective Liouville superoperator (3.147) within the vector-potential gauge via the broad spectral function (3.152) has no physical justification.

At this point, it is crucial to stress that the applicability limits of the Markov approximation just recognized do not contrast with the request of a gauge-invariant formulation of the problem. To this end, by denoting with

$$U_{\overline{\alpha}\alpha} = \langle\overline{\alpha}|\alpha\rangle \tag{3.158}$$

the matrix elements of the unitary transformation connecting the time-independent ($\eta = 0$) Wannier–Stark states $|\alpha\rangle$ to the time-dependent states $|\overline{\alpha}\rangle$ corresponding to the generic gauge η, it is easy to show that the single-particle density matrix (3.116) will gauge transform according to

$$\rho^{\mathrm{sp}}_{\overline{\alpha}_1\overline{\alpha}_2} = \sum_{\alpha_1\alpha_2} U_{\overline{\alpha}_1\alpha_1} U^*_{\overline{\alpha}_2\alpha_2} \rho^{\mathrm{sp}}_{\alpha_1\alpha_2} . \tag{3.159}$$

If we now apply the above gauge transformation to the single-particle equation (3.146), it is easy to verify that the Markov superoperator $\mathcal{L}^{\mathrm{sp}}$ will gauge transform according to

[7] As discussed extensively in [87, 88], such extra-term – induced by the possible time variation of our basis states – describes field-induced interband transitions, usually referred to as "Zener tunneling."

$$\mathcal{L}^{\mathrm{sp}}_{\overline{\alpha}_1\overline{\alpha}_2,\overline{\alpha}'_1\overline{\alpha}'_2} = \sum_{\alpha_1\alpha_2,\alpha'_1\alpha'_2} U_{\overline{\alpha}_1\alpha_1} U^*_{\overline{\alpha}_2\alpha_2} \mathcal{L}^{\mathrm{sp}}_{\alpha_1\alpha_2,\alpha'_1\alpha'_2} U^*_{\overline{\alpha}'_1\alpha'_1} U_{\overline{\alpha}'_2\alpha'_2} \,. \tag{3.160}$$

This is the gauge-invariant formulation of high-field transport we were looking for: starting from the Markov superoperator evaluated within the scalar-potential basis $\{|\alpha\rangle\}$, we are able to move to the generic gauge η via the transformation in (3.160). We stress once again that such gauge-invariant prescription differs from the usual vector-potential treatments.

In order to compare quantitatively the usual treatment of the ICFE previously mentioned with the rigorous gauge-invariant formulation presented so far, it is useful to review the results of a fully three-dimensional calculation of high-field charge transport in state-of-the-art semiconductor superlattices presented in [87, 88].

Figure 3.1 shows the steady-state carrier drift velocity as a function of the applied field for a 45/45Å GaAs/$Al_{0.3}Ga_{0.7}As$ superlattice at room temperature in the low-density limit. Here, the usual ICFE model (triangles) is compared to the result of the proposed gauge-invariant approach (squares). At low fields the two curves exhibit a similar behavior, but they tend to separate as the field increases. In particular, the drift velocity corresponding to the usual ICFE treatment within the accelerated-Bloch-state picture at high fields is by far higher than the gauge-invariant one; this is the result of the erroneous estimation of the ICFE due to the neglect of the Zener-tunneling term (3.157) induced by the time variation of the vector-potential basis states. The peak at about 40 kV/cm, in both curves, corresponds to the phonon resonance, i.e., for this value of the applied field E the Wannier–Stark or Bloch

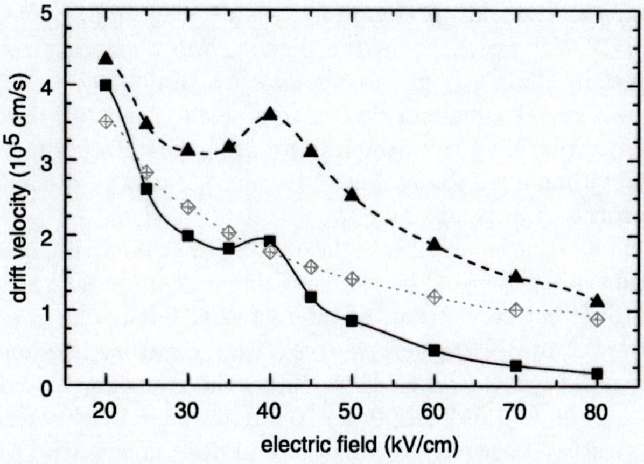

Fig. 3.1. Steady-state carrier drift velocity as a function of the applied field for a 45/45 Å GaAs/$Al_{0.3}Ga_{0.7}As$ superlattice at room temperature in the low-density limit. Gauge-invariant result (*squares*), conventional ICFE model (*triangles*), and Boltzmann limit (*diamonds*) (see text). *Lines* are a guide to the eye. Reprinted from [88]

energy eEd is equal to the optical-phonon energy. Let us finally compare the two quantum-mechanical results (squares and triangles) with the purely semiclassical (Boltzmann) one (diamonds). The latter shows a good agreement with the gauge-invariant one (squares) for a wide field range (20–50 kV/cm), while it differs significantly from the standard ICFE model (triangles). As anticipated, this may account for the surprisingly good agreement between semiclassical and rigorous quantum-transport calculations reported, e.g., in [108], as well as for the anomalous carrier heating typical of standard ICFE models [186].

3.5 Alternative Formulation of the Markov Limit: The "Quantum Fermi's Golden Rule"

As already pointed out, the global-description Markov superoperator (3.32) as well as its reduced-description version in (3.88) are not Lindblad like. It is imperative to stress that this serious limitation was originally evidenced by Herbert Spohn and co-workers [187] about three decades ago; in particular, they clearly pointed out that the choice of the adiabatic-decoupling strategy is definitely not unique (see below), and only one among the available possibilities, developed in the pioneering work by Edward B. Davies [188], could be shown to preserve positivity: it was the case of a "small" subsystem of interest interacting with a thermal environment, and selected through a partial-trace reduction. Unfortunately, the theory was restricted to finite-dimensional subsystems only (i.e., N-level atoms) and to the particular projection scheme of the partial trace.

Aim of this section is to present an alternative and more general adiabatic procedure [172–174] which (i) in the discrete-spectrum case reduces to the Davies' model in [188], (ii) in the semiclassical limit (see Sect. 3.3.1) gives the well-known Fermi's golden rule (see, e.g., [189]), and (iii) describes a genuine Lindblad evolution also in the infinite-dimensional/continuous-spectrum case, thus providing a reliable/robust treatment of energy-dissipation and decoherence processes in semiconductor quantum devices. As we shall see, by means of such alternative adiabatic-decoupling approach, different Markovian approximations are generated by choosing different projection schemes (corresponding to different subsystems of interest, e.g., many-electron description, single-particle picture). However, we stress that, contrary to standard master-equation formulations (see, e.g., [171, 190]), the resulting Liouville superoperators are always of Lindblad type; it follows that in this new adiabatic-decoupling strategy positivity is intrinsic and does not depend on the chosen subsystem of interest.

In order to introduce this alternative formulation of the Markov limit, let us go back to the integro-differential equation (3.23), and let us consider the following time symmetrization: given the two times t' and t, we shall introduce the "average" or "macroscopic" time $T = \frac{t+t'}{2}$ and the "relative" time

$\tau = t - t'$[8]: the basic idea is that the relevant time characterizing/describing our effective system evolution is the macroscopic time T; this suggests to replace the second-order contribution in (3.23) with the following "macroscopic" version[9]:

$$\frac{\mathrm{d}}{\mathrm{d}T}\hat{\rho}^i(T) = -\int_0^{T-T_0} \mathrm{d}\tau \left[\hat{\mathcal{H}}^i\left(T + \frac{\tau}{2}\right), \left[\hat{\mathcal{H}}^i\left(T - \frac{\tau}{2}\right), \hat{\rho}^i\left(T - \frac{\tau}{2}\right)\right]\right] . \tag{3.161}$$

In the spirit of the adiabatic approximation previously recalled, the density-matrix operator $\hat{\rho}^i$ can be taken out of the time integral and evaluated at the current time T. As already stressed, this important approximation is valid when the interaction-picture density matrix $\hat{\rho}^i$ varies slowly, i.e., in the so-called weak-coupling limit.

As described in [173], it is now convenient to rewrite the finite-domain time integration over τ by introducing a corresponding Gaussian correlation function $\mathrm{e}^{-\frac{\tau^2}{2\bar{t}^2}}$ whose width $\bar{t}$ may be regarded as a safe overestimation of the collision duration introduced in Chap. 2 , which is typically shorter than the macroscopic evolution time-scale $T - T_0$. More specifically, in the completed-collision limit (i.e., for $T - T_0$ greater than the collision duration) the time integration in (3.161) may be safely extended up to infinity. Moreover, for the sake of simplicity, here we shall neglect again energy-renormalization contributions[10]; As discussed in [172], this amounts to a symmetrization of the τ integration domain.

Within such approximation scheme, the result in (3.161) reduces to

$$\frac{\mathrm{d}}{\mathrm{d}T}\hat{\rho}^i(T) = -\frac{1}{2}\int_{-\infty}^{\infty} \mathrm{d}\tau\, \mathrm{e}^{-\frac{\tau^2}{2\bar{t}^2}} \left[\hat{\mathcal{H}}^i\left(T + \frac{\tau}{2}\right), \left[\hat{\mathcal{H}}^i\left(T - \frac{\tau}{2}\right), \hat{\rho}^i\left(T\right)\right]\right] . \tag{3.162}$$

We stress how the proposed time symmetrization gives rise to a fully symmetric superoperator, compared to the strongly asymmetric result of the conventional Markov treatment in (3.32).

The second crucial step in order to get a genuine Lindblad superoperator for the global dynamics is to exploit once again the slowly varying character of the density-matrix operator $\hat{\rho}^i$ on the right-hand side of (3.162). The key idea

---

8 This change of variables has very solid bases, as it is common and well established in a wide variety of contests, such as phase-space (or Wigner) formulation of quantum mechanics (see, e.g., [191]), standard quantum-kinetics Green's functions (see, e.g., [102]), and even classical radiation theory (e.g., in the treatment of "Bremsstrahlung").

9 The second-order term (3.161) may be readily derived from the original version in (3.23) by rewriting the double commutator in terms of the new variables ($t = T + \frac{\tau}{2}, t' = T - \frac{\tau}{2}$) and replacing on both sides the microscopic time t with the macroscopic time T.

10 A more general derivation accounting for energy-renormalization terms can be found in [172].

is to perform on both sides of (3.162) a so-called temporal coarse graining, i.e., a weighted time average on a so-called microscopic scale, a scale over which the variation of $\hat{\rho}^i(T)$ is negligible. Since for small and intermediate coupling regimes such time-scale is fully compatible with the collision-duration time-scale $\bar{t}$, we shall perform such time average employing once again a Gaussian correlation function of width $\frac{\bar{t}}{2}$, i.e.,

$$\frac{\mathrm{d}}{\mathrm{d}t}\hat{\rho}^i(T) = -\frac{1}{\sqrt{2\pi}\bar{t}}\int_{-\infty}^{\infty}\mathrm{d}T'\mathrm{e}^{-\frac{2T'^2}{\bar{t}^2}} \cdot \\ \cdot \int_{-\infty}^{\infty}\mathrm{d}\tau\mathrm{e}^{-\frac{\tau^2}{2\bar{t}^2}}\left[\hat{\mathcal{H}}^i\left(T+T'+\frac{\tau}{2}\right),\left[\hat{\mathcal{H}}^i\left(T+T'-\frac{\tau}{2}\right),\hat{\rho}^i\left(T\right)\right]\right] . \tag{3.163}$$

Moving back to the original Schrödinger picture, combining the two Gaussian distributions, and relabeling the macroscopic time T with t, the above equation can be rewritten in the following compact form

$$\frac{\mathrm{d}\hat{\rho}}{\mathrm{d}t} = -\frac{1}{2}\left[\hat{A},\left[\hat{A},\hat{\rho}\right]\right] \tag{3.164}$$

with

$$\hat{A} = \left(\frac{2}{\pi\bar{t}^2}\right)^{\frac{1}{4}}\int_{-\infty}^{\infty}\mathrm{d}t'\hat{\mathcal{H}}^i(t')\mathrm{e}^{-\frac{t'^2}{\bar{t}^2}} . \tag{3.165}$$

This is the genuine Lindblad-like superoperator we were looking for; indeed, the operator $\hat{A}$ is always Hermitian, and thus the resulting dynamics is positive-definite (see (3.37)).

If we now write the new Markov superoperator (3.164) in our noninteracting carrier-plus-quasiparticle basis λ, we obtain an effective equation of motion of the form

$$\frac{\mathrm{d}\rho_{\lambda_1\lambda_2}}{\mathrm{d}t} = \frac{1}{2}\sum_{\lambda'_1\lambda'_2}\left[\tilde{\mathcal{P}}_{\lambda_1\lambda_2,\lambda'_1\lambda'_2}\rho_{\lambda'_1\lambda'_2} - \tilde{\mathcal{P}}^*_{\lambda'_1\lambda'_1,\lambda_1\lambda'_2}\rho_{\lambda'_2\lambda_2}\right] + \mathrm{H.c.} \tag{3.166}$$

with

$$\tilde{\mathcal{P}}_{\lambda_1\lambda_2,\lambda'_1\lambda'_2} = A_{\lambda_1\lambda'_1}A^*_{\lambda_2\lambda'_2} . \tag{3.167}$$

Taking into account that the matrix elements $A_{\lambda\lambda'}$ of the Lindblad operator (3.165) are given by

$$A_{\lambda\lambda'} = \sqrt{\frac{2\pi}{\hbar}}H'_{\lambda\lambda'}\tilde{D}_{\lambda\lambda'} \tag{3.168}$$

with

$$\tilde{D}_{\lambda\lambda'} = \frac{\mathrm{e}^{-\left(\frac{\epsilon_\lambda-\epsilon_{\lambda'}}{2\bar{\epsilon}}\right)^2}}{\left(2\pi\bar{\epsilon}^2\right)^{\frac{1}{4}}} , \tag{3.169}$$

the explicit form of the scattering superoperator (3.167) comes out to be

$$\tilde{\mathcal{P}}_{\lambda_1\lambda_2,\lambda'_1\lambda'_2} = \frac{2\pi}{\hbar} H'_{\lambda_1\lambda'_1} H'^*_{\lambda_2\lambda'_2} \frac{\mathrm{e}^{-\frac{\left(\epsilon_{\lambda_1}-\epsilon_{\lambda'_1}\right)^2+\left(\epsilon_{\lambda_2}-\epsilon_{\lambda'_2}\right)^2}{4\overline{\epsilon}^2}}}{\sqrt{2\pi\overline{\epsilon}}} . \tag{3.170}$$

Here, $\overline{\epsilon} = \frac{\hbar}{\overline{t}}$ is a measure of the energy uncertainty in the interaction process induced by our temporal coarse graining.

It is imperative to stress that the new quantum-transport equation (3.166) has exactly the same structure of the corresponding result in (3.40), provided to replace the highly asymmetric scattering superoperator $\mathcal{P}$ of the conventional Markov treatment in (3.49) with its new symmetrized version in (3.170); indeed, the latter can be regarded as a generalization of the conventional Fermi's golden rule to the density-matrix formalism; indeed, in the semiclassical limit ($\lambda_1 = \lambda_2, \lambda'_1 = \lambda'_2$) the new scattering superoperator (3.170) reduces to what could be considered a dressed-vertex/smoothed version of the conventional Fermi's golden rule, i.e.,

$$\tilde{P}_{\lambda\lambda'} = \tilde{\mathcal{P}}_{\lambda\lambda,\lambda'\lambda'} = \frac{2\pi}{\hbar} |H'_{\lambda\lambda'}|^2 \frac{\mathrm{e}^{-\frac{(\epsilon_\lambda-\epsilon_{\lambda'})^2}{2\overline{\epsilon}^2}}}{\sqrt{2\pi\overline{\epsilon}}} , \tag{3.171}$$

and in the infinite collision time limit ($\overline{\epsilon} \to 0$) the standard Fermi's golden rule (3.54) is readily recovered.[11]

In passing, we stress that the generalized transition rates (3.170) could be regarded as a "quantum version" of the celebrated Fermi's golden rule. This should not generate confusion: of course the Fermi's golden rule is itself a quantum-mechanical result, but the latter is then employed within a semiclassical picture as generator of a "classical Markov process." Instead, the generalized transition rates (3.170) do not describe a classical Markov process, but rather its quantum analogue: a so-called quantum dynamical semigroup (see, e.g., [169]) for the full density matrix. Moreover, it is worth stressing that there are many proposed quantum generalizations of the well-known Fermi's golden rule (see, e.g., [192]), which are robust as well as physically and mathematically meaningful. However, all these generalizations consider a bipartite system, one of which is finally traced over, the other being an N-level atom (a rather particular case). In contrast, the alternative adiabatic strategy discussed so far is by far more general and refers to a closed quantum system, similarly to the original idea by Enrico Fermi.

---

[11] It may be argued that this approximation scheme – derived employing a specific Gaussian smoothing function – suffers from lack of generality. However, one has to look at asymptotic features only, since the approximation is valid in the weak-coupling regime plus longtime limit ($\overline{\epsilon} \to 0$); in this case, one does not appreciate features of this Gaussian smoothing, but rather the (much more universal) asymptotic character of our dynamics; as such, we could say that the selected Gaussian correlation is a good representative among all the possible asymptotic Markovian approximations to the exact dynamics.

As pointed out previously, the above theoretical scheme becomes meaningful and applicable only when a well-defined subsystem of interest is identified (together with a corresponding infinite-dimensional environment), so that its characteristic time-scale $\overline{t}$ can be estimated, and the corresponding (irreversible) semigroup dynamics can correctly describe the projected – but fully reversible (see, e.g., [193]) – exact Hamiltonian dynamics.

Following this spirit, let us consider again the reduced- or electronic-description picture introduced in Sect. 3.3.2; more specifically, let us consider a many-electron subsystem interacting with quasiparticles via the microscopic Hamiltonian (3.86). Starting from the new global (i.e., carrier-plus-quasiparticle) Markov superoperator in (3.164) and following all the conceptual steps and basic approximations discussed in Sect. 3.3.2 as well as in Appendix C, it is possible to derive an alternative version of the effective Markov superoperator in (3.88). More specifically, one gets

$$\frac{\mathrm{d}\hat{\rho}^c}{\mathrm{d}t} = -\frac{1}{2}\sum_{\boldsymbol{q}\pm}\left(N^\circ_{\boldsymbol{q}} + \frac{1}{2} \pm \frac{1}{2}\right)\left(\hat{A}^{\pm\dagger}_{\boldsymbol{q}}\hat{A}^{\pm}_{\boldsymbol{q}}\hat{\rho}^c - \hat{A}^{\pm}_{\boldsymbol{q}}\hat{\rho}^c\hat{A}^{\pm\dagger}_{\boldsymbol{q}}\right) + \mathrm{H.c.} \tag{3.172}$$

with

$$\hat{A}^{\pm}_{\boldsymbol{q}} = \left(\frac{2}{\pi\overline{t}^2}\right)^{\frac{1}{4}}\int_{-\infty}^{\infty}\mathrm{d}t'\mathrm{e}^{\frac{\hat{H}^c t'}{\mathrm{i}\hbar}}\hat{\mathcal{H}}^{\pm}_{\boldsymbol{q}}\mathrm{e}^{-\frac{\hat{H}^c t'}{\mathrm{i}\hbar}}\mathrm{e}^{\pm\frac{\epsilon_{\boldsymbol{q}} t'}{\mathrm{i}\hbar}}\mathrm{e}^{-\frac{t'^2}{\overline{t}^2}}\ . \tag{3.173}$$

Due to the reduction procedure, the electronic Lindblad operators (3.173) are not Hermitian anymore, and the original double-commutator structure of the global description is lost. However, it is easy to verify that, contrary to the standard electronic Markov superoperator (3.88), this new version in (3.172) is Lindblad like (see the general Lindblad form in (3.36)).

Let us now discuss the explicit form of the new electronic scattering superoperator within our many-electron basis ξ. To this aim, by expanding the various terms in (3.172) it is easy to obtain the following equation of motion for our electronic density matrix:

$$\frac{\mathrm{d}\rho^c_{\xi_1\xi_2}}{\mathrm{d}t} = \frac{1}{2}\sum_{\xi'_1\xi'_2}\left[\tilde{\mathcal{P}}_{\xi_1\xi_2,\xi'_1\xi'_2}\rho^c_{\xi'_1\xi'_2} - \tilde{\mathcal{P}}^*_{\xi'_1\xi'_1,\xi_1\xi'_2}\rho^c_{\xi'_2\xi_2}\right] + \mathrm{H.c.} \tag{3.174}$$

with generalized scattering rates given by

$$\tilde{\mathcal{P}}_{\xi_1\xi_2,\xi'_1\xi'_2} = \sum_{\boldsymbol{q}\pm}\left(N^\circ_{\boldsymbol{q}} + \frac{1}{2} \pm \frac{1}{2}\right)A^{\boldsymbol{q}\pm}_{\xi_1\xi'_1}A^{\boldsymbol{q}\pm *}_{\xi_2\xi'_2}\ , \tag{3.175}$$

where

$$A^{\boldsymbol{q}\pm}_{\xi\xi'} = \sqrt{\frac{2\pi}{\hbar}}H^{\boldsymbol{q}\pm}_{\xi\xi'}\tilde{D}^{\boldsymbol{q}\pm}_{\xi\xi'} \tag{3.176}$$

are the matrix elements of the electronic Lindblad operator (3.173) and

$$\tilde{D}^{q\pm}_{\xi\xi'} = \frac{e^{-\left(\frac{\epsilon_\xi - \epsilon_{\xi'} \pm \epsilon_q}{2\bar{\epsilon}}\right)^2}}{\left(2\pi\bar{\epsilon}^2\right)^{\frac{1}{4}}} . \tag{3.177}$$

Once again, we get exactly the generalized in- minus out-scattering structure in (3.100) provided to replace the conventional scattering operator (3.101) with its symmetrized version in (3.175). In the semiclassical limit ($\xi_1\xi_1' = \xi_2\xi_2'$) and for infinite collision durations ($\bar{\epsilon} \to 0$), it is easy to verify that the generalized rates (3.175) reduce to the conventional Fermi's golden rule (3.107).

Let us consider again the single-particle picture discussed in Sect. 3.3.3. More specifically, by inserting into the equation of motion (3.119) the new version of the global-description Markov superoperator in (3.164) and employing once again the mean-field approximation previously introduced, it is possible to derive a modified version of the nonlinear single-particle equations (3.132) and (3.138). In the low-density limit – when applicable (see Chap. 9) – such new single-particle equations may be expressed again via linear scattering superoperators. In particular, for the case of the generic carrier–quasiparticle interaction (3.137) we get

$$\left.\frac{d\rho^{sp}_{\alpha_1\alpha_2}}{dt}\right|_{c-qp} = \frac{1}{2}\sum_{\alpha_1'\alpha_2'}\left[\tilde{\mathcal{P}}_{\alpha_1\alpha_2,\alpha_1'\alpha_2'}\rho^{sp}_{\alpha_1'\alpha_2'} - \tilde{\mathcal{P}}^*_{\alpha_1'\alpha_1,\alpha_1\alpha_2'}\rho^{sp}_{\alpha_2'\alpha_2}\right] + \text{H.c.} \tag{3.178}$$

with generalized single-particle scattering rates

$$\tilde{\mathcal{P}}_{\alpha_1\alpha_2,\alpha_1'\alpha_2'} = \sum_{q\pm}\left(N^\circ_q + \frac{1}{2} \pm \frac{1}{2}\right) A^\pm_{\alpha_1\alpha_1',q} A^{\pm *}_{\alpha_2\alpha_2',q} , \tag{3.179}$$

where

$$A^\pm_{\alpha\alpha',q} = \sqrt{\frac{2\pi}{\hbar}} g^\pm_{\alpha\alpha',q} \tilde{D}^\pm_{\alpha\alpha',q} \tag{3.180}$$

and

$$\tilde{D}^\pm_{\alpha\alpha',q} = \frac{e^{-\left(\frac{\epsilon_\alpha - \epsilon_{\alpha'} \pm \epsilon_q}{2\bar{\epsilon}}\right)^2}}{\left(2\pi\bar{\epsilon}^2\right)^{\frac{1}{4}}} . \tag{3.181}$$

As for the reduced-description picture, in the semiclassical limit ($\alpha_1\alpha_1' = \alpha_2\alpha_2'$) and for an infinite collision duration ($\bar{\epsilon} \to 0$) the single-particle scattering operator (3.179) reduces to the conventional carrier–quasiparticle scattering rates (3.140).

As key result of the alternative Markov approximation presented in this section, the conclusion is that the low-density single-particle dynamics is still described via the equation of motion (3.146), provided to replace in (3.148) the original scattering superoperator $\mathcal{P}$ in (3.144) with its new symmetrized version in (3.179).

As final crucial step, it is important to stress that the positivity properties discussed so far remain valid no matter how the subsystem of interest is chosen (e.g., many-electron description, single-particle picture, or any other non-conventional projection scheme). To this end, by adopting a more mathematical/abstract language [172], we notice that the usual partial-trace projection $\Pi_{\rm pt}$ – employed, e.g., in our reduced-description picture – is of the form

$$\Pi_{\rm pt}\hat{a} = \hat{a}^{\rm s} \otimes \hat{\mathcal{I}}^{\rm env} \ , \qquad \hat{a}^{\rm s} = {\rm tr}\left\{\hat{a}\hat{\rho}^{\rm env}\right\} \ , \tag{3.182}$$

where $\hat{a}$ is a generic operator belonging to our system (s) $\otimes$ environment (env) global space and $\hat{\rho}^{\rm env}$ is the "environment" density-matrix operator (see, e.g., $\hat{\rho}^{\circ}$ in our reduced description). In view of the general properties of the above projection class, it follows that the partial trace $\Pi_{\rm pt}$ is a completely positive map (see, e.g., [169]). Moreover, the projected observables, all being of the form $\hat{a}^{\rm s} \otimes \hat{\mathcal{I}}^{\rm env}$, constitute a subalgebra of the global-observable algebra. Based upon these two crucial remarks, let us now consider a generic projection Π on a given subalgebra of observables, which is also a completely positive map. More precisely, given a global observable $\hat{a}$, such a projection is defined via the corresponding set of Kraus operators $\{\hat{V}_i\}$ [194] according to

$$\Pi\hat{a} = \sum_i \hat{V}_i^{\dagger}\hat{a}\hat{V}_i \ . \tag{3.183}$$

Then one could easily verify that such a subalgebra is made by observables that commute with each of the $\hat{V}_i$ and $\hat{V}_i^{\dagger}$.

As discussed in [172], by projecting the new global Markov dynamics in (3.164) over our subalgebra, one finally gets

$$\frac{d\hat{\rho}^{\rm s}}{dt} = -\sum_{ii'}\left(\frac{1}{2}\{\hat{A}_{ii'}^{\dagger}\hat{A}_{ii'}, \hat{\rho}^{\rm s}\} - \hat{A}_{ii'}\hat{\rho}^{\rm s}\hat{A}_{ii'}^{\dagger}\right) \tag{3.184}$$

with

$$\hat{A}_{ii'} = \hat{V}_i\hat{A}\hat{V}_{i'} \ . \tag{3.185}$$

The Lindblad-like structure (see (3.36)) of the above projected Markov dynamics shows indeed that – also for the wide projection class in (3.183) – the new global Markov superoperator (3.164) allows one to obtain a positive-definite density-matrix evolution for the subsystem/subalgebra of interest.

As we shall discuss extensively in Chap. 4, the possibility of getting a positive-definite dynamics for the rather general projection class in (3.183) is paving the way for the introduction of novel simulation strategies aimed at investigating quantum-transport phenomena in systems with open boundaries (see Sect. 4.3).

4

Generalization to Systems with Open Boundaries

In this chapter we shall discuss how to extend the density-matrix approach previously introduced to quantum systems with open spatial boundaries, which corresponds to the case of a generic quantum device inserted into an electric circuit.

4.1 Semiconductor Bloch Equations for Open Systems

As anticipated in Sect. 2.6, in spite of the quantum-mechanical nature of carrier dynamics in the core region of typical nanostructured devices – like semiconductor superlattices and double-barrier structures – the overall behavior of such quantum systems is often the result of a complex interplay between phase coherence and energy dissipation/decoherence (see, e.g., [110]), the latter being primarily due to the presence of spatial boundaries (see, e.g., [85]). It follows that a proper treatment of such novel nanoscale devices requires a theoretical modeling able to properly account for both coherent – i.e., scattering-free – and incoherent – i.e., phase-breaking – processes on the same footing. Following this spirit, a generalization to systems with open boundaries of the well-known semiconductor Bloch equations (see Chap. 2) has been originally proposed in [195]; however, such theoretical analysis was primarily related to the interplay between phase coherence and energy relaxation within the device active region, and – apart from its abstract formulation – no detailed investigation of the carrier-injection process (from the electric contacts into the device active region) has been performed.

Aim of this section is to review such open-system generalization, originally introduced in [195] and further developed in [80]; the latter provides a quantum-mechanical description of the coupling dynamics between the device active region and external charge reservoirs and allows one to account for the semi-phenomenological injection models commonly employed in state-of-the-art simulations of realistic one- and two-dimensional open quantum devices (see, e.g., [56] and references therein). Among such simulation strategies it is worth mentioning the approach proposed by Massimo V. Fischetti [196]; more specifically, as previously discussed in Sect. 1.4 (see (1.83)), the open-system transport equation proposed in [196] is of the form

$$\frac{\partial f_\alpha}{\partial t} = \sum_{\alpha'} (P_{\alpha\alpha'} f_{\alpha'} - P_{\alpha'\alpha} f_\alpha) - \frac{f_\alpha - f_\alpha^\circ}{\tau_\alpha^{\mathrm{t}}} . \tag{4.1}$$

Here, f_α° denotes the equilibrium carrier distribution of the device electric contacts while τ_α^{t} can be regarded as the device transit time for an electron in state α. As anticipated, in spite of a rigorous treatment of the scattering dynamics (via the standard Boltzmann collision term involving microscopic scattering rates $P_{\alpha\alpha'}$), the last (relaxation-time-like) term describes carrier injection/loss on a partially phenomenological level and does not depend on the real position of the device spatial boundaries. Indeed, although the transit time τ_α^{t} is related to the device dimensions, the semiclassical distribution function f_α alone does not provide a genuine quantum-mechanical real-space description.

In order to provide a fully microscopic real-space formulation of the carrier–injection process, we shall revisit the theoretical approach originally proposed in [195]. To this aim, the starting point is the conventional set of semiconductor Bloch equations, i.e., the equations of motion for the single-particle density matrix (3.116); as shown in Sect. 3.3.3, for a closed system in the low-density limit the latter may be written according to (3.146), i.e.,

$$\frac{\mathrm{d}\rho_{\alpha_1\alpha_2}}{\mathrm{d}t} = \sum_{\alpha_1'\alpha_2'} \mathcal{L}_{\alpha_1\alpha_2,\alpha_1'\alpha_2'} \rho_{\alpha_1'\alpha_2'} , \tag{4.2}$$

where the effective single-particle Liouville operator $\mathcal{L}$ in (3.147) is the sum of two terms: coherent (i.e., scattering-free) single-particle evolution plus energy-dissipation/decoherence dynamics.

In order to incorporate within such density-matrix picture the presence of spatial boundaries, the key idea is first to move to the Wigner phase-space picture introduced in Sect. 2.3 via the standard Weyl–Wigner transform (2.55); more specifically, by denoting with $\boldsymbol{k} = \frac{\boldsymbol{p}}{\hbar}$ the carrier momentum in units of $\hbar$, the Wigner function corresponding to our single-particle density matrix is given by

$$f_{\mathrm{W}}(\boldsymbol{r}, \boldsymbol{k}) = \sum_{\alpha_1\alpha_2} \mathcal{W}_{\alpha_1\alpha_2}(\boldsymbol{r}, \boldsymbol{k}) \rho_{\alpha_1\alpha_2} \tag{4.3}$$

with

$$\mathcal{W}_{\alpha_1\alpha_2}(\boldsymbol{r}, \boldsymbol{k}) = \int \mathrm{d}^3 r' \psi_{\alpha_1}\left(\boldsymbol{r} + \frac{\boldsymbol{r}'}{2}\right) \frac{\mathrm{e}^{-i\boldsymbol{k}\cdot\boldsymbol{r}'}}{(2\pi)^{\frac{3}{2}}} \psi_{\alpha_2}^*\left(\boldsymbol{r} - \frac{\boldsymbol{r}'}{2}\right) . \tag{4.4}$$

As anticipated in Sect. 2.3, such Wigner-function picture allows for a direct comparison between semiclassical and quantum-mechanical results. Indeed, by applying the Weyl–Wigner transform to the average-value equation (3.115), one gets

$$a = \int \mathrm{d}^3 r\, \mathrm{d}^3 k a_{\mathrm{W}}(\boldsymbol{r}, \boldsymbol{k}) f_{\mathrm{W}}(\boldsymbol{r}, \boldsymbol{k}) \;, \tag{4.5}$$

where

$$a_{\mathrm{W}}(\boldsymbol{r}, \boldsymbol{k}) = \sum_{\alpha_1 \alpha_2} \mathcal{W}_{\alpha_1 \alpha_2}(\boldsymbol{r}, \boldsymbol{k}) a_{\alpha_1 \alpha_2} \tag{4.6}$$

is the Weyl–Wigner transform of the single-particle operator $\hat{a}$.[1] The new average-value equation (4.5) involves the usual phase-space integral of classical statistics; however, the (real) Wigner function f_{W} is not positive-definite, and therefore the latter cannot be regarded as a probability distribution.

Starting from the explicit form of the Weyl–Wigner transform in (4.4), it is possible to rewrite the definition of the Wigner function (4.3) in a fully operatorial way according to

$$f_{\mathrm{W}}(\boldsymbol{r}, \boldsymbol{k}) = \mathrm{tr}\left\{ \hat{\mathcal{W}}(\boldsymbol{r}, \boldsymbol{k}) \hat{\rho} \right\} \tag{4.8}$$

with

$$\hat{\mathcal{W}}(\boldsymbol{r}, \boldsymbol{k}) = \int \mathrm{d}^3 r' \left| \boldsymbol{r} - \frac{\boldsymbol{r}'}{2} \right\rangle \frac{\mathrm{e}^{-i\boldsymbol{k}\cdot\boldsymbol{r}'}}{(2\pi)^{\frac{3}{2}}} \left\langle \boldsymbol{r} + \frac{\boldsymbol{r}'}{2} \right| \;. \tag{4.9}$$

This result shows that the Wigner function f_{W} may be regarded as the average value of the above Wigner operator $\hat{\mathcal{W}}$.

By applying the Weyl–Wigner transform (4.3) to the original semiconductor Bloch equation (4.2), the latter are transformed into the following closed-system Wigner-function equation:

$$\frac{\partial f_{\mathrm{W}}(\boldsymbol{r}, \boldsymbol{k})}{\partial t} = \int \mathrm{d}^3 r' \, \mathrm{d}^3 k' \mathcal{L}(\boldsymbol{r}, \boldsymbol{k}; \boldsymbol{r}', \boldsymbol{k}') f_{\mathrm{W}}(\boldsymbol{r}', \boldsymbol{k}') \;, \tag{4.10}$$

where

$$\mathcal{L}(\boldsymbol{r}, \boldsymbol{k}; \boldsymbol{r}', \boldsymbol{k}') = \sum_{\alpha_1 \alpha_2, \alpha'_1 \alpha'_2} \mathcal{W}_{\alpha_1 \alpha_2}(\boldsymbol{r}, \boldsymbol{k}) \mathcal{L}_{\alpha_1 \alpha_2, \alpha'_1 \alpha'_2} \mathcal{W}^*_{\alpha'_1 \alpha'_2}(\boldsymbol{r}', \boldsymbol{k}') \tag{4.11}$$

is the original Liouville superoperator written in the new Wigner picture $\boldsymbol{r}, \boldsymbol{k}$.

At this point, the crucial step is to impose the desired spatial boundary conditions on the Wigner-function equation (4.10). This is commonly realized by employing the very same u boundary-condition scheme (see Fig. 1.22)

---

[1] In order to show that the phase-space integration (4.5) is equivalent to the original average-value expression in (3.115), it is crucial to employ the Hermiticity property $\mathcal{W}_{\alpha_1 \alpha_2}(\boldsymbol{r}, \boldsymbol{k}) = \mathcal{W}^*_{\alpha_2 \alpha_1}(\boldsymbol{r}, \boldsymbol{k})$ together with the orthonormality relation

$$\int \mathrm{d}^3 r \, \mathrm{d}^3 k \mathcal{W}_{\alpha_1 \alpha_2}(\boldsymbol{r}, \boldsymbol{k}) \mathcal{W}^*_{\alpha'_1 \alpha'_2}(\boldsymbol{r}, \boldsymbol{k}) = \delta_{\alpha_1 \alpha_2, \alpha'_1 \alpha'_2} \,. \tag{4.7}$$

discussed in Appendix B and already applied in Sect. 1.4 to the Boltzmann equation; indeed, following exactly the same prescription adopted in the semiclassical picture, we may fix the value of the Wigner function f_W entering the device active region Ω_d, which in turn will fix the flux of incoming particles.[2] Similar to the semiclassical case, such a first-order boundary condition may be successfully imposed by adding to the closed-system Wigner-function equation (4.10) a boundary source term (see (B.6))

$$S(\boldsymbol{r},\boldsymbol{k}) = \delta(\boldsymbol{r}-\boldsymbol{r}^b)v^\mathrm{in}(\boldsymbol{k})f^b_\mathrm{W}(\boldsymbol{k}) \tag{4.12}$$

and a corresponding (negative) boundary loss term (see (B.7))

$$-\Lambda(\boldsymbol{r},\boldsymbol{k}) = -\delta(\boldsymbol{r}-\boldsymbol{r}^b)v^\mathrm{in}(\boldsymbol{k})f_\mathrm{W}(\boldsymbol{r},\boldsymbol{k}) \;, \tag{4.13}$$

where $v^\mathrm{in}(\boldsymbol{k})$ denotes again the incoming part of the carrier group velocity $\boldsymbol{v}(\boldsymbol{k})$ normal to the boundary surface (see Appendix B).

As a result, the open-system version of the Wigner-function equation (4.10) is given by

$$\frac{\partial f_\mathrm{W}(\boldsymbol{r},\boldsymbol{k})}{\partial t} = \int \mathrm{d}^3r'\,\mathrm{d}^3k'\,\mathcal{L}(\boldsymbol{r},\boldsymbol{k};\boldsymbol{r}',\boldsymbol{k}')f_\mathrm{W}(\boldsymbol{r}',\boldsymbol{k}') + \left.\frac{\partial f_\mathrm{W}(\boldsymbol{r},\boldsymbol{k})}{\partial t}\right|_b \tag{4.14}$$

with

$$\left.\frac{\partial f_\mathrm{W}(\boldsymbol{r},\boldsymbol{k})}{\partial t}\right|_b = S(\boldsymbol{r},\boldsymbol{k}) - \Lambda(\boldsymbol{r},\boldsymbol{k}) \;. \tag{4.15}$$

As shown in Appendix B, the effect of the loss contribution Λ in (4.13) may also be described/regarded as a renormalization of the original (closed-system) Liouville superoperator in (4.11); more specifically, the Wigner-function equation (4.14) may also be written as

$$\frac{\partial f_\mathrm{W}(\boldsymbol{r},\boldsymbol{k})}{\partial t} = \int \mathrm{d}^3r'\,\mathrm{d}^3k'\,\mathcal{L}^\mathrm{open}(\boldsymbol{r},\boldsymbol{k};\boldsymbol{r}',\boldsymbol{k}')f_\mathrm{W}(\boldsymbol{r}',\boldsymbol{k}') + S(\boldsymbol{r},\boldsymbol{k}) \tag{4.16}$$

with

$$\mathcal{L}^\mathrm{open}(\boldsymbol{r},\boldsymbol{k};\boldsymbol{r}',\boldsymbol{k}') = \mathcal{L}(\boldsymbol{r},\boldsymbol{k};\boldsymbol{r}',\boldsymbol{k}') + \Delta\mathcal{L}(\boldsymbol{r},\boldsymbol{k};\boldsymbol{r}',\boldsymbol{k}') \tag{4.17}$$

and

$$\Delta\mathcal{L}(\boldsymbol{r},\boldsymbol{k};\boldsymbol{r}',\boldsymbol{k}') = -\delta(\boldsymbol{r}-\boldsymbol{r}')\delta(\boldsymbol{k}-\boldsymbol{k}')\Lambda(\boldsymbol{r}',\boldsymbol{k}') \;. \tag{4.18}$$

The final step is to move back to the original density-matrix formalism: by applying to the open-system Wigner-function equation (4.16) the inverse Weyl–Wigner transform

[2] It is imperative to stress that the application of the conventional U scheme in Appendix B is limited to first-order Liouville superoperators only (see (B.2)); such requirement is fulfilled by adopting an effective-mass band model (see (1.14)) and by neglecting non-local contributions ($\boldsymbol{r} \neq \boldsymbol{r}'$) in the Weyl–Wigner transform of the scattering superoperator.

$$\rho_{\alpha_1\alpha_2} = \int d^3r\, d^3k \mathcal{W}^*_{\alpha_1\alpha_2}(\boldsymbol{r},\boldsymbol{k}) f_W(\boldsymbol{r},\boldsymbol{k}) \;, \tag{4.19}$$

we finally get

$$\frac{d\rho_{\alpha_1\alpha_2}}{dt} = \sum_{\alpha'_1\alpha'_2} \mathcal{L}^{open}_{\alpha_1\alpha_2,\alpha'_1\alpha'_2} \rho_{\alpha'_1\alpha'_2} + S_{\alpha_1\alpha_2} \;, \tag{4.20}$$

where once again the open-system Liouville superoperator $\mathcal{L}^{open}$ corresponds to the original (closed-system) superoperator $\mathcal{L}$ in (4.2) renormalized by

$$\Delta\mathcal{L}_{\alpha_1\alpha_2,\alpha'_1\alpha'_2} = -\int d^3r\, d^3k \mathcal{W}^*_{\alpha_1\alpha_2}(\boldsymbol{r},\boldsymbol{k}) \Lambda(\boldsymbol{r},\boldsymbol{k}) \mathcal{W}_{\alpha'_1\alpha'_2}(\boldsymbol{r},\boldsymbol{k}) \;, \tag{4.21}$$

and

$$S_{\alpha_1\alpha_2} = \int d^3r\, d^3k \mathcal{W}^*_{\alpha_1\alpha_2}(\boldsymbol{r},\boldsymbol{k}) S(\boldsymbol{r},\boldsymbol{k}) \tag{4.22}$$

is the Weyl–Wigner antitransform of the source term (4.12).

Equation (4.20) is the desired generalization to open systems of the semiconductor Bloch equations in (4.2). It is imperative to stress that the open character of the system manifests itself via the non-Hermitian correction $\Delta\mathcal{L}$, whose effect – exactly as for the semiclassical case discussed in Sect. 1.4 – is again equivalent to dissipation/decoherence processes within the simulated region, as originally pointed out in [85].

In order to illustrate the power and flexibility of this theoretical approach, we shall start by reviewing a few simulated experiments originally presented in [195].

As a first example, let us consider an electron wavepacket entering the double-barrier structure of a GaAs/AlGaAs resonant-tunneling diode. Figure 4.1 shows the time evolution of the wavepacket in the absence of carrier–phonon scattering as a function of position (a) and energy (b). It is easy to recognize the well-established resonance scenario typical of any purely coherent dynamics: as the wavepacket enters our semiconductor nanostructure, a part of it is transmitted and a part is reflected (see Fig. 4.1a). Since in this simulation carrier–phonon scattering is not included, the wavepacket central energy is conserved, i.e., no energy dissipation occurs (see Fig. 4.1b). In contrast, in the presence of nonelastic scattering processes, the resonance dynamics of Fig. 4.1a is strongly modified by the scattering itself, as shown in Fig. 4.2a. In particular, the presence of phase-breaking scattering events is found to reduce both the interference peaks and the transmitted wavepacket. This is confirmed by the corresponding energy distribution in Fig. 4.2b, where we clearly recognize the granular nature of the dissipation process through the formation of so-called phonon replica. Indeed, this is the fingerprint of any microscopic treatment of energy relaxation (see, e.g., [90]).

As a second testbed, let us consider electrically injected Bloch oscillations in semiconductor superlattices. More specifically, the system under investigation consists of a biased GaAs/AlGaAs superlattice surrounded by

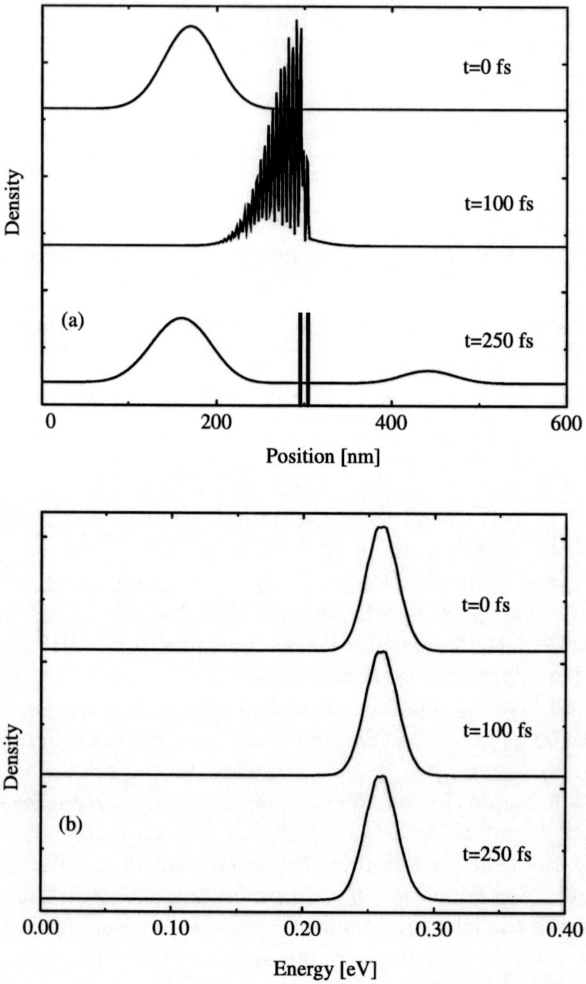

Fig. 4.1. Carrier distribution at different times as a function of position (**a**) and energy (**b**) corresponding to an electron wavepacket injected into a resonant-tunneling nanostructure in the absence of scattering processes. Here, the barrier height is 0.5 eV while the barrier width and separation are, respectively, 20 and 60 Å. The two barriers are schematically depicted as *black vertical lines* (see text). Reprinted from [195]

two semi-infinite GaAs regions. In this simulated experiment an electronic wavepacket is injected from the left contact (GaAs region) into the superlattice (see Fig. 4.3). Figures 4.3 and 4.4 illustrate the time evolution of the electronic wavepacket with and without carrier–phonon scattering, respectively. When the wavepacket reaches the superlattice structure most of it gets reflected backward, while some portion of it tunnels into the Wannier–Stark

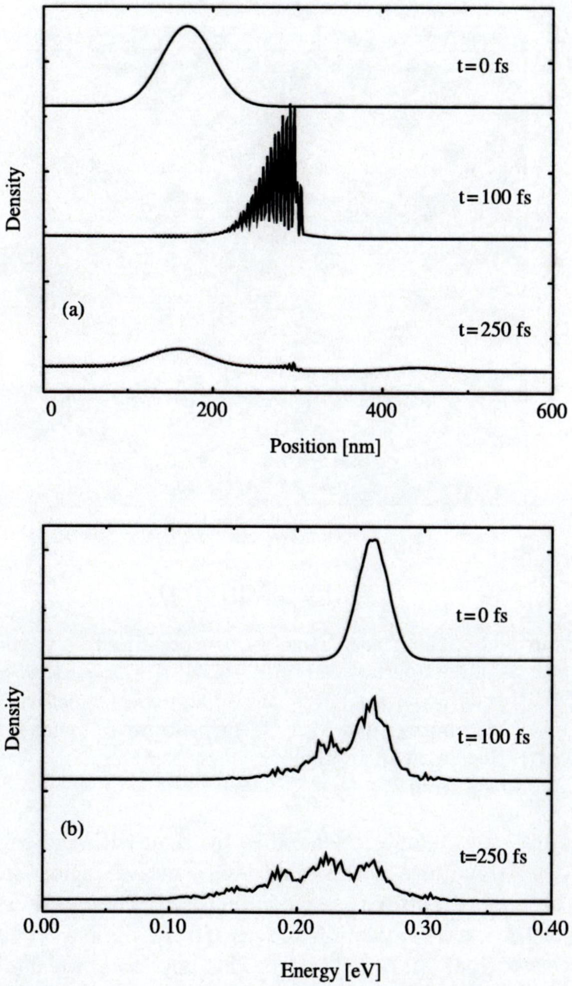

Fig. 4.2. Same as in Fig. 4.1, but in the presence of scattering processes (see text). Reprinted from [195]

ladder associated with the superlattice (see, e.g., [178]) and starts to oscillate at a frequency of about 3.5 THz. Every time the wavepacket reaches the boundary of the oscillation region, a part of it gets transmitted via Zener tunneling. We should notice, however, that such Zener processes do not destroy the Bloch oscillation dynamics, but simply reduce the charge density within the superlattice region. Indeed, in the scattering-free case (see Fig. 4.3) Bloch oscillations are found to persist on a picosecond time-scale. In contrast, once scattering mechanisms are considered (see Fig. 4.4), the phonon-induced decoherence drastically reduces their life-time.

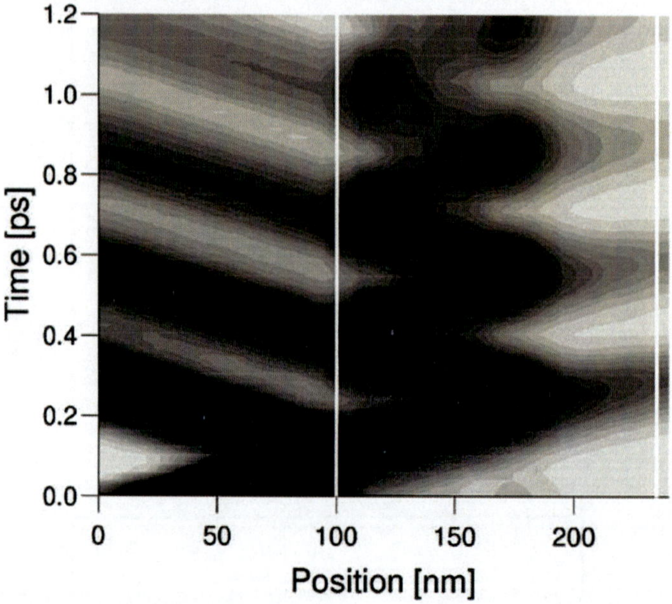

Fig. 4.3. Contour plot of the charge density corresponding to a wavepacket electrically injected into a finite superlattice region (marked with *vertical white lines*) in the absence of scattering processes. Here, the superlattice structure is made by 21 periods of a 8 Å AlGaAs barrier and 57 Å GaAs well unit cell with a barrier height of 0.4 eV (see text). Reprinted from [195]

The simulated experiments reviewed so far deal with the coherent versus incoherent carrier dynamics within the device active region only; however, primary goal of the open-system semiconductor Bloch equation (4.20) is the investigation of the carrier-injection process (from the electric contacts into the device active region). In order to face this key issue, we shall now review a few simulated experiments originally presented in [80]; to this end, we shall focus on a very simple semiconductor nanostructure: a single-barrier equidistant from the device contacts (see Fig. 4.5). As envelope-function basis states we adopt the scattering states of the device potential profile (see, e.g., [197]); moreover, to better identify the role played by carrier injection, we shall neglect all other sources of energy dissipation/decoherence in the device active region, like carrier–phonon and carrier–carrier scattering.

Figure 4.5 shows results for a GaAs-based single-barrier potential profile when carriers are primarily injected from left. Here, the steady-state real-space charge distribution obtained from the phenomenological injection model (4.1) (dashed curves) is compared to that of the microscopic model in (4.20) (solid curves). As we can see, the two models give completely different results. The phenomenological model gives basically what we expect: Since we have

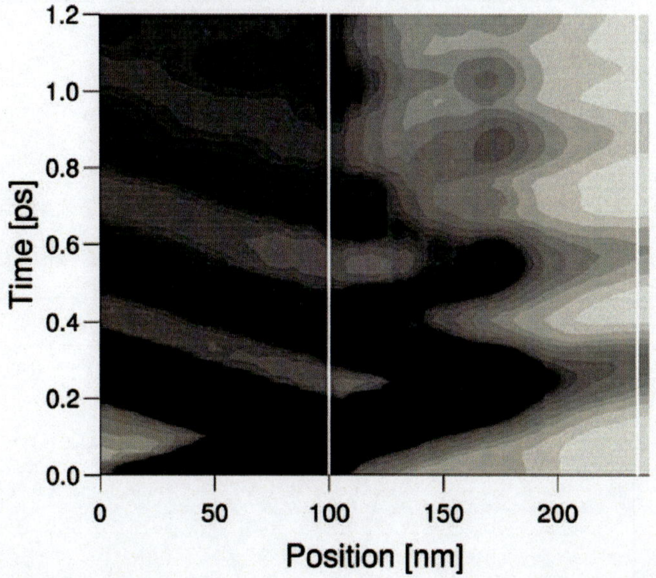

Fig. 4.4. Same as in Fig. 4.3, but in the presence of scattering processes (see text). Reprinted from [195]

significant carrier injection from left only and since the potential barrier is relatively high, the carrier distribution is mainly located on the left side. In contrast, the microscopic model gives an almost symmetric charge distribution. In order to understand the origin of this unphysical result, let us focus on the nature of the source term in (4.20). Contrary to the phenomenological injection/loss term in (4.1), the latter is intrinsically non-diagonal, i.e., the injection of a carrier with well-defined wavevector $\boldsymbol{k}$ (see (4.12)) is described by a non-diagonal source contribution $S_{\alpha_1\alpha_2}$. In other words, we inject into the device active region a coherent superposition of states α_1 and α_2, in clear contrast with the idea of injection from a thermal – i.e., diagonal – charge reservoir. More specifically, in this case the generic scattering state α on the left comes out to be an almost equally weighted superposition of $+k$ and $-k$:

$$\psi_k(z) = a_k \mathrm{e}^{\mathrm{i}kz} + b_k \mathrm{e}^{-\mathrm{i}kz} \ . \tag{4.23}$$

This, in turn, tells us that the generic plane-wave state k injected from the left contact is also an almost equally weighted superposition of the left and right scattering states. This is why the charge distribution (solid curve in Fig. 4.5) is almost symmetric: any electron injected from left couples to left as well as to right scattering states. The anomaly of the microscopic model is even more pronounced if we look at the carrier distribution in momentum space (see inset in Fig. 4.5). While for the phenomenological model (dashed curve) we get a positive-definite distribution showing, as expected, the two

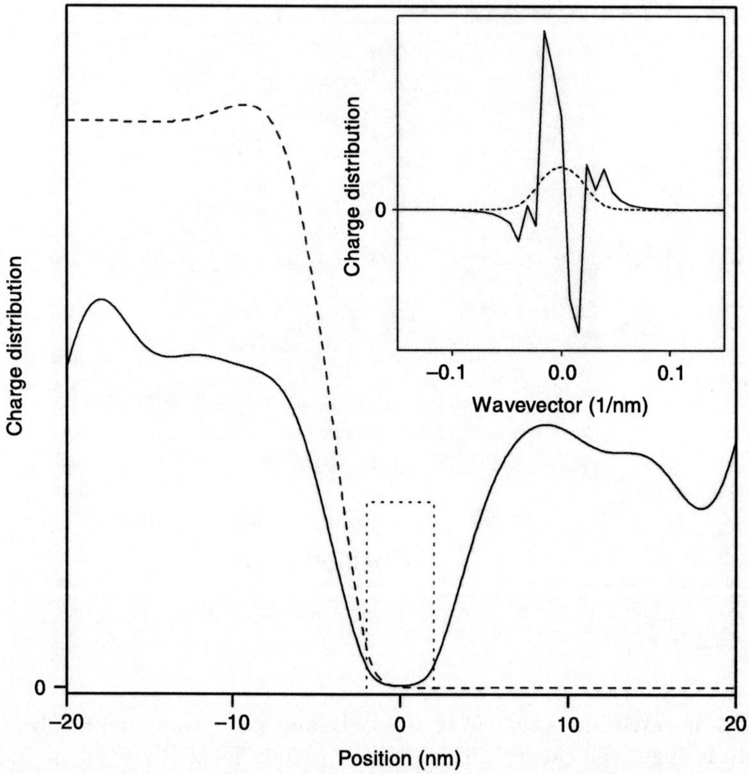

Fig. 4.5. Comparison between the real-space charge distribution obtained from the phenomenological injection model (4.1) (*dashed curves*) and the microscopic model in (4.20) (*solid curves*) for a GaAs-based single-barrier structure (height $V_\circ = 0.5\,\text{eV}$ and width $a = 4\,\text{nm}$) equidistant from the electric contacts. In this room-temperature simulation, due to a misalignment $\Delta\mu = 0.2\,\text{eV}$ of the left and right chemical potential, carriers are primarily injected from left (total carrier concentration $n \sim 10^{17}\,\text{cm}^{-3}$). The corresponding charge distribution in momentum space is also reported in the inset (see text). Reprinted from [80]

symmetric wavevector components of the scattering state, the microscopic result is not positive-definite; this tells us that for the particular problem under examination the theoretical approach considered so far does not provide a "good" Wigner function.

Generally speaking, the scenario previously discussed is highly non-physical; this unexpected behavior is mainly ascribed to the boundary-condition scheme employed so far, which implies injection of plane-wave electrons (see the source term in (4.12)), regardless of the shape of the device potential profile. It is then clear that, in order to overcome this limitation, one needs an alternative real-space description corresponding to a diagonal injection over the scattering states α of the device potential profile. To this end, as originally proposed in [80], it is possible to consider a generalization of the Wigner-function formulation employed so far. The key idea is to extend the Weyl–Wigner transform (4.4) from the plane-wave states $\boldsymbol{k}$ to a generic basis set $\{|\beta\rangle\}$ according to

$$\overline{\mathcal{W}}^{\beta_1\beta_2}_{\alpha_1\alpha_2}(\boldsymbol{r}) = \int \mathrm{d}^3r' \psi_{\alpha_1}\left(\boldsymbol{r}+\frac{\boldsymbol{r}'}{2}\right) \frac{\chi^*_{\beta_1}\left(\boldsymbol{r}+\frac{\boldsymbol{r}'}{2}\right)\chi_{\beta_2}\left(\boldsymbol{r}-\frac{\boldsymbol{r}'}{2}\right)}{\Omega_{\mathrm{d}}^{-1}} \psi^*_{\alpha_2}\left(\boldsymbol{r}-\frac{\boldsymbol{r}'}{2}\right) , \tag{4.24}$$

where $\chi_\beta(\boldsymbol{r}) = \langle r|\beta\rangle$ and Ω_{d} denotes again the volume of the simulated region. In analogy to (4.3), such generalized Wigner function is given by

$$\overline{f}^{\beta_1\beta_2}_{\mathrm{W}}(\boldsymbol{r}) = \sum_{\alpha_1\alpha_2} \overline{\mathcal{W}}^{\beta_1\beta_2}_{\alpha_1\alpha_2}(\boldsymbol{r})\rho_{\alpha_1\alpha_2} . \tag{4.25}$$

As described in [80], by adopting as basis states $|\beta\rangle$ again the scattering states of the device potential profile $|\alpha\rangle$, and assuming a source term diagonal in this basis, it is possible to incorporate the spatial boundary conditions on the new Wigner function (4.25), and finally to get a generalized/modified version of the open-system semiconductor Bloch equation (4.20).

Figure 4.6 shows again results for the GaAs-based single-barrier potential profile previously considered. Here, the steady-state simulation based on the phenomenological injection model (4.1) (dashed curves) is now compared to the results of this alternative approach (solid curves) based on the new Wigner function (4.25). As we can see, the highly non-physical behaviors of Fig. 4.5 (solid curves) have been completely removed. Indeed, the momentum distribution in the inset is always positive-definite and the two models exhibit a very similar behavior. As discussed in [80], one deals with relatively small deviations close to the device spatial boundaries, ascribed to interlevel-coupling mechanisms not present in the phenomenological injection model. This is clearly a fingerprint of a genuine real-space description, where the point-like carrier injection is located at the device spatial boundaries; however, when the device active region is relatively far from its electric contacts these deviations can be safely neglected, and the phenomenological model (4.1) provides reliable results.

The analysis presented so far has clearly pointed out intrinsic limitations of the open-system semiconductor Bloch equation (4.20) in describing the crucial coupling between device active region and its electric contacts. In particular, the anomalous behavior reported in Fig. 4.5 has been primarily ascribed to the standard U boundary-condition scheme (see Appendix B) applied to the conventional Wigner function, since this corresponds to the injection of plane-wave electrons, and therefore to a non-diagonal injection within the device scattering-state basis. As a confirmation of this interpretation, the introduction in [80] of the generalized Wigner function (4.25) – and of a corresponding density-matrix equation – has allowed to remove such non-physical behavior.

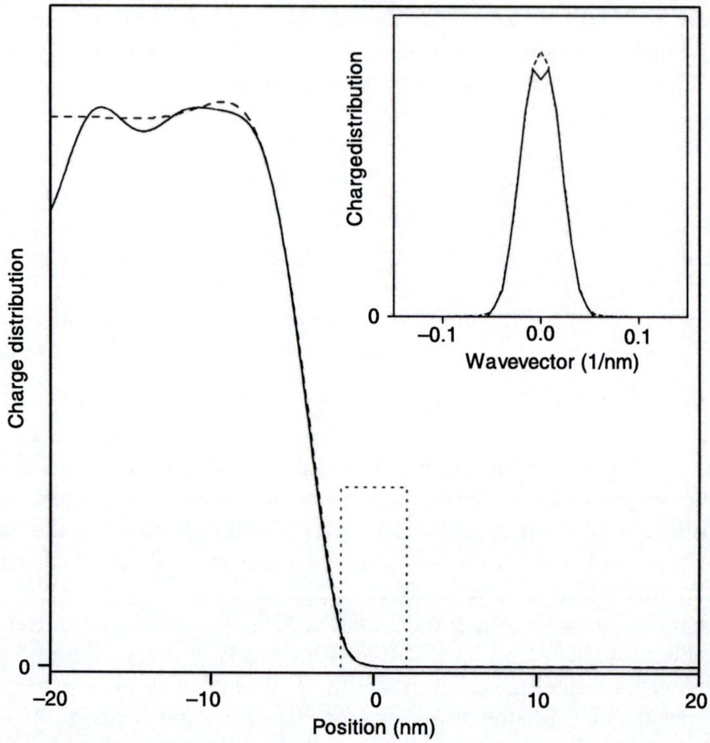

Fig. 4.6. Same as in Fig. 4.5, but for the alternative microscopic model based on the generalized Wigner function in (4.25) (see text). Reprinted from [80]

4.2 Failure of the Conventional Wigner-Function Formalism

Motivated by the anomalous results (see Fig. 4.1) obtained via the open-system semiconductor Bloch equation (4.20), in this section we shall try to better identify the physical origin of such unexpected behavior, and its link with the Wigner-function boundary-condition scheme previously discussed.

For the case of a one-dimensional problem, the closed-system Wigner-function equation in (4.10) is of the form

$$\frac{\partial f_{\mathrm{W}}(z,k)}{\partial t} = \int \mathrm{d}z'\,\mathrm{d}k'\,\mathcal{L}(z,k;z',k') f_{\mathrm{W}}(z',k') \,. \tag{4.26}$$

Different approaches for the study of quantum-transport phenomena in semiconductor nanomaterials and nanodevices based on the Wigner-function formalism have been proposed [198–212]. On the one hand, starting from the pioneering work by William R. Frensley [198], a few groups (see, e.g.,

[201]) have performed quantum-transport simulations based on a direct numerical solution of the Wigner-function equation (4.26) via finite-difference approaches (see Sect. 5.1.1) by imposing to $f_{\mathrm{W}}(z,k)$ the standard U boundary-condition scheme (see Fig. 1.22) reviewed in Appendix B; on the other hand, as discussed in the previous section, a generalization to systems with open boundaries of the semiconductor Bloch equations has been also proposed [80, 195]. In addition to these two alternative simulation strategies – both based on effective treatments of relevant interaction mechanisms – Carlo Jacoboni and co-workers have proposed a fully quantum-mechanical simulation scheme for the study of electron–phonon interaction based on the so-called Wigner paths [164].[3]

The generalized semiconductor Bloch-equation approach previously reviewed seems to suggest an intrinsic limitation of the conventional Wigner-function formalism in describing quantum-transport phenomena through systems with open boundaries. On the other hand, no clear evidence of such limitations has been reported so far via Wigner-function simulations based on finite-difference treatments.

Aim of this section is to investigate this apparent contradiction, thus shedding light on the real limitations of the conventional Wigner-function picture applied to open-device modeling. Our analysis, originally presented in [213], will show that the artificial spatial separation between device active region and external reservoirs is intrinsically incompatible with the non-local character of quantum mechanics.

In order to gain more insight into this highly non-trivial problem, let us start considering the explicit form of the Wigner-function equation (4.26) in the absence of any scattering mechanism; the latter, derived in Appendix D and often referred to as Wigner transport equation, may be written as (see, e.g., [85, 155])

$$\frac{\partial f_{\mathrm{W}}(z,k)}{\partial t} + v(k)\,\frac{\partial f_{\mathrm{W}}(z,k)}{\partial z} + \int \mathrm{d}k' \mathcal{V}(z,k-k') f_{\mathrm{W}}(z,k') = 0 \,, \tag{4.27}$$

where

$$\mathcal{V}(z,k-k') = \frac{\mathrm{i}}{\hbar}\int \mathrm{d}z' \frac{\mathrm{e}^{-\mathrm{i}(k-k')z'}}{2\pi}\left[V\left(z+\frac{z'}{2}\right) - V\left(z-\frac{z'}{2}\right)\right] \tag{4.28}$$

is a non-local Wigner superoperator corresponding to the device potential profile $V(z)$ (see Appendix D) while $v(k) = \frac{\hbar k}{m^*}$ denotes the electron group velocity within the usual effective-mass approximation (see (1.14)). In order to better understand the role played by the potential superoperator (4.28) as

[3] This approach is intrinsically able to overcome the standard approximations of conventional quantum-transport models, namely the Markov approximation and the completed-collision limit (see Chap. 3); however, due to the huge amount of computation required, its applicability is often limited to short time-scales and extremely simplified situations.

well as the link with its semiclassical counterpart, let us consider the particular case of the quadratic potential profile

$$V(z) = \frac{1}{2}az^2 + bz + c \tag{4.29}$$

corresponding to the classical force

$$F(z) = -\frac{\mathrm{d}V(z)}{\mathrm{d}z} = -(az + b). \tag{4.30}$$

In this case we have

$$V\left(z + \frac{z'}{2}\right) - V\left(z - \frac{z'}{2}\right) = -F(z)z' , \tag{4.31}$$

which allows us to rewrite the potential superoperator as

$$\mathcal{V}(z, k - k') = -\frac{\mathrm{i}F(z)}{\hbar} \int \mathrm{d}z' \frac{z'\mathrm{e}^{-\mathrm{i}(k-k')z'}}{2\pi} = \frac{F(z)}{\hbar}\frac{\partial}{\partial k}\delta(k - k') . \tag{4.32}$$

By inserting this last result into the original Wigner equation (4.27), we finally get

$$\left(\frac{\partial}{\partial t} + v(k)\frac{\partial}{\partial z} + \frac{F(z)}{\hbar}\frac{\partial}{\partial k}\right) f_{\mathrm{W}}(z, k) = 0 . \tag{4.33}$$

This clearly shows that for the case of a quadratic potential the (non-local) potential superoperator $\mathcal{V}$ reduces to the conventional (local) drift term of the Boltzmann transport theory, and the deterministic dynamics of the classical physics is recovered (see (1.64) and (1.67)). This can be regarded as a formal proof of the fact that, for a particle subjected to a quadratic potential, classical and quantum mechanics coincide, a fundamental result originally pointed out by Richard P. Feynman via his "path integral" formulation of quantum mechanics (see, e.g., [214] and references therein). In contrast, for a non-quadratic potential profile the action of the Wigner superoperator (4.28) is definitely non-local, thus introducing quantum corrections to the local drift term of classical mechanics.[4] This quantum versus classical analysis confirms the crucial role of the Wigner function as a direct link between semiclassical and quantum treatments.

Let us now focus on the steady-state version of the Wigner equation (4.27):

[4] It is worth stressing that, also in the presence of non-quadratic potential profiles, the Wigner transport equation (4.27) contains again the very same diffusion term of classical mechanics (see (1.64)); however, this is due to our effective-mass approximation (see (1.14)), which corresponds to assuming a kinetic term quadratic in the carrier momentum (see (D.1) in Appendix D). Indeed, for the case of a non-parabolic band model, also the classical diffusion term is replaced by a corresponding non-local operator (see, e.g., [117] and references therein).

$$v(k)\,\frac{\partial f_{\mathrm{W}}(z,k)}{\partial z} + \int \mathrm{d}k' \mathcal{V}(z,k-k') f_{\mathrm{W}}(z,k') = 0\,. \tag{4.34}$$

Following the spirit of a standard first-order boundary-condition problem (see Appendix B), we shall now impose the desired spatial boundary conditions for f_{W} at the left ($z = -\frac{l}{2}$) and right ($z = +\frac{l}{2}$) electric contacts, specifying the "incoming" Wigner distribution $f^b_{\mathrm{W}}(k) = f_{\mathrm{W}}\left(z^b(k), k\right)$, where $z^b(k) = \frac{l}{2}\left(\theta(-k) - \theta(+k)\right)$ denote the left and right boundary coordinate corresponding, respectively, to positive and negative carrier wavevectors k (θ being the usual Heaviside step function). By integrating the transport equation (4.34) from the spatial boundary $z^b(k)$ to the current point z we get[5]

$$f_{\mathrm{W}}(z,k) = f^b_{\mathrm{W}}(k) + \int \mathrm{d}z'\mathrm{d}k' \mathcal{U}(z,k;z',k') f_{\mathrm{W}}(z',k') \tag{4.35}$$

with

$$\mathcal{U}(z,k;z',k') = -\theta(z' - z^b(k))\theta(z - z')\frac{\mathcal{V}(z',k-k')}{v(k)}\,. \tag{4.36}$$

In a compact notation we have

$$f_{\mathrm{W}} = f^b_{\mathrm{W}} + \mathcal{U} f_{\mathrm{W}} \quad \text{or} \quad f_{\mathrm{W}} = \frac{1}{1-\mathcal{U}} f^b_{\mathrm{W}}\,. \tag{4.37}$$

By expanding the above formal solution in powers of the interaction superoperator/propagator $\mathcal{U}$ – and therefore of the potential $\mathcal{V}$ – we get the well-known Neumann series

$$f_{\mathrm{W}} = \sum_{n=0}^{\infty} \mathcal{U}^n f^b_{\mathrm{W}}\,. \tag{4.38}$$

Let us now focus on the case of a symmetric potential profile ($V(z) = V(-z)$), which in turn corresponds to an antisymmetric potential superoperator, i.e., $\mathcal{V}(z,k) = -\mathcal{V}(-z,k)$. Using this property together with the symmetric nature of our spatial boundaries, i.e., $z^b(k) = -z^b(-k)$, it is possible to show that the interaction propagator $\mathcal{U}$ is also preserving the potential symmetry.[6] This result is extremely important: it implies that for any symmetric potential profile and spatial boundaries the Neumann expansion (4.38) gives always a Wigner function symmetric in the spatial coordinate: $f_{\mathrm{W}}(z,k) = f_{\mathrm{W}}(-z,k)$.

---

[5] This formal solution may also be regarded as a sort of Schwinger–Dyson equation linking the desired Wigner function f_{W} to the free-particle one ($f^\circ_{\mathrm{W}}(z,k) = f^b_{\mathrm{W}}(k)$) through the potential superoperator $\mathcal{V}$.

[6] More specifically, by applying the interaction superoperator $\mathcal{U}$ to any symmetric function ($a(z,k) = a(-z,k)$), we get as a result again a symmetric function of z, i.e.,

$$b(z,k) = -\int_{z^b(k)}^{z} \mathrm{d}z' \int \mathrm{d}k' \frac{\mathcal{V}(z',k-k')}{v(k)} a(z',k') = b(-z,k). \tag{4.39}$$

Therefore, in total agreement with the numerical results of the generalized semiconductor Bloch equations presented in Fig. 4.5, the spatial charge density ($n(z) = \int f_W(z,k)\,dk$) is always symmetric, no matter which is the shape of the injected carrier distribution $f_W^b(k)$. As anticipated, such symmetric behavior – which is an exact result of the treatment presented so far – has never been observed via finite-difference calculations.

In order to solve this apparent contradiction, let us now focus on a particular choice of the device potential profile:

$$V(z) = \frac{V_\circ}{2}\left[1 + \cos(2\kappa z)\right] \ . \tag{4.40}$$

The corresponding superoperator in (4.28) is simply given by

$$\mathcal{V}(z,k) = \frac{V_\circ}{2\hbar}\sin(2\kappa z)[\delta(k-\kappa) - \delta(k+\kappa)] \ . \tag{4.41}$$

For this particular potential superoperator – characterized by a factorization/decoupling of position (z) and momentum (k) coordinates – it is possible to obtain the spatial charge distribution analytically:

$$n(z) = \int dk f_W^b(k)\, {}_1F_2\left[\frac{1}{2};\, 1+\alpha(k),\, 1-\alpha(k);\, q(z)\right] \tag{4.42}$$

with

$$q(z) = \frac{2m^* V_\circ}{\hbar^2\kappa^2}\sin\left[\kappa\left(z - \frac{l}{2}\right)\right]\sin\left[\kappa\left(z + \frac{l}{2}\right)\right] \ . \tag{4.43}$$

Here, m^* is the electron effective mass, $\alpha(k) = \frac{k}{\kappa}$, and ${}_1F_2$ denotes the generalized hypergeometric function of type $(1,2)$.

In order to investigate the main features of the analytical results obtained so far, let us start by considering extremely simple spatial boundary conditions: a monoenergetic carrier injection from left only, i.e., $f_W^b(k) \propto \delta(k - k_\circ)$. Moreover, we choose a potential periodicity $\kappa = \frac{\pi}{l}$ corresponding to just one maximum within the device active region, so as to mimic the single-barrier device considered in [80] (see Fig. 4.5). Figure 4.7 shows a comparison between the analytical spatial charge distribution in (4.42) and the phenomenological result obtained via conventional scattering-state calculations. As we can see, the two curves differ significantly: While the phenomenological charge distribution (dashed curve) – describing an extremely small tunneling dynamics through the potential maximum – exhibits a clear and non-ambiguous asymmetric behavior, the analytical result of the Wigner-function approach (solid curve) is always symmetric, in total agreement with the analysis proposed in [80]. This implies that, also for an infinitely high-potential barrier and for a monoenergetic injection from left only, carriers are "instantaneously" present also on the right part of the device, a typical non-local feature of our quantum-mechanical calculation. More important, within the Wigner-function treatment (solid curve) we deal with a spatial charge/probability distribution

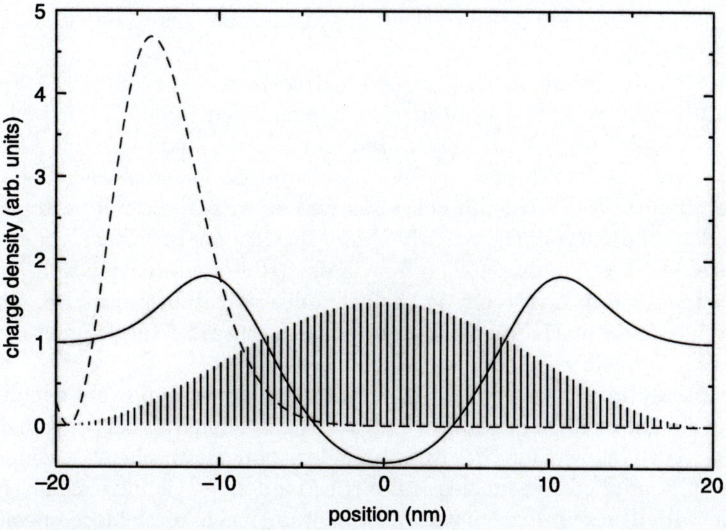

Fig. 4.7. Quantum transport through a GaAs-based device active region ($l = 40$ nm) with a cosine-like potential profile ($V_0 = 150$ meV) sandwiched between its electric contacts: comparison between the analytical charge distribution (*solid curve*) and the corresponding phenomenological result (*dashed curve*) for the case of a monoenergetic carrier injection from left ($\epsilon_\circ = 70$ meV) (see text). Reprinted from [213]

with negative values, which tells us that the analytical solution $f_W(z,k)$ of the differential equation (4.27) is not necessarily a Wigner function.[7]

Let us finally come to potential discrepancies between exact and finite-difference results. Figure 4.8a shows a comparison between the conventional finite-difference solution of the Wigner equation (4.34) proposed in [85] (dashed curve) and a numerical (finite-difference) iterative solution of the result in (4.35) (solid curve); for both cases the same 80×80 phase-space discretization has been employed. As we can see, while the iterative solution coincides with the analytical result (see solid curve in Fig. 4.7), the finite-difference result comes out to be strongly asymmetric. A closer inspection reveals that such anomalous behavior is mainly ascribed to the improper discretization scheme adopted in [85]: More specifically, the usual non-symmetric discretization (left or right derivative according to the sign of the wavevector k) of the kinetic/diffusion term in (4.34) comes out to be incompatible with the symmetric spatial discretization of the potential superoperator[8]; such se-

[7] Indeed, it is well known that a "good" Wigner function, i.e., obtained as the Weyl–Wigner transform of a density matrix, will never give rise to negative charge distributions.

[8] We stress that this problem does not affect the iterative solution of the integral equation (4.35), since in this case the kinetic part of the dynamics is treated exactly.

rious inconsistency gives rise to the fictitious decoherence/damping dynamics reported in Fig. 4.8a.[9]

As a confirmation, in [213] a modified/corrected version of the Frensley finite-difference scheme has been also implemented, where both terms (kinetic and potential) are now discretized on the same spatial grid.[10] As we can see, the new result, reported in Fig. 4.8b, coincides exactly with the analytical solution in Fig. 4.7 (solid curve) as well as with the iterative solution in Fig. 4.8a (solid curve). This clearly shows that the more regular – and physically sound – results obtained so far via finite-difference calculations may be ascribed to such inconsistency in the usual discretization procedure, which in turn tends to limit the highly non-physical features of the Wigner-function formalism applied to systems with open boundaries.

The analysis reviewed so far allows us to draw a few important conclusions. First of all, the results of the analytically solvable device model previously considered clearly show that the usual boundary-condition scheme – widely and successfully applied to semiclassical device modeling – is intrinsically incompatible with the non-local nature of quantum mechanics. More specifically, from a strictly mathematical point of view it is true that the Wigner equation (4.34) is a first-order equation in z parameterized by the wavevector k; on the other hand, from a physical point of view the separation between active region and external reservoirs/contacts is only apparent, since the action of the potential superoperator $\mathcal{V}$ in (4.34) is local in space, but its value inside the device depends on the properties of the potential $V(z)$ both inside and outside the device active region (see (4.28)). We are then forced to conclude that the application of the standard boundary-condition scheme to the Wigner equation (4.34) is not physically justified, since it may provide solutions which are not Wigner functions, i.e., which do not correspond to the state of a quantum system. A clear and unambiguous proof of such non-physical outcomes are the negative values of the electron probability distribution reported in Figs. 4.7 and 4.8.[11] Generally speaking, what is intrinsically wrong in the

---

[9] It is worth mentioning that Frensley himself pointed out that different active-region/reservoir coupling schemes (i.e., non-symmetric versus symmetric derivatives) may lead to different physical pictures and may also potentially lead to negative charge distributions.

[10] Within such modified discretization scheme, the kinetic part is still evaluated via left $(f_j - f_{j-1})$ or right $(f_{j+1} - f_j)$ derivatives (according to the sign of the wavevector), but the potential superoperator $\mathcal{V}_j$ is now replaced by its left $((\mathcal{V}_j + \mathcal{V}_{j-1})/2)$ or right $((\mathcal{V}_{j+1} + \mathcal{V}_j)/2)$ spatial average.

[11] As discussed in Appendix D, in deriving the Wigner transport equation (4.27) it is crucial to assume the spatial-decoherence limit $\rho(+\infty, -\infty) = 0$, which is physically justified in the presence of one or more phase-breaking interaction processes, namely carrier–carrier, carrier–phonon, etc. Strictly speaking, however, this assumption is not compatible with the fully coherent regime corresponding to the one-electron Hamiltonian (D.1); it follows that the validity limits of the Wigner transport equation (4.27) are not straightforward.

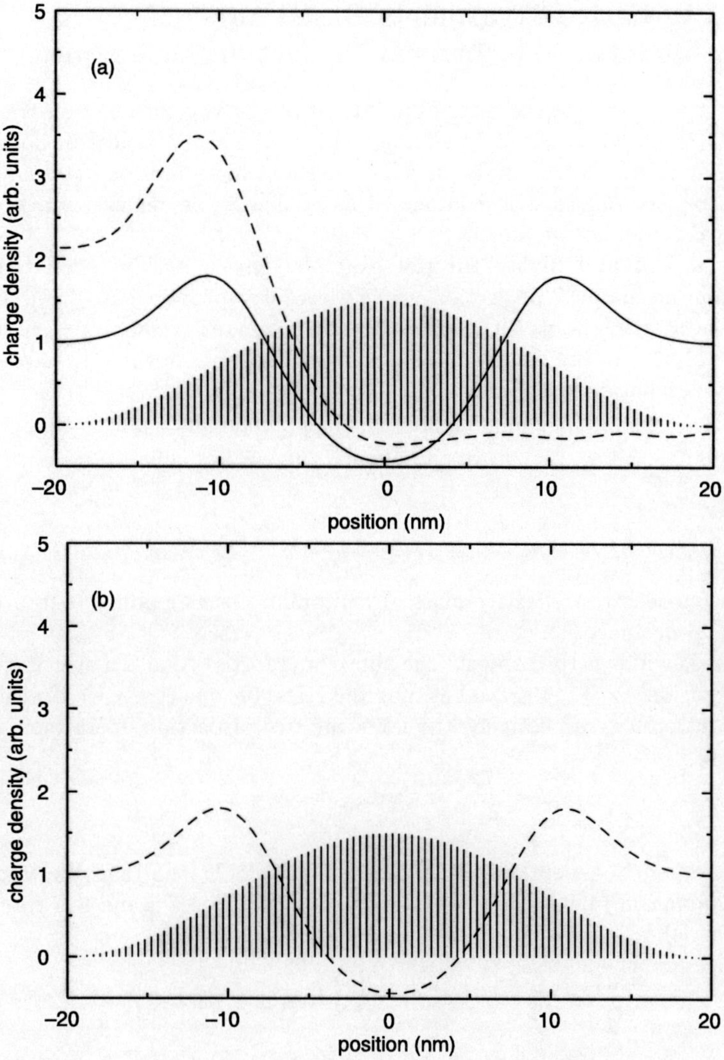

Fig. 4.8. (**a**) Comparison between the conventional finite-difference result (*dashed curve*) and a corresponding iterative numerical solution of the integral equation (4.35). (**b**) Finite-difference calculation based on the modified/corrected discretization scheme (see text). Reprinted from [213]

usual Wigner-function treatment of open devices is the classical-like spatial separation between active region and reservoirs. This is similar to isolate a portion of a given energy spectrum and try to treat such subset of energy levels as an independent subsystem.

4.3 Alternative Treatments Based on Fully Quantum-Mechanical Projection Techniques

In order to overcome the severe limitations of the Wigner-function transport models reviewed so far, it is vital to replace the semiclassical U boundary-condition scheme with a fully quantum-mechanical separation between device active region and electric contacts; this goal may be realized via properly designed projection strategies.

As a first and highly tutorial example, let us briefly recall the fully quantum-mechanical projection scheme recently proposed in [215]. In this case, the starting point is the Liouville–von Neumann equation describing the time evolution of the one-electron density-matrix operator $\hat{\rho}$ in the absence of energy dissipation/decoherence, i.e.,

$$\frac{\mathrm{d}\hat{\rho}}{\mathrm{d}t} = \frac{1}{\mathrm{i}\hbar}\left[\hat{H}, \hat{\rho}\right] , \tag{4.44}$$

where

$$\hat{H} = \frac{\hat{p}^2}{2m^*} + \hat{v} \tag{4.45}$$

is the one-electron effective-mass Hamiltonian corresponding to the device potential profile $V(z)$.[12]

The key idea is to "project" the above operatorial equation over the device active region ($|z| \leq \frac{l}{2}$) as well as over the reservoir one ($|z| > \frac{l}{2}$). To this end, let us introduce, respectively, the following two projection operators:

$$\hat{d} = \int_{-\frac{l}{2}}^{+\frac{l}{2}} |z\rangle\langle z|\mathrm{d}z , \qquad \hat{r} = \hat{\mathcal{I}} - \hat{d} . \tag{4.46}$$

By denoting, in general, with $\hat{O}_{\mathrm{ab}}$ the quantity $\hat{a}\hat{O}\hat{b}$, the Liouville–von Neumann equation (4.44) projected over the two subspaces d and r is translated into the following system of coupled equations:

$$\begin{aligned}
\frac{\mathrm{d}\hat{\rho}_{\mathrm{dd}}}{\mathrm{d}t} &= \frac{1}{\mathrm{i}\hbar}\left[\hat{H}_{\mathrm{dd}}, \hat{\rho}_{\mathrm{dd}}\right] + \frac{1}{\mathrm{i}\hbar}\left\{\hat{H}_{\mathrm{dr}}\hat{\rho}_{\mathrm{rd}} - \hat{\rho}_{\mathrm{dr}}\hat{H}_{\mathrm{rd}}\right\} \\
\frac{\mathrm{d}\hat{\rho}_{\mathrm{rr}}}{\mathrm{d}t} &= \frac{1}{\mathrm{i}\hbar}\left[\hat{H}_{\mathrm{rr}}, \hat{\rho}_{\mathrm{rr}}\right] + \frac{1}{\mathrm{i}\hbar}\left\{\hat{H}_{\mathrm{rd}}\hat{\rho}_{\mathrm{dr}} - \hat{\rho}_{\mathrm{rd}}\hat{H}_{\mathrm{dr}}\right\} \\
\frac{\mathrm{d}\hat{\rho}_{\mathrm{dr}}}{\mathrm{d}t} &= \frac{1}{\mathrm{i}\hbar}\left\{\hat{H}_{\mathrm{dd}}\hat{\rho}_{\mathrm{dr}} - \hat{\rho}_{\mathrm{dr}}\hat{H}_{\mathrm{rr}} + \hat{H}_{\mathrm{dr}}\hat{\rho}_{\mathrm{rr}} - \hat{\rho}_{\mathrm{dd}}\hat{H}_{\mathrm{dr}}\right\} \\
\frac{\mathrm{d}\hat{\rho}_{\mathrm{rd}}}{\mathrm{d}t} &= \frac{1}{\mathrm{i}\hbar}\left\{\hat{H}_{\mathrm{rr}}\hat{\rho}_{\mathrm{rd}} - \hat{\rho}_{\mathrm{rd}}\hat{H}_{\mathrm{dd}} + \hat{H}_{\mathrm{rd}}\hat{\rho}_{\mathrm{dd}} - \hat{\rho}_{\mathrm{rr}}\hat{H}_{\mathrm{rd}}\right\} .
\end{aligned} \tag{4.47}$$

Taking into account that $\hat{O}_{\mathrm{ab}} = \hat{O}^{\dagger}_{\mathrm{ba}}$, we deal with three independent equations only, for the three reduced density matrices $\hat{\rho}_{\mathrm{dd}}$, $\hat{\rho}_{\mathrm{rr}}$, and $\hat{\rho}_{\mathrm{dr}} = \hat{\rho}^{\dagger}_{\mathrm{rd}}$;

[12] As shown in Appendix D, the Wigner transport equation (4.27) may be obtained by applying to the real-space version of the above Liouville–von Neumann equation (see (D.2)) the Weyl–Wigner transform in (D.6).

the first two quantities are genuine density-matrix operators describing, respectively, device and reservoir subspaces, while the quantity $\hat{\rho}_{\rm dr}$ describes quantum coupling effects between device active region and environment. The system of equations (4.47) is formally equivalent to the structure of the optical Bloch equations in (2.34); indeed, the operator $\hat{\rho}_{\rm dr}$ may be regarded as a device-reservoir polarization field.

Exactly as for the case of a two-level system (see Sect. 2.2), it is possible to perform a so-called adiabatic elimination of the polarization $\hat{\rho}_{\rm dr}$. As usual, this is performed by formally integrating the equation of motion for $\hat{\rho}_{\rm dr}$ from $-\infty$ to the current time t, and by assuming that device and reservoir are initially uncorrelated ($\hat{\rho}_{\rm dr}(t=-\infty)=0$); the result is a set of two coupled integro-differential equations for the two reduced density-matrix operators $\hat{\rho}_{\rm dd}$ and $\hat{\rho}_{\rm rr}$.

Finally, taking into account that

$$\hat{H} = \hat{H}_\circ + \hat{H}' = \left(\hat{H}_{\rm dd} + \hat{H}_{\rm rr}\right) + \left(\hat{H}_{\rm dr} + \hat{H}_{\rm rd}\right) , \tag{4.48}$$

when the device-reservoir coupling Hamiltonian $\hat{H}' = \hat{H}_{\rm dr} + \hat{H}_{\rm rd}$ is small (compared to $\hat{H}_\circ = \hat{H}_{\rm dd} + \hat{H}_{\rm rr}$), it is possible to perform a suitable Markov approximation (see Chap. 3), and thus to get a set of local equations of the form

$$\begin{aligned} \frac{{\rm d}\hat{\rho}_{\rm dd}}{{\rm d}t} &= \frac{1}{{\rm i}\hbar}\left[\hat{H}_{\rm dd}, \hat{\rho}_{\rm dd}\right] + \left\{\hat{S}_{\rm d} - \Lambda_{\rm d}\left(\hat{\rho}_{\rm dd}\right)\right\} \\ \frac{{\rm d}\hat{\rho}_{\rm rr}}{{\rm d}t} &= \frac{1}{{\rm i}\hbar}\left[\hat{H}_{\rm rr}, \hat{\rho}_{\rm rr}\right] + \left\{\hat{S}_{\rm r} - \Lambda_{\rm r}\left(\hat{\rho}_{\rm rr}\right)\right\} . \end{aligned} \tag{4.49}$$

Here, Λ denotes loss superoperators (acting on the reduced density matrices) while $\hat{S}$ denotes corresponding source/injection terms; their explicit form depends on the specific adiabatic procedure adopted in performing the Markov approximation. In particular, by employing the new adiabatic-decoupling strategy proposed in Sect. 3.5, it is possible to show that the two density matrices – solutions of the coupled set in (4.49) – are always positive-definite; this is indeed a distinguished feature of the present projection strategy, compared to the Wigner-function treatments previously reviewed.[13]

We stress that the projection strategy proposed so far is still global: the coupled set (4.49) describes the dynamical evolutions of both device and reservoir subsystems. In order to move from a global (closed) to a reduced (open) treatment of the device subsystem only, the crucial step is to fix the state $\hat{\rho}_{\rm rr}$ of the external environment, e.g., assuming that – due to additional dissipation/decoherence mechanisms – the (two or more) charge reservoirs are always

[13] More specifically, the above set of equations describes the evolution of two coupled quantum dynamical semigroups, thus preserving the total (device + reservoir) trace ${\rm tr}\left\{\hat{\rho}_{\rm dd} + \hat{\rho}_{\rm rr}\right\}$; each equation, in contrast, may be regarded as a nonhomogeneous (i.e., trace nonpreserving) Markovian master equation.

in thermal-equilibrium conditions specified by their (independent) chemical potentials. As a result, the coupled set (4.49) reduces to the device equation only:

$$\frac{\mathrm{d}\hat{\rho}_{\mathrm{dd}}}{\mathrm{d}t} = \frac{1}{\mathrm{i}\hbar}\left[\hat{H}_{\mathrm{dd}}, \hat{\rho}_{\mathrm{dd}}\right] + \left\{\hat{S}_{\mathrm{d}}^{\circ} - \Lambda_{\mathrm{d}}\left(\hat{\rho}_{\mathrm{dd}}\right)\right\} , \tag{4.50}$$

where the $\hat{S}_{\mathrm{d}}^{\circ}$ is now a genuine source term evaluated assuming the reservoirs always at thermal equilibrium.

We stress the strong formal similarity between the above active-region transport equation and the open-system semiconductor Bloch equations (4.20). However, in spite of the very same injection versus loss structure, it is imperative to point out once again that the present projection strategy ensures the positive-definite character of the reduced/projected density-matrix operator $\hat{\rho}_{\mathrm{dd}}$.

At this point, a few crucial comments are in order. In spite of the simple and highly tutorial nature of this approach, its concrete implementation is not straightforward. Indeed, the net spatial separation between device and reservoir (see (4.46)) borrowed from the semiclassical picture translates into singular behaviors of the projected Hamiltonians in (4.48) on the device boundary surface. Moreover, such a one-particle formulation does not allow for a consistent incorporation of many-body contributions.

In order to overcome the intrinsic limitations of the above one-electron projection scheme, it is imperative to introduce a clear separation between the degrees of freedom of the subsystem of interest (i.e., electrons primarily localized within the device active region) and those of the external reservoirs (i.e., carriers primarily localized within the electric contacts). Given such separation, it is possible to derive an effective transport equation for the subsystem of interest via a suitable statistical average over the coordinates of the "environment."

To this aim, let us consider a generic many-body Hamiltonian of the form

$$\hat{H} = \hat{H}^{\mathrm{d}} + \hat{H}^{\mathrm{r}} + \hat{H}^{\mathrm{dr}} , \tag{4.51}$$

where

$$\hat{H}^{\mathrm{d}} = \hat{H}_{\circ}^{\mathrm{d}} + \hat{H}_{\mathrm{d}}' = \sum_{\alpha_{\mathrm{d}}} \epsilon_{\alpha_{\mathrm{d}}} \hat{c}_{\alpha_{\mathrm{d}}}^{\dagger} \hat{c}_{\alpha_{\mathrm{d}}} + \hat{H}_{\mathrm{d}}' \tag{4.52}$$

and

$$\hat{H}^{\mathrm{r}} = \hat{H}_{\circ}^{\mathrm{r}} + \hat{H}_{\mathrm{r}}' = \sum_{\alpha_{\mathrm{r}}} \epsilon_{\alpha_{\mathrm{r}}} \hat{c}_{\alpha_{\mathrm{r}}}^{\dagger} \hat{c}_{\alpha_{\mathrm{r}}} + \hat{H}_{\mathrm{r}}' \tag{4.53}$$

are, respectively, the device and reservoir Hamiltonians, while

$$\hat{H}^{\mathrm{dr}} = \sum_{\alpha_{\mathrm{d}}\alpha_{\mathrm{r}}} \left(\gamma_{\alpha_{\mathrm{d}}\alpha_{\mathrm{r}}}^{-} \hat{c}_{\alpha_{\mathrm{d}}}^{\dagger} \hat{c}_{\alpha_{\mathrm{r}}} + \gamma_{\alpha_{\mathrm{d}}\alpha_{\mathrm{r}}}^{+} \hat{c}_{\alpha_{\mathrm{d}}} \hat{c}_{\alpha_{\mathrm{r}}}^{\dagger}\right) \tag{4.54}$$

describes the device–reservoir coupling. Here, α_{d} (α_{r}) denotes the generic device (reservoir) single-particle state, while $\hat{H}_{\mathrm{d}}'$ ($\hat{H}_{\mathrm{r}}'$) describes additional interaction mechanisms (e.g., carrier–carrier as well as carrier–quasiparticle couplings) acting within the device (reservoir) only. In contrast, the Hamiltonian $\hat{H}^{\mathrm{dr}}$ in (4.54) accounts for device–reservoir interaction processes: the first contribution describes carrier injection ($\alpha_{\mathrm{r}} \to \alpha_{\mathrm{d}}$) via the destruction of a carrier in state α_{r} and the creation of a carrier in state α_{d}, while the second one describes carrier loss ($\alpha_{\mathrm{d}} \to \alpha_{\mathrm{r}}$) via the inverse process.

It is important to stress that, opposite to the projection scheme in (4.46), within the present many-body formulation there is no net spatial separation between device region and reservoirs: indeed, both device and reservoir single-particle states may extend over the whole coordinate space. Moreover, the physical properties of the device–reservoir interaction Hamiltonian (4.54) are dictated by the explicit form of the coupling matrix elements $\gamma^{\pm}_{\alpha_{\mathrm{d}}\alpha_{\mathrm{r}}} = \gamma^{\mp *}_{\alpha_{\mathrm{d}}\alpha_{\mathrm{r}}}$; the latter, in general, are given by a properly weighted spatial overlap between device and reservoir single-particle wavefunctions.

In order to better define the proposed reduction/projection procedure, it is useful to rewrite the device–reservoir Hamiltonian (4.54) as

$$\hat{H}^{\mathrm{dr}} = \hbar \sum_{\alpha_{\mathrm{r}}} \left(\hat{\mathcal{H}}^{-}_{\alpha_{\mathrm{r}}} \hat{c}_{\alpha_{\mathrm{r}}} + \hat{\mathcal{H}}^{+}_{\alpha_{\mathrm{r}}} \hat{c}^{\dagger}_{\alpha_{\mathrm{r}}} \right) , \tag{4.55}$$

where

$$\hat{\mathcal{H}}^{-}_{\alpha_{\mathrm{r}}} = \frac{1}{\hbar} \sum_{\alpha_{\mathrm{d}}} \gamma^{-}_{\alpha_{\mathrm{d}}\alpha_{\mathrm{r}}} \hat{c}^{\dagger}_{\alpha_{\mathrm{d}}} = \hat{\mathcal{H}}^{+\dagger}_{\alpha_{\mathrm{r}}} \tag{4.56}$$

are operators (parameterized by the reservoir states α_{r}) acting on the device subsystem only. It is imperative to stress the strong similarity between the carrier–quasiparticle Hamiltonian (3.86) and the above device–reservoir one. Indeed, here the reservoir states α_{r} play the role of the quasiparticle modes $\boldsymbol{q}$, and the two terms in (4.55) – corresponding to carrier injection and loss – may be regarded as device-absorption and -emission processes.

By applying the Markov approximation scheme proposed in Sect. 3.5 to the device–reservoir interaction Hamiltonian (4.54), the resulting contribution to the global (device plus reservoir) dynamics is given by the following Lindblad-like superoperator[14]:

$$\left.\frac{\mathrm{d}\hat{\rho}}{\mathrm{d}t}\right|_{\mathrm{dr}} = -\frac{1}{2} \left[\hat{A}^{\mathrm{dr}}, \left[\hat{A}^{\mathrm{dr}}, \hat{\rho} \right] \right] \tag{4.57}$$

with

$$\hat{A}^{\mathrm{dr}} = \sum_{\alpha_{\mathrm{d}}\alpha_{\mathrm{r}}} \left(A^{-}_{\alpha_{\mathrm{d}}\alpha_{\mathrm{r}}} \hat{c}^{\dagger}_{\alpha_{\mathrm{d}}} \hat{c}_{\alpha_{\mathrm{r}}} + A^{+}_{\alpha_{\mathrm{d}}\alpha_{\mathrm{r}}} \hat{c}_{\alpha_{\mathrm{d}}} \hat{c}^{\dagger}_{\alpha_{\mathrm{r}}} \right) \tag{4.58}$$

[14] The explicit form of the device–reservoir Lindblad operator $\hat{A}^{\mathrm{dr}}$ in (4.58) may be easily obtained inserting into the general prescription (3.165) the explicit form of the unperturbed Hamiltonian $\hat{H}_{\circ} = \hat{H}^{\mathrm{d}}_{\circ} + \hat{H}^{\mathrm{r}}_{\circ}$ as well as of the interaction Hamiltonian $\hbar\hat{\mathcal{H}} = \hat{H}^{\mathrm{dr}}$ and employing the anticommutation relations of our (fermionic) creation and destruction operators.

and

$$A^-_{\alpha_d\alpha_r} = \sqrt{\frac{2\pi}{\hbar}}\gamma^-_{\alpha_d\alpha_r}\frac{e^{-\left(\frac{\epsilon_{\alpha_d}-\epsilon_{\alpha_r}}{2\overline{\epsilon}}\right)^2}}{\left(2\pi\overline{\epsilon}^2\right)^{\frac{1}{4}}} = A^{+*}_{\alpha_d\alpha_r} \ . \tag{4.59}$$

In order to derive the desired equation for the device subsystem, the key approximation is to fix the state of the external environment, assuming that – due to additional dissipation/decoherence mechanisms – the (two or more) charge reservoirs are always in thermal-equilibrium conditions specified by their (independent) chemical potentials. Within such approximation scheme, the global (device plus reservoirs) density-matrix operator $\hat{\rho}$ can be written as the product of the device density-matrix operator times a time-independent density-matrix operator describing the collection of external reservoirs[15]:

$$\hat{\rho} = \hat{\rho}^{\rm d}\hat{\rho}^{\rm r} \ . \tag{4.60}$$

Moreover, in view of the local-equilibrium conditions in the various reservoirs, the environment single-particle density matrix (see (3.116)) is always diagonal:

$$\rho_{\alpha_r\alpha'_r} = \mathrm{tr}\left\{\hat{c}^\dagger_{\alpha'_r}\hat{c}_{\alpha_r}\hat{\rho}^{\rm r}\right\} = f^\circ_{\alpha_r}\delta_{\alpha_r\alpha'_r} \ . \tag{4.61}$$

Applying to the Lindblad superoperator (4.57) the very same reduction scheme presented in Sect. 3.3.2 for the carrier–quasiparticle coupling (see also Appendix C), and employing the device–reservoir factorization (4.60) together with the equilibrium condition (4.61), it is possible to derive an effective equation describing the time evolution of the device density-matrix operator $\hat{\rho}^{\rm d}$:

$$\frac{d\hat{\rho}^{\rm d}}{dt} = \left.\frac{d\hat{\rho}^{\rm d}}{dt}\right|_{\hat{H}^{\rm d}} + \left.\frac{d\hat{\rho}^{\rm d}}{dt}\right|_{\rm dr} \tag{4.62}$$

with

$$\left.\frac{d\hat{\rho}^{\rm d}}{dt}\right|_{\rm dr} = -\frac{1}{2}\sum_{\alpha_r\pm}\left(\frac{1}{2}\pm\frac{1}{2}\mp f^\circ_{\alpha_r}\right)\left(\hat{A}^{\pm\dagger}_{\alpha_r}\hat{A}^{\pm}_{\alpha_r}\hat{\rho}^{\rm d} - \hat{A}^{\pm}_{\alpha_r}\hat{\rho}^{\rm d}\hat{A}^{\pm\dagger}_{\alpha_r}\right) + \mathrm{H.c.} \tag{4.63}$$

and

$$\hat{A}^-_{\alpha_r} = \sum_{\alpha_d}\left(A^-_{\alpha_d\alpha_r}\hat{c}^\dagger_{\alpha_d} + A^+_{\alpha_d\alpha_r}\hat{c}_{\alpha_d}\right) \ . \tag{4.64}$$

The time evolution of $\hat{\rho}^{\rm d}$ in (4.62) is the sum of a contribution due to the device Hamiltonian only, plus the device–reservoir term (4.63)[16]; the latter is

---

[15] This approximation is formally equivalent to the carrier–quasiparticle factorization scheme (3.76) and may be physically justified assuming that the dissipation/decoherence processes induced by the interaction Hamiltonian $\hat{H}'_{\rm r}$ is so strong to maintain each carrier reservoir at thermal equilibrium.

[16] Comparing the Lindblad-like term (4.63) with the corresponding carrier–quasiparticle one in (3.172), the only differences are due to the peculiar nature

Lindblad like (see the general Lindblad form in (3.36)), thus preserving the positive-definite character of the density-matrix operator $\hat{\rho}^{\mathrm{d}}$.

In spite of the homogeneous nature of the above quantum equation, it is imperative to recall its many-body character: it is true that the global trace of the many-body density-matrix operator $\hat{\rho}^{\mathrm{d}}$ is preserved; however, in view of the open character of the device–reservoir Hamiltonian (4.54), the average number of electrons within the device,

$$N_{\mathrm{d}} = \mathrm{tr}\left\{\sum_{\alpha_{\mathrm{d}}} \hat{c}^{\dagger}_{\alpha_{\mathrm{d}}} \hat{c}_{\alpha_{\mathrm{d}}} \hat{\rho}\right\} , \tag{4.65}$$

is not preserved. In other words, in spite of the apparent absence of injection/loss terms, the Lindblad superoperator (4.63) couples many-body states of the device characterized by a different number of electrons; it follows that a direct numerical solution of the many-electron equation (4.62) is computationally prohibitive.

In order to circumvent the limitations of the above many-electron treatment, the key strategy is to introduce a single-particle description of the device subsystem (see Sect. 3.3.3). More specifically, following the general prescription in (3.116), the device single-particle density matrix is defined as

$$\rho_{\alpha_{\mathrm{d}}\alpha'_{\mathrm{d}}} = \mathrm{tr}\left\{\hat{c}^{\dagger}_{\alpha'_{\mathrm{d}}} \hat{c}_{\alpha_{\mathrm{d}}} \hat{\rho}\right\} . \tag{4.66}$$

Combining the Lindblad-like global evolution (4.57) with the definition in (4.66), the resulting single-particle dynamics is given by

$$\frac{d\rho_{\alpha_{\mathrm{d}}\alpha'_{\mathrm{d}}}}{dt} = \left.\frac{d\rho_{\alpha_{\mathrm{d}}\alpha'_{\mathrm{d}}}}{dt}\right|_{\hat{H}^{\mathrm{d}}} + \left.\frac{d\rho_{\alpha_{\mathrm{d}}\alpha'_{\mathrm{d}}}}{dt}\right|_{\mathrm{dr}} \tag{4.67}$$

with

$$\left.\frac{d\rho_{\alpha_{\mathrm{d}}\alpha'_{\mathrm{d}}}}{dt}\right|_{\mathrm{dr}} = -\frac{1}{2}\mathrm{tr}\left\{\hat{c}^{\dagger}_{\alpha'_{\mathrm{d}}} \hat{c}_{\alpha_{\mathrm{d}}} \left[\hat{A}^{\mathrm{dr}}, \left[\hat{A}^{\mathrm{dr}}, \hat{\rho}\right]\right]\right\} . \tag{4.68}$$

As shown in Sect. 3.3.3, employing the cyclic property of the trace in (C.7) it is possible to rewrite the above device–reservoir contribution as

$$\left.\frac{d\rho_{\alpha_{\mathrm{d}}\alpha'_{\mathrm{d}}}}{dt}\right|_{\mathrm{dr}} = -\frac{1}{2}\mathrm{tr}\left\{\left[\left[\hat{c}^{\dagger}_{\alpha'_{\mathrm{d}}} \hat{c}_{\alpha_{\mathrm{d}}}, \hat{A}^{\mathrm{dr}}\right], \hat{A}^{\mathrm{dr}}\right] \hat{\rho}\right\} . \tag{4.69}$$

Starting from the explicit form of the Lindblad operator $\hat{A}^{\mathrm{dr}}$ in (4.58) and employing once again the fermionic anticommutation relations, it is possible

(fermionic versus bosonic) of their thermal environments: while for the quasiparticle case absorption and emission processes are proportional to $N^{\circ}_{\boldsymbol{q}}$ and $N^{\circ}_{\boldsymbol{q}} + 1$, respectively, here carrier injection/absorption and loss/emission processes are proportional, respectively, to the thermal occupation $f^{\circ}_{\alpha_{\mathrm{r}}}$ of the initial state α_{r} and to the Pauli factor $1 - f^{\circ}_{\alpha_{\mathrm{r}}}$ describing the degree of occupation of the final state α_{r}.

to evaluate the double commutator in (4.69), and eventually to perform the trace reduction assuming the device–reservoir factorization (4.60). The result of this straightforward derivation (not reported here) is a closed equation of motion for the single-particle density matrix of the device

$$\left.\frac{d\rho_{\alpha_d\alpha'_d}}{dt}\right|_{dr} = \frac{1}{2}\sum_{\alpha_r\alpha''_d}\left[f^\circ_{\alpha_r} A^-_{\alpha_d\alpha_r} A^{-*}_{\alpha''_d\alpha_r}\left(\delta_{\alpha''_d\alpha'_d} - \rho_{\alpha''_d\alpha'_d}\right)\right.$$
$$\left. - \left(1 - f^\circ_{\alpha_r}\right) A^{+*}_{\alpha_d\alpha_r} A^{+}_{\alpha''_d\alpha_r}\rho_{\alpha''_d\alpha'_d}\right] + \text{H.c.}\,. \quad (4.70)$$

In order to properly identify the physical origin of the different contributions, let us consider once again the semiclassical limit ($\rho_{\alpha_d\alpha'_d} = f_{\alpha_d}\delta_{\alpha_d\alpha'_d}$) of the above single-particle quantum equation:

$$\left.\frac{df_{\alpha_d}}{dt}\right|_{dr} = \sum_{\alpha_r}\left[f^\circ_{\alpha_r}\left|A^-_{\alpha_d\alpha_r}\right|^2\left(1 - f_{\alpha_d}\right) - \left(1 - f^\circ_{\alpha_r}\right)\left|A^+_{\alpha_d\alpha_r}\right|^2 f_{\alpha_d}\right]\,. \quad (4.71)$$

As we can see, the two terms describe, respectively, device carrier absorption ($\alpha_r \to \alpha_d$) and emission ($\alpha_d \to \alpha_r$); for each process the corresponding contribution is proportional to the occupation f of the initial state and to the Pauli factor $1 - f$ of the final state. As anticipated in Sect. 3.3.3, this shows that the factor $\left(\delta_{\alpha_d\alpha'_d} - \rho_{\alpha_d\alpha'_d}\right)$ in (4.70) is the quantum-mechanical generalization of the Pauli factor $(1 - f_{\alpha_d})$.

A closer inspection of the quantum-transport equation (4.70) shows that, in spite of our single-particle description, the latter is still linear; this feature is ascribed to the fact that – from the device point of view – the device–reservoir Hamiltonian (4.54) is simply linear in the creation and destruction operators. Moreover, opposite to the many-electron equation (4.63), the single-particle equation (4.70) is not homogeneous anymore; indeed, the latter may be simply rewritten as a source minus a loss contribution according to

$$\left.\frac{d\rho_{\alpha_d\alpha'_d}}{dt}\right|_{dr} = S^d_{\alpha_d\alpha'_d} - \frac{1}{2}\sum_{\alpha''_d}\left(\Gamma^d_{\alpha_d\alpha''_d}\rho_{\alpha''_d\alpha'_d} + \rho_{\alpha_d\alpha''_d}\Gamma^d_{\alpha''_d\alpha'_d}\right) \quad (4.72)$$

with

$$S^d_{\alpha_d\alpha'_d} = \sum_{\alpha_r} f^\circ_{\alpha_r} A^-_{\alpha_d\alpha_r} A^{-*}_{\alpha'_d\alpha_r} \quad (4.73)$$

and

$$\Gamma^d_{\alpha_d\alpha'_d} = \sum_{\alpha_r}\left(1 - f^\circ_{\alpha_r}\right)\left(A^{+*}_{\alpha_d\alpha_r} A^{+}_{\alpha'_d\alpha_r} + A^-_{\alpha_d\alpha_r} A^{-*}_{\alpha'_d\alpha_r}\right)\,. \quad (4.74)$$

We stress the strong formal similarity between the above single-particle transport equation and the open-system semiconductor Bloch equations (4.20). However, in spite of the very same injection versus loss structure, it is imperative to point out that – as for the one-electron model previously discussed – the present many-body projection strategy ensures the positive-definite character of the single-particle density matrix $\rho_{\alpha_d\alpha'_d}$; indeed, this is guaranteed by the positive-definite character of the source term (4.73) as well as of the loss term (4.74). Moreover, in the semiclassical limit ($\rho_{\alpha_d\alpha'_d} = f_{\alpha_d}\delta_{\alpha_d\alpha'_d}$) the injection/loss structure in (4.72) reduces to the standard phenomenological model in (1.84): the usual injection and loss rates (S_α and Γ^b_α) are simply given by the diagonal elements of the corresponding matrices in (4.73) and (4.74).

As a final remark, we stress once again that the open-system (i.e., injection versus loss) character of the quantum-transport formulation presented so far is intrinsically related to the open-system structure of the device–reservoir Hamiltonian (4.54). Indeed, adopting as alternative version of the device–reservoir coupling a two-body interaction Hamiltonian of the form (see (3.131))

$$\hat{H}^{\mathrm{dr}} = \frac{1}{2} \sum_{\alpha_d\alpha_r,\alpha'_d\alpha'_r} \gamma_{\alpha_d\alpha_r,\alpha'_d\alpha'_r} \hat{c}^\dagger_{\alpha_d} \hat{c}^\dagger_{\alpha_r} \hat{c}_{\alpha'_d} \hat{c}_{\alpha'_r} , \tag{4.75}$$

the number of electrons within the device is never changed; indeed, opposite to the original device–reservoir interaction Hamiltonian (4.54), here the energy dissipation and decoherence on the device subsystem is induced by carrier–carrier processes $\alpha'_d\alpha'_r \to \alpha_d\alpha_r$, where each carrier performs a single-particle transition ($\alpha'_d \to \alpha_d$ and $\alpha'_r \to \alpha_r$) within the corresponding subsystem, thus preventing any device–reservoir charge transfer.[17] Such a "closed-system paradigm" is the starting point of the alternative kinetic approach presented in the following section.

4.4 A Simple Kinetic Model Based on a Closed-System Paradigm

Aim of this section is to review an extremely simple and radically different approach for the study of quantum devices with spatial open boundaries originally proposed in [216, 217]; the key idea – grounded on a "closed-system paradigm" – is to replace the usual modeling of open quantum devices (based on phenomenological injection/loss rates) with a kinetic description of the system–reservoir thermalization process.

In order to introduce this alternative modeling strategy, let us start by recalling the conventional treatment of the problem. As discussed in Sect. 1.4, within a semiclassical framework the carrier dynamics of a quantum device is usually described via an effective transport equation of the form (see (1.83))

[17] It is possible to show that in the presence of the alternative device–reservoir Hamiltonian (4.75) the single-particle injection/loss result (4.72) is replaced by a homogeneous Lindblad-like superoperator describing two-body scattering processes between device and reservoir electrons (see Sect. 4.4).

$$\frac{\partial f_\alpha}{\partial t} = \left.\frac{\partial f_\alpha}{\partial t}\right|_{\rm d} + \left.\frac{\partial f_\alpha}{\partial t}\right|_{\rm dr} . \tag{4.76}$$

As anticipated in Sect. 4.1, the first term describes scattering dynamics within the device active region and is usually treated at a kinetic level via a Boltzmann-like collision operator of the form

$$\left.\frac{\partial f_\alpha}{\partial t}\right|_{\rm d} = \sum_{\alpha'} \left(P^{\rm d}_{\alpha\alpha'} f_{\alpha'} - P^{\rm d}_{\alpha'\alpha} f_\alpha\right) , \tag{4.77}$$

where $P^{\rm d}_{\alpha\alpha'}$ is the total (i.e., summed over all relevant interaction mechanisms) scattering rate from state α' to state α. In contrast, the last term in (4.76) accounts for the open character of the system and describes injection/loss contributions from/to the (at least two) external carrier reservoirs. These processes are usually modeled by a relaxation-time-like term of the form (see (1.84))

$$\left.\frac{\partial f_\alpha}{\partial t}\right|_{\rm dr} = -\Gamma^b_\alpha \left(f_\alpha - f^\circ_\alpha\right) . \tag{4.78}$$

In spite of the kinetic nature of the scattering dynamics in (4.77), the contribution (4.78) describes carrier injection and loss processes on a partially phenomenological level; as discussed in Chap. 5, this formulation of the problem, in turn, requires hybrid simulation strategies combining, e.g., a Monte Carlo sampling of the scattering dynamics with a direct numerical integration of injection/loss terms (see Sect. 5.3).

The key idea proposed in [216, 217] is to replace the conventional relaxation-time term (4.78) with a Boltzmann-like operator of the form

$$\left.\frac{\partial f_\alpha}{\partial t}\right|_{\rm dr} = \sum_{\alpha'} \left(P^{\rm dr}_{\alpha\alpha'} f_{\alpha'} - P^{\rm dr}_{\alpha'\alpha} f_\alpha\right) . \tag{4.79}$$

This contribution has indeed the same structure of the scattering operator in (4.77); however, the new scattering rates $P^{\rm dr}_{\alpha\alpha'}$ describe electronic transitions within the simulated region induced by the coupling to the external carrier reservoirs. It is worthwhile to stress that, contrary to the conventional injection/loss term (4.78), in this case there is no particle exchange between device active region and thermal reservoirs. The total number of simulated carriers is therefore conserved.

Let us now discuss the explicit form of the scattering rates $P^{\rm dr}_{\alpha\alpha'}$ entering the new collision operator (4.79). In the absence of scattering processes ($P^{\rm d}_{\alpha\alpha'} = 0$), the steady-state solution of the conventional injection/loss model (4.78) is $f_\alpha = f^\circ_\alpha$, i.e., the carrier distribution inside the device coincides with the distribution in the external carrier reservoirs. As a first requirement, we therefore impose the same steady-state solution ($f_\alpha = f^\circ_\alpha$) to the new collision operator in (4.79). This, in turn, will impose conditions on the explicit form of the scattering rates $P^{\mathrm{dr}}_{\alpha\alpha'}$. More specifically, from the detailed-balance principle (see, e.g., [72]) we get[18]:

$$\frac{P^{\mathrm{dr}}_{\alpha\alpha'}}{P^{\mathrm{dr}}_{\alpha'\alpha}} = \frac{f^\circ_\alpha}{f^\circ_{\alpha'}} . \tag{4.80}$$

It follows that our transition rates should be of the form

$$P^{\mathrm{dr}}_{\alpha\alpha'} = \overline{P}_{\alpha\alpha'} f^\circ_\alpha , \tag{4.81}$$

where $\overline{P}$ can be any positive and symmetric transition matrix ($\overline{P}_{\alpha\alpha'} = \overline{P}_{\alpha'\alpha} > 0$). What is important in steady-state conditions is the ratio of the scattering rates entering (4.80) and not their absolute values which are, in contrast, crucial in determining the transient nonequilibrium response of the system. Since our aim is to replace the injection/loss term (4.78) with the Boltzmann-like term (4.79), as second requirement we ask that the relaxation dynamics induced by the new collision term corresponds to the phenomenological relaxation times in (4.78). This amounts to imposing that the total out-scattering rate – summed over all possible final states – coincides with the relaxation rates Γ^b_α:

$$\sum_{\alpha'} P^{\mathrm{dr}}_{\alpha'\alpha} = \Gamma^b_\alpha . \tag{4.82}$$

By assuming as simplest form of the symmetric transition matrix in (4.81)

$$\overline{P}_{\alpha\alpha'} = p_\alpha p_{\alpha'} , \tag{4.83}$$

the requirement in (4.82) reduces to the following system of equations for the unknown quantities p_α:

$$\sum_{\alpha'} p_{\alpha'} f^\circ_{\alpha'} = \frac{\Gamma^b_\alpha}{p_\alpha} . \tag{4.84}$$

Since the sum on the left is α-independent, this tells us that p_α is simply proportional to Γ^b_α:

$$p_\alpha = C\Gamma^b_\alpha . \tag{4.85}$$

By inserting the proportionality relation (4.85) into the system of equations (4.84), it is easy to express the constant quantity C in terms of the loss rates Γ^b_α:

$$C = \frac{1}{\sqrt{\sum_{\alpha''} \Gamma^b_{\alpha''} f^\circ_{\alpha''}}} . \tag{4.86}$$

Recalling the result in (4.85), we obtain

[18] For the case of a nonequilibrium carrier distribution f°_α, the detailed-balance principle should be replaced by a total-balance one. However, by inserting (4.81) into the effective collision operator in (4.79) it is easy to verify that its steady-state solution is $f_\alpha = f^\circ_\alpha$.

$$\overline{P}_{\alpha\alpha'} = p_\alpha p_{\alpha'} = \frac{\Gamma^b_\alpha \Gamma^b_{\alpha'}}{\sum_{\alpha''} \Gamma^b_{\alpha''} f^\circ_{\alpha''}} . \tag{4.87}$$

By inserting this result into (4.81), the explicit form of the desired device–reservoir scattering rates entering the Boltzmann-like collision term (4.79) comes out to be

$$P^{\mathrm{dr}}_{\alpha\alpha'} = \frac{\Gamma^b_\alpha \Gamma^b_{\alpha'}}{\sum_{\alpha''} \Gamma^b_{\alpha''} f^\circ_{\alpha''}} f^\circ_\alpha . \tag{4.88}$$

As discussed in Chap. 5, the present kinetic formulation in terms of Boltzmann-like collision operators only is particularly suited for a standard Monte Carlo simulation approach (see Sect. 5.2), where one deals with a fixed number of carriers. In this respect, contrary to the phenomenological model in (4.76), in the present closed-system formulation the total carrier density is not fixed by the external reservoirs and the resulting transport equation is homogeneous.

The above closed-system strategy has been extensively tested in [216, 217]. To this end, a fully three-dimensional Monte Carlo simulation scheme has been developed (see Chap. 5), employing as envelope-function basis states the product of scattering states along the field/growth direction times two-dimensional plane waves accounting for the in-plane dynamics. Moreover, in order to properly describe phonon-induced energy and momentum relaxation within the device active region, in addition to the new scattering-like thermalization mechanism in (4.79), carrier–phonon scattering in a fully three-dimensional fashion has been also included.[19]

We start our analysis considering an extremely simple transport problem: a GaAs mesoscopic bulk system of length $l = 200\,\mathrm{nm}$ sandwiched between two reservoirs with different chemical potentials ($\mu_{\mathrm{left}} - \mu_{\mathrm{right}} = 50\,\mathrm{meV}$). We have applied to this problem the simulation strategy previously described (see (4.79)), comparing the results with those of the conventional simulation approach (see (4.78)).

Figure 4.9a displays the transient carrier dynamics resulting from the conventional injection/loss model in (4.78). Here, we show the time evolution of the carrier distribution in momentum space at steps of 1 ps. Since in this model we start at time $t = 0$ with an empty-device configuration, the simulated experiment shows a progressive increase of the carrier distribution, which from the very beginning exhibits a strong left–right asymmetry due to the chemical-potential misalignment. This scenario manifests the open nature of the conventional approach, which does not allow the direct use of a standard Monte Carlo procedure (see Sect. 5.3).

Figure 4.9b shows again the transient evolution of the carrier distribution in momentum space, but obtained via the proposed simulation approach. In this case we deal with a fixed number of particles which at time $t = 0$ are arbitrarily chosen to be equally distributed in the three-dimensional momentum

[19] Since in the simulated experiments discussed below we shall focus on low-density conditions, carrier–carrier scattering has not been considered.

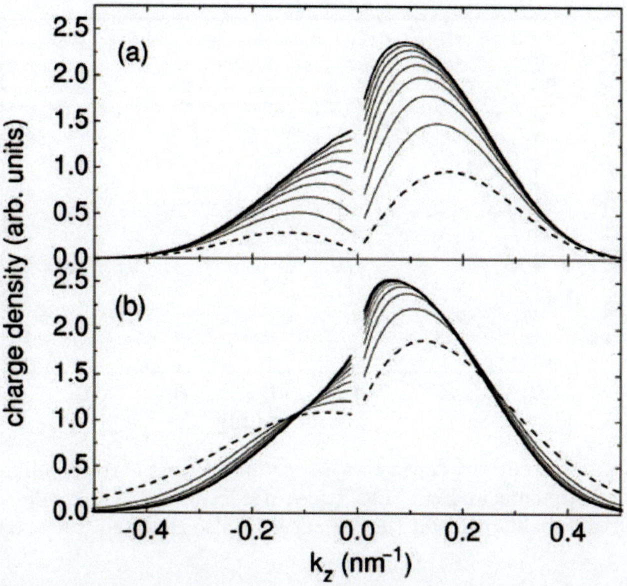

Fig. 4.9. Room-temperature transport properties of a GaAs mesoscopic bulk system of length $l = 200\,\mathrm{nm}$ sandwiched between two reservoirs with different chemical potentials ($\mu_{\mathrm{left}} - \mu_{\mathrm{right}} = 50\,\mathrm{meV}$). Transient dynamics of the carrier distribution in momentum space – from 1 ps (*dashed curve*) to 9 ps (*thick solid curve*) at intervals of 1 ps (*thin solid curves*) – as obtained (**a**) from the conventional injection/loss model (4.78) and (**b**) from the proposed simulation strategy (see text). Reprinted from [217]

space. Moreover, the total carrier density, which is now a free parameter, has been set equal to the steady-state value in Fig. 4.9a (which can be directly evaluated from the thermal distributions f°_α). Contrary to the time evolution in Fig. 4.9a, here at very short times the device region is already occupied and its charge distribution in momentum space is almost symmetric. Only at later times, due to the effective scattering mechanism in (4.81), we recover the asymmetric distribution of Fig. 4.9a (see solid curve).

Figure 4.10 shows the charge current density as a function of time corresponding to the two simulated experiments in Fig. 4.9a (dashed curve) and Fig. 4.9b (solid curve). At time $t = 0$ the current is in both cases equal to zero; this is, however, ascribed to different reasons: in Fig. 4.9a at $t = 0$ the carrier density is equal to zero while the mean velocity is different from zero; in contrast, in Fig. 4.9b the mean velocity is equal to zero while the carrier density is different from zero. In spite of a slightly different transient, both

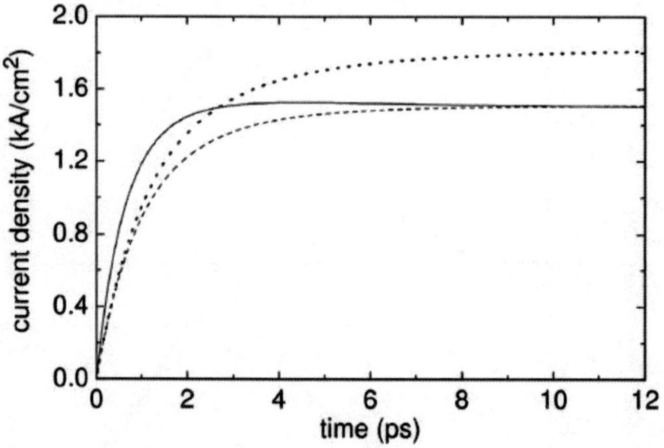

Fig. 4.10. Charge current density as a function of time corresponding to the two simulated experiments in Figs. 4.9a (*dashed curve*) and 4.9b (*solid curve*). The scattering-free or ballistic result (*dotted curve*) is also reported (see text). Reprinted from [217]

curves reach almost the same steady-state value, confirming the validity of the proposed simulation strategy.[20]

The steady-state regime is the result of a strong interplay between the thermalization induced by the external reservoirs and the phonon-induced momentum relaxation within the device active region. Indeed, in the phonon-free case (dotted curve in Fig. 4.10) the steady-state current – which is fully ballistic – reaches significantly higher values. The momentum-relaxation dynamics previously mentioned is clearly visible in Fig. 4.9a, where the peaks of the injected carrier distribution are progressively shifted toward lower wavevectors.

As a second testbed, a prototypical semiconductor quantum device has been considered: a GaAs/AlGaAs resonant-tunneling diode with a barrier height of 0.24 eV and a barrier width and separation of 2.8 and 4.4 nm, respectively. Figure 4.11 shows the current–voltage characteristics obtained from the proposed closed-system simulation scheme with (solid curve) and without (dashed curve) carrier–phonon scattering. The results confirm the validity of the present approach. Indeed, the latter describes properly the typical resonance scenario: as expected, in the presence of phase-breaking processes, like carrier–phonon scattering, the resonance peak is significantly reduced; however, also in this more realistic case, the proposed simulation strategy comes out to properly describe the key phenomena under investigation.

[20] To compare the transient response of the two models it is useful to assume identical initial conditions. To this end we have also simulated the open system starting from the same initial distribution considered in Fig. 4.9b. The resulting current density as a function of time (not reported here) coincides within a few percents with the dashed curve.

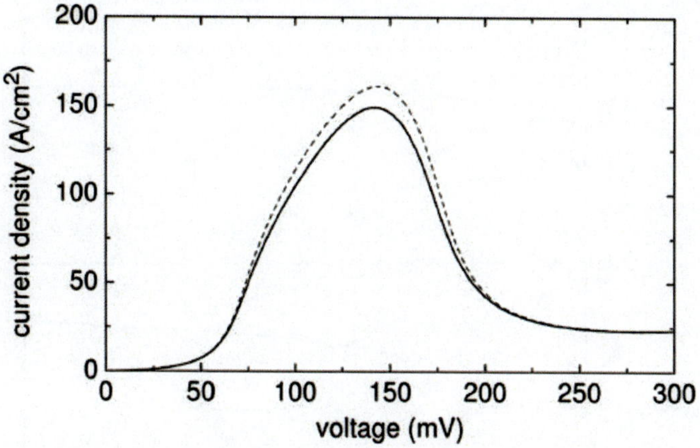

Fig. 4.11. Room-temperature current–voltage characteristics of a GaAs/AlGaAs resonant-tunneling diode (with barrier height 0.24 eV and barrier width and separation of 2.8 and 4.4 nm, respectively) as obtained via the proposed closed-system simulation scheme with (*solid curve*) and without (*dashed curve*) carrier–phonon scattering (see text). Reprinted from [217]

As a final example, we have considered a typical nonequilibrium situation, an electrically driven quantum-dot device. To this end, we have adopted an extremely simplified model: an electronic two-level system coupled to the phonon modes of the host material as well as to two external (injecting and extracting) charge reservoirs (see inset in Fig. 4.12).

According to the general prescription previously introduced (see (4.77)), by denoting with a and b the ground and excited states of our two-level system, the phonon-induced scattering dynamics will be described in terms of two interlevel rates corresponding to phonon absorption and emission:

$$P^{\mathrm{d}}_{ba} = W^{\mathrm{d}} N^{\circ} \ , \qquad P^{\mathrm{d}}_{ab} = W^{\mathrm{d}} \left(N^{\circ} + 1 \right) \ , \tag{4.89}$$

where $N^{\circ} = \left(\mathrm{e}^{\frac{\Delta\epsilon}{k_B T}} - 1 \right)^{-1}$ is the Bose occupation number (see (3.90)) corresponding to the interlevel energy splitting $\Delta\epsilon = \epsilon_b - \epsilon_a$. In the absence of coupling to the external reservoirs, regardless of the value of the carrier–phonon coupling constant W^{d}, the steady-state solution is the thermal equilibrium one:

$$\mathcal{R}^{\mathrm{eq}} = \frac{f^{\mathrm{eq}}_b}{f^{\mathrm{eq}}_a} = \frac{P^{\mathrm{d}}_{ba}}{P^{\mathrm{d}}_{ab}} = \frac{N^{\circ}}{N^{\circ} + 1} = \mathrm{e}^{-\frac{\Delta\epsilon}{k_B T}} \ . \tag{4.90}$$

In contrast, in the presence of external reservoirs characterized by different values of their chemical potentials, μ_{left} and μ_{right} (see inset in Fig. 4.12), the population ratio $\mathcal{R}$ may differ significantly from the thermal-equilibrium value in (4.90). More specifically, in the absence of phonon-induced interlevel scattering ($W^{\mathrm{d}} = 0$) the population ratio $\mathcal{R}^{\circ}$ is fully dictated by the

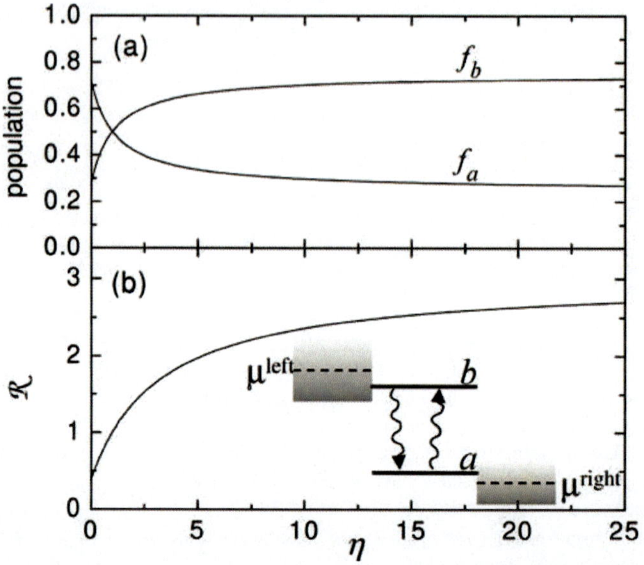

Fig. 4.12. Electrically driven nonequilibrium carrier distribution in a quantum-dot device: (**a**) excited- and ground-state populations f_b and f_a and (**b**) their ratio $\mathcal{R}$ as a function of the coupling-constant ratio η, for an interlevel energy splitting $\Delta\epsilon =$ 25 meV and for a reservoir-induced population ratio $\mathcal{R}_\circ = 3$ at room temperature (see text). A schematics of the electrically driven two-level system is also reported in the inset. Reprinted from [217]

carrier distributions within the reservoirs and may be greater than 1, i.e., a population-inversion regime may be established (see Chap. 8).

Following again the proposed closed-system simulation strategy, we shall describe the effect of the system–reservoir coupling via the effective Boltzmann collision term (4.79). More specifically, for the case of our two-level system we have

$$P^{\mathrm{dr}}_{ba} = W^{\mathrm{dr}} \frac{\mathcal{R}^\circ}{\mathcal{R}^\circ + 1} , \qquad P^{\mathrm{dr}}_{ab} = W^{\mathrm{dr}} \frac{1}{\mathcal{R}^\circ + 1} , \tag{4.91}$$

where W^{dr} denotes a suitable system–reservoir coupling constant. Analogous to the result in (4.90), it is easy to see that, in the absence of carrier–phonon scattering, such effective rates provide as steady-state solution the desired population ratio $\mathcal{R}^\circ$, i.e.,

$$\frac{f_b}{f_a} = \frac{P^{\mathrm{dr}}_{ba}}{P^{\mathrm{dr}}_{ab}} = \mathcal{R}^\circ . \tag{4.92}$$

In the presence of carrier–phonon scattering as well as coupling to the external reservoirs, the actual value of the population ratio $\mathcal{R}$ is the result of a non-trivial interplay between carrier–phonon interlevel scattering and device–reservoir coupling.

To better understand this nonequilibrium behavior, in [217] a few simulated experiments based on the above electrically driven quantum-dot model have been performed, varying the ratio $\eta = \frac{W^{\mathrm{dr}}}{W^{\mathrm{d}}}$ between carrier reservoir and phonon-scattering coupling constants. Figure 4.12 shows (a) the excited- and ground-state carrier populations f_b and f_a and (b) their ratio $\mathcal{R}$ as a function of the coupling-constant ratio η, for an interlevel energy splitting $\Delta\epsilon = 25\,\mathrm{meV}$ and for a reservoir-induced population ratio $\mathcal{R}^\circ = 3$ at room temperature. As we can see, for $\eta = 0$ (closed-system limit) the thermal-equilibrium value $\mathcal{R}^{\mathrm{eq}} = \frac{1}{e}$ is recovered. For increasing values of η we see a progressive increase of the population ratio, which becomes greater than 1 (population-inversion regime). For $\eta = 25$ the resulting population ratio is very close to the scattering-free value $\mathcal{R}^\circ = 3$, which tells us that in such regime the effect of interlevel phonon scattering is negligible.

As a final step, we shall now extend the closed-system approach previously introduced to the quantum-mechanical case. To this end, the Boltzmann-like equation (4.77) of the semiclassical theory needs to be replaced by the corresponding set of semiconductor Bloch equations, i.e., the equations of motion for the single-particle density matrix in (3.116); as shown in Sect. 3.3.3, for a closed system in the low-density limit the latter may be written according to (3.146), i.e.,

$$\frac{\mathrm{d}\rho_{\alpha_1\alpha_2}}{\mathrm{d}t} = \sum_{\alpha_1'\alpha_2'} \mathcal{L}_{\alpha_1\alpha_2,\alpha_1'\alpha_2'}\rho_{\alpha_1'\alpha_2'} \,. \tag{4.93}$$

Here, the effective single-particle Liouville superoperator $\mathcal{L}$ is the sum of two terms (see (3.147)): a coherent (i.e., scattering-free) single-particle contribution plus a scattering superoperator Γ^{d} describing energy dissipation and decoherence within the device active region.

Following again the paradigm of the present closed-system picture, the idea is to account for device–reservoir dissipation/dechoerence processes via an additional "ad hoc" scattering superoperator Γ^{dr} describing on a kinetic level the device–reservoir thermalization dynamics. More specifically, in analogy to (4.79), for the quantum-mechanical case we may write

$$\left.\frac{\mathrm{d}\rho_{\alpha_1\alpha_2}}{\mathrm{d}t}\right|_{\mathrm{dr}} = \sum_{\alpha_1'\alpha_2'} \Gamma^{\mathrm{dr}}_{\alpha_1\alpha_2,\alpha_1'\alpha_2'}\rho_{\alpha_1'\alpha_2'} \,. \tag{4.94}$$

Let us now try to identify the simplest form of the device–reservoir scattering superoperator Γ^{dr}. As a first requirement, we shall ask that in the semiclassical limit ($\rho_{\alpha_1\alpha_2} = f_{\alpha_1}\delta_{\alpha_1\alpha_2}$) the quantum equation (4.94) will reduce to the semiclassical result (4.79). This requires that

$$\Gamma^{\mathrm{dr}}_{\alpha\alpha,\alpha'\alpha'} = P^{\mathrm{dr}}_{\alpha\alpha'} - \delta_{\alpha\alpha'}\sum_{\alpha''} P^{\mathrm{dr}}_{\alpha''\alpha} \,. \tag{4.95}$$

Following the spirit of the well-known T_1/T_2 model introduced in Chap. 2 (see (2.40)), in addition to the semiclassical or T_1 terms in (4.95), the scattering superoperator Γ^{dr} should also contain decoherence or T_2 contributions;

the latter describe, in general, phase-breaking effects induced by the external reservoir on the carrier subsystem and will produce a damping of the non-diagonal density-matrix elements. According to the simplest formulation of the T_1/T_2 model, for a given non-diagonal term $\rho_{\alpha_1 \neq \alpha_2}$, the corresponding decoherence or damping rate is given by the average of the total out-scattering rates for states α_1 and α_2, i.e.,

$$\Gamma^{\mathrm{dr}}_{\alpha_1\alpha_2,\alpha_1\alpha_2} = -\frac{1}{2}\left(\sum_{\alpha'} P^{\mathrm{dr}}_{\alpha'\alpha_1} + \sum_{\alpha'} P^{\mathrm{dr}}_{\alpha'\alpha_2}\right) . \tag{4.96}$$

As discussed in [168], in addition to T_1 and T_2 terms, a generic scattering superoperator may also contain additional contributions describing non-trivial couplings between diagonal and non-diagonal density-matrix elements. Such extra-terms may lead to non-diagonal steady-state solutions (see Chap. 3). However, since in the absence of interaction mechanisms inside the simulated region we require a quasi-thermal, i.e., diagonal, steady-state solution, these extra-terms in the scattering superoperator Γ^{dr} are set equal to zero.

Combining the T_1 terms in (4.95) with the T_2 terms in (4.96), we finally get

$$\begin{aligned}\Gamma^{\mathrm{dr}}_{\alpha_1\alpha_2,\alpha_1'\alpha_2'} = {} & \delta_{\alpha_1\alpha_1',\alpha_2\alpha_2'}\left(P^{\mathrm{dr}}_{\alpha_1\alpha_1'} - \delta_{\alpha_1\alpha_1'}\sum_{\alpha''} P^{\mathrm{dr}}_{\alpha''\alpha_1}\right) \\ & -\frac{1}{2}\delta_{\alpha_1\alpha_2,\alpha_1'\alpha_2'}\left(\sum_{\alpha'} P^{\mathrm{dr}}_{\alpha'\alpha_1} + \sum_{\alpha'} P^{\mathrm{dr}}_{\alpha'\alpha_2}\right) .\end{aligned} \tag{4.97}$$

As we can see, the only ingredients entering the above scattering superoperator are the device–reservoir effective scattering rates (4.81).

We finally stress that – opposite to the open-system density-matrix formulation presented in Sect. 4.1 – the particle-preserving scattering superoperator Γ^{dr} ensures the positive-definite character of our single-particle density matrix, thus providing a reliable framework for the simulation of open quantum devices. This is confirmed by the fully quantum-mechanical investigation (not reviewed here) of the electrically driven quantum-dot system of Fig. 4.12 presented in [217].

5

Simulation Strategies

In this chapter we shall introduce basic concepts as well as key instruments related to the numerical modeling of semiconductor nanomaterials and nanodevices. The large variety of available numerical instruments may be subdivided into two major classes: (i) deterministic techniques and (ii) stochastic approaches. As we shall see, while the former are based on deterministic discretization algorithms, the latter are strongly linked to the use of random numbers. As anticipated in Sect. 2.6, the proper choice of the optimal modeling technique depends strongly on the problem under examination, i.e., semiclassical versus quantum-mechanical regimes described via phenomenological versus microscopic treatments (see Fig. 2.9); it follows that for specific problems, a proper combination of deterministic and stochastic algorithms is also required (see Sect. 5.3).

Generally speaking, irrespective of the nature of the specific problem under investigation, many of the key equations in semiconductor physics are integro-differential equations of the form

$$\frac{\partial F(X)}{\partial t} = \int^{\Omega} \mathrm{d}X' \mathcal{L}(X, X') F(X') + S(X) , \tag{5.1}$$

where X denotes a generic point in a n-dimensional real and/or momentum space Ω, and $F(X)$ may describe any complex field (as, e.g., the quantum-mechanical wavefunction) or any physical phase-space distribution (like, e.g., the distribution function of the Boltzmann theory). The operator $\mathcal{L}$ entering (5.1) is, in general, a nonlinear operator over the X-space, while the possible source term S may account for spatial boundary conditions related to a given simulated volume Ω_d (see Chaps. 1 and 4).

For the particular case of stationary solutions, i.e.,

$$F(X, t) = F(X) \mathrm{e}^{-\mathrm{i}\omega t} , \tag{5.2}$$

the time-dependent equation (5.1) reduces to the following eigenfrequency problem:

$$\int^{\Omega} \mathrm{d}X' \mathcal{L}(X, X') F(X') + S(X) = -\mathrm{i}\omega F(X) . \tag{5.3}$$

5.1 Direct or Deterministic Integration Techniques

In this section we shall recall the basic numerical techniques commonly employed for the direct or deterministic solution of the generic semiconductor equation (5.1) as well as of its frequency-domain counterpart (5.3), e.g., Schrödinger, Poisson, Wigner, and optical Bloch equations.

Generally speaking, a direct numerical solution of (5.1) within the continuous X-space is not the most convenient strategy; in contrast, as a general prescription, it is highly preferable to discretize our equation via the introduction of a proper set of orthonormal basis functions $\chi_j(X)$. More specifically, the key step is to expand the unknown solution $F(X)$ as linear combination of such basis functions:

$$F(X) = \sum_j F_j \chi_j(X) . \tag{5.4}$$

By inserting the above expansion into (5.1) and employing the orthonormality relation

$$\int^{\Omega} \mathrm{d}X \chi_j^*(X) \chi_{j'}(X) = \delta_{jj'} , \tag{5.5}$$

we finally get

$$\frac{\mathrm{d}F_j}{\mathrm{d}t} = \sum_{j'} \mathcal{L}_{jj'} F_{j'} + S_j \tag{5.6}$$

with

$$\mathcal{L}_{jj'} = \int^{\Omega} \mathrm{d}X \, \mathrm{d}X' \chi_j^*(X) \mathcal{L}(X, X') \chi_{j'}(X') \tag{5.7}$$

and

$$S_j = \int^{\Omega} \mathrm{d}X \chi_j^*(X) S(X) . \tag{5.8}$$

The general discretization procedure outlined so far allows us to transform the original integro-differential equation (5.1) into a discrete coupled set of ordinary differential equations.

For the particular case of stationary solutions (see (5.2)), by inserting the linear combination (5.4) into the eigenfrequency problem (5.3) we obtain

$$\sum_{j'} \mathcal{L}_{jj'} F_{j'} + S_j = -\mathrm{i}\omega F_j . \tag{5.9}$$

For the particular case of a linear operator $\mathcal{L}$ in the absence of source terms ($S_j = 0$) the set of coupled (5.9) reduces to the following eigenvalue problem:

$$\mathrm{i}\sum_{j'} \mathcal{L}_{jj'} F_{j'} = \omega F_j \,. \tag{5.10}$$

The optimal choice of the basis functions $\chi_j(X)$ depends strongly on the specific problem under examination and gives rise to the distinct discretization algorithms discussed below.

5.1.1 The Finite-Element/Finite-Difference Method

For spatially dependent problems in the presence of boundary source terms, the most effective discretization strategy is given by the so-called finite-element method [218–231]. The latter – sometimes referred to as finite-element analysis – is a numerical technique for finding approximate solutions of partial integro-differential equations. The method is based either on eliminating the original integro-differential equation completely (see the stationary problem (5.9)) or rendering the original integro-differential problem into an approximating system of ordinary differential equations (see the time-dependent problem (5.6)), which are then numerically integrated using standard techniques like, e.g., various Runge–Kutta methods (see Sect. 5.1.3).

Generally speaking, in solving partial integro-differential equations the primary challenge is to create an equation that approximates the equation to be studied, but is numerically stable, meaning that errors in the input data and intermediate calculations do not accumulate and cause the resulting output to be meaningless. There are many ways of doing this, all with advantages and disadvantages. The finite-element method is a good choice for solving partial differential equations over complicated domains (as, e.g., cars and oil pipelines), when the domain changes (as, e.g., during a solid-state reaction with a moving boundary), when the desired precision varies over the entire domain, or when the solution lacks smoothness.[1]

A particular but extremely relevant version of the finite-element method recalled so far is given by the well-known finite-difference technique [232–239]; the latter approximates the solutions to differential equations by replacing derivative expressions with approximately equivalent difference quotients. It is possible to show that the finite-difference approach is formally equivalent to choose as basis set $\chi_j(X)$ in (5.4) a number of Dirac delta functions centered on the points of a properly designed spatial grid:

$$\chi_j(X) = \delta(X - X_j) \,. \tag{5.11}$$

A detailed discussion of the finite-element as well as of the finite-difference methods is out of the scope of the present book; in what follows we shall

[1] For instance, in a frontal-crash simulation it is possible to increase prediction accuracy in "important" areas like the front of the car and reduce it in its rear (thus reducing the cost of the simulation). Another example would be the simulation of the weather pattern on the Earth, where it is more important to have accurate predictions over land than over the wide-open sea.

just underline the main features of the finite-difference scheme discussing its application to the one-dimensional version of the Boltzmann equation (1.67) written within the relaxation-time approximation (1.71):

$$\frac{\partial f}{\partial t} + v_k \frac{\partial f}{\partial z} + \frac{F}{\hbar}\frac{\partial f}{\partial k} = -\Gamma^{\rm coll}(f - f^\circ) \ . \tag{5.12}$$

As already pointed out in Sect. 1.4, this is a standard first-order boundary-condition problem in real space parameterized by the value of the wavevector k (see Appendix B), which requires to impose the desired spatial boundary conditions for the carrier distribution f at the left ($z = -\frac{l}{2}$) and right ($z = +\frac{l}{2}$) electric contacts (see Fig. 1.22), specifying, e.g., the value of the "incoming" carrier distribution.

According to the general prescription of the finite-difference method previously recalled, the continuous description z, k is replaced by a suitable phase-space discretization; more specifically, by adopting a uniform-grid scheme corresponding to an even number of points N_z and N_k, we have

$$z_{j_z} = \Delta z \left(j_z - \frac{N_z + 1}{2} \right) , \qquad k_{j_k} = \Delta k \left(j_k - \frac{N_k + 1}{2} \right) , \tag{5.13}$$

with $\Delta z = \frac{l}{N_z+1}$ and $\Delta k = \frac{2\pi}{N_k a_\circ}$, $a_\circ$ denoting the lattice period along z; the original distribution function $f(z, k)$ is then translated into its discretized version

$$f_{j_z, j_k} = f(z_{j_z}, k_{j_k}) \ . \tag{5.14}$$

According to the previous finite-difference discretization scheme, the original Boltzmann equation (5.12) is transformed into the following set of coupled equations:

$$\frac{{\rm d}f_{j_z,j_k}}{{\rm d}t} + v_{j_k} Q^z_{j_z,j_k} + \frac{F}{\hbar} Q^k_{j_z,j_k} = -\Gamma^{\rm coll}_{j_z,j_k}(f_{j_z,j_k} - f^\circ_{j_z,j_k}) \ , \tag{5.15}$$

where Q^z and Q^k denote so-called difference quotients, i.e., the discretized versions of the partial derivatives of f with respect to z and k.

More specifically, in order to properly include the spatial boundary values according to the general scheme of Fig. 1.22, the form of the spatial difference quotient Q^z depends on the sign of the wavevector k_{j_k} (left- or right-derivative scheme):

$$Q^z_{j_z,j_k} = \begin{cases} \frac{f_{j_z,j_k} - f_{j_z-1,j_k}}{\Delta z} & \text{for } k_{j_k} > 0 \\ \frac{f_{j_z+1,j_k} - f_{j_z,j_k}}{\Delta z} & \text{for } k_{j_k} < 0 \end{cases} \ . \tag{5.16}$$

Indeed, for positive values of k the difference quotient involves the left boundary value ($z_{j_z=0} = -\frac{l}{2}$), while for negative values the latter involves the right one ($z_{j_z=N_z+1} = +\frac{l}{2}$).

In contrast, in view of the intrinsic Brillouin zone periodicity $f_{j_z,j_k} = f_{j_z,j_k \pm N_k}$, the difference quotient Q^k is typically written via the following symmetrized-derivative scheme:

$$Q^k_{j_z,j_k} = \frac{f_{j_z,j_k+1} - f_{j_z,j_k-1}}{2\Delta k} . \tag{5.17}$$

The coupled set of ordinary differential equations (5.15) together with the explicit form of the difference quotients in (5.16) and (5.17) constitutes the basis of the finite-difference method applied to the one-dimensional Boltzmann equation (5.12). In particular, for steady-state conditions all the time derivatives entering (5.15) vanish, and the problem reduces to a coupled set of N_z times N_k linear equations for the unknown quantities f_{j_z,j_k}.

The discretization scheme discussed so far may be easily extended to the superlattice version of the Boltzmann equation introduced in Sect. 6.1 (see (6.9)). Indeed, this is exactly the numerical technique employed for the study of the quantum-well infrared photodetectors presented in Chap. 7.

Moreover, as already underlined in Sect. 4.2, the one-dimensional Wigner transport equation (4.27) may be regarded as the quantum-mechanical generalization of the semiclassical Boltzmann equation (5.12) in the ballistic limit (see also Appendix D); this suggests to employ also for the Wigner equation the same finite-difference scheme previously introduced. In particular, in view of the fact that the Wigner equation (4.27) contains the very same diffusion term of the Boltzmann equation (5.12), the latter may be treated via the same k-dependent difference quotient given in (5.16); in contrast, in view of the non-local character of the potential term (see Appendix D), the difference quotient (5.17) is replaced by a corresponding non-local operator (see, e.g., [85]). As originally pointed out in [213] and discussed extensively in Sect. 4.2, in order to obtain a correct solution of the Wigner transport equation via finite-difference calculations, it is crucial to adopt the very same real-space discretization scheme for both diffusion and potential contributions.

5.1.2 The Plane-Wave Expansion

As alternative discretization procedure – compared to the finite-element/difference method – let us now introduce the so-called plane-wave expansion; the latter is particularly suited for the solution of differential equations describing physical systems with spatial periodicity. To this end, let us rewrite the generic integro-differential equation (5.1) for the particular case of a three-dimensional real space ($X = \boldsymbol{r}$):

$$\frac{\partial F(\boldsymbol{r})}{\partial t} = \int^{\Omega} d^3r' \mathcal{L}(\boldsymbol{r}, \boldsymbol{r}') F(\boldsymbol{r}') + S(\boldsymbol{r}) . \tag{5.18}$$

As anticipated, the function $F(\boldsymbol{r})$ may describe either a quantum-mechanical amplitude or a physically measurable quantity.

For the first case, the system periodicity requires that for any symmetry translation $\boldsymbol{R}$ the squared modulus of the amplitude F be preserved, i.e., $|F|^2(\boldsymbol{r} + \boldsymbol{R}) = |F|^2(\boldsymbol{r})$, which is fully equivalent to impose the following generalized Bloch theorem (see (1.7) and (1.8)):

$$F_{\boldsymbol{k}}(\boldsymbol{r}+\boldsymbol{R}) = \mathrm{e}^{\mathrm{i}\boldsymbol{k}\cdot\boldsymbol{R}} F_{\boldsymbol{k}}(\boldsymbol{r}) \;, \tag{5.19}$$

$\boldsymbol{k}$ denoting any wavevector within the first Brillouin zone corresponding to the system periodicity volume $\Omega_\circ$.[2]

For the second case – that of a physically measurable quantity – the requirement dictated by the system periodicity is $F(\boldsymbol{r}+\boldsymbol{R}) = F(\boldsymbol{r})$; the latter corresponds to the Bloch condition (5.19) for $\boldsymbol{k} = \boldsymbol{0}$. It follows that for field amplitudes – e.g., quantum-mechanical wavefunctions – any wavevector $\boldsymbol{k}$ in (5.19) is allowed, while for genuine physical quantities – e.g., charge or field distributions – the only allowed Bloch wavevector is $\boldsymbol{k} = \boldsymbol{0}$.

Following the general prescription in (5.4), we shall now expand our real-space solution according to

$$F_{\boldsymbol{k}}(\boldsymbol{r}) = \sum_{\boldsymbol{G}} F_{\boldsymbol{G}+\boldsymbol{k}} \chi_{\boldsymbol{G}+\boldsymbol{k}}(\boldsymbol{r}) \;, \tag{5.20}$$

where

$$\chi_{\boldsymbol{G}+\boldsymbol{k}}(\boldsymbol{r}) = \frac{1}{\sqrt{\Omega}} \mathrm{e}^{\mathrm{i}(\boldsymbol{G}+\boldsymbol{k})\cdot\boldsymbol{r}} \tag{5.21}$$

is a set of plane waves corresponding to the reciprocal-lattice vectors $\boldsymbol{G}$, parameterized by the Bloch wavevector $\boldsymbol{k}$. It is easy to verify that the above plane-wave expansion fulfills the Bloch theorem (5.19).

By inserting the plane-wave expansion (5.20) into the integro-differential problem (5.18), we get the following set of coupled equations:

$$\frac{\mathrm{d}F_{\boldsymbol{G}+\boldsymbol{k}}}{\mathrm{d}t} = \sum_{\boldsymbol{G}'} \mathcal{L}_{\boldsymbol{G}\boldsymbol{G}',\boldsymbol{k}} F_{\boldsymbol{G}'+\boldsymbol{k}} + S_{\boldsymbol{G}+\boldsymbol{k}} \tag{5.22}$$

with

$$\mathcal{L}_{\boldsymbol{G}\boldsymbol{G}',\boldsymbol{k}} = \int^{\Omega} \mathrm{d}^3 r \, \mathrm{d}^3 r' \chi^*_{\boldsymbol{G}+\boldsymbol{k}}(\boldsymbol{r}) \mathcal{L}(\boldsymbol{r},\boldsymbol{r}') \chi_{\boldsymbol{G}'+\boldsymbol{k}}(\boldsymbol{r}') \tag{5.23}$$

and

$$S_{\boldsymbol{G}+\boldsymbol{k}} = \int^{\Omega} \mathrm{d}^3 r \chi^*_{\boldsymbol{G}+\boldsymbol{k}}(\boldsymbol{r}) S(\boldsymbol{r}) \;. \tag{5.24}$$

For the particular case of stationary solutions, i.e.,

$$F_{\boldsymbol{G}+\boldsymbol{k}}(t) = F_{\boldsymbol{G}+\boldsymbol{k}} \mathrm{e}^{-\mathrm{i}\omega t} \;, \tag{5.25}$$

the coupled set of differential equations (5.22) reduces to the following eigenfrequency problem:

$$\sum_{\boldsymbol{G}'} \mathcal{L}_{\boldsymbol{G}\boldsymbol{G}',\boldsymbol{k}} F_{\boldsymbol{G}'+\boldsymbol{k}} + S_{\boldsymbol{G}+\boldsymbol{k}} = -\mathrm{i}\omega F_{\boldsymbol{G}+\boldsymbol{k}} \;. \tag{5.26}$$

[2] Such system periodicity may correspond either to that of a real crystal lattice or to a so-called superperiodicity as for the case of periodically repeated nanostructures (see Sect. 1.2.2) called superlattices (see Figs. 1.17 and 1.18).

Moreover, for the case of a linear operator $\mathcal{L}$ and in the absence of source terms ($S = 0$), the original integro-differential equation (5.18) reduces to a standard eigenvalue problem:

$$\mathrm{i} \sum_{\boldsymbol{G}'} \mathcal{L}_{\boldsymbol{G}\boldsymbol{G}',\boldsymbol{k}} F_{\boldsymbol{G}'+\boldsymbol{k}} = \omega F_{\boldsymbol{G}+\boldsymbol{k}} \; . \tag{5.27}$$

As typical and extremely relevant application of the plane-wave expansion reviewed so far, we shall now discuss the numerical solution of the envelope-function Schrödinger equation introduced in Sect. 1.2.2 (see also Appendix A); for the case of a periodically repeated nanostructure, the latter is given by

$$\hat{H}^{\mathrm{env}} \overline{\psi}_{\boldsymbol{k}\overline{\nu}}(\boldsymbol{r}) \equiv \left[-\frac{\hbar^2 \nabla_{\boldsymbol{r}}^2}{2m^*} + \overline{V}(\boldsymbol{r}) \right] \overline{\psi}_{\boldsymbol{k}\overline{\nu}}(\boldsymbol{r}) = \epsilon_{\boldsymbol{k}\overline{\nu}} \overline{\psi}_{\boldsymbol{k}\overline{\nu}}(\boldsymbol{r}) \; . \tag{5.28}$$

Following the general prescription in (5.20), let us expand the unknown envelope function $\overline{\psi}$ as a linear combination of the plane-wave set (5.21):

$$\overline{\psi}_{\boldsymbol{k}\overline{\nu}}(\boldsymbol{r}) = \sum_{\boldsymbol{G}} c_{\boldsymbol{G}+\boldsymbol{k},\overline{\nu}} \chi_{\boldsymbol{G}+\boldsymbol{k}}(\boldsymbol{r}) \; . \tag{5.29}$$

By inserting this plane-wave expansion into the Schrödinger equation (5.28), the latter is readily transformed into the following eigenvalue problem:

$$\sum_{\boldsymbol{G}'} H^{\mathrm{env}}_{\boldsymbol{G}\boldsymbol{G}',\boldsymbol{k}} c_{\boldsymbol{G}'+\boldsymbol{k},\overline{\nu}} = \epsilon_{\boldsymbol{k}\overline{\nu}} c_{\boldsymbol{G}+\boldsymbol{k},\overline{\nu}} \tag{5.30}$$

with

$$H^{\mathrm{env}}_{\boldsymbol{G}\boldsymbol{G}',\boldsymbol{k}} = \int^{\Omega} \mathrm{d}^3 r \chi^*_{\boldsymbol{G}+\boldsymbol{k}}(\boldsymbol{r}) \hat{H}^{\mathrm{env}} \chi_{\boldsymbol{G}'+\boldsymbol{k}}(\boldsymbol{r}) \; . \tag{5.31}$$

By inserting into (5.31) the explicit form of the envelope-function Hamiltonian $\hat{H}^{\mathrm{env}}$, its plane-wave matrix elements come out to be:

$$H^{\mathrm{env}}_{\boldsymbol{G}\boldsymbol{G}',\boldsymbol{k}} = \frac{\hbar^2 |\boldsymbol{G}+\boldsymbol{k}|^2}{2m^*} \delta_{\boldsymbol{G}\boldsymbol{G}'} + \overline{V}_{\boldsymbol{G}-\boldsymbol{G}'} \tag{5.32}$$

with

$$\overline{V}_{\boldsymbol{G}-\boldsymbol{G}'} = \frac{1}{\Omega} \int^{\Omega} \mathrm{d}^3 r \overline{V}(\boldsymbol{r}) \mathrm{e}^{-\mathrm{i}(\boldsymbol{G}-\boldsymbol{G}')\cdot\boldsymbol{r}} \; . \tag{5.33}$$

The plane-wave solution of the original envelope-function Schrödinger equation (5.28) proposed so far is then equivalent to solve the eigenvalue problem (5.30), which amounts to a numerical diagonalization of the Hamiltonian matrix $H^{\mathrm{env}}_{\boldsymbol{G}\boldsymbol{G}',\boldsymbol{k}}$ in (5.32). More specifically, such diagonalization procedure needs to be performed for a suitable number of $\boldsymbol{k}$ values, which will allow one to obtain the desired electronic miniband structure $\epsilon_{\boldsymbol{k}\overline{\nu}}$ of the periodic nanostructure under examination (matrix eigenvalues) together with the corresponding wavefunction coefficients $c_{\boldsymbol{G}+\boldsymbol{k},\overline{\nu}}$ (matrix eigenvectors).

A closer inspection of the Hamiltonian matrix (5.32) shows that the basic ingredient entering the present plane-wave calculation is the potential matrix $\overline{V}_{G-G'}$ in (5.33), corresponding to the three-dimensional Fourier transform of the nanostructure confinement potential $\overline{V}(\boldsymbol{r})$ evaluated over the generic reciprocal-lattice vector $\boldsymbol{G} - \boldsymbol{G}'$.

At this point a few comments are in order.

Any numerical diagonalization procedure applied to the plane-wave Hamiltonian (5.32) requires to deal with a finite-size matrix; this, in turn, corresponds to a proper truncation of our basis set. To this end, the original plane-wave expansion (5.29) is limited to reciprocal-lattice values $\boldsymbol{G}$ energetically smaller than a given energy cutoff: $\epsilon_{\boldsymbol{G}} = \frac{\hbar^2 G^2}{2m^*} < \epsilon_{\text{c.o.}}$. It is therefore crucial to verify that the results (band diagrams and envelope functions) obtained via such a truncation are nearly independent from the energetic cutoff; more precisely, it is imperative to verify that the deviations from the exact results induced by this plane-wave truncation are within the desired numerical precision. Generally speaking, as for many other numerical strategies, the optimal choice of the truncated basis set is the result of a compromise between computational cost (which increases significantly with the size of the basis) and required numerical precision.

As discussed extensively in Sect. 1.2.2, the effective dimensionality of the problem depends strongly on the specific nanostructure under investigation: while for genuine quantum-dot systems a fully three-dimensional calculation is imperative, for nanomaterials of higher dimensionality – like quantum wires and wells – one may benefit from a factorization of the original three-dimensional real space into a parallel ($\boldsymbol{r}_{\parallel}$) times a perpendicular ($\boldsymbol{r}_{\perp}$) subspace (see (1.43)), and the previous plane-wave expansion is applied to the perpendicular/confinement subspace $\boldsymbol{r}_{\perp}$ only. It follows that systems of higher dimensionality require a much smaller plane-wave set, thus limiting significantly the computational effort, compared to the fully three-dimensional case of quantum-dot structures.

We finally stress that for nanostructures characterized by relatively high doping concentrations, the effective potential entering the Schrödinger equation (5.28) is the sum of the confinement potential previously considered and of an electrostatic contribution due to the specific doping profile and to the corresponding charge distribution, $\overline{V}(\boldsymbol{r}) = \overline{V}^c(\boldsymbol{r}) + \overline{V}^e(\boldsymbol{r})$; such electrostatic potential is obtained by solving the Poisson equation

$$-\frac{\varepsilon}{4\pi} \nabla_{\boldsymbol{r}}^2 \overline{V}^e(\boldsymbol{r}) = \rho(\boldsymbol{r}) \ , \tag{5.34}$$

where ε is the static dielectric constant and ρ denotes the total spatial charge density given by the doping profile plus the corresponding electronic charge distribution, i.e.,

$$\rho(\boldsymbol{r}) = \rho^{\mathrm{d}}(\boldsymbol{r}) - e \sum_{\boldsymbol{k}\overline{\nu}} f_{\boldsymbol{k}\overline{\nu}} \left|\overline{\psi}_{\boldsymbol{k}\overline{\nu}}(\boldsymbol{r})\right|^2 \ ; \tag{5.35}$$

here, $f_{\boldsymbol{k}\overline{\nu}}$ denotes the carrier distribution over our envelope-function states $\boldsymbol{k}\overline{\nu}$.[3] It follows that in the presence of a significant doping level, the evaluation of the nanostructure envelope functions $\overline{\psi}_{\boldsymbol{k}\overline{\nu}}$ as well as of the corresponding miniband diagram $\epsilon_{\boldsymbol{k}\overline{\nu}}$ requires a self-consistent solution of the following set of coupled (Schrödinger + Poisson) equations:

$$\left[-\frac{\hbar^2\nabla^2_{\boldsymbol{r}}}{2m^*}+\overline{V}^{\mathrm{c}}(\boldsymbol{r})+\overline{V}^{\mathrm{e}}(\boldsymbol{r})\right]\overline{\psi}_{\boldsymbol{k}\overline{\nu}}(\boldsymbol{r})=\epsilon_{\boldsymbol{k}\overline{\nu}}\overline{\psi}_{\boldsymbol{k}\overline{\nu}}(\boldsymbol{r})$$

$$-\frac{\varepsilon}{4\pi}\nabla^2_{\boldsymbol{r}}\overline{V}^{\mathrm{e}}(\boldsymbol{r})=\rho^{\mathrm{d}}(\boldsymbol{r})-e\sum_{\boldsymbol{k}\overline{\nu}}f_{\boldsymbol{k}\overline{\nu}}\left|\overline{\psi}_{\boldsymbol{k}\overline{\nu}}(\boldsymbol{r})\right|^2 . \tag{5.36}$$

For the case of a periodic nanostructure, both the electrostatic potential and the doping charge profile may be expanded via the $\boldsymbol{k}=\boldsymbol{0}$ plane-wave set as well:

$$\overline{V}^{\mathrm{e}}(\boldsymbol{r})=\sum_{\boldsymbol{G}}\overline{V}^{\mathrm{e}}_{\boldsymbol{G}}\mathrm{e}^{\mathrm{i}\boldsymbol{G}\cdot\boldsymbol{r}} , \qquad \rho^{\mathrm{d}}(\boldsymbol{r})=\sum_{\boldsymbol{G}}\rho^{\mathrm{d}}_{\boldsymbol{G}}\mathrm{e}^{\mathrm{i}\boldsymbol{G}\cdot\boldsymbol{r}} . \tag{5.37}$$

By inserting the plane-wave expansions (5.29) and (5.37) into the Schrödinger/Poisson set (5.36), the latter is transformed into the following set of coupled linear equations for the plane-wave coefficients $c_{\boldsymbol{G}+\boldsymbol{k},\overline{\nu}}$:

$$\frac{\hbar^2|\boldsymbol{G}+\boldsymbol{k}|^2}{2m^*}c_{\boldsymbol{G}+\boldsymbol{k},\overline{\nu}}+\sum_{\boldsymbol{G}'}\left(\overline{V}^{\mathrm{c}}_{\boldsymbol{G}-\boldsymbol{G}'}+\overline{V}^{\mathrm{e}}_{\boldsymbol{G}-\boldsymbol{G}'}\right)c_{\boldsymbol{G}'+\boldsymbol{k},\overline{\nu}}=\epsilon_{\boldsymbol{k}\overline{\nu}}c_{\boldsymbol{G}+\boldsymbol{k},\overline{\nu}},$$

$$\overline{V}^{\mathrm{e}}_{\boldsymbol{G}}=\frac{4\pi}{\varepsilon G^2}\left(\rho^{\mathrm{d}}_{\boldsymbol{G}}-\frac{e}{\Omega}\sum_{\boldsymbol{k}\overline{\nu}}f_{\boldsymbol{k}\overline{\nu}}\sum_{\boldsymbol{G}'}c^*_{\boldsymbol{G}'+\boldsymbol{k},\overline{\nu}}c_{\boldsymbol{G}+\boldsymbol{G}'+\boldsymbol{k},\overline{\nu}}\right) . \tag{5.38}$$

The above system of coupled equations consists of an eigenvalue problem (which provides carrier energies and envelope functions) plus an algebraic version of the Poisson equation obtained via our plane-wave picture. They are typically solved by means of the following iterative procedure. As initial condition, one evaluates the carrier eigenstates by solving the eigenvalue problem in the absence of internal potential ($\overline{V}^{\mathrm{e}}_{\boldsymbol{G}}=0$); the resulting zero-order eigensolutions (i.e., wavefunction coefficients $c_{\boldsymbol{G}+\boldsymbol{k},\overline{\nu}}$) are then inserted into the Poisson equation which, in turn, provides the first-order internal potential $\overline{V}^{\mathrm{e}}_{\boldsymbol{G}}$. The above two-step procedure is repeated until convergence is achieved.

Compared to more conventional real-space Schrödinger–Poisson calculations (see, e.g., [28] and references therein), the use of the above plane-wave description – possible, however, for periodic nanosystems only – is rather stable and efficient; indeed, this is confirmed by the vertical-transport analysis

[3] For equilibrium or quasiequilibrium problems the carrier distribution $f_{\boldsymbol{k}\overline{\nu}}$ coincides with the usual Fermi–Dirac distribution, while for nonequilibrium regimes the latter is given by the instantaneous solution of a corresponding Boltzmann equation (see, e.g., [28] and references therein).

originally proposed in [240] and reviewed in Chap. 6 as well as by the investigation of quantum-well infrared photodetectors and quantum-cascade lasers presented in Chaps. 7 and 8.

As additional application, let us finally show the power and flexibility of the plane-wave expansion for the study of Coulomb-correlation effects in low-dimensional nanostructures. As discussed extensively in Chap. 9, the key ingredients for the investigation of such many-body phenomena are the two-body Coulomb matrix elements in (9.7):

$$V^{\mathrm{cc}}_{\alpha_1\alpha_2,\alpha_1'\alpha_2'} = \int \mathrm{d}^3r \int \mathrm{d}^3r' \overline{\psi}^*_{\overline{\alpha}_1}(\boldsymbol{r})\overline{\psi}^*_{\overline{\alpha}_2}(\boldsymbol{r}')V^{\mathrm{cc}}(\boldsymbol{r}-\boldsymbol{r}')\overline{\psi}_{\overline{\alpha}_2'}(\boldsymbol{r}')\overline{\psi}_{\overline{\alpha}_1'}(\boldsymbol{r}) \,. \tag{5.39}$$

Generally speaking, a real-space numerical evaluation of the above six-dimensional integral is extremely demanding. As alternative strategy, let us Fourier expand the two-body Coulomb potential V^{cc} with respect to the relative coordinate $\boldsymbol{r}-\boldsymbol{r}'$:

$$V^{\mathrm{cc}}(\boldsymbol{r}-\boldsymbol{r}') = \int \mathrm{d}^3q V^{\mathrm{cc}}(\boldsymbol{q})\mathrm{e}^{\mathrm{i}\boldsymbol{q}\cdot(\boldsymbol{r}-\boldsymbol{r}')} \,. \tag{5.40}$$

By inserting the above Fourier integral into (5.39), our Coulomb matrix elements may be rewritten as

$$V^{\mathrm{cc}}_{\alpha_1\alpha_2,\alpha_1'\alpha_2'} = \int \mathrm{d}^3q F^*_{\overline{\alpha}_1'\overline{\alpha}_1}(\boldsymbol{q})V^{\mathrm{cc}}(\boldsymbol{q})F_{\overline{\alpha}_2\overline{\alpha}_2'}(\boldsymbol{q}) \tag{5.41}$$

with

$$F_{\overline{\alpha}\overline{\alpha}'}(\boldsymbol{q}) = \int \mathrm{d}^3r\overline{\psi}^*_{\overline{\alpha}}(\boldsymbol{r})\mathrm{e}^{-\mathrm{i}\boldsymbol{q}\cdot\boldsymbol{r}}\overline{\psi}_{\overline{\alpha}'}(\boldsymbol{r}) \,. \tag{5.42}$$

Thanks to the Fourier expansion (5.40) of the two-body Coulomb potential, we realize a factorization of the two real-space coordinates $\boldsymbol{r}$ and $\boldsymbol{r}'$; indeed, in this way the evaluation of the Coulomb matrix elements (5.39) reduces to the evaluation of the single-particle form factors in (5.42).

For the case of a periodic nanostructure, the single-particle label $\overline{\alpha}$ is given by the three-dimensional Bloch vector $\boldsymbol{k}$ plus the miniband index $\overline{\nu}$ ($\overline{\alpha} \equiv \boldsymbol{k}\overline{\nu}$) and the corresponding envelope function ($\overline{\psi}_{\overline{\alpha}} = \overline{\psi}_{\boldsymbol{k}\overline{\nu}}$) may be expressed again as a linear combination of the $\boldsymbol{k}$-dependent plane-wave set (5.21). More specifically, by inserting the plane-wave expansion (5.29) into the single-particle form factor (5.42), it is easy to get the following result

$$F_{\boldsymbol{k}\overline{\nu},\boldsymbol{k}'\overline{\nu}'}(\boldsymbol{q}) = \sum_{\boldsymbol{G},\boldsymbol{G}'} c^*_{\boldsymbol{G}+\boldsymbol{k},\overline{\nu}}c_{\boldsymbol{G}'+\boldsymbol{k}',\overline{\nu}'}\delta((\boldsymbol{G}'+\boldsymbol{k}')-(\boldsymbol{G}+\boldsymbol{k})-\boldsymbol{q}) \,. \tag{5.43}$$

This clearly shows that the evaluation of the single-particle form factors (5.42) – and therefore of the two-body Coulomb matrix elements (5.39) – may be directly performed in terms of the plane-wave coefficients $c_{\boldsymbol{G}+\boldsymbol{k},\overline{\nu}}$ only.

5.1.3 The Time-Step Integration

As anticipated, while for stationary conditions (see (5.2)) the original integro-differential problem is transformed into the eigenfrequency equation (5.9), for transient regimes the result of the discretization procedure previously introduced (see (5.4)) is the coupled set of ordinary differential equations in (5.6); the latter may also be written in a compact form as

$$\frac{\mathrm{d}F_j(t)}{\mathrm{d}t} = A_j\left(\{F_j(t)\}, t\right) , \tag{5.44}$$

where

$$A_j\left(\{F_j(t)\}, t\right) = \sum_{j'} \mathcal{L}_{jj'}(t)F_{j'}(t) + S_j(t) \tag{5.45}$$

is, in general, a nonlinear and nonhomogeneous function of all the F_j.

In order to perform a numerical solution of this set of ordinary differential equations, the standard approach commonly employed is the so-called time-step integration (see, e.g., [241] and references therein). In its simplest formulation, the latter is based on a uniform time discretization. More specifically, by denoting with Δt the uniform elementary time step, the corresponding integration grid is given by: $t_n = n\Delta t$.

By employing a simple first-order integration scheme, one gets

$$F_j(t_n) = F_j(t_{n-1}) + \Delta t\, A_j\left(\{F_j(t_{n-1})\}, t_{n-1}\right) . \tag{5.46}$$

Within such basic integration procedure the value of F_j at time t_n is approximated by F_j at the previous time t_{n-1} plus the time-step Δt multiplied by the function A_j evaluated at the previous time t_{n-1}.

While for norm-nonpreserving operators A_j (as, e.g., time-dependent Schrödinger equations) the first-order integration scheme (5.46) may be highly inadequate, for norm-preserving operators (as, e.g., Boltzmann as well as Wigner transport equations) the latter is well suited, provided to choose appropriate time-step values. Indeed, the simple first-order prescription (5.46) is the time-step integration scheme employed for many of the simulated experiments discussed in this book (see Chaps. 4, 8, and 10).

For norm-nonpreserving problems as well as for extremely demanding simulation tasks, the first-order integration previously introduced may be conveniently replaced by a variety of higher-order integration schemes known as Runge–Kutta methods (see, e.g., [241] and references therein); in particular, the most popular time-step integration scheme is the so-called fourth-order Runge–Kutta (RK4) method.

5.2 Monte Carlo or Stochastic Sampling

In this section we shall recall the fundamentals of the Monte Carlo method. More specifically, we shall introduce the so-called Weighted Monte Carlo method, a quite general simulation strategy for the stochastic solution of any set of differential equations.

5.2.1 Two Different Points of View About Monte Carlo

The name Monte Carlo was originally introduced by scientists of the Manhattan Project in the 1940s. The essence of the Monte Carlo method [242–253] is the invention of games of chance whose behavior and outcome can be used to study some interesting phenomena. While there is no essential link to computers, the effectiveness of numerical or simulated gambling as a serious scientific pursuit has been enormously enhanced by the availability of modern digital computers. It is interesting, and may strike some as remarkable, that carrying out games of chance or random sampling will produce anything worthwhile. Indeed, in the past some authors have claimed that Monte Carlo will never be a method of choice for other than rough estimates of numerical quantities. Such criticisms have been definitely removed by the enormous success of the Monte Carlo method in many diverse research areas [254–271], mainly stimulated by its cross-disciplinal nature and rather intuitive implementation.

As a first and extremely popular example of Monte Carlo sampling, let us consider a circle and its circumscribed square. The ratio of the area of the circle to the area of the square is $\frac{\pi}{4}$. It is plausible that if points were placed at random in the square, a fraction $\frac{\pi}{4}$ would also lie inside the circle. If this is true (and we shall prove later that in a certain sense it is), then one could measure $\frac{\pi}{4}$ by putting a round cake pan with diameter L inside a square cake pan with side L and collecting rain in both. It is also possible to program a computer to generate random pairs of cartesian coordinates to represent random points in the square and count the fraction that lie in the circle. This fraction as determined from many experiments should be close to $\frac{\pi}{4}$, and the latter is called an estimate for $\frac{\pi}{4}$.

This tutorial example illustrates that random sampling may be used to solve a well-defined mathematical problem, in this case the evaluation of the area of a circle with unitary diameter. This is the so-called mathematical point of view about Monte Carlo: as discussed in Sect. 5.2.3, a deterministic problem, e.g., the evaluation of sums and integrals, may be performed via a stochastic algorithm. The estimations obtained in this way are statistical in nature and subjected to the laws of chance. This aspect of Monte Carlo is a drawback, but not a fatal one, since we can always determine how accurate the answer is (see Sect. 5.2.3).

A second and complementary example is the estimation of the chances of winning at solitaire. On the one hand, assuming the deck is perfectly shuffled before laying out the cards and choosing a particular strategy for placing one pile of cards on another, the problem is a straightforward one in elementary probability theory, but it is also a very tedious one. On the other hand, it is straightforward to program a computer to randomize lists representing the 52 cards of a deck, prepare lists representing the different piles, and then simulate the playing of the game to completion. Collecting statistics over many repetitions would lead to a Monte Carlo estimate of the chance of success. This method would in fact be the easiest way of making any such estimate. In this case we may regard the computer gambling as a "faithful" simulation of the real random process under examination, namely, the card shuffling. This is the so-called physical point of view about Monte Carlo: such a stochastic sampling may be regarded as a computer simulation of a real/natural stochastic process; as such, the latter comes out to be highly cross-disciplinal and intuitive, but its application is limited to a classical or semiclassical description of genuine stochastic phenomena (see (5.2.4)).

As a final remark, we stress that the above subdivision – between mathematical and physical Monte Carlo samplings – is somewhat artificial and arbitrary. Indeed, as we shall see in Sect. 5.2.5, it is possible to introduce a quite general simulation framework – often referred to as Weighted Monte Carlo – for the solution of any given differential equation, regardless of its physical interpretation. If the equation under investigation describes a natural stochastic process (as for the case of Boltzmann-like treatments), this mathematical sampling reduces to the direct simulation of such phenomenon (see Sect. 5.2.4); in contrast, if the differential equation under examination does not correspond to a natural stochastic process (as for the case of Schrödinger-like problems), it is still possible to solve it via the Weighted Monte Carlo method, but in this case there is no physical counterpart of the random process employed in the simulation, and the latter is by far less intuitive.

5.2.2 A Bit of Probability Theory

Generally speaking, any Monte Carlo calculation is a numerical stochastic process, i.e., it is a sequence of random events. While we shall not discuss the philosophical question of what random events are, we shall assume that they do exist and that it is possible and useful to organize a computer code to produce effective equivalents of natural random processes. The most popular examples of random events are the flip of a coin and the roll of a die. When the output of a random event is a number, the latter is called random variable.

Let us start by considering a discrete random variable g, i.e., a random variable characterized by a discrete set of possible outcomes g_i; we shall associate to each outcome a given probability p_i such that

$$0 < p_i \leq 1 , \qquad \sum_i p_i = 1 . \tag{5.47}$$

The most important statistical property of such a random variable is its expectation value:

$$\langle g \rangle = \sum_i g_i p_i . \tag{5.48}$$

For the case of a continuous random variable, i.e., a random variable characterized by a continuous set of possible outcomes $g(x)$, we shall associate to the generic continuous outcome a given probability density $f(x)$ such that

$$f(x) \geq 0 \;, \qquad \int f(x)dx = 1 \;. \tag{5.49}$$

More precisely, the probability density $f(x)$ is defined as the derivative of the so-called cumulative probability distribution $F(x)$ (see, e.g., [252]), i.e.,

$$f(x) = \frac{\mathrm{d}F(x)}{\mathrm{d}x} \;; \tag{5.50}$$

the latter, in turn, is defined as the probability of getting as random outcome any value smaller than x. The expectation value of a continuous random variable is obtained by replacing in (5.48) the discrete index i with the continuous variable x, i.e.,

$$\langle g \rangle = \int g(x)\mathrm{d}F(x) = \int g(x)f(x)\mathrm{d}x \;. \tag{5.51}$$

For both discrete and continuous random variables, the magnitude of the statistical fluctuations around their expectation value $\langle g \rangle$ is described by the so-called statistical variance; the latter is defined as the expectation value of the square of the fluctuation $\Delta g = g - \langle g \rangle$, i.e.,

$$\mathrm{var}(g) = \langle (g - \langle g \rangle)^2 \rangle = \langle g^2 \rangle - \langle g \rangle^2 \;. \tag{5.52}$$

The square root of the variance is typically referred to as the standard deviation.

For many concrete statistical problems, both expectation values and corresponding variances are not known; it is thus imperative to obtain a numerical estimation of these quantities. To this aim, let us consider the arithmetic average (or mean) of N random samplings generated according to the corresponding probability distribution:

$$G_N = \frac{1}{N}\sum_{l=1}^{N} g_l \tag{5.53}$$

(Here, for discrete random variables $g_l \equiv g_{i_l}$, while for continuous ones $g_l \equiv g(x_l)$). Such arithmetic average is itself a random variable, and its expectation value coincides with that of the original one: $\langle G_N \rangle = \langle g \rangle$; this suggests to regard the quantity in (5.53) as an estimate of the theoretical expectation value $\langle g \rangle$; it is then crucial to establish the degree of accuracy of G_N, quantitatively described by its statistical variance. Taking into account that the N samplings in (5.53) are statistically uncorrelated (see, e.g., [252]), we have

$$\mathrm{var}(G_N) = \mathrm{var}\left(\frac{1}{N}\sum_{l=1}^{N} g_l\right) = \frac{1}{N^2}\sum_{l=1}^{N} \mathrm{var}(g) = \frac{\mathrm{var}(g)}{N} \;. \tag{5.54}$$

As the number N of samplings increases, the variance of G_N decreases as $\frac{1}{N}$; this tells us that for any random variable with finite variance, the statistical error $(G_N - \langle g \rangle)$ may be progressively reduced by simply increasing the sampling size. In the limit $N \to \infty$ we have

$$\lim_{N\to\infty} \mathrm{var}(G_N) = 0 \ , \qquad \lim_{N\to\infty} G_N = \langle g \rangle \ . \tag{5.55}$$

This constitutes the essence of the well-known central-limit theorem (see, e.g., [252]). More specifically, for any given N it is possible to define a probability distribution $P_N(G_N)$ which describes the probability to get from N samplings the arithmetic average G_N; it is possible to show that in the limit $N \to \infty$ such probability P_N approaches a Gaussian (or normal) distribution

$$\lim_{N\to\infty} P_N(G_N) \propto \mathrm{e}^{-\frac{|G_N - \langle g \rangle|^2}{2\sigma_N^2}} \tag{5.56}$$

with a standard error[4]

$$\sigma_N = \sqrt{\mathrm{var}(G_N)} = \sqrt{\frac{\mathrm{var}(g)}{N}} \ . \tag{5.57}$$

It is important to stress that this error is statistical in nature, i.e., the best one can do is to predict the probability that a certain sampling will produce an estimation error $|G_N - \langle g \rangle|$ smaller than a given precision value.[5]

Let us now focus on the main task related to stochastic calculations: the numerical algorithms for the generation of random variables. As a matter of fact, any modern digital computer provides a routine for the generation of random numbers[6] uniformly distributed over the interval $(0,1)$ (see, e.g., [28, 252] and references therein). The main question is then how to transform this available source of uniformly distributed numbers into random numbers distributed according to a given probability profile.

To this end, let us start by considering a continuous random variable x distributed according to the probability density $f(x)$. In this case, the most efficient algorithm is the so-called direct-generation technique. As discussed in detail in [252], by denoting with R a random number uniformly distributed over $(0,1)$, the general prescription is

$$x(R) = F^{-1}(R) \ , \tag{5.58}$$

---

[4] In general, the variance $\mathrm{var}(g)$ – and therefore the standard error σ_N – is not known; however, as discussed in more detail in Sect. 5.2.3, the latter can be estimated within the same random sampling as well.

[5] Indeed, according to elementary statistics, the probability of getting an estimation error less than σ_N, $2\sigma_N$, and $3\sigma_N$ is, respectively, about 68.3, 95.4, and 99.7%.

[6] In general, such a routine produces "pseudo-random numbers," i.e., starting from a given initial entry (called "seed") the routine produces a well-established periodic sequence of uncorrelated numbers uniformly distributed over $(0,1)$; it is also possible to work with genuine random numbers – generated employing, e.g., current time/date – but pseudo-random numbers are highly preferable since the whole simulation process remains deterministic like.

where F^{-1} denotes the inverse of the cumulative probability distribution $F(x)$ introduced in (5.50).

A typical and highly relevant example of such direct-generation technique is the random generation of scattering times distributed according to an exponential profile, i.e.,

$$f(t) = \Gamma e^{-\Gamma t} , \qquad F(t) = \int_0^t f(t')dt' = 1 - e^{-\Gamma t} . \tag{5.59}$$

By inverting the cumulative distribution $F(t)$ and taking into account that – in view of its uniform character – $R = 1 - R$, it is easy to get the following prescription for the generation of random scattering times parameterized by the inverse life-time Γ:

$$t(R) = -\frac{1}{\Gamma} \log(R) . \tag{5.60}$$

This result plays a central role for the direct Monte Carlo simulation of stochastic processes discussed in Sect. 5.2.4.

The application of the direct-generation method requires to know analytically the cumulative distribution $F(x)$ as well as its inverse. If this is not the case, a second – less efficient but more flexible – strategy is the well-known rejection technique. Provided to work with a non-singular probability density $f(x)$ characterized by a maximum value f^{max}, this second generation strategy proceeds as follows:

(i) We generate with uniform probability a random number $\overline{x}$ within our probability domain.
(ii) We generate with uniform probability a second random number $\overline{f}$ between 0 and f^{max}.[7]
(iii) We compare $\overline{f}$ with $f(\overline{x})$:
  a) if $\overline{f} \le f(\overline{x})$ the number $\overline{x}$ is accepted.
  b) if $\overline{f} > f(\overline{x})$ the latter is rejected and a new pair $\overline{x}, \overline{f}$ is generated (see steps (i) and (ii)).

For many problems characterized by multi-dimensional simulation domains with non-trivial shapes, the most efficient random-generation strategy comes out to be a proper combination of both direct and rejection techniques (see, e.g., [28, 252] and references therein).

Let us finally come to the generation of discrete random variables. To this end, it is important to notice that, given a random number R uniformly distributed over the interval $(0, 1)$, the probability that the latter will lie within a generic subinterval (x_1, x_2) is equal to its length $(x_2 - x_1)$. Keeping this in mind, for any discrete random variable i characterized by a probability

[7] It is not necessary to know the maximum of $f(x)$; the method works also with any overestimation f^{max} of it.

distribution p_i we shall associate to each outcome i a subinterval of length p_i. It follows that, in order to sample the discrete number i, it is enough to generate a random number R (uniformly distributed over $(0,1)$) and to determine the selected subinterval. More specifically, by defining the cumulative probability[8]

$$F_i = \sum_{i'=1}^{i} p_i \; , \tag{5.61}$$

we shall look for the outcome i such that

$$F_{i-1} \leq R \leq F_i \; . \tag{5.62}$$

This is typically accomplished by performing a loop over the possible outcomes i up to the fulfillment of the above inequality. In order to improve the efficiency of this algorithm, it is convenient to order the possible outcomes according to their probabilities: $p_1 \geq p_2 \geq p_3 \ldots$ (see, e.g., [28]).

5.2.3 Monte Carlo Sampling of Sums and Integrals

The simplest example of Monte Carlo sampling within the mathematical point of view previously recalled (see Sect. 5.2.1) is the Monte Carlo evaluation of a sum:

$$S = \sum_i a_i \; . \tag{5.63}$$

To this end, let us consider a given probability distribution p_i such to fulfill the general requirements in (5.47). The key idea is to rewrite the sum in (5.63) as

$$S = \sum_i a_i = \sum_i \left(\frac{a_i}{p_i} \right) p_i \; . \tag{5.64}$$

If we now compare this last expression for the sum with the definition of the expectation value of a discrete random variable in (5.48), our original sum (5.63) may also be written as the expectation value of a new random variable e:

$$S = \langle e \rangle = \sum_i e_i p_i \; , \qquad e_i = \frac{a_i}{p_i} \; . \tag{5.65}$$

Recalling that the theoretical expectation value of any random variable may be approximated by a mathematical average of N random samplings (see (5.53), i.e.,

$$\langle e \rangle \approx E_N = \frac{1}{N} \sum_{l=1}^{N} e_{i_l} \tag{5.66}$$

$(e_{i_1}, e_{i_2}, \ldots, e_{i_N}$ denoting N random samplings of e),[9] we finally obtain

[8] The cumulative probability (5.61) may be regarded as the discrete version of the cumulative distribution $F(x)$ introduced in (5.50).

[9] We stress that the label l runs over the total number of samplings (from 1 to N) while the label i runs over the possible outcomes of our probability distribution p_i, corresponding to the addenda entering the sum in (5.63).

$$S = \sum_i a_i \approx \frac{1}{N} \sum_{l=1}^{N} e_{i_l} = \frac{1}{N} \sum_{l=1}^{N} \frac{a_{i_l}}{p_{i_l}} \ . \tag{5.67}$$

Given the above prescription, the central question is to estimate how big is the error performed within such a statistical sampling. From the central-limit theorem recalled in Sect. 5.2.2 we know that the estimated value is distributed according to a Gaussian (or normal) distribution (see (5.56); its deviation from the theoretical value S is typically expressed via the so-called statistical error (see (5.57)):

$$\Delta S = \sqrt{\mathrm{var}(E_N)} = \sqrt{\frac{\mathrm{var}(e)}{N}} \ . \tag{5.68}$$

In general, the variance var(e) of our estimator is not known[10]; however, the latter can be estimated within our Monte Carlo sampling as well. More specifically, taking into account that

$$\left\langle \frac{1}{N} \sum_l e_{i_l}^2 - \left[\frac{1}{N} \sum_l e_{i_l} \right]^2 \right\rangle = \langle e^2 \rangle - \frac{1}{N^2} \left\langle \sum_l e_{i_l}^2 + \sum_{l \neq l'} e_{i_l} e_{i_{l'}} \right\rangle \tag{5.69}$$

and recalling that $\langle e_{i_l} e_{i_{l'}} \rangle = \langle e_{i_l} \rangle \langle e_{i_{l'}} \rangle$, we get

$$\begin{aligned} \left\langle \frac{1}{N} \sum_l e_{i_l}^2 - \left[\frac{1}{N} \sum_l e_{i_l} \right]^2 \right\rangle &= \left(1 - \frac{1}{N}\right) \langle e^2 \rangle - \frac{N(N-1)}{N^2} \langle e \rangle^2 \\ &= \frac{N-1}{N} \mathrm{var}(e) = (N-1) \Delta S^2 \ . \end{aligned} \tag{5.70}$$

A good estimator for ΔS is then given by

$$\Delta S \approx \left\{ \frac{1}{N-1} \left[\frac{1}{N} \sum_l e_{i_l}^2 - \left(\frac{1}{N} \sum_l e_{i_l} \right)^2 \right] \right\}^{\frac{1}{2}} \ . \tag{5.71}$$

The general structure of a typical computer code for the Monte Carlo evaluation of a sum may be schematized as follows:

(a) Initialization of the two estimators s_1 and s_2 for S and ΔS.
(b) Loop over the random samplings ($l = 1, 2, \ldots, N$):
   (i) random generation of the index i_l with probability p_{i_l};
   (ii) evaluation of the estimator $e_{i_l} = \frac{a_{i_l}}{p_{i_l}}$;
   (iii) updating of the two estimators, i.e.,

$$s_1 = s_1 + e_{i_l} \ , \qquad s_2 = s_2 + e_{i_l}^2 \ .$$

[10] It is imperative to point out that, in order to rely on the central-limit theorem, the variance of our estimator needs to be finite (see, e.g., [252]); this requirement, always fulfilled for a finite number of addenda, may be problematic for the case of an infinite summation.

(c) Final averages

$$S = \frac{s_1}{N}, \qquad \Delta S = \left(\frac{\frac{s_2}{N} - \left(\frac{s_1}{N}\right)^2}{N-1} \right)^{\frac{1}{2}} . \tag{5.72}$$

Let us now come to the role played by the choice of the probability distribution. According to the general prescription, any probability distribution p_i gives a correct result S within the statistical error ΔS; however, a good choice of p_i may strongly reduce the statistical fluctuations. As a general prescription – formally derived at the end of this section – in order to reduce the statistical error it is crucial to limit the fluctuations of our random outcomes $e_i = \frac{a_i}{p_i}$ by choosing p_i as close as possible to $|a_i|$.

In order to show effectiveness versus limitations of such a strategy, let us consider the particular case of positive-definite addenda ($a_i > 0$). By choosing $p_i \propto a_i$ we get

$$e_i = \frac{a_i}{p_i} = \text{constant} , \qquad \text{var}(e) = 0 . \tag{5.73}$$

In this case we have no fluctuations at all: a single random sampling gives the right answer. However, this is just an illusion, since the evaluation of such "ideal" probability distribution

$$p_i = \frac{a_i}{\sum_i a_i} = \frac{a_i}{S} \tag{5.74}$$

requires itself the evaluation of the original sum S in (5.63). A realistic choice, in contrast, is performed by adopting a probability distribution p_i whose expression is known analytically and is as close as possible to the distribution of the a_i.

In order to provide a concrete and highly tutorial implementation of such a Monte Carlo sampling, let us consider the simplest example: the sum of two numbers. As a starting point let us consider the sum of two positive-definite addenda:

$$S = 9 + 1 . \tag{5.75}$$

Following the general prescription outlined above,

(i) we choose the probabilities $p^{(9)}$ and $p^{(1)} = 1 - p^{(9)}$ corresponding to the two elements 9 and 1;
(ii) we compute the two possible outcomes of the estimator $e^{(9)} = \frac{9}{p^{(9)}}$ and $e^{(1)} = \frac{1}{p^{(1)}}$;
(iii) we obtain an estimate of S by performing N random samplings of e.

More specifically, let us consider three different choices of the probability distribution, choice 1, 2, and 3, detailed in Table 5.1; for each of them we have performed a corresponding Monte Carlo sampling, whose results are reported in Figs. 5.1, 5.2, 5.3, and 5.4. As expected, for all three choices our

Table 5.1. Three different choices of the probability distribution employed for the Monte Carlo sampling of the sum in (5.75) reported in Figs. 5.1, 5.2, 5.3, and 5.4 (see text)

choice	$p^{(9)}$	$p^{(1)}$	$e^{(9)}$	$e^{(1)}$
1	$\frac{1}{2}$	$\frac{1}{2}$	18	2
2	$\frac{8}{9}$	$\frac{1}{9}$	$\frac{81}{8}$	9
3	$\frac{1}{10}$	$\frac{9}{10}$	90	$\frac{10}{9}$

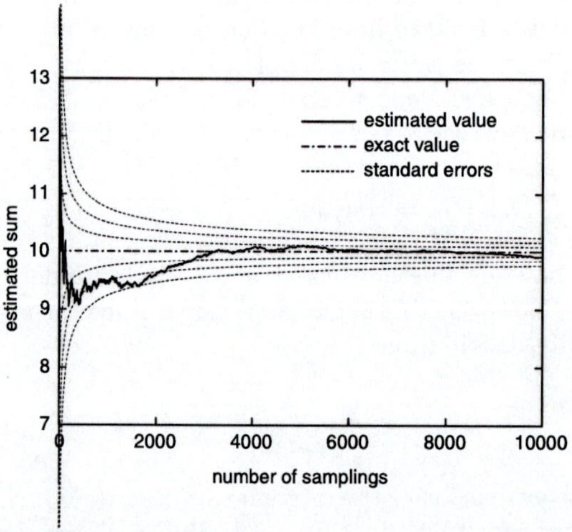

Fig. 5.1. Monte Carlo estimation of the sum in (5.75) as a function of the number of samplings corresponding to the choice 1 in Table 5.1 (see text)

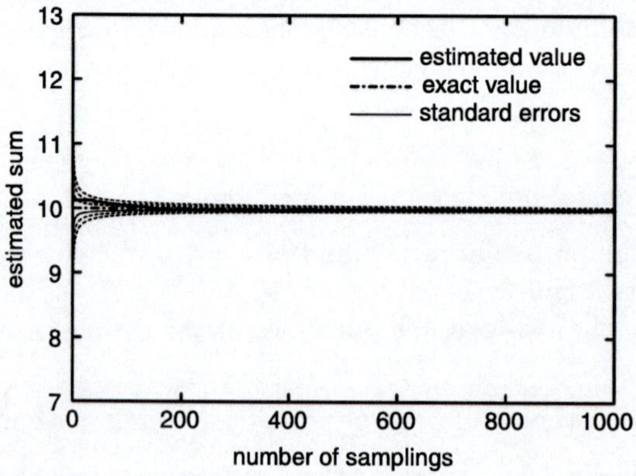

Fig. 5.2. Monte Carlo estimation of the sum in (5.75) as a function of the number of samplings corresponding to the choice 2 in Table 5.1 (see text)

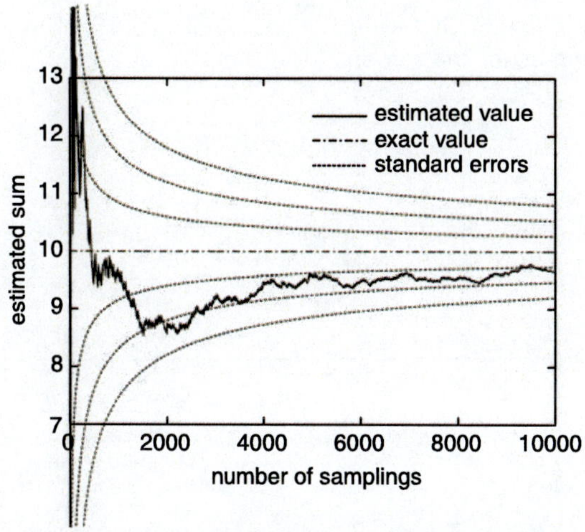

Fig. 5.3. Monte Carlo estimation of the sum in (5.75) as a function of the number of samplings corresponding to the choice 3 in Table 5.1 (see text)

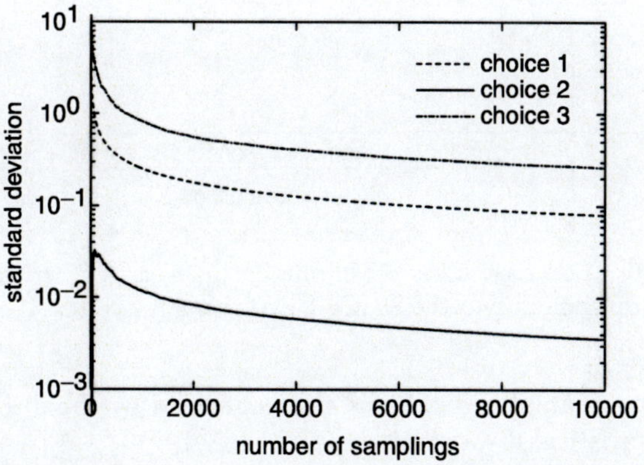

Fig. 5.4. Standard error as a function of the number of samplings corresponding to the three Monte Carlo estimations reported in Figs. 5.1, 5.2, and 5.3 (see text)

Monte Carlo sampling provides the correct result $(9 + 1 = 10)$ within the corresponding statistical error; the latter, in contrast, depends strongly on the choice of the distribution probability. In particular, choice 1 – corresponding to an equiprobable random generation of the two addenda – shows an intermediate level of statistical fluctuations (see Fig. 5.1), while choices 2 and 3 display extreme and opposite behaviors: in total agreement with the general prescription previously discussed, for choice 2 – corresponding to a probability distribution nearly proportional to the two addenda – we realize a significant suppression of the statistical fluctuations (see Fig. 5.2), while for choice 3 the

Table 5.2. Two different choices of the probability distribution employed for the Monte Carlo sampling of the sum in (5.76) reported in Figs. 5.5, 5.6, and 5.7 (see text)

choice	$p^{(10)}$	$p^{(-9)}$	$e^{(10)}$	$e^{(-9)}$
1	$\frac{1}{4}$	$\frac{3}{4}$	40	−12
2	$\frac{10}{19}$	$\frac{9}{19}$	19	−19

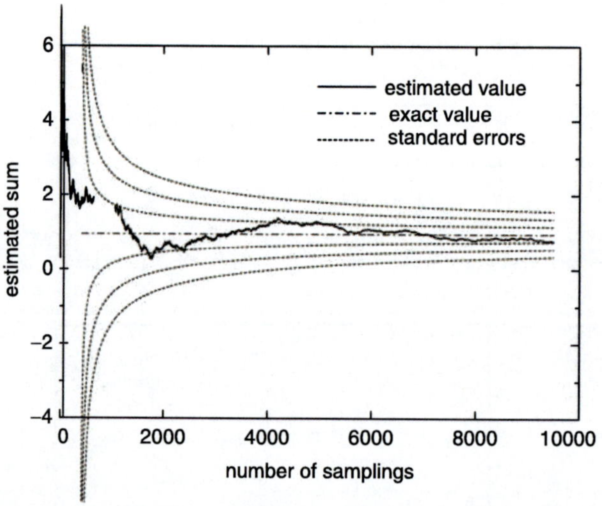

Fig. 5.5. Monte Carlo estimation of the sum in (5.76) as a function of the number of samplings corresponding to the choice 1 in Table 5.2 (see text)

latter are strongly amplified. This is also confirmed by the direct comparison of the three statistical errors reported in Fig. 5.4.

This basic example has clearly shown that a proper choice of the probability distribution is imperative in order to optimize our Monte Carlo sampling. This conclusion, however, is valid provided to work with positive-definite addenda. In order to better clarify this crucial point, let us consider as second example the sum

$$S = 10 - 9 \; . \tag{5.76}$$

For this second example we shall consider two different choices of the probability distribution, choice 1 and 2, detailed in Table 5.2; for each of them we have performed a corresponding Monte Carlo sampling, whose results are reported in Figs. 5.5, 5.6, and 5.7. Again, as expected, for both choices (see Figs. 5.5 and 5.6) our Monte Carlo sampling provides the correct result $(10 - 9 = 1)$ within the corresponding statistical errors; however, opposite to the case of positive-definite addenda in (5.75), the level of sta-

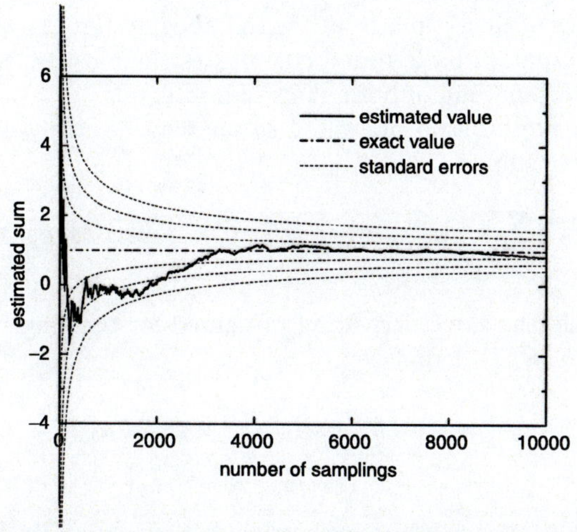

Fig. 5.6. Monte Carlo estimation of the sum in (5.76) as a function of the number of samplings corresponding to the choice 2 in Table 5.2 (see text)

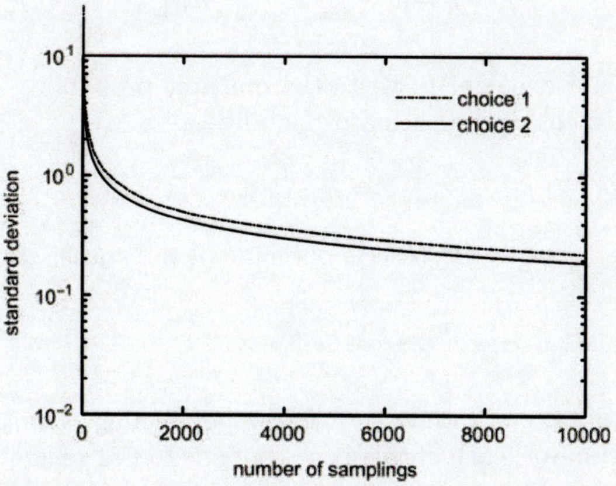

Fig. 5.7. Standard error as a function of the number of samplings corresponding to the two Monte Carlo estimations reported in Figs. 5.5 and 5.6 (see text)

tistical fluctuations is not so sensitive to the choice of the probability distribution (see Fig. 5.7). This intrinsic limitation – usually referred to as the "sign problem" (see, e.g., [252] and references therein) – may be qualitatively understood realizing that in this case we estimate a quantity of the order of 1 by subtracting numbers of the order of 10. More precisely, while for the case of positive-definite addenda the choice 2 in Table 5.1 allows us to deal with nearly constant (and positive) values of the estimator, in this case – also for the most convenient probability distribution (choice 2 in Table 5.2) – we deal with equal values of the estimator but with opposite sign. As we shall see, this unavoidable fluctuation background constitutes the most severe limitation of the Monte Carlo sampling of both sums and integrals.

The general prescription presented so far may be easily extended to the case of a multiple sum:

$$S = \sum_{i_1,i_2,\ldots,i_n} a_{i_1,i_2,\ldots,i_n} = \sum_{\lambda} a_{\lambda} \, , \quad \lambda \equiv i_1, i_2, \ldots, i_n \, . \tag{5.77}$$

Following exactly the same derivation presented for the single sum in (5.63), the key result in (5.67) is replaced by

$$S = \sum_{\lambda} a_{\lambda} \approx \frac{1}{N}\sum_{l=1}^{N} e_{\lambda_l} = \frac{1}{N}\sum_{l=1}^{N} \frac{a_{\lambda_l}}{p_{\lambda_l}} \, . \tag{5.78}$$

Again, the performance level of such a Monte Carlo sampling may be strongly influenced by the choice of the multiple probability distribution $p_{\lambda} = p_{i_1,i_2,\ldots,i_n}$.

A typical example of multiple sum is the evaluation of the mth power of a matrix:

$$S_{ll'} \equiv \{A^m\}_{ll'} = \sum_{i_1,i_2,\ldots,i_{m-1}} A_{li_1} A_{i_1 i_2} \ldots A_{i_{m-1}l'} \, . \tag{5.79}$$

In this case it is convenient to write the multiple probability distribution as a product of elementary transition probabilities, i.e.,

$$p_{i_1,i_2,\ldots,i_{m-1}} = P_{li_1} P_{i_1 i_2} \ldots P_{i_{m-1}l'} \, , \tag{5.80}$$

which allows us to write the generic outcome of the statistical estimator as

$$e_{i_1,i_2,\ldots,i_{m-1}} = E_{li_1} E_{i_1 i_2} \ldots E_{i_{m-1}l'} \, , \qquad E_{ij} = \frac{A_{ij}}{P_{ij}} \, . \tag{5.81}$$

In this way we obtain an estimate of $S_{ll'}$ by performing N random samplings of the above estimator; each of them corresponds to the generation of a sort of "random walk" $i_1, i_2, \ldots, i_{m-1}$ connecting the initial and final indices (l and l') according to the transition probability P_{ij}.

Let us now come to the Monte Carlo evaluation of an integral:

$$S = \int a(x)\mathrm{d}x \, . \tag{5.82}$$

Comparing the sum (5.63) with the above integral, we see that the discrete index i is simply replaced by the continuous variable x; this suggests to introduce a "continuous version" of the Monte Carlo procedure discussed so far. To this end, let us consider a given probability density $f(x)$ such to fulfill the general requirements in (5.49); this allows us to rewrite the integral in (5.82) as

$$S = \int a(x)\mathrm{d}x = \int \left(\frac{a(x)}{f(x)}\right) f(x)\mathrm{d}x \; . \tag{5.83}$$

By comparing this last expression for the integral with the definition of the expectation value of a continuous random variable in (5.51), our original integral may also be written as the expectation value of a continuous random variable e:

$$S = \langle e \rangle = \int e(x) f(x)\mathrm{d}x \; , \qquad e(x) = \frac{a(x)}{f(x)} \; . \tag{5.84}$$

Exactly as for the case of the sum (see (5.67)), recalling that the theoretical expectation value of any random variable may be approximated by a mathematical average of N random samplings (see (5.53)), i.e.,

$$\langle e \rangle \approx E_N = \frac{1}{N}\sum_{l=1}^{N} e(x_l) \tag{5.85}$$

$(e(x_1), e(x_2), \ldots, e(x_N)$ denoting N random samplings of e), we finally get

$$S = \int a(x)\mathrm{d}x \approx \frac{1}{N}\sum_{l=1}^{N} e(x_l) = \frac{1}{N}\sum_{l=1}^{N} \frac{a(x_l)}{f(x_l)} \; . \tag{5.86}$$

Exactly as for the case of the sum, the central-limit theorem allows us to estimate the statistical error associated with our Monte Carlo calculation (see (5.68)), and the latter may be readily sampled within the same simulation as well via a continuous version of the prescription in (5.71). Again, the choice of the probability distribution may play a crucial role for the optimization of such a Monte Carlo sampling; according to the general prescription previously introduced, in order to minimize the statistical fluctuations of the estimator, one should employ a probability density $f(x)$ as close as possible to $|a(x)|$.

As basic example of Monte Carlo integration, let us consider the following problem:

$$S = \int_0^1 \sin(x)\mathrm{d}x \; . \tag{5.87}$$

To this aim, let us consider two different choices of the probability density, choice 1 and 2, detailed in Table 5.3; for each of them we have performed a

Table 5.3. Two different choices of the probability density employed for the Monte Carlo samplings of the integral in (5.87) reported in Figs. 5.8, 5.9, and 5.10 (see text)

choice	$a(x)$	$f(x)$	$e(x)$
1	$\sin(x)$	1	$\sin(x)$
2	$\sin(x)$	$2x$	$\frac{\sin(x)}{2x}$

corresponding Monte Carlo sampling, whose results are reported in Figs. 5.8, 5.9, and 5.10. Again, for both choices (see Figs. 5.8 and 5.9) our Monte Carlo sampling provides the correct result ($1-\cos(1) \approx 0.46$) within the corresponding statistical error; the latter, in contrast, depends significantly on the choice of the probability density $f(x)$ (see Table 5.3), as can be seen in Fig. 5.10.

As second and more intriguing example, let us consider the following integral:

$$S = \int_0^1 \frac{\cos(x)}{\sqrt{x}} \mathrm{d}x \ . \tag{5.88}$$

By employing again a uniform probability density (see choice 1 in Table 5.4), we get the sampling profile reported in Fig. 5.11. As usual, by

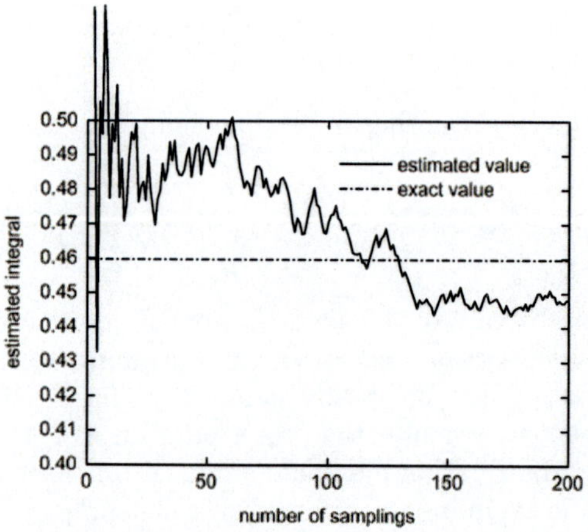

Fig. 5.8. Monte Carlo estimation of the integral in (5.87) as a function of the number of samplings corresponding to the choice 1 in Table 5.3 (see text)

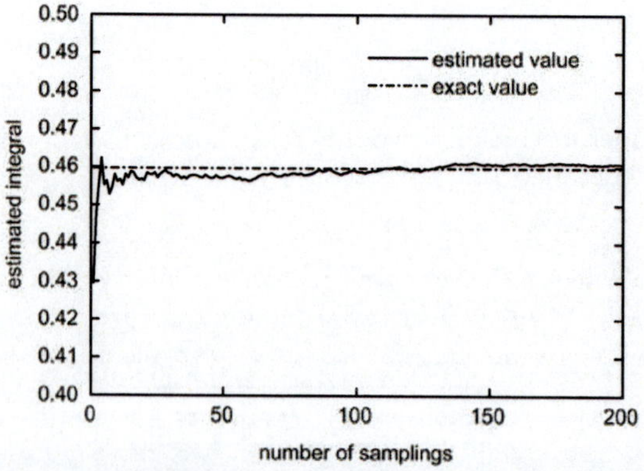

Fig. 5.9. Monte Carlo estimation of the integral in (5.87) as a function of the number of samplings corresponding to the choice 2 in Table 5.3 (see text)

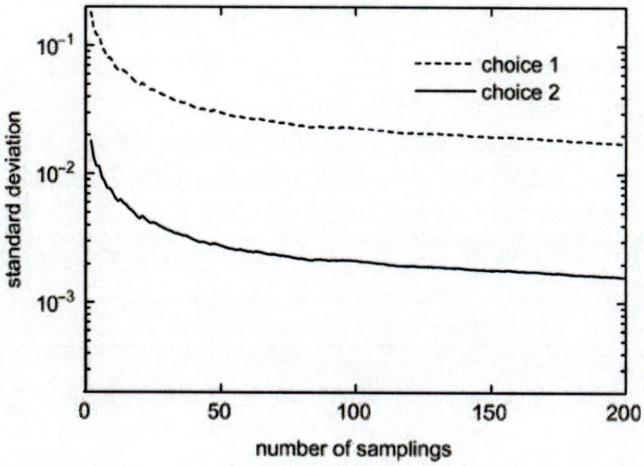

Fig. 5.10. Standard error as a function of the number of samplings corresponding to the two Monte Carlo estimations reported in Figs. 5.8 and 5.9 (see text)

Table 5.4. Two different choices of the probability density employed for the Monte Carlo samplings of the integral in (5.88) reported in Figs. 5.11, 5.12, and 5.13 (see text)

choice	$a(x)$	$f(x)$	$e(x)$
1	$\frac{\cos(x)}{\sqrt{x}}$	1	$\frac{\cos(x)}{\sqrt{x}}$
2	$\frac{\cos(x)}{\sqrt{x}}$	$\frac{1}{2\sqrt{x}}$	$2\cos(x)$

increasing the number of samplings the statistical fluctuations decrease; however, opposite to all previous cases, here the estimated value deviates significantly from the exact one (corresponding to the horizontal line). Such anomalous behavior suggests that the Monte Carlo sampling of Fig. 5.11 is

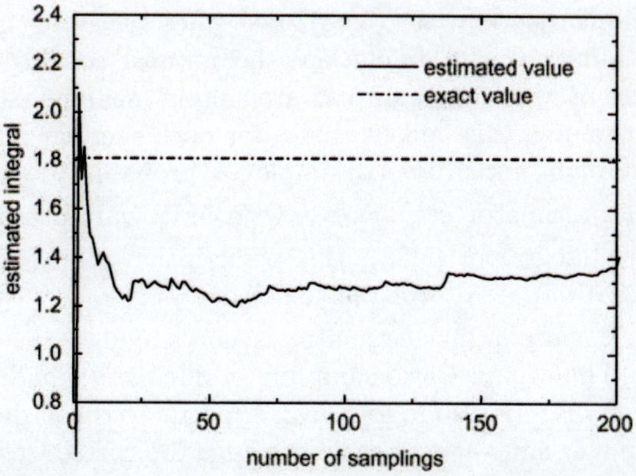

Fig. 5.11. Monte Carlo estimation of the integral in (5.88) as a function of the number of samplings corresponding to the choice 1 in Table 5.4 (see text)

ill-defined; this is confirmed by extending the very same simulation over a larger number of samplings (see Fig. 5.12): opposite to the partial sampling profile in Fig. 5.11, the new result, corresponding to a much longer simulation path, shows unambiguously that the anomalous statistical fluctuations do not scale according to the central-limit theorem. This can be readily understood noticing that for the case of a constant probability density (choice 1 in Table 5.4) the variance of the corresponding estimator is not finite:

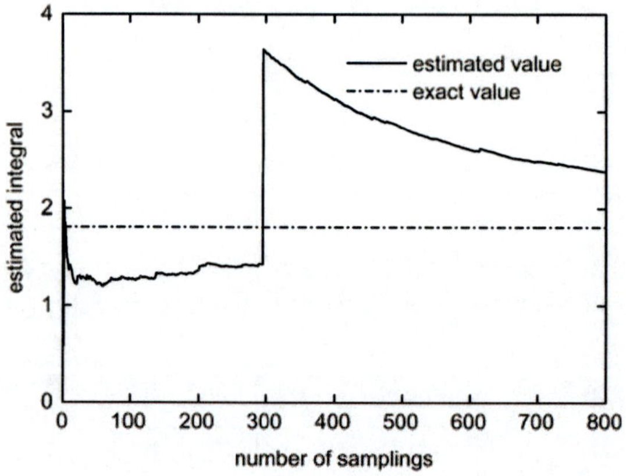

Fig. 5.12. Same as in Fig. 5.11 but for a larger number of samplings (see text)

$$\mathrm{var}(e) = \langle e^2 \rangle - \langle e \rangle^2 = \int_0^1 e^2(x)\mathrm{d}x - S^2 = \int_0^1 \frac{\cos^2(x)}{x}\mathrm{d}x - S^2 = \infty \,. \quad (5.89)$$

Since the finite value of var(e) is an essential prerequisite for the application of the central-limit theorem (see (5.68)) and since the latter is the "building block" of any Monte Carlo simulation, the natural conclusion is that the sampling profiles of Figs. 5.11 and 5.12 are simply meaningless.

Generally speaking, this tells us that – for both singular and non-singular functions $a(x)$ – it is imperative to employ a probability density $f(x)$ corresponding to an estimator $e(x) = \frac{a(x)}{f(x)}$ with finite variance. For the particular case of the integral in (5.88), this prescription may be easily realized by employing as alternative probability density $f(x) = \frac{1}{2\sqrt{x}}$ (see choice 2 in Table 5.4); the corresponding estimator is non-singular and its variance is certainly finite. The Monte Carlo sampling profile corresponding to this new probability density is reported in Fig. 5.13; opposite to the ill-defined sampling of Fig. 5.12, the new simulation profile provides the correct result (horizontal line) within the statistical error, and the latter scales down according to the prescription of the central-limit theorem.

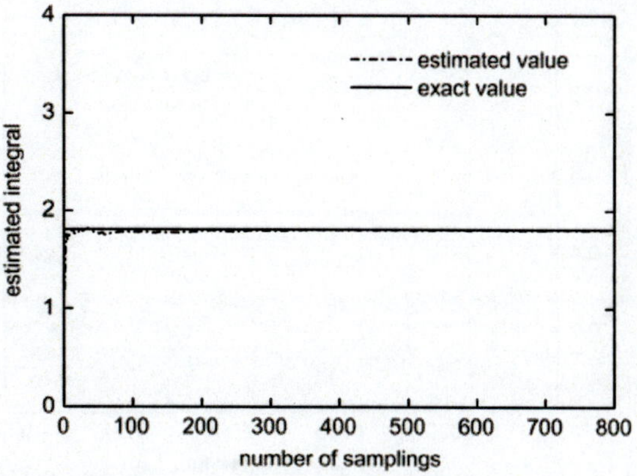

Fig. 5.13. Monte Carlo estimation of the integral in (5.88) as a function of the number of samplings corresponding to the choice 2 in Table 5.4 (see text)

So far, our analysis has been limited to the integration of positive-definite functions (see (5.87) and (5.88)). As for the statistical sampling of a sum (see (5.76)), in the presence of oscillating functions (i.e., characterized by both positive and negative values) the effectiveness of the proposed Monte Carlo method is strongly reduced. In order to better elucidate this crucial point, let us consider the following time integral[11]:

$$S = \Re\left(\int_0^\infty \mathrm{e}^{\mathrm{i}(\omega+\mathrm{i}\Gamma)t}\mathrm{d}t\right) = \frac{\Gamma}{\omega^2+\Gamma^2} \,. \tag{5.90}$$

Following again the general prescription, in this case we shall employ a probability density proportional to the modulus of the exponential function to be integrated, i.e., $f(t) = \Gamma\mathrm{e}^{-\Gamma t}$. As detailed in Table 5.5, by setting $\Gamma =$

Table 5.5. Three different parameter sets employed for the Monte Carlo samplings of the integral in (5.90) reported in Figs. 5.14, 5.15, 5.16, 5.17, and 5.18 (see text)

case	ω	Γ	Exact value
A	1	1	$S = \frac{1}{2}$
B	5	1	$S = \frac{1}{26}$
C	25	1	$S = \frac{1}{626}$

[11] As already pointed out in Chap. 2, such time integrals are always present in any treatment of phase coherence versus dissipation in semiconductor quantum devices (see (2.42)): the frequency ω is typically related to some characteristic energy of the system, while the inverse life-time Γ describes dissipation/decoherence effects.

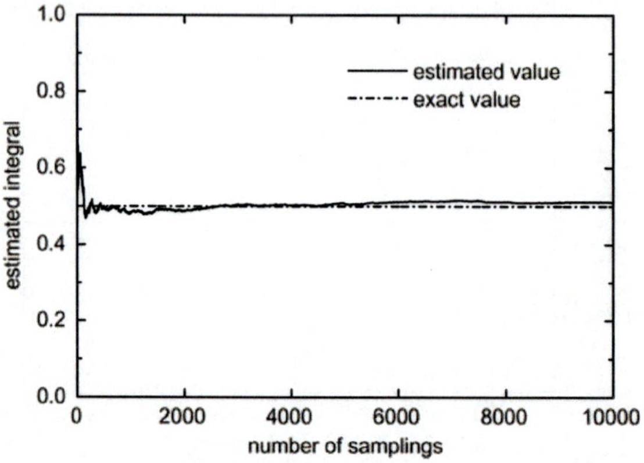

Fig. 5.14. Monte Carlo estimation of the integral in (5.90) as a function of the number of samplings corresponding to case A in Table 5.5 (see text)

1, we have performed three different simulations corresponding to $\omega = 1$, $\omega = 5$, and $\omega = 25$, whose results are reported, respectively, in Figs. 5.14, 5.15, and 5.16. As we can see, for increasing values of ω the level of statistical fluctuations increases, and the approach becomes less and less effective; this can be understood as follows: while the absolute error ΔS (see Fig. 5.17) is not significantly changed, the relative error $\frac{\Delta S}{S}$ (see Fig. 5.18) is progressively increased. This is basically ascribed to the fact that the value of the integral in (5.90) scales down as $\frac{1}{\omega^2}$ (see Table 5.5), while the statistical fluctuations of the estimator are nearly ω-independent and always of the order of 1; therefore, the relative error increases as ω^2. Such intrinsic limitation may be regarded as a continuous version of the "sign problem" previously faced for the case of the sum.

The main conclusion is that for $\omega \ll \Gamma$ – usually referred to as incoherent regime (see Sect. 2.2) – the Monte Carlo sampling comes out to be quite effective; in contrast, for $\omega \gg \Gamma$ – usually referred to as coherent regime (see again Sect. 2.2) – the latter is definitely inadequate, and alternative integration schemes, e.g., deterministic (see Sect. 5.1) or deterministic-plus-stochastic approaches (see Sect. 5.3), are highly preferable.

The Monte Carlo strategy discussed so far may be readily extended to the case of a multi-dimensional integral defined over a generic volume Ω:

$$S = \int^{\Omega} a(X)\mathrm{d}X \ , \qquad X \equiv x_1, x_2, \ldots, x_n \ . \tag{5.91}$$

To this end, by introducing a multi-dimensional probability density $f(X) = f(x_1, x_2, \ldots, x_n)$, the prescription (5.86) is replaced by

$$S = \int^{\Omega} a(X)\mathrm{d}X \approx \frac{1}{N}\sum_{l=1}^{N} e(X_l) = \frac{1}{N}\sum_{l=1}^{N} \frac{a(X_l)}{f(X_l)} \ . \tag{5.92}$$

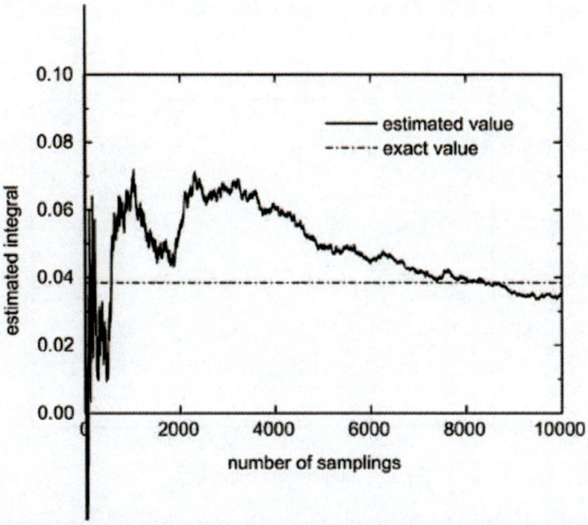

Fig. 5.15. Monte Carlo estimation of the integral in (5.90) as a function of the number of samplings corresponding to case B in Table 5.5 (see text)

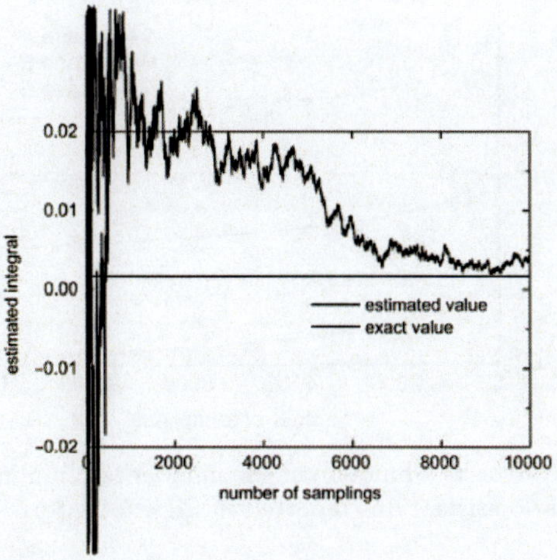

Fig. 5.16. Monte Carlo estimation of the integral in (5.90) as a function of the number of samplings corresponding to case C in Table 5.5 (see text)

As anticipated in the introductory part of this section, the most effective Monte Carlo sampling is realized by choosing a probability density proportional to the absolute value of the function to be integrated: $f(X) \propto |a(X)|$. This fundamental result – also known as the "importance-sampling theorem" (see, e.g., [252]) – may be derived as follows. By recalling that the square of the statistical error ΔS is simply proportional to var(e) (see (5.68)), the goal is to look for the probability density that minimizes the variance of our

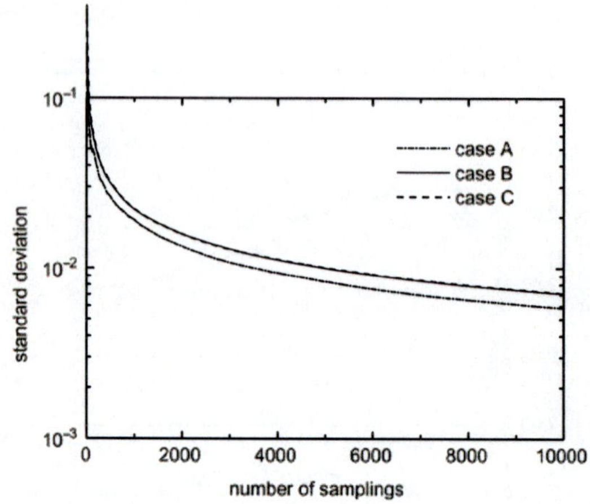

Fig. 5.17. Absolute error as a function of the number of samplings corresponding to the three Monte Carlo estimations reported in Figs. 5.14, 5.15, and 5.16 (see text)

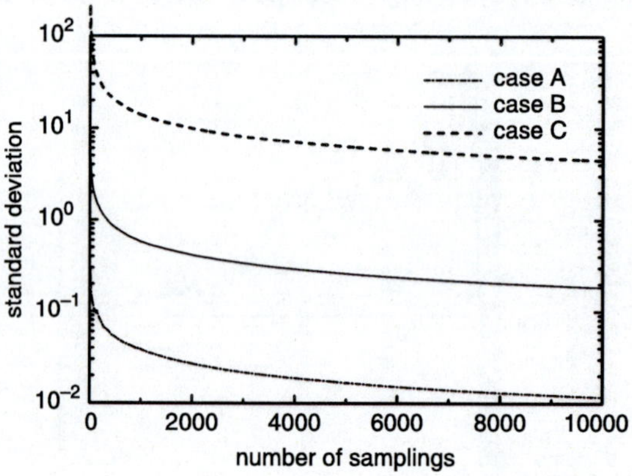

Fig. 5.18. Relative error as a function of the number of samplings corresponding to the three Monte Carlo estimations reported in Figs. 5.14, 5.15, and 5.16 (see text)

estimator. For the case of the multi-dimensional integral (5.91) the estimator is given by $e(X) = \frac{a(X)}{f(X)}$, and its variance may be written as

$$\begin{aligned} \mathrm{var}(e) &= \langle e^2 \rangle - \langle e \rangle^2 \\ &= \int^{\Omega} (e(X))^2 f(X) \mathrm{d}X - \langle e \rangle^2 \\ &= \int^{\Omega} \frac{a^2(X)}{f(X)} \mathrm{d}X - S^2 \, . \end{aligned} \tag{5.93}$$

In order to get the probability density $f(X)$ that minimizes the variance, we shall evaluate the functional derivative of $\mathrm{var}(e)$ with respect to $f(X)$ taking into account the normalization constraint of the probability density; this is accomplished by introducing a Lagrange multiplier λ:

$$\frac{\delta}{\delta f}\left[\mathrm{var}(e) + \lambda \int^{\Omega} f\,\mathrm{d}X\right] = 0\,. \tag{5.94}$$

By inserting the explicit form of the variance (5.93) into (5.94) we have

$$\frac{\delta}{\delta f}\left[\left(\int^{\Omega} \frac{a^2}{f}\mathrm{d}X - S^2\right) + \lambda \int^{\Omega} f\,\mathrm{d}X\right]$$

$$= \int^{\Omega} \mathrm{d}X \frac{\delta}{\delta f}\left[\frac{a^2}{f} + \lambda f\right] = \int^{\Omega} \mathrm{d}X \left[\lambda - \left(\frac{a}{f}\right)^2\right] = 0 \tag{5.95}$$

and therefore[12]

$$f(X) = \frac{1}{\sqrt{\lambda}}|a(X)| \propto |a(X)|\,. \tag{5.96}$$

Adopting this "ideal" probability density, it is easy to verify that when $a(X)$ is positive-definite the estimator is constant ($e(X) = \sqrt{\lambda}$) and the corresponding variance is simply equal to zero, i.e., no fluctuations at all; in contrast, for oscillating functions the absolute value of the estimator is still constant but its sign changes randomly, giving rise to an unavoidable level of statistical fluctuations, i.e., the "sign problem" previously mentioned.

Opposite to the direct/deterministic approaches reviewed in Sect. 5.1, the Monte Carlo sampling allows one to treat integration domains Ω of arbitrary shape. Moreover, as discussed extensively in [252], while for deterministic integration schemes the number of discretization cells – and thus the computation effort – increases rapidly with the number of dimensions n, the Monte Carlo sampling comes out to be nearly independent from the dimensionality of our integration. It follows that for problems involving a significantly high number of dimensions/coordinates, the Monte Carlo strategy is always preferable.

As we shall see in Sect. 5.2.5, the Monte Carlo solution of a large class of integro-differential equations reduces to the stochastic sampling of an infinite sum of nested time integrals[13]:

$$S = \sum_{n=0}^{\infty} c_n \int_{t_0}^{t} \mathrm{d}t_n \ldots \int_{t_0}^{t_2} \mathrm{d}t_1 a(t_n, \ldots, t_1)\,. \tag{5.97}$$

This can be performed by combining the two Monte Carlo algorithms for the evaluation of sums and integrals previously described. More specifically, we choose a given multi-dimensional probability density $f_n(t_n, \ldots, t_1)$ such that $t_0 \leq t_1 \leq \cdots \leq t_n \leq t$, and we compute the corresponding estimator

[12] We stress that the above result applies to the case of complex functions $a(X)$ as well.

[13] A typical example is the Neumann series (3.24) introduced in Sect. 3.3.1.

$$e_n(t_n,\ldots,t_1) = \frac{c_n a(t_n,\ldots,t_1)}{f_n(t_n,\ldots,t_1)} . \tag{5.98}$$

As usual, we shall obtain an estimate of the infinite sum of nested time integrals in (5.97) by performing N random samplings of the above estimator. As anticipated, this sampling strategy constitutes the key ingredient employed for the Monte Carlo solution of differential equations (see Sect. 5.2.5).

5.2.4 Direct Monte Carlo Simulation of Stochastic Processes

As pointed out in Sect. 5.2.1, the Monte Carlo method may be viewed from two different perspectives. So far, we have discussed the so-called mathematical point of view, according to which the Monte Carlo method is a stochastic technique for the numerical evaluation of sums and integrals (see Sect. 5.2.3), i.e., a deterministic problem is solved via a stochastic approach. In this section we shall discuss the so-called physical point of view, according to which a Monte Carlo sampling is a "faithful" simulation of a natural stochastic process.

Let us start our analysis by considering a generic physical system described (within a classical or semiclassical picture) by a point X within a multi-dimensional phase-space Ω. On the one hand, the system evolution will be governed by its own internal dynamics; on the other hand, the latter will also be influenced by possible interactions with its environment. While the former gives rise to a deterministic system evolution, the latter may be conveniently described via stochastic transition processes. More specifically, by denoting with X the state of the system, its time evolution may be expressed as

$$\frac{\mathrm{d}X}{\mathrm{d}t} = \left.\frac{\mathrm{d}X}{\mathrm{d}t}\right|_{\mathrm{det}} + \left.\frac{\mathrm{d}X}{\mathrm{d}t}\right|_{\mathrm{stoch}} . \tag{5.99}$$

The explicit form of the first contribution – accounting for the (deterministic) internal system dynamics – depends on the specific problem under investigation (see below). The stochastic term – describing within a scattering picture the effect of the system–environment coupling – may be written as

$$\left.\frac{\mathrm{d}X}{\mathrm{d}t}\right|_{\mathrm{stoch}} = \sum_n \delta(t - t_n)\Delta X_n . \tag{5.100}$$

Here, the system–environment interaction is described via instantaneous processes (taking place at the generic time t_n) where the system undergoes an instantaneous transition from the current state X to $X'_n = X + \Delta X_n$.

Within such effective picture, the resulting system dynamics may be regarded as a random walk given by a sequence of deterministic system evolutions, usually referred to as "scattering-free flights" or simply "free flights," interrupted by instantaneous phase-space transitions, usually referred to as "scattering events." While the free-flight dynamics – being in general relatively simple – can be evaluated via analytical or quasi-analytical methods, the random sequence of instantaneous transition processes in (5.100) obeys well-precise system–environment scattering probabilities. More specifically, we shall denote with $W^s(X', X)$ the probability per time unit that the sth interaction mechanism will induce a scattering process from the initial state X to the final state X'[14]; the corresponding scattering rate is then given by the integral of $W^s(X', X)$ over all possible final states X', i.e.,

$$\Gamma^s(X) = \int^{\Omega} W^s(X', X) \mathrm{d}X' \,. \tag{5.101}$$

The latter may be regarded as a sort of inverse life-time induced by the interaction mechanism s. It follows that the total scattering rate, and therefore the total inverse life-time, is obtained summing the scattering rate (5.101) over all possible interaction mechanisms:

$$\Gamma(X) = \sum_s \Gamma^s(X) \,. \tag{5.102}$$

Indeed, from a physical point of view, $\frac{1}{\Gamma(X)}$ corresponds to the average time spent by the system within the deterministic trajectory $X(t)$, which in turn corresponds to the average duration of the free flight, i.e., the time elapsed between a generic scattering event and the following one.

An extremely simple example of random walk – i.e., a sequence of free flights interrupted by scattering events – is reported in Figs. 5.19 and 5.20. In

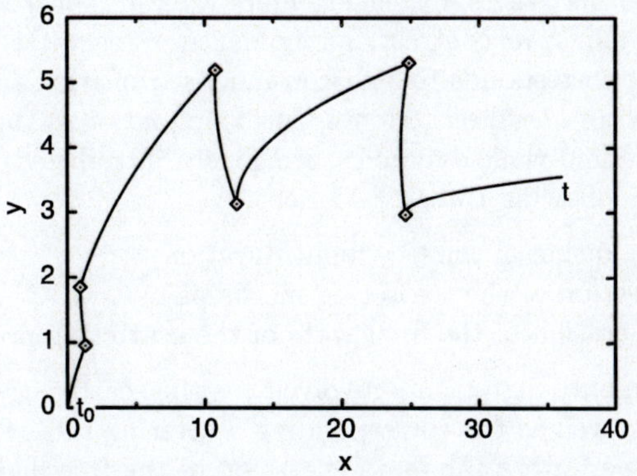

Fig. 5.19. Example of a bidimensional random-walk profile in real space for the case of a classical particle accelerated by an external force/field along the horizontal axis. In view of the local character of the scattering processes, the resulting random walk is continuous; the latter is formed by parabolic trajectories interrupted by point-like scattering events

[14] Generally speaking, the scattering probability $W^s(X', X)$ depends both on initial and final states and not only on their difference $\Delta X = X' - X$.

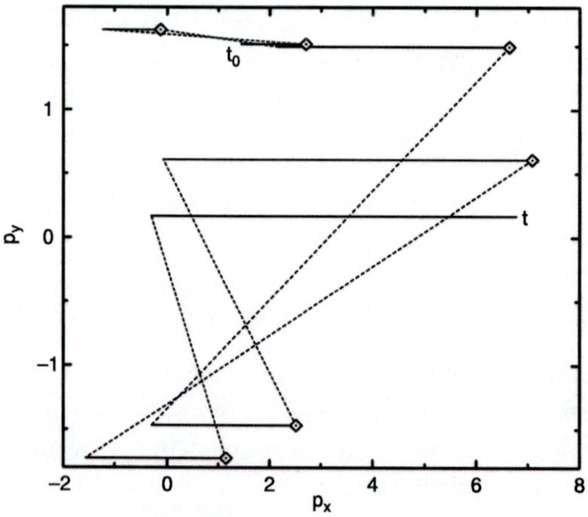

Fig. 5.20. Example of a bidimensional random-walk profile in momentum space for the case of a classical particle accelerated by an external force/field along the horizontal axis. The free-flight dynamics corresponds to a linear increase of the particle momentum along the direction of the applied force; opposite to the real-space random walk in Fig. 5.19, here each scattering event corresponds to a discontinuous transition

this case the physical system under investigation is a classical particle within a bidimensional space subjected to an external force/field along the horizontal axis. As for any classical or semiclassical model (see Sect. 1.3), here we deal with local scattering events, i.e., the random walk in real space (see Fig. 5.19) is always continuous; in contrast, in momentum space (see Fig. 5.20) each scattering event corresponds to a discontinuous transition.

Primary goal of any direct Monte Carlo simulation is the statistical sampling of the random-walk dynamics previously introduced; this is typically realized via the following three basic steps:

(i) random generation of the free-flight duration;
(ii) random selection of the scattering mechanism;
(iii) random generation of the final state of the scattering process.

To this aim, the only ingredients involved are the scattering probabilities per time unit W^s as well as the corresponding scattering rates Γ^s (see (5.101)).

In order to perform a random generation of the free-flight duration, it is crucial to determine its probability distribution. Let us denote with $P\left(X(t)\right)$ the probability of finding the system at time t within the trajectory $X(t)$; recalling the definition of total scattering rate in (5.102), we have

$$\frac{\mathrm{d}}{\mathrm{d}t} P\left(X(t)\right) = -\Gamma\left(X(t)\right) P\left(X(t)\right) . \tag{5.103}$$

Given the system in the state $X(t_a)$ at time t_a (i.e., $P\left(X(t_a)\right) = 1$) and integrating the above equation over the free flight, i.e., over the deterministic system trajectory from t_a to t_b, the probability of finding the system within the same trajectory at time t_b is given by

$$P\left(X(t_b)\right) = \mathrm{e}^{-\int_{t_a}^{t_b} \Gamma(X(t'))\mathrm{d}t'} . \tag{5.104}$$

It follows that the probability density associated with the free-flight duration $t = t_b - t_a$ is of the form

$$f^{\mathrm{f.f.}}(t_a, t) \propto \mathrm{e}^{-g(t_a,t)} , \qquad g(t_a, t) = \int_{t_a}^{t_a+t} \Gamma(t')\mathrm{d}t' . \tag{5.105}$$

Since the total scattering rate Γ is in general a function of the state of the system X, the quantity g is usually a non-trivial function of time; it follows that the cumulative distribution of the probability density $f^{\mathrm{f.f.}}$ is generally not known analytically and the direct-generation method introduced in Sect. 5.2.2 (see (5.58)) cannot be employed.

As discussed extensively in [28], this tedious problem may be greatly simplified by adopting the so-called self-scattering technique originally proposed by Rees in 1968 [272]. Denoting with $\Gamma^{\circ}(X)$ the maximum of the total scattering rate (5.102) over our simulation domain Ω (or any overestimation of it), the self-scattering technique proceeds as follows: we introduce a fictitious scattering mechanism called "self-scattering" such that

$$\Gamma^{\mathrm{s.s.}}(X) = \Gamma^{\circ} - \Gamma(X) , \qquad W^{\mathrm{s.s.}}(X', X) \propto \delta(X - X') . \tag{5.106}$$

It is straightforward to realize that this new scattering mechanism produces no real effects on the system dynamics; indeed, in a self-scattering event the initial and final states of the scattering coincide ($X = X'$) and the deterministic dynamics of the free flight is not modified. This is clearly shown in Figs. 5.21 and 5.22, where the bidimensional Monte Carlo simulation of Figs. 5.19 and 5.20 has been repeated in the presence of self-scattering.

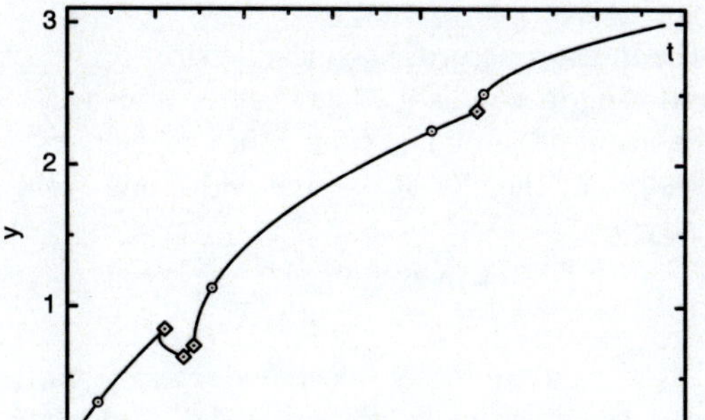

Fig. 5.21. Same as in Fig. 5.19 but in the presence of self-scattering processes; in this simulation $\Gamma^{\mathrm{s.s.}} = \frac{2}{3}\Gamma^{\circ}$ (see text)

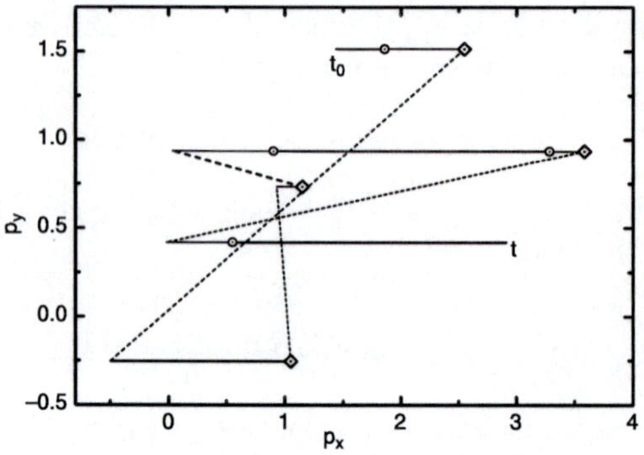

Fig. 5.22. Same as in Fig. 5.20 but in the presence of self-scattering processes; in this simulation $\Gamma^{\mathrm{s.s.}} = \frac{2}{3}\Gamma^\circ$ (see text)

The big advantage of the self-scattering technique is that the total (real + self) scattering rate Γ° is constant, and the corresponding free-flight probability density is simply given by

$$f^{\mathrm{f.f.}}(t) \propto \mathrm{e}^{-\Gamma^\circ t} \ . \tag{5.107}$$

This allows us to employ the standard free-flight generation introduced in (5.60):

$$t(R) = -\frac{1}{\Gamma^\circ} \log(R) \ . \tag{5.108}$$

Given the free-flight duration, and thus the following scattering time, the next step is the random selection of the scattering mechanism s. To this aim, given the distribution probability (including self scattering)

$$p_s = \frac{\Gamma^s}{\Gamma^\circ} \ , \tag{5.109}$$

we shall select the index s according to the general prescription in (5.62).

Once the scattering mechanism s has been established, the missing step is the random generation of the final state. More specifically, given the current state X of the system at the end of the free flight, one evaluates the following probability density:

$$f^{\mathrm{f.s.}}(X') = \frac{W^s(X', X)}{\Gamma^s(X)} \ . \tag{5.110}$$

The final state X' is then randomly selected starting from the above distribution profile; due to the complexity of the simulation domain as well as of the probability density $f^{\mathrm{f.s.}}(X')$, such generation process requires, in general, a proper combination of both direct and rejection techniques (see Sect. 5.2.2).

The general simulation scheme presented so far – usually referred to as "Ensemble Monte Carlo" (see, e.g., [28]) – is based on the generation of a large number of independent random walks, each of them describing a possible system evolution in the presence of random disturbances. The estimated value of a physical quantity g at a given time is then obtained by performing a corresponding ensemble average over all the random walks/samplings:

$$\langle g \rangle \approx \frac{1}{N} \sum_{l=1}^{N} g(X_l) \; . \tag{5.111}$$

It is worth stressing that, in virtue of the well-known "ergodic theorem" (see, e.g., [273] and references therein), for the case of homogeneous and steady-state phenomena such average over many random walks may be safely replaced by a time average over a single random walk:

$$\overline{g} = \lim_{T \to \infty} \frac{1}{T} \int_0^T g\left(X(t)\right) \mathrm{d}t \; . \tag{5.112}$$

This kind of Monte Carlo simulation is often referred to as "one-particle Monte Carlo" (see also Sect. 5.2.5).

The most relevant application of the above direct Monte Carlo strategy in the context of the present book is the analysis of charge transport in optoelectronic semiconductor devices, where the system under investigation is the generic charge carrier traveling within our semiconductor crystal. As discussed in Sect. 1.3, the latter may be safely described via the so-called semiclassical picture in terms of the usual phase-space coordinates: $X \equiv \boldsymbol{r}, \boldsymbol{k}$. In this case, the deterministic carrier free flights are governed by the effective Hamilton equations (1.66), while the scattering events are commonly described via point-like scattering probabilities derived within the well-known Fermi's golden rule (see, e.g., [28]); in addition to interparticle (carrier–carrier) scattering, the simulated electrons/holes will undergo stochastic transitions induced by a number of interaction mechanisms ascribed to the host material, as, e.g., carrier–phonon, carrier–plasmon, carrier–impurity scattering. As shown extensively in Chaps. 4, 6, and 8, the direct Monte Carlo simulation of charge-transport phenomena is a key instrument for the design and optimization of new-generation optoelectronic devices.

5.2.5 Sampling of Differential Equations: The Weighted Monte Carlo Method

As discussed in the introductory part of this chapter, many of the physical phenomena governing the behavior of semiconductor quantum devices are described by the generic integro-differential equation (5.1); by limiting our analysis to closed-system problems ($S = 0$), the latter may be written in a compact form as

$$\frac{\partial F}{\partial t} = \mathcal{L} F \; . \tag{5.113}$$

If we now integrate the above equation from t_0 to t we get

$$F(t) = F(t_0) + \int_{t_0}^{t} \mathrm{d}t' \mathcal{L}(t') F(t') \ . \tag{5.114}$$

Starting from this formal solution, it is possible to perform the following iterative substitution:

$$\begin{aligned} F(t) &= F(t_0) + \int_{t_0}^{t} \mathrm{d}t' \mathcal{L}(t') F(t') \\ &= F(t_0) + \int_{t_0}^{t} \mathrm{d}t' \mathcal{L}(t') F(t_0) \\ &\quad + \int_{t_0}^{t} \mathrm{d}t' \mathcal{L}(t') \int_{t_0}^{t'} \mathrm{d}t'' \mathcal{L}(t'') F(t'') \\ &= F(t_0) + \int_{t_0}^{t} \mathrm{d}t' \mathcal{L}(t') F(t_0) \\ &\quad + \int_{t_0}^{t} \mathrm{d}t' \mathcal{L}(t') \int_{t_0}^{t'} \mathrm{d}t'' \mathcal{L}(t'') F(t_0) + \ldots \\ &= \Delta F^{(0)}(t) + \Delta F^{(1)}(t) + \Delta F^{(2)}(t) + \ldots \ . \end{aligned} \tag{5.115}$$

The general form of the previous expansion is then given by

$$F(t) = \sum_{n=0}^{\infty} \Delta F^{(n)}(t) \tag{5.116}$$

with

$$\Delta F^{(n)}(t) = \int_{t_0}^{t} \mathrm{d}t_n \ldots \int_{t_0}^{t_2} \mathrm{d}t_1 \mathcal{L}(t_n) \ldots \mathcal{L}(t_1) F(t_0) \ . \tag{5.117}$$

Recalling the definition of the non-local operator $\mathcal{L}$ in terms of the generic coordinate set X (see (5.1)), the explicit form of the nth order contribution (5.117) is given by

$$\int_{t_0}^{t} \mathrm{d}t_n \ldots \int_{t_0}^{t_2} \mathrm{d}t_1 \int \mathrm{d}X_n \ldots \int \mathrm{d}X_1 \mathcal{L}(X, X_n; t_n) \ldots \mathcal{L}(X_2, X_1; t_1) F(X_1, t_0) \ . \tag{5.118}$$

In terms of the random-walk picture introduced in Sect. 5.2.4, the nth order contribution may be regarded as a sequence of unperturbed evolutions (i.e., free flights) $t_0 \to t_1$, $t_1 \to t_2$, $\ldots$, $t_n \to t$ interrupted by interaction processes (i.e., scattering events) $\mathcal{L}(X_2, X_1; t_1)$, $\mathcal{L}(X_3, X_2; t_2)$, $\ldots$, $\mathcal{L}(X, X_n; t_n)$; the corresponding random walk within our simulation domain is schematically depicted in Fig. 5.23.

The above iterative expansion for $F(X, t)$ is an infinite sum of nested time integrals; as such, the latter can be numerically sampled by applying

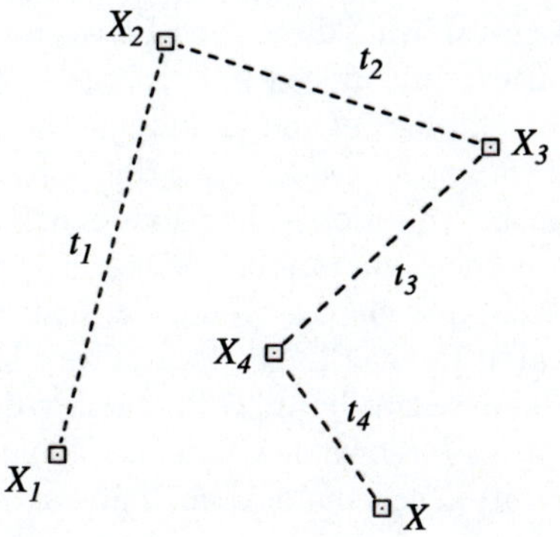

Fig. 5.23. Schematic representation of the random walk corresponding to the generic term in (5.117) (see text)

the corresponding Monte Carlo method introduced in Sect. 5.2.3 (see (5.97) and (5.98)). In particular, in addition to the nested time integrals, here we deal with an additional multi-dimensional integration over the coordinate set $X_1, X_2, \ldots, X_n$. It follows that, in order to perform the desired Monte Carlo sampling, we need to introduce a corresponding multi-dimensional probability density $f_n(X_n, t_n; \ldots; X_1, t_1)$. Moreover, on the basis of the random-walk representation discussed above (see Fig. 5.23), it is convenient to write this multi-dimensional probability density as a product of elementary transition probabilities T:

$$f_n(X_n, t_n; \ldots; X_1, t_1) = p_n T(X, X_n; t_n) \ldots T(X_2, X_1; t_1) f_\circ(X_1) \ , \quad (5.119)$$

where p_n is the probability to generate the nth order term, $f_\circ(X_1)$ is the probability density corresponding to the initial point (X_1) of the random walk, and $T(X', X; t)$ denotes the transition probability for the generic interaction process $X \to X'$ at time t.

It is imperative to stress that, opposite to the direct Monte Carlo sampling of Sect. 5.2.4, these probability distributions – employed for the generation of the random walk – have no physical meaning; indeed, similar to the Monte Carlo sampling of sums and integrals, any choice of these artificial probability profiles will produce a correct result within the corresponding statistical error; the latter, in turn, may be significantly reduced following the importance-sampling prescription in (5.96).

The Monte Carlo scheme introduced so far applies to a generic integro-differential equation, irrespective of its physical interpretation; it follows that, opposite to the direct Monte Carlo simulation, the latter is not limited to the study of classical (or semiclassical) stochastic phenomena and may be employed to investigate a variety of quantum-mechanical solid-state problems (see below).

We shall now try to establish a direct link between the two different points of view about Monte Carlo introduced in Sect. 5.2.1; to this end, let us consider the Monte Carlo sampling of the Boltzmann equation (see (1.64)). Indeed, this case plays a very special role, since the Boltzmann equation – being a particular type of master equation – describes exactly a physical system interacting with its environment via random scattering processes; it follows that its mathematical-like sampling via the Monte Carlo algorithm described above should be strongly related to the direct Monte Carlo simulation presented in Sect. 5.2.4. As discussed extensively in [274], starting from an integral version of the Boltzmann equation – known as Chambers equation [275] – and following the iterative-substitution scheme described above, it is possible to expand the semiclassical distribution function in powers of the scattering probabilities of the Boltzmann theory. Each term of this expansion may be regarded as a random walk within the semiclassical phase space, and its Monte Carlo sampling corresponds to a random generation of free flights interrupted by scattering events. In this case, however, the artificial probability profiles employed for the generation of the random walk have, in general, nothing to do with the physical probabilities entering the direct Monte Carlo simulation of Sect. 5.2.4. If, as a very special case, one adopts as artificial probabilities for the generation of the random walk exactly the physical probabilities entering the direct-simulation scheme, the Monte Carlo sampling of the Boltzmann equation and the direct Monte Carlo algorithm coincide; it follows that in this particular context there is no way to distinguish between the mathematical and the physical point of view.

The possibility of sampling the Boltzmann equation employing probability profiles different from the physical/natural ones is an important feature of the general Monte Carlo scheme previously introduced. As originally proposed by Carlo Jacoboni and co-workers (see [274] and references therein), within such generalized Monte Carlo simulation scheme – often referred to as Weighted Monte Carlo – scattering events occur with arbitrary probabilities, and the "weight" of the "simulative particle" (corresponding to the estimator of the random walk) is modified accordingly. In this way it is possible to better investigate the presence and physical relevance of "rare events," i.e., events that in direct simulations would occur too rarely; within such a Weighted Monte Carlo scheme this is realized by employing artificial probability distributions such to concentrate a large fraction of random walks toward a particular phase-space region of interest.

As anticipated, in addition to the weighted simulation of genuine stochastic phenomena, the Monte Carlo sampling of integro-differential equations previously discussed allows one to study a variety of different physical problems. In the context of the present book, the most relevant application is the Monte Carlo sampling of the global (electrons plus phonons) Liouville–von Neumann equation as well as of its Wigner-function version; such Quantum Monte Carlo (QMC) approach – originally proposed by Carlo Jacoboni and co-workers [108] – has allowed for a fully quantum-mechanical treatment of high-field transient-transport phenomena in semiconductor bulk and nanostructures [161–167], thus overcoming some of the basic limitations of conventional kinetic treatments, e.g., the completed-collision limit and the Markov approximation (see Chap. 3).

In spite of the generality of the previous sampling scheme, it is crucial to stress that its computational effectiveness is strongly related to the physical problem under examination. As already pointed out in discussing the Monte Carlo integration of oscillating functions (see Figs. 5.14, 5.15, 5.16, 5.17, and 5.18), the presence of strongly oscillating behaviors may reduce significantly the performance of our Monte Carlo sampling, i.e., the well-known sign problem. In particular, while in the so-called incoherent regime (see Fig. 5.14) the statistical error is small, in the so-called coherent regime (see Fig. 5.16) the latter increases significantly and the Monte Carlo sampling becomes highly problematic. It follows that the application of the Monte Carlo method to the study of purely coherent phenomena is unavoidably limited to extremely short time-scales (corresponding to a few quantum-mechanical oscillations within the simulated time domain); in contrast, the investigation of the non-trivial interplay between quantum coherence and dissipation/decoherence on much longer time-scales requires a proper combination of direct/deterministic techniques and stochastic approaches (see below).

5.3 Proper Combinations of Direct and Monte Carlo Schemes

In this section we shall briefly discuss how direct/deterministic integration techniques and stochastic simulation strategies may be properly combined in order to model physical systems with open spatial boundaries as well as phenomena governed by both phase coherence and energy-dissipation/decoherence processes.

As shown in Sect. 5.2, irrespective of the physical problem under investigation, the Monte Carlo sampling is by far the most convenient numerical approach for the simulation of scattering-like processes, but hardly applicable to phase-coherence phenomena. In contrast, finite-element/difference strategies (see Sect. 5.1.1) are able to properly treat quantum-mechanical phase coherence (see, e.g., [85]), but less effective in describing purely stochastic processes. This suggests to combine deterministic and stochastic algorithms, assigning to each of them well-precise and separate tasks.

Following this spirit, the key idea is to rewrite the integro-differential equation (5.1) by splitting the global time evolution of the unknown function $F(X)$ into a deterministic contribution and a stochastic one:

$$\frac{\partial F(X)}{\partial t} = \left.\frac{\partial F(X)}{\partial t}\right|_{\mathrm{det}} + \left.\frac{\partial F(X)}{\partial t}\right|_{\mathrm{stoch}} . \tag{5.120}$$

To this end, we shall define as stochastic contribution the sum over all relevant interaction mechanisms s of generalized Boltzmann collision terms:

$$\left.\frac{\partial F(X)}{\partial t}\right|_{\mathrm{stoch}} = \sum_s \int \mathrm{d}X' \left[W^s_{\mathrm{in}}(X, X')F(X') - W^s_{\mathrm{out}}(X', X)F(X)\right] . \quad (5.121)$$

Compared to the traditional collision term of the Boltzmann theory (see (1.65)), here the function $F(X)$ may also describe negative or complex quantities, as, e.g., Wigner functions or density-matrix elements (see Chaps. 2 and 3); moreover, the scattering probabilities W^s_{in} and W^s_{out} – corresponding to generalized in- and out-scattering processes – do not necessarily coincide. In spite of such remarkable differences, the Monte Carlo sampling of the stochastic contribution (5.121) is by far the most convenient strategy for its numerical evaluation (see, e.g., [276–279]).

In addition to the stochastic contribution (5.121), the system under examination will be governed by its own internal dynamics; the latter will manifest itself via the drift and diffusion terms of the semiclassical theory (see (1.64)) or via coherent oscillations of its quantum-mechanical state (see (3.128)). Moreover, in the presence of open spatial boundaries (see Sects. 1.4 and 2.4), the system evolution is also influenced by source versus loss terms (see, e.g., (1.80) and (1.84)). All these contributions may be conveniently treated via finite-element/difference methods, thus contributing to the deterministic term in (5.120).

The combination of direct and Monte Carlo calculations is typically realized via the standard time-step integration introduced in Sect. 5.1.3. To this end, the deterministic contribution is evaluated via a proper phase-space discretization (realized by expanding the unknown function $F(X)$ in terms of a suitable set of basis functions $\chi_j(X)$ (see (5.4))); at the same time, the stochastic contribution (5.121) is sampled via the Weighted Monte Carlo method introduced in Sect. 5.2.5 by performing a field-to-particle mapping and vice versa.[15] More specifically, in the spirit of the time-step integration scheme of Sect. 5.1.3 we have

$$F_j(t_n) = F_j(t_{n-1}) + \Delta F_j^{\mathrm{det}} + \Delta F_j^{\mathrm{stoch}} . \quad (5.122)$$

Here, at each time step the deterministic term $\Delta F_j^{\mathrm{det}}$ is evaluated via the first-order integration scheme in (5.46) (or via an improved version of it), while the stochastic one is sampled via a Weighted Monte Carlo simulation from time t_{n-1} to time t_n.

A first example of such combined – deterministic versus stochastic – calculation is the simulation of optoelectronic semiconductor devices with open

[15] As discussed extensively in [279], at each time step such deterministic versus stochastic simulation approach requires to map our initial field F_j into "simulative" particles – corresponding to the various random walks – and to translate them back to the same discretization grid j.

spatial boundaries (see Chaps. 4 and 6). In particular, the latter has been extensively applied to the design and optimization of mid-infrared quantum-cascade lasers discussed in Sect. 8.2 (see, e.g., [280] and references therein). In view of its open character, i.e., source versus loss terms, this problem requires a Monte Carlo simulation where the total number of particles (in the active region of the device) is not constant. As discussed in Chap. 4, to overcome this limitation it is also possible to adopt an alternative simulation strategy based on a closed-system paradigm (see Sect. 4.4).

A second and particularly relevant application of the above numerical approach is the microscopic simulation of ultrafast optical quantum processes (see, e.g., [276–279]). As pointed out in Chap. 2, the most appropriate instrument for the description of coherent optical phenomena in semiconductor bulk and nanostructures is the electronic single-particle density matrix introduced in Sect. 3.3.3; the latter will evolve according to the well-known semiconductor Bloch equations, whose explicit form depends on the most relevant interaction mechanisms involved. As discussed extensively in [279], the time evolution of the single-particle density matrix may be regarded as a coherent/deterministic term plus an incoherent/stochastic contribution, in total agreement with the separation (5.120). In particular, for both diagonal (carrier populations) as well as non-diagonal density-matrix elements (carrier polarizations) the corresponding incoherent/stochastic contributions have exactly the Boltzmann-like structure (5.121); this implies that, in spite of their complex values, also for the polarization fields it is possible to employ a Weighted Monte Carlo sampling; the latter amounts once again to a generation of random walks, where the free-flight duration is dictated by the out-scattering rates in (5.121) and the complex nature of the simulated quantity is accounted for "attaching" to each random trajectory a phase factor proportional to the initial value of the polarization. Within such generalized simulation scheme, the decay of the electronic polarization is the result of a non-trivial phase-interference process among all possible random walks, a physical scenario qualitatively similar to the well-known Feynman path integral [214].

The above generalized Monte Carlo method has proven to be quite successful in understanding the fundamental physical processes that govern the ultrafast optical response of semiconductor materials; indeed, the latter has allowed to explain a wide variety of state-of-the-art coherent optical experiments (see, e.g., [281–284]).

Part II

State-of-the-Art Unipolar Quantum Devices: General Properties and Key Examples

6

Modeling of Unipolar Semiconductor Nanodevices

In this chapter we shall discuss in very general terms the most effective approaches for the study of unipolar transport in nanodevices; to this end, we shall address separately the low- and the high-field regimes, and for both regimes we shall provide a semiclassical treatment of the problem as well as its quantum-mechanical generalization.

As for any other semiconductor-based optoelectronic device, the modeling of unipolar nanodevices may be performed at different levels, each corresponding to a different degree of accuracy/approximation (see, e.g., [280]).

Within a strictly semiclassical picture, a first level of description is by analogy with n-level atomic systems and therefore in terms of global quantities. This approach is often grounded on purely macroscopic models, the only relevant physical quantities being the various carrier concentrations/populations N_n, within each device nanostructure subband n.[1] The time evolution of the macroscopic quantities N_n is typically described by a set of phenomenological rate equations of the form

$$\frac{dN_n}{dt} = \sum_{n'} \left(W_{nn'} N_{n'} - W_{n'n} N_n\right) + \left(S_n - \Gamma_n^b N_n\right) \quad . \tag{6.1}$$

This is the sum of an interlevel-scattering term ($W_{nn'}$ denoting phenomenological scattering rates connecting levels n' and n) plus an injection/loss contribution associated with the spatial boundaries of our unipolar device (S_n and Γ_n^b denoting, respectively, carrier injection and loss rates for level n).

A second and more refined (i.e., kinetic) semiclassical level of description is provided in terms of the single-particle distribution function f_α over the fully three-dimensional electronic states α of our nanodevice. As discussed in

[1] Such a macroscopic nanostructure description is analogous to the well-established hydrodynamic and drift-diffusion models for conventional microelectronic device simulations (see, e.g., [79]), where the key ingredients, i.e., total carrier density, average drift velocity, and average kinetic energy, correspond to the various momenta of the semiclassical distribution function.

Sect. 1.4 (see (1.83)), the equation of motion for f_α may be schematically written as

$$\frac{\partial f_\alpha}{\partial t} = \left.\frac{\partial f_\alpha}{\partial t}\right|_{\mathrm{d}} + \left.\frac{\partial f_\alpha}{\partial t}\right|_{\mathrm{dr}} . \tag{6.2}$$

Here, the first term describes scattering dynamics within the device active region and is usually treated at a kinetic level via a Boltzmann-like collision operator of the form

$$\left.\frac{\partial f_\alpha}{\partial t}\right|_{\mathrm{d}} = \sum_{\alpha'} \left[P_{\alpha\alpha'} f_{\alpha'} - P_{\alpha'\alpha} f_\alpha\right] , \tag{6.3}$$

where $P_{\alpha\alpha'}$ is the total scattering rate (i.e., summed over all relevant interaction mechanisms including light–matter coupling) from state α' to state α. The second term in (6.2) accounts for the open character of the system and describes injection/loss contributions from/to the (at least two) external carrier reservoirs. According to (1.84), these processes are usually modeled by a relaxation-time-like term of the form

$$\left.\frac{\partial f_\alpha}{\partial t}\right|_{\mathrm{dr}} = -\Gamma^b_\alpha \left(f_\alpha - f^\circ_\alpha\right) = S_\alpha - \Gamma^b_\alpha f_\alpha , \tag{6.4}$$

where Γ^b_α may be interpreted as the inverse device transit time for an electron in state α, while f°_α is the single-particle carrier distribution in the external reservoirs. The latter may describe distinct quasiequilibrium distributions corresponding to left and right chemical potentials or may account for a generic nonequilibrium carrier distribution within the external reservoirs.

Employing this kinetic approach, let us now show that the physical parameters entering the effective equation (6.1) are unavoidably global or macroscopic quantities, i.e., their microscopic evaluation would require the knowledge of the carrier distribution function f_α. Indeed, for the relevant case of a quasi-two-dimensional unipolar nanodevice, the average carrier occupation/population in a given subband n is the sum/integral over the in-plane (or parallel) wavevector $\boldsymbol{k}_\parallel$ of the single-particle distribution f_α; more specifically, in terms of the conventional envelope-function picture introduced in Sect. 1.2.2 we have $\alpha = \nu, \overline{\alpha} \equiv \nu, \boldsymbol{k}_\parallel n$, i.e., the generic envelope-function state α is specified by its band index ν plus its envelope-function index $\overline{\alpha} = \boldsymbol{k}_\parallel n$. For unipolar transport problems, the band index ν may be disregarded, and the link between the macroscopic carrier populations in (6.1) and the microscopic carrier distribution is given by

$$N_n = \sum_{\boldsymbol{k}_\parallel} f_{\boldsymbol{k}_\parallel n} . \tag{6.5}$$

Let us now reconsider the Boltzmann-like equation (6.3) written in this envelope-function representation ($\boldsymbol{k}_\parallel n$), i.e.,

$$\left.\frac{\partial f_{\boldsymbol{k}_\parallel n}}{\partial t}\right|_{\mathrm{d}} = \sum_{\boldsymbol{k}'_\parallel n'} \left[P_{\boldsymbol{k}_\parallel n, \boldsymbol{k}'_\parallel n'} f_{\boldsymbol{k}'_\parallel n'} - P_{\boldsymbol{k}'_\parallel n', \boldsymbol{k}_\parallel n} f_{\boldsymbol{k}_\parallel n}\right] . \tag{6.6}$$

By summing the above three-dimensional transport equation over the in-plane wavevector $\boldsymbol{k}_{\|}$, we get

$$\left.\frac{\mathrm{d}N_n}{\mathrm{d}t}\right|_{\mathrm{d}} = \sum_{n'} \left(W_{nn'}N_{n'} - W_{n'n}N_n\right) \tag{6.7}$$

with

$$W_{nn'} = \frac{\sum_{\boldsymbol{k}_{\|},\boldsymbol{k}'_{\|}} P_{\boldsymbol{k}_{\|}n,\boldsymbol{k}'_{\|}n'} f_{\boldsymbol{k}'_{\|}n'}}{\sum_{\boldsymbol{k}'_{\|}} f_{\boldsymbol{k}'_{\|}n'}} \,. \tag{6.8}$$

The effective Boltzmann collision term (6.7) has indeed the same form of the one in (6.1); however, as clearly shown by (6.8), the effective interlevel scattering rates $W_{nn'}$ are obtained by averaging the microscopic transition probabilities $P_{\boldsymbol{k}_{\|}n,\boldsymbol{k}'_{\|}n'}$ over the initial and final in-plane carrier wavevectors $\boldsymbol{k}'_{\|}$ and $\boldsymbol{k}_{\|}$. It follows that their values are strongly affected by the actual nature and shape of the unknown in-plane carrier distribution.

For this reason, the phenomenological rate-equation scheme in (6.1) can only operate as an "a posteriori" fitting procedure. Indeed, rather than within a many-level system, the electron dynamics in semiconductor-based quantum devices generally occurs within a many-subband structure, and the existence of transverse, i.e., in-plane, degrees of freedom should be properly taken into account.

In view of their Boltzmann-like structures, the kinetic equations (6.3) is commonly "sampled" via the direct Monte Carlo simulation presented in Sect. 5.2.4; indeed, one of the main advantages of this technique is that it allows us to include, on equal footing, a large variety of scattering mechanisms.

At this point, it is imperative to stress that – in spite of the kinetic nature of the scattering dynamics in (6.3) – the contribution (6.4) still treats carrier-injection and loss processes on a partially phenomenological level. On the other hand, it is crucial to notice that state-of-the-art quantum devices often present an active region which consists of a periodic repetition of identical "stages." This is the case, for example, of the quantum-well infrared photodetectors discussed in Chap. 7 as well as of the quantum-cascade devices discussed in Chap. 8. In this case, the ending stages of the sequence are the only ones coupled to the external carrier reservoirs (via electric contacts), while each internal stage actually behaves as an injector/collector of carriers for the following/previous one. In the presence of a large number of stages (typical values are in the range 10–70) the charge dynamics, and therefore the nanodevice operation, is mainly determined by inter-stage coupling rather than by finite-size effects. It follows that such multistage nanostructures may be safely treated as a periodic repetition of a given device building block, neglecting any spatial-boundary effects.

By exploiting such device periodicity, fully kinetic approaches – alternative to the partially phenomenological one in (6.2) – may be introduced; this is typically realized by imposing proper boundary conditions that allow us to "close the circuit" and thus to employ a particle-conserving Monte Carlo simulation (see Chap. 5). More specifically, two different kinetic treatments may be considered, depending on the strength of the applied electric bias, i.e., low- versus high-field regimes.

6.1 Vertical Transport in the Low-Field Regime

Let us consider a periodic nanodevice subjected to a small/moderate applied bias. As discussed in Sect. 1.2.2, the presence of a periodic confinement potential gives rise to a superlattice structure, whose electronic spectrum exhibits the well-known miniband character (see Fig. 1.18); more specifically, within the standard envelope-function picture (see Sect. 1.2.2), the electronic-state label is specified via the three-dimensional carrier wavevector ($\boldsymbol{k} = \boldsymbol{k}_{\parallel}, \boldsymbol{k}_{\perp}$) plus the miniband index $\overline{\nu}$ (see (1.51)), i.e., $\overline{\alpha} = \boldsymbol{k}\overline{\nu}$.

For a wide class of nanodevices, including the quantum-well infrared photodetectors presented in Chap. 7, the relatively small applied bias (corresponding to a given electric field $\boldsymbol{E}$) may be safely described via the conventional drift term of the semiclassical theory; in particular, for spatially homogeneous problems (along $\boldsymbol{r}_{\parallel}$), the corresponding semiclassical transport equation is formally identical to the bulk one in (1.67), provided one neglects the diffusion term and replaces the conventional band index ν with the miniband index $\overline{\nu}$:

$$\left(\frac{\partial}{\partial t} - \frac{e\boldsymbol{E}}{\hbar} \cdot \boldsymbol{\nabla}_{\boldsymbol{k}}\right) f_{\boldsymbol{k}\overline{\nu}} = \sum_{\boldsymbol{k}'\overline{\nu}'} \left[P_{\boldsymbol{k}\overline{\nu},\boldsymbol{k}'\overline{\nu}'} f_{\boldsymbol{k}'\overline{\nu}'} - P_{\boldsymbol{k}'\overline{\nu}',\boldsymbol{k}\overline{\nu}} f_{\boldsymbol{k}\overline{\nu}}\right] \tag{6.9}$$

with

$$P_{\boldsymbol{k}\overline{\nu},\boldsymbol{k}'\overline{\nu}'} = (1 - f_{\boldsymbol{k}\overline{\nu}}) \sum_{s} P^{s}_{\boldsymbol{k}\overline{\nu},\boldsymbol{k}'\overline{\nu}'} \ . \tag{6.10}$$

As anticipated, the system periodicity allows us to investigate the current–voltage characteristics of the nanodevice under examination, without resorting to partially phenomenological injection/loss models; indeed, the unipolar current density may be obtained directly from the solution $f_{\boldsymbol{k}\overline{\nu}}$ of the Boltzmann equation (6.9) according to

$$\boldsymbol{J}(\boldsymbol{E}) = -\frac{e}{\Omega} \sum_{\boldsymbol{k}\overline{\nu}} \boldsymbol{v}_{\boldsymbol{k}\overline{\nu}} f_{\boldsymbol{k}\overline{\nu}} \ , \qquad \boldsymbol{v}_{\boldsymbol{k}\overline{\nu}} = \frac{1}{\hbar} \boldsymbol{\nabla}_{\boldsymbol{k}} \epsilon_{\boldsymbol{k}\overline{\nu}} \ . \tag{6.11}$$

As discussed extensively in [240], the transport equation (6.9) provides a fully three-dimensional description of both localized (bound) and extended (continuum) states, crucial ingredient for a fully microscopic treatment of ionization versus capture processes in quantum-well-based optoelectronic devices (see, e.g., [64] and references therein). To better elucidate this crucial feature, let us consider a modified version of the superlattice structure of Fig. 1.17: by

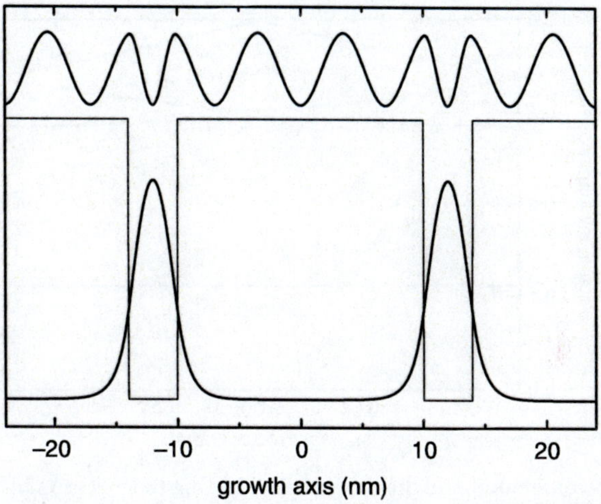

Fig. 6.1. Electronic confinement-potential profile and charge distributions for a GaAs/AlGaAs multiple quantum well with well (GaAs) and barrier (AlGaAs) widths of 4 and 20 nm, respectively, and with a band-offset of 0.28 eV. Here, the two spatial charge distributions correspond to the squared moduli of the carrier envelope wavefunction (evaluated at the minizone center) for the bound-state miniband 1 and continuum-state miniband 4 of Fig. 6.2

increasing the AlGaAs barrier width by a factor 10 (from 2 to 20 nm), we obtain the GaAs/AlGaAs multiple quantum-well structure depicted in Fig. 6.1. Since our primary interest is the interplay between localized and extended states, the corresponding electronic miniband diagram reported in Fig. 6.2 includes also a significant portion of the energy continuum.

By comparing Figs. 1.17 and 1.18 with Figs. 6.1 and 6.2, we notice that the significant increase of the barrier width gives rise to a strong charge confinement of the bound states accompanied by a nearly total suppression of interwell tunneling (see Fig. 6.1); the latter, in turn, manifests itself via a nearly dispersionless character of the bound-state miniband 1 (see Fig. 6.2). In contrast, the charge distribution corresponding to continuum states in miniband 4 is only partially influenced by the confinement profile and extends over a large fraction of our multiple-well structure; this is confirmed by the dispersive character of the corresponding minibands in the continuum (see Fig. 6.2).

The above comparison tells us that while for the superlattice structure of Fig. 1.17 vertical-transport phenomena may easily take place thanks to the dispersive low-energy miniband of Fig. 1.18 (see, e.g., [285] and references therein), for the multiple quantum well of Fig. 6.1 transport within the bound-state miniband is nearly suppressed due to its dispersionless character,[2] and

[2] Indeed, in the dispersionless-miniband limit, the perpendicular component of the carrier group velocity $\boldsymbol{v}_{\boldsymbol{k}\overline{\nu}}$ in (6.11) goes to zero and therefore the miniband current density $\boldsymbol{J}_{\perp}$ vanishes.

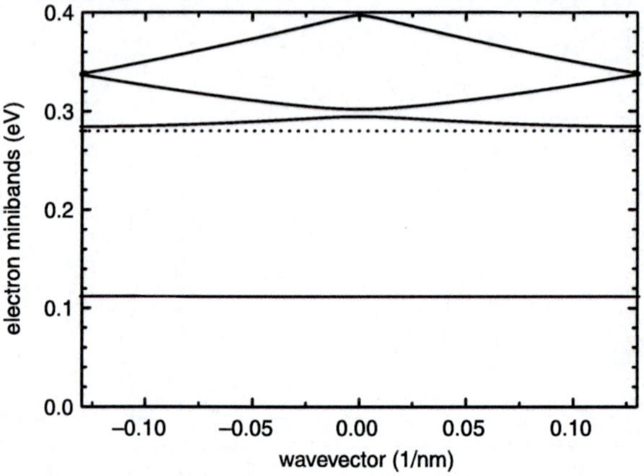

Fig. 6.2. Electronic miniband diagram corresponding to the GaAs/AlGaAs multiple quantum-well structure of Fig. 6.1. Here, we deal with a single-bound-state miniband (lower than the barrier height) plus a number of continuous-state minibands separated by small minigaps

vertical transport is basically determined/controlled by charge flow within the continuum-state minibands of Fig. 6.2.

It is then clear that the current–voltage characteristics of such kind of nanostructures are strongly determined/influenced by the fraction of carriers within the continuum as well as by their average group velocity. This peculiar feature has made such class of nanodevices candidates-of-choice for infrared detection (see, e.g., [64] and references therein); indeed, a closer inspection of the typical energy scale of the miniband diagram in Fig. 6.2 shows that the presence of electromagnetic radiation in the infrared spectral region may give rise to a significant fraction of photon-induced current, usually referred to as photocurrent, due to bound-to-continuum transitions induced by interminiband photon absorption.

As shown in [240], the actual shape of the carrier distribution over the continuum states is the result of a non-trivial interplay between phonon- and photon-induced ionization versus capture processes in the presence of an applied vertical bias. It is exactly for this kind of transport problems that the fully kinetic approach in (6.9) has revealed all its power and flexibility. Indeed, the latter has allowed to perform parameter-free simulated experiments [240] via the Weighted Monte Carlo simulation strategy discussed in Sect. 5.2.5; to this end, the envelope-function single-particle states $\boldsymbol{k}\overline{\nu}$ have been computed via a self-consistent solution of Schrödinger and Poisson equations (see (5.36)) via a suitable plane-wave expansion (see Sect. 5.1.2). Such a fully three-dimensional description of carrier wavefunctions and energy bands has allowed for a consistent treatment of all relevant electron–phonon scattering rates for both intra- and interminiband transitions, thus providing a quantitative (i.e., parameter-free) evaluation of scattering-induced coupling between extended (bulk-like) and localized (quantum-well-like) states in the vertical carrier transport across a generic heterostructure. In particular, many of the simulated experiments in [240] have been devoted to the investigation of the thermionic-emission current in multiple quantum-well structures; indeed, this phenomenon has been widely studied since the end of the 1980s owing to its role in determining the "dark current" in quantum-well infrared photodetectors (see, e.g., [64] and references therein).

To better clarify the role and relevance of the parameter-free kinetic approach reviewed so far, it is worth mentioning that many of the transport models employed for the investigation of quantum-well infrared photodetectors (see, e.g., [42] and references therein) are based on strongly simplified electronic band diagrams: the continuum portion of the spectrum is typically described via three-dimensional bulk states, while bound levels are simply treated as purely two-dimensional states. Such approximated treatments are intrinsically unable to provide a microscopic (i.e., parameter-free) description of ionization (bound-to-continuum) versus capture (continuum-to-bound) processes, since within such simplified scheme their initial and final states are not treated on equal footing (i.e., two- versus three-dimensional wavefunctions); it follows that all bound-to-continuum as well as continuum-to-bound transition rates cannot be evaluated microscopically and thus enter the simulation as fitting parameters. Therefore, in spite of their high computational efficiency, such simplified transport models – exactly as for the case of the phenomenological rate-equation scheme in (6.1) – can only operate as an "a posteriori" fitting procedure.

As a final step, let us discuss the quantum-mechanical generalization of the superlattice transport model presented so far. To this end, the semiclassical equation (6.9) needs to be replaced by a corresponding quantum equation for the single-particle density matrix (3.116); in the low-density regime (see Sect. 3.3.3) the desired density-matrix equation reduces to the linear problem in (3.146). More specifically, within the superlattice envelope-function basis employed so far ($\overline{\alpha} = \boldsymbol{k}\overline{\nu}$) and assuming a macroscopically homogeneous intraminiband charge distribution, the single-particle density matrix comes out to be diagonal with respect to the $\boldsymbol{k}$ coordinate (see, e.g., [110]):

$$\rho_{\overline{\alpha}\overline{\alpha}'} = \rho_{\boldsymbol{k},\overline{\nu}\overline{\nu}'}\delta_{\boldsymbol{k}\boldsymbol{k}'} \,. \tag{6.12}$$

Moreover, as discussed in [161], in the low-bias regime the field-induced variation of the single-particle density matrix (6.12) may be expressed again via the drift term of the semiclassical theory. It follows that in this case the quantum-transport equation (3.146) reduces to

$$\left(\frac{\partial}{\partial t} - \frac{e\boldsymbol{E}}{\hbar}\cdot\boldsymbol{\nabla}_{\boldsymbol{k}}\right)\rho_{\boldsymbol{k},\overline{\nu}_1\overline{\nu}_2} = \sum_{\boldsymbol{k}',\overline{\nu}'_1\overline{\nu}'_2} \mathcal{L}_{\boldsymbol{k},\overline{\nu}_1\overline{\nu}_2;\boldsymbol{k}',\overline{\nu}'_1\overline{\nu}'_2}\rho_{\boldsymbol{k}',\overline{\nu}'_1\overline{\nu}'_2} \,, \tag{6.13}$$

where $\mathcal{L}_{\boldsymbol{k},\overline{\nu}_1\overline{\nu}_2;\boldsymbol{k}',\overline{\nu}'_1\overline{\nu}'_2}$ are the ($\boldsymbol{k}$-diagonal) matrix elements of the Liouville superoperator (3.147) within the field-free superlattice basis $\overline{\alpha} = \boldsymbol{k}\overline{\nu}$. We stress the strong similarity between the quantum-transport equation (6.13) and the semiclassical result in (6.9); indeed, the latter may be readily obtained by inserting into the quantum equation the usual semiclassical approximation:

$$\rho_{\mathbf{k},\overline{\nu}\overline{\nu}'} = f_{\mathbf{k}\overline{\nu}} \delta_{\overline{\nu}\overline{\nu}'} \,. \tag{6.14}$$

This amounts to neglecting interminiband ($\overline{\nu} \neq \overline{\nu}'$) quantum-mechanical phase coherence. It follows that, in order to describe coherent phenomena in superlattice-based nanodevices (governing, in particular, ultrafast electro-optical processes (see Chap. 2)), the inclusion of such interminiband density-matrix elements is imperative; this is realized by replacing the semiclassical treatment in (6.9) with its quantum-mechanical generalization in (6.13).

As a final remark, it is important to recall that the proposed quantum-transport equation applies to the low-field regime only; in contrast, as discussed extensively in Sect. 3.4, in the presence of strong applied fields the superlattice field-free basis needs to be replaced by the corresponding (field-induced) Wannier–Stark states (see [87, 88]).

6.2 Vertical Transport in the High-Field Regime

As anticipated, state-of-the-art semiconductor-based optoelectronic devices at the nanoscale are often based on periodically repeated core regions, for which spatial-boundary effects may be safely neglected. This strategy is exploited, for example, to increase either the photocurrent in infrared detectors (see Chap. 7) or the emitted power in coherent-light sources like quantum-cascade lasers (see Chap. 8). This feature turns out to be crucial also in the proper modeling of these unipolar devices, since it allows one to consistently "close the circuit" without the need of any phenomenological parameter accounting for the coupling of the device with its electric contacts. While for small/moderate applied fields (see Sect. 6.1) it is possible to exploit the fully three-dimensional superlattice periodicity of the nanostructure under examination, in the presence of a significant applied bias the latter cannot be treated via the conventional drift term of the semiclassical theory (see, e.g., [161] and references therein). In this case, the applied field needs to be added to the nanostructure potential profile; as a result, the total potential acting on the carriers is not periodic anymore, i.e., the usual superlattice periodicity is lost. Nevertheless, by assuming a uniform bias drop, the effective field profile across the nanostructure is still periodic, which implies that also in the presence of a strong homogeneous external field all relevant physical properties (e.g., the semiclassical distribution function f_α) should display the periodicity of the nanostructure under investigation.

More specifically, by neglecting finite-size effects, the semiclassical approach providing a fully three-dimensional description of the whole (periodically repeated) nanodevice under examination is given by the Boltzmann-like equation (6.6), where the generic carrier state is specified by its parallel or in-plane wavevector $\boldsymbol{k}_\parallel$ plus the subband index n. Since such nanostructures are obtained as periodic repetition of a given building block usually referred to as "device stage," it follows that in the presence of an applied uniform electric field the Wannier–Stark-like states localized in the different stages are physically equivalent (i.e., their envelope wavefunctions differ by a phase factor only) (see, e.g., [178]). Based on such translational symmetry, the generic subband index n – running over the whole nanostructure – may be conveniently replaced by a stage index ℓ plus an intra–stage subband index $\tilde{n}$, i.e., $\overline{\alpha} = \boldsymbol{k}_\parallel n = \boldsymbol{k}_\parallel \ell \tilde{n}$. Employing this new notation, the Boltzmann-like transport equation (6.6) comes out to be

$$\frac{\partial f_{\boldsymbol{k}_\parallel \ell \tilde{n}}}{\partial t} = \sum_{\boldsymbol{k}'_\parallel \ell' \tilde{n}'} \left[P_{\boldsymbol{k}_\parallel \ell \tilde{n}, \boldsymbol{k}'_\parallel \ell' \tilde{n}'} f_{\boldsymbol{k}'_\parallel \ell' \tilde{n}'} - P_{\boldsymbol{k}'_\parallel \ell' \tilde{n}', \boldsymbol{k}_\parallel \ell \tilde{n}} f_{\boldsymbol{k}_\parallel \ell \tilde{n}} \right] . \tag{6.15}$$

In view of the spatial periodicity previously mentioned, the carrier distribution function should also be stage independent:

$$f_{\boldsymbol{k}_\parallel \ell \tilde{n}} = f_{\boldsymbol{k}_\parallel \ell' \tilde{n}} . \tag{6.16}$$

Thanks to this crucial symmetry property, it is possible to reduce the global transport equation (6.15) to the following single-stage version:

$$\frac{\partial f_{\boldsymbol{k}_\parallel \tilde{n}}}{\partial t} = \sum_{\boldsymbol{k}'_\parallel \tilde{n}'} \left[\tilde{P}_{\boldsymbol{k}_\parallel \tilde{n}, \boldsymbol{k}'_\parallel \tilde{n}'} f_{\boldsymbol{k}'_\parallel \tilde{n}'} - \tilde{P}_{\boldsymbol{k}'_\parallel \tilde{n}', \boldsymbol{k}_\parallel \tilde{n}} f_{\boldsymbol{k}_\parallel \tilde{n}} \right] \tag{6.17}$$

with

$$\tilde{P}_{\boldsymbol{k}_\parallel \tilde{n}, \boldsymbol{k}'_\parallel \tilde{n}'} = \sum_{\ell'} P_{\boldsymbol{k}_\parallel \ell \tilde{n}, \boldsymbol{k}'_\parallel \ell' \tilde{n}'} . \tag{6.18}$$

This constitutes a strong simplification of the problem, since within the present single-stage scheme our kinetic description of the device is limited, e.g., to the $\ell = 0$ stage only (see, for example, the quantum-cascade structure depicted in Fig. 8.1) and thus to a limited number of subbands $\tilde{n}$ (typically of the order of 10); all the required information regarding the inter-stage carrier coupling is provided by the reduced (or single-stage) scattering rates $\tilde{P}$ in (6.18).[3,4]

As anticipated, within such single-stage picture each device stage actually behaves as an injector/collector of carriers for the following/previous one; this

[3] In view of the spatial periodicity, all inter-stage scattering rates depend on the stage difference $\ell' - \ell$ only: $P_{\boldsymbol{k}_\parallel, \ell, \tilde{n}; \boldsymbol{k}'_\parallel, \ell', \tilde{n}'} = P_{\boldsymbol{k}_\parallel, 0, \tilde{n}; \boldsymbol{k}'_\parallel, \ell' - \ell, \tilde{n}'}$; it follows that the single-stage rates $\tilde{P}$ in (6.18) are ℓ-independent as well.

[4] Within a conventional Monte Carlo simulation (see Sect. 5.2), the evaluation of the single-stage scattering rates $\tilde{P}$ is not particularly demanding, since the latter is performed just once at the beginning of our simulated experiment.

suggests to rewrite the global transport equation (6.15) splitting the scattering term into intra-stage ($\ell = \ell'$) and inter-stage ($\ell \neq \ell'$) contributions:

$$\frac{\partial f_{\mathbf{k}_\parallel \tilde{n}}}{\partial t} = \left.\frac{\partial f_{\mathbf{k}_\parallel \tilde{n}}}{\partial t}\right|_{\text{intra}} + \left.\frac{\partial f_{\mathbf{k}_\parallel \tilde{n}}}{\partial t}\right|_{\text{inter}} \tag{6.19}$$

with

$$\left.\frac{\partial f_{\mathbf{k}_\parallel \tilde{n}}}{\partial t}\right|_{\text{intra}} = \sum_{\mathbf{k}'_\parallel \tilde{n}'} \left[\tilde{P}^0_{\mathbf{k}_\parallel \tilde{n},\mathbf{k}'_\parallel \tilde{n}'} f_{\mathbf{k}'_\parallel \tilde{n}'} - \tilde{P}^0_{\mathbf{k}'_\parallel \tilde{n}',\mathbf{k}_\parallel \tilde{n}} f_{\mathbf{k}_\parallel \tilde{n}}\right] \tag{6.20}$$

and

$$\left.\frac{\partial f_{\mathbf{k}_\parallel \tilde{n}}}{\partial t}\right|_{\text{inter}} = \tilde{S}_{\mathbf{k}_\parallel \tilde{n}} - \tilde{\Gamma}_{\mathbf{k}_\parallel \tilde{n}} f_{\mathbf{k}_\parallel \tilde{n}} \,. \tag{6.21}$$

Here, the first term describes intra-stage scattering processes only (induced by the corresponding rates $\tilde{P}^0_{\mathbf{k}_\parallel \tilde{n},\mathbf{k}'_\parallel \tilde{n}'} = P_{\mathbf{k}_\parallel 0 \tilde{n},\mathbf{k}'_\parallel 0 \tilde{n}'}$) while the second contribution has been intentionally expressed in terms of a source/injection term

$$\tilde{S}_{\mathbf{k}_\parallel \tilde{n}} = \sum_{\ell \neq 0} \sum_{\mathbf{k}'_\parallel \tilde{n}'} P_{\mathbf{k}_\parallel 0 \tilde{n},\mathbf{k}'_\parallel \ell \tilde{n}'} f_{\mathbf{k}'_\parallel \tilde{n}'} \tag{6.22}$$

and of a loss/escape rate

$$\tilde{\Gamma}_{\mathbf{k}_\parallel \tilde{n}} = \sum_{\ell \neq 0} \sum_{\mathbf{k}'_\parallel \tilde{n}'} P_{\mathbf{k}'_\parallel \ell \tilde{n}',\mathbf{k}_\parallel 0 \tilde{n}} \,. \tag{6.23}$$

We stress the strong formal similarity between the partially phenomenological approach in (6.2) and the single-stage formulation in (6.19): the injection/loss term (6.21) describes carriers entering/leaving our reference stage ($\ell = 0$) from/to any other stage ($\ell \neq 0$). It follows that the rigorous single-stage result (6.19) may be regarded as a formal derivation/justification of the partially phenomenological treatment in (6.2). However, it is imperative to notice that, strictly speaking, the term (6.22) is not a genuine source contribution, since the latter depends on the actual carrier distribution f within the external stages, whose knowledge, in turn, requires to solve the single-stage Boltzmann equation (6.17).

As anticipated, the fully kinetic approach in (6.17) provides a parameter-free description of vertical-transport phenomena across our periodic nanostructure. Indeed, given the steady-state solution $f_{\mathbf{k}_\parallel \tilde{n}}$, the device current density may be written as

$$J = -\frac{e}{\Omega_\parallel} \sum_{\mathbf{k}_\parallel \tilde{n},\mathbf{k}'_\parallel \tilde{n}'} \left(\tilde{P}^+_{\mathbf{k}'_\parallel \tilde{n}',\mathbf{k}_\parallel \tilde{n}} - \tilde{P}^-_{\mathbf{k}'_\parallel \tilde{n}',\mathbf{k}_\parallel \tilde{n}}\right) f_{\mathbf{k}_\parallel \tilde{n}} \tag{6.24}$$

with

$$\tilde{P}^\pm_{\mathbf{k}'_\parallel \tilde{n}',\mathbf{k}_\parallel \tilde{n}} = \sum_{\ell > 0} P_{\mathbf{k}'_\parallel,\pm\ell,\tilde{n}';\mathbf{k}_\parallel,0,\tilde{n}} \,. \tag{6.25}$$

Here, the $\pm$ sign refers to left- and right-side stages, respectively.[5]

A closer inspection of the current-density equation (6.24) shows that within the present semiclassical picture $\overline{\alpha} = \boldsymbol{k}_\parallel \tilde{n}$ we deal with scattering-induced transport only; indeed, in the absence of any scattering mechanism ($P_{\boldsymbol{k}_\parallel \ell \tilde{n}, \boldsymbol{k}'_\parallel \ell' \tilde{n}'} = 0$) the current density is always equal to zero. This non-conventional behavior – compared to the standard bulk-transport scenario – may also be ascribed to our semiclassical treatment, based on the neglect of so-called coherent-transport phenomena, i.e., phenomena for which the quantum-mechanical phase relation between different basis states plays a dominant role. As we shall discuss in Chap. 8, the latter are essential for the understanding of ultrafast electro-optical processes in quantum-cascade devices.

In order to properly describe coherent-transport phenomena, the basic step is, once again, to replace the semiclassical transport equation (6.17) with a corresponding quantum equation for the single-particle density matrix in (3.116). More specifically, assuming a homogeneous in-plane charge distribution and neglecting any inter-stage ($\ell \neq \ell'$) phase coherence,[6] the single-particle density matrix written in our periodic-nanostructure basis comes out to be diagonal with respect to the $\boldsymbol{k}_\parallel$ coordinate (see, e.g., [110]) as well as with respect to the stage index ℓ:

$$\rho_{\overline{\alpha}\overline{\alpha}'} = \rho_{\boldsymbol{k}_\parallel \ell, \tilde{n}\tilde{n}'} \delta_{\boldsymbol{k}_\parallel \boldsymbol{k}'_\parallel} \delta_{\ell\ell'} \ . \tag{6.26}$$

Moreover, in view of the spatial periodicity previously mentioned (see (6.16)), the above single-particle density matrix should also be stage independent:

$$\rho_{\boldsymbol{k}_\parallel \ell, \tilde{n}\tilde{n}'} = \rho_{\boldsymbol{k}_\parallel \ell', \tilde{n}\tilde{n}'} \ . \tag{6.27}$$

As discussed extensively in Chap. 3, the derivation of a closed equation of motion for the single-particle density matrix is a highly non-trivial task involving a number of interconnected approximations (see Sect. 3.3.3). In particular, in the low-density regime it is possible to derive the linear quantum equation (3.146) involving the Liouville superoperator in (3.147). In view of the specific properties of the single-particle density matrix in (6.26) and (6.27), for the case of our periodic-nanodevice problem the quantum-transport equation (3.146) comes out to be of the form

$$\frac{d\rho_{\boldsymbol{k}_\parallel, \tilde{n}_1\tilde{n}_2}}{dt} = \frac{\epsilon_{\boldsymbol{k}_\parallel \tilde{n}_1} - \epsilon_{\boldsymbol{k}_\parallel \tilde{n}_2}}{i\hbar} \rho_{\boldsymbol{k}_\parallel, \tilde{n}_1\tilde{n}_2} + \sum_{\boldsymbol{k}'_\parallel, \tilde{n}'_1\tilde{n}'_2} \tilde{\Gamma}_{\boldsymbol{k}_\parallel, \tilde{n}_1\tilde{n}_2; \boldsymbol{k}'_\parallel, \tilde{n}'_1\tilde{n}'_2} \rho_{\boldsymbol{k}'_\parallel, \tilde{n}'_1\tilde{n}'_2} \tag{6.28}$$

with

$$\tilde{\Gamma}_{\boldsymbol{k}_\parallel, \tilde{n}_1\tilde{n}_2; \boldsymbol{k}'_\parallel, \tilde{n}'_1\tilde{n}'_2} = \sum_\ell \Gamma_{\boldsymbol{k}_\parallel, 0, \tilde{n}_1\tilde{n}_2; \boldsymbol{k}'_\parallel, \ell, \tilde{n}'_1\tilde{n}'_2} \ . \tag{6.29}$$

[5] The current density (6.24) corresponds to the out-scattering charge flux across the right stage boundary (+ sign) minus the out-scattering charge flux across the left one (– sign).

[6] This amounts to saying that the electron coherence length is significantly shorter than the period of our nanostructure.

Exactly as for the semiclassical transport equation (6.19), the quantum equation (6.28) is based on a single-stage description. Here, the key ingredient is the single-stage scattering superoperator $\tilde{\Gamma}$ in (6.29). The latter may be regarded as the quantum-mechanical generalization of the single-stage scattering rates (6.18); indeed, in the usual semiclassical limit,

$$\rho_{\boldsymbol{k}_\parallel,\tilde{n}\tilde{n}'} = f_{\boldsymbol{k}_\parallel\tilde{n}}\delta_{\tilde{n}\tilde{n}'} \;, \tag{6.30}$$

the nanodevice quantum equation (6.28) reduces to the single-stage Boltzmann equation (6.19).

At this point, an important comment is in order. The previous quantum modeling based on the linear transport equation (3.146) applies to nanodevices operating in the low-density regime only. This tells us that, in order to properly model quantum devices characterized by a significant subband population – and thus by a relevant carrier–carrier interaction dynamics (see Chap. 8) – the low-density linear superoperator $\mathcal{L}$ in (6.28) should be replaced by nonlinear in- and out-scattering superoperators describing all relevant interaction mechanisms, e.g., carrier–phonon, carrier–carrier (see (3.132) and (3.138)). The resulting nonlinear quantum equation written again within our single-stage picture ($\overline{\alpha} = \boldsymbol{k}_\parallel\tilde{n}$) is of the form

$$\frac{d\rho_{\boldsymbol{k}_\parallel,\tilde{n}_1\tilde{n}_2}}{dt} = \frac{\epsilon_{\boldsymbol{k}_\parallel\tilde{n}_1} - \epsilon_{\boldsymbol{k}_\parallel\tilde{n}_2}}{i\hbar}\rho_{\boldsymbol{k}_\parallel,\tilde{n}_1\tilde{n}_2} + \sum_s \left[\mathcal{F}^s_{\text{in}}(\rho)_{\boldsymbol{k}_\parallel,\tilde{n}_1\tilde{n}_2} - \mathcal{F}^s_{\text{out}}(\rho)_{\boldsymbol{k}_\parallel,\tilde{n}_1\tilde{n}_2}\right] \;, \tag{6.31}$$

where the label s denotes the generic interaction mechanism.

As discussed in Chap. 5, in spite of the potential nonlinear character of the nanodevice quantum equation (6.31) as well as of its semiclassical counterpart in (6.19), the latter may be instantaneously linearized via suitable time-step discretization techniques, and within such approximation schemes they may be numerically solved via a proper combination of finite-difference and Monte Carlo strategies (see Sect. 5.3).

6.3 Investigation of Coupled Carrier–Quasiparticle Nonequilibrium Regimes

According to the conventional treatment of carrier–quasiparticle interaction presented in Chap. 3, the typical assumption is to consider the quasiparticle subsystem (e.g., photons, phonons, plasmons) as characterized by a huge number of degrees of freedom (compared to the carrier subsystem). In other words this amounts to saying that the $\boldsymbol{q}$ subsystem has an infinitely high heat capacity, i.e., it behaves as a thermal bath; this allows one to consider the quasiparticle subsystem always in thermal equilibrium, i.e., not significantly perturbed by the carrier dynamics. Such approximation scheme allows us to write the global (carrier plus quasiparticle) density-matrix operator according to the factorization (3.76). In turn, this amounts to saying that the single-particle density matrix of the quasiparticle subsystem is always diagonal, i.e.,

$$\rho_{\boldsymbol{q}\boldsymbol{q}'} = \mathrm{tr}\left\{\hat{b}^\dagger_{\boldsymbol{q}'}\hat{b}_{\boldsymbol{q}}\hat{\rho}\right\} = N^\circ_{\boldsymbol{q}}\delta_{\boldsymbol{q}\boldsymbol{q}'} \; , \tag{6.32}$$

where $N^\circ_{\boldsymbol{q}}$ is the usual Bose occupation number in (3.90).

For many ultrafast and/or nonequilibrium carrier phenomena in nanomaterials and nanodevices, however, this assumption is highly questionable, and the carrier–quasiparticle treatment presented in Chap. 3 should be extended in order to account for such quasiparticle nonequilibrium regimes. Generally speaking, the latter may be classified according to the following qualitative subdivision. On the one hand, we deal with coherent quasiparticle phenomena, where the non-diagonal density-matrix elements in (6.32) play a crucial role; in addition to various coherent-light effects introduced in Chap. 2 as well as to quantum-optics phenomena (see, e.g., [286–288] and references therein), a typical example, discussed extensively in [110], is that of the so-called coherent phonons. On the other hand, also in the absence of quasiparticle phase coherence, the diagonal density-matrix elements in (6.32) may deviate significantly from the equilibrium Bose distribution $N^\circ_{\boldsymbol{q}}$ in (3.90); two relevant examples of such nonequilibrium phenomena, both discussed in Chap. 8, are photonic cavity–mode amplification and hot-phonon effects (see below).

Since for many of the quantum devices considered in this book coherent quasiparticle effects play a very minor role,[7] in the reminder of this section we shall mainly focus on the second class of nonequilibrium phenomena, whose treatment may be performed within a purely semiclassical picture; moreover, for the sake of simplicity, we shall also neglect carrier phase coherence. Within such a fully semiclassical scenario, our kinetic description will simply involve the carrier distribution

$$f_\alpha = \mathrm{tr}\left\{\hat{c}^\dagger_\alpha \hat{c}_\alpha \hat{\rho}\right\} \tag{6.33}$$

as well as the quasiparticle distribution

$$N_{\boldsymbol{q}} = \mathrm{tr}\left\{\hat{b}^\dagger_{\boldsymbol{q}}\hat{b}_{\boldsymbol{q}}\hat{\rho}\right\} \; . \tag{6.34}$$

Following the conventional single-particle correlation expansion of the quantum-kinetic theory (recalled in Sect. 3.3.3 and reviewed extensively in [110]) and neglecting all non-diagonal carrier as well as quasiparticle density-matrix elements, it is possible to derive a coupled set of nonlinear equations for the above carrier as well as quasiparticle distributions (see below).

As discussed in Chap. 2, in the presence of strong electro-optical excitations one deals with a nonequilibrium carrier distribution f_α, also referred to

[7] Indeed, the only quasiparticle coherent phenomena addressed extensively in this book are those related to coherent-light–matter coupling induced by laser sources; however, in this case the light field is typically treated within a purely classical framework.

as hot-carrier regime[8]; in such conditions the carrier subsystem may transfer a relevant fraction of its excess energy to the quasiparticle one via carrier–quasiparticle interaction. Qualitatively speaking, within the conventional scenario presented in Chap. 3 energy dissipation is a single-step process: the quasiparticle subsystem is assumed to have an infinitely high heat capacity, thus absorbing a large fraction of carrier energy without changing its temperature (thermal-bath behavior). In contrast, for particularly strong carrier–quasiparticle couplings and/or quasiparticle subsystems with a small heat capacity, the latter may be easily driven out of equilibrium, exhibiting so-called hot-quasiparticle phenomena; in this case, energy dissipation comes out to be a two-step process: the carrier subsystem transfers again a significant fraction of energy to the quasiparticle one; in turn, the quasiparticle subsystem will transfer its excess energy to additional degrees of freedom characterized by a much larger heat capacity (given, e.g., by other quasiparticle excitations).

More specifically, adopting again as prototypical carrier–quasiparticle interaction Hamiltonian the general form in (3.137), the time evolution induced by such carrier–quasiparticle coupling is described by the following set of coupled kinetic equations:

$$\begin{aligned}
\left.\frac{\mathrm{d}f_\alpha}{\mathrm{d}t}\right|_{\mathrm{c-qp}} &= \sum_{\boldsymbol{q}\pm}\left(N_{\boldsymbol{q}}+\frac{1}{2}\pm\frac{1}{2}\right)\sum_{\alpha'}\left[(1-f_\alpha)\,P^{\boldsymbol{q}\pm}_{\alpha\alpha'}f_{\alpha'}-(1-f_{\alpha'})\,P^{\boldsymbol{q}\pm}_{\alpha'\alpha}f_\alpha\right] \\
\left.\frac{\mathrm{d}N_{\boldsymbol{q}}}{\mathrm{d}t}\right|_{\mathrm{c-qp}} &= \sum_{\pm}\pm\left(N_{\boldsymbol{q}}+\frac{1}{2}\pm\frac{1}{2}\right)\sum_{\alpha\neq\alpha'}(1-f_\alpha)\,P^{\boldsymbol{q}\pm}_{\alpha\alpha'}f_{\alpha'}-\frac{N_{\boldsymbol{q}}-N^\circ_{\boldsymbol{q}}}{\tau^{\mathrm{th}}}
\end{aligned} \tag{6.35}$$

with

$$P^{\boldsymbol{q}\pm}_{\alpha\alpha'}=\frac{2\pi}{\hbar}\left|g^{\pm}_{\alpha\alpha',\boldsymbol{q}}\right|^2\delta(\epsilon_\alpha-\epsilon_{\alpha'}\pm\epsilon_{\boldsymbol{q}})=P^{\boldsymbol{q}\mp}_{\alpha'\alpha}\,. \tag{6.36}$$

The first equation in (6.35) is still Boltzmann-like (i.e., in- minus out-scattering): it describes the effect on the carrier distribution f_α due to quasiparticle emission (+) and absorption (−) processes. The second equation, in contrast, is not Boltzmann like: the first term describes the effect on the quasiparticle distribution $N_{\boldsymbol{q}}$ due again to quasiparticle emission and absorption, while the second contribution accounts for thermalization effects on the quasiparticle subsystem induced by additional degrees of freedom; the latter are described via a standard relaxation-time model (see (1.71)) in terms of a phenomenological relaxation time τ^{th}.

[8] The origin of the term "hot carriers" is related to the fact that, in the presence of strong applied fields, the average kinetic energy of the electrons is significantly higher than the thermal one; in other words the effective electron temperature is significantly higher than the lattice one (see, e.g., [28]). It is important to stress, however, that by means of quasi-resonant optical excitations it is also possible to generate nonequilibrium carrier distributions characterized by extremely low electronic temperatures (see, e.g., [90]); it follows that the term "hot carriers" has a well-precise meaning only in high-field transport regimes.

In the strong-thermalization limit ($\tau^{\mathrm{th}} \to 0$), the quasiparticle distribution tends to its thermal value in (3.90): $N_{\boldsymbol{q}} \to N_{\boldsymbol{q}}^{\circ}$. In this case, the coupled set in (6.35) reduces to the carrier equation only, provided to replace the quasiparticle population $N_{\boldsymbol{q}}$ with its equilibrium value $N_{\boldsymbol{q}}^{\circ}$; indeed, it is easy to verify that in this limit the resulting transport equation coincides with the standard Boltzmann result in (3.139).

For many realistic situations, however, the thermalization time-scale τ^{th} is comparable to the typical carrier–quasiparticle scattering times. In this case, the resulting dynamics is characterized by a non-trivial interplay between carrier–quasiparticle energy dissipation and quasiparticle thermalization processes; it follows that a detailed modeling based on the coupled set of equations (6.35) is imperative.[9]

As anticipated, two prototypical examples of such two-step coupled dissipation dynamics will be addressed in Chap. 8. The first one is related to cavity–mode lasing (see, e.g., [66] and references therein): due to strong carrier–photon coupling in semiconductor-based electromagnetic cavities, the population of a given photonic mode may be strongly amplified with respect to its equilibrium value; in turn, such out-of-equilibrium population will be limited by cavity-loss processes. The second example is the so-called hot-phonon regime[10]: as originally pointed out by Paolo Lugli and Stephen M. Goodnick [121], photoexcited carriers in semiconductor nanostructures are found to emit a large fraction of optical phonons, thus driving the corresponding distribution out of equilibrium; such "hot" population, in turn, is limited/controlled via decay of optical phonons into acoustic ones (see, e.g., [105] and references therein).

We finally stress that in the absence of nonequilibrium sources like, e.g., strong electro-optical excitations, the global evolution of the carrier as well as of the quasiparticle distribution is fully dictated by the coupled set in (6.35). In this case, it is easy to verify that its steady-state solution coincides with the thermal-equilibrium one; indeed, the detailed-balance principle (see, e.g., [72]) applied to the carrier equation in (6.35) requires that

$$\frac{(1 - f_{\alpha})\, f_{\alpha'}}{(1 - f_{\alpha'})\, f_{\alpha}} = \frac{N_{\boldsymbol{q}}}{N_{\boldsymbol{q}} + 1} \,. \tag{6.37}$$

It is straightforward to check that the detailed-balance condition (6.37) is fulfilled by the following equilibrium carrier and quasiparticle distributions at the same temperature T:

---

[9] As discussed in Chap. 5, this can be conveniently performed via a proper Monte Carlo sampling based on a time-step discretization procedure (see Sect. 5.1.3).

[10] Exactly as for the carriers, the term "hot phonons" refers to the fact that the nonequilibrium phonon distribution – typically induced by strong carrier–phonon coupling – may be qualitatively described in terms of a higher effective temperature.

$$f_\alpha = f_\alpha^\circ = \frac{1}{e^{\frac{\epsilon_\alpha - \mu}{k_B T}} + 1}, \qquad N_{\boldsymbol{q}} = N_{\boldsymbol{q}}^\circ = \frac{1}{e^{\frac{\epsilon_{\boldsymbol{q}}}{k_B T}} - 1}. \tag{6.38}$$

Here, the carrier subsystem is described by the usual Fermi–Dirac distribution corresponding to its chemical potential μ (see (1.12)), while the quasiparticle subsystem is described by the usual Bose–Einstein distribution (see (3.90)).

In the low-density limit discussed in Sect. 3.3.3, the Pauli factors $(1 - f_\alpha)$ may be safely neglected, and the detailed-balance condition (6.37) reduces to

$$\frac{f_{\alpha'}}{f_\alpha} = \frac{N_{\boldsymbol{q}}}{N_{\boldsymbol{q}} + 1}. \tag{6.39}$$

This new condition is still fulfilled by the equilibrium solution (6.38) provided to replace the Fermi–Dirac distribution with its classical (i.e., Maxwell–Boltzmann) counterpart:

$$f_\alpha^\circ = e^{-\frac{\epsilon_\alpha - \mu}{k_B T}}. \tag{6.40}$$

At this point, it is worth stressing that the description of carrier–quasiparticle interaction presented in Chap. 3 – as well as its nonequilibrium generalization previously recalled – is based on a fully quantum-mechanical treatment of the quasiparticle subsystem; the latter may describe a variety of additional degrees of freedom, e.g., photons, phonons, plasmons. Within this treatment, in addition to quasiparticle absorption processes (proportional to $N_{\boldsymbol{q}}$), we deal with emission processes proportional to $N_{\boldsymbol{q}}+1$ (see (6.35)): the latter, in turn, may be regarded as a stimulated-emission contribution (i.e., proportional to $N_{\boldsymbol{q}}$) plus a so-called spontaneous-emission term (i.e., $N_{\boldsymbol{q}}$ independent). Such classification in terms of stimulated versus spontaneous emission processes comes out to play an important role in describing carrier–phonon as well as carrier-light processes in the so-called quantum limit, i.e., when the average quasiparticle occupation $N_{\boldsymbol{q}}$ is much smaller than (or of the order of) one; this applies, in particular, to a variety of photoluminescence phenomena not addressed in this book (see, e.g., [90]). In contrast, for average quasiparticle occupations $N_{\boldsymbol{q}}$ much larger than one, the spontaneous-emission term may be safely neglected, and both absorption and emission processes become equally probable, being both proportional to the quasiparticle occupation $N_{\boldsymbol{q}}$; this is indeed the typical scenario introduced in Chap. 2 to describe the interaction of a simple electronic two-level system with a so-called classical-light field, like, e.g., an intense electromagnetic source (see (2.6)). A typical example of such a classical-light behavior is the strong amplification regime in quantum-cascade lasers discussed in Chap. 8.

By neglecting spontaneous-emission contributions, the detailed-balance requirement in (6.39) reduces to

$$\frac{f_{\alpha'}}{f_\alpha} = 1 \quad \rightarrow \quad f_{\alpha'} = f_\alpha. \tag{6.41}$$

This result seems to be highly unphysical: it tells us that for a generic carrier–quasiparticle interaction (e.g., carrier–photon, carrier–phonon, carrier–plasmon) the neglect of spontaneous-emission contributions – which corresponds to a purely classical treatment of the quasiparticle subsystem – gives, as equilibrium solution, a constant, i.e., α-independent, carrier distribution ($f_{\alpha'} = f_\alpha$); this, in turn, amounts to saying that the carrier subsystem is characterized by an infinitely high temperature. In other words, the fully quantum-mechanical treatment of the quasiparticle subsystem seems to be an essential prerequisite in order to get the correct thermal-equilibrium distribution of the carrier system in (6.38).

A closer inspection of the problem reveals that the apparent anomaly of the detailed-balance result (6.41) is related to the fact that the carrier indices α and α' are always linked to the quasiparticle mode $\boldsymbol{q}$ via the energy-conservation relation:

$$\epsilon_{\alpha'} - \epsilon_\alpha = \epsilon_{\boldsymbol{q}} \; . \tag{6.42}$$

Taking into account that the classical-quasiparticle limit $N_{\boldsymbol{q}} \gg 1$ corresponds to state that the quasiparticle energy $\epsilon_{\boldsymbol{q}}$ is much smaller than the thermal energy $k_b T$, in view of the energy-conservation condition (6.42) the energy difference $(\epsilon_{\alpha'} - \epsilon_\alpha)$ between initial and final carrier states is also much smaller than $k_{\mathrm{B}}T$, which implies that the corresponding carrier populations ($f_{\alpha'}$ and f_α) nearly coincide. It follows that the condition (6.41) is not valid in general; in contrast, it applies to well-precise quasiparticle phase-space regions $\Omega_{\boldsymbol{q}}$.

As a general conclusion, it is imperative to stress that, before adopting a classical treatment of quasiparticle degrees of freedom, it is important to verify the fulfillment of the classical-limit condition $N_{\boldsymbol{q}} \gg 1$ for all the quasiparticle modes $\boldsymbol{q}$ involved in the phenomenon under examination. As we shall see, this condition is fulfilled in the analysis of the electro-optical response of infrared photodetectors to an intense monochromatic radiation presented in Sect. 7.2 as well as in the investigation of quantum-cascade lasers presented in Chap. 8. In contrast, such classical-quasiparticle approximation does not apply to the investigation of the photocurrent response of infrared photodetectors to a room-temperature black-body radiation presented in Sect. 7.3; indeed, in this case the central black-body photon energy is comparable to the typical intersubband splitting, and in order to properly reproduce a black-body thermalization of the carrier subsystem a quantum-mechanical treatment of the electromagnetic radiation is imperative.

7

Quantum-Well Infrared Photodetectors

In this chapter we shall discuss the basic physical processes as well as open technological problems related to the design and optimization of new-generation quantum-well infrared photodetectors. In particular, we shall focus on the development of efficient quantum devices for the terahertz spectral region.

7.1 Fundamentals of Semiconductor-Based Infrared Detection

The recent development of reliable far-infrared semiconductor-based coherent-light sources (see, e.g., [66] and references therein) together with their potential applications in imaging, communication, and medicine identifies terahertz-radiation detection as a crucial technological milestone (see, e.g., [289, 290] and references therein). To this end, many approaches have been proposed in the last years, which aim at accessing the 1–10 THz region of the electromagnetic spectrum. Currently proposed solutions encompass a variety of different approaches, each with its own peculiar characteristics.

From the electronics world, field-effect transistors are extending their operation frequency into the sub-terahertz and terahertz regions exploiting plasmonic-resonance effects (see, e.g., [291–293]). On the other hand, optoelectronics technology benefiting from electro-optical properties of $LiTaO_3$, $LiNbO_3$, and ZnTe crystals has been proposed (see, e.g., [294, 295].

As discussed in Chap. 6, various semiconductor nanodevices also play a fundamental role in this context, quantum-cascade structures (see, e.g., [296, 297]) as well as quantum-well infrared photodetectors (see, e.g., [64] and references therein) being among the most promising directions. Concerning the latter, infrared-radiation detection via conventional multiple quantum-well designs (see Fig. 7.1) resorts on direct bound-to-continuum electronic transitions (see Figs. 6.1 and 6.2), which allowed to achieve remarkable levels

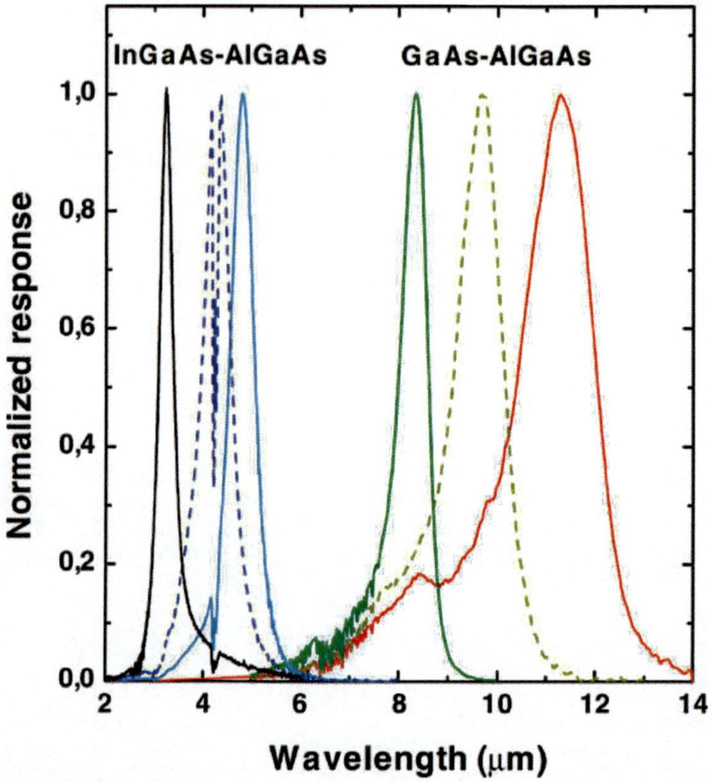

Fig. 7.1. Spectral-response curves of six quantum-well infrared photodetectors covering the two atmospheric transmission windows of 3–5 and 8–12μm wavelength regions. The dip in the second curve from the left at about 4.2 μm is due to the CO_2 absorption, and the noise from 5.5 to 7.5 μm is due to the water absorption. Reprinted with permission from [64]

of performance in the mid-infrared spectral region. Recently, the use of multilevel architectures – opening up to bound-to-bound electronic transitions – has been proposed and studied, focusing both on their intrinsic nonlinear character and on their wide-band absorption spectra. While the latter feature allows for multi-color (see, e.g., [298]) or wide-band detection (see, e.g., [299–302]), second-order nonlinearities of two-level systems have been studied and experimentally demonstrated with the idea of using the devices for second-order autocorrelation measurements (see, e.g., [303–306]).

The extension of the conventional (bound-to-continuum) photodetection principle into the far-infrared range is not straightforward. In particular, one of the main issues in terahertz-operating devices is the huge dark-current values that cause the so-called background-limited infrared-photodetection temperature (see Sect. 7.3) to be in the range 10–15 K (see, e.g., [307, 308]), i.e., much lower than that of state-of-the-art mid-infrared detectors.

The theoretical analysis originally proposed in [309] and reviewed in Sect. 7.2 addresses specifically the application of a multi-level architecture in terahertz quantum-well-based designs and concludes that a bound-to-bound-to-continuum scheme may efficiently face the above-mentioned dark-current issue. A more recent investigation [310, 311], reviewed in Sect. 7.3, analyzes the performances of such novel architectures, focusing on the characteristic figure of merit previously mentioned, namely the background-limited infrared-photodetection temperature. The results presented in [310, 311] suggest the possibility to achieve a consistent improvement of the operational temperature of semiconductor-based terahertz detection by means of these multi-level design strategies.

7.2 Single- Versus Multi-photon Strategies

Standard quantum-well infrared photodetectors are based on single-bound-state quantum wells (see Figs. 6.1 and 6.2), and electronic transitions between such bound state and the continuum are used for photon detection. Incident photons excite bound electrons into the continuum contributing to what is called the photocurrent, which is the detection signal "measured" by such devices. The well width and composition, and thus the depth of the bound state, are designed to match with the energy of the photons to be detected.

Such semiconductor-based detectors were already demonstrated as successful and reliable devices to cover the mid-infrared region of the electromagnetic spectrum. As previously stressed, however, the scaling down of the operation frequency into the terahertz range (1–10 THz) is not straightforward and still work has to be done after the first demonstrations. The main obstacle is the fact that at these energies (4.1–41 meV) the dark current, mainly due to the high-energy tail of the electron distribution function, may become predominant over the photocurrent signal.

In what follows we review an alternative terahertz-detection design [309–311]; the latter, instead of resorting on a direct bound-to-continuum transition (as for conventional mid-infrared detectors), is based on a ladder of equally spaced bound states, whose step is tuned to the target frequency of operation. As we shall see, this kind of photodetector is intrinsically less responsive to light than the single-bound-state system since many photons have to be absorbed to excite an electron into the continuum. However, the reduction of the dark current yields an overall better signal-to-noise ratio. Indeed, the latter is the figure of merit usually adopted for the characterization of this device and depends on the number N_{b} of bound states in the design. More specifically, the signal-to-noise ratio is defined as

$$R_{N_{\mathrm{b}}} = \frac{J_{N_{\mathrm{b}}}^{\mathrm{ph}}}{J_{N_{\mathrm{b}}}^{\mathrm{dark}}} , \tag{7.1}$$

where $J_{N_b}^{ph}$ is the photocurrent density and $J_{N_b}^{dark}$ is the dark-current density. Since we are focusing on how the number of bound states may affect the detector response, the relevant quantity to consider is indeed

$$R_{N_b}^{rel} = \frac{R_{N_b}}{R_1} , \tag{7.2}$$

i.e., the signal-to-noise ratio of the proposed many-level design normalized to that of the conventional (i.e., single-bound-state) one.

To get a first insight into the general idea of the proposed architecture, we shall start considering an extremely simple device model originally introduced in [309]: a one-dimensional parabolic system; the latter would formally produce an infinite sequence of evenly spaced bound states along the growth direction. The key approximation of this model is to limit our analysis to the first $N_b + N_c$ levels, the former being the bound states and the latter representing/describing the continuum. The set of discrete N_c wavefunctions is chosen to properly model the real continuum, i.e., their number is so that the electron population in the continuum is basically N_c independent. As already pointed out in Chap. 6, we stress that within this simplified one-dimensional model any in-plane carrier dissipation/thermalization process is necessarily treated via effective/phenomenological interlevel-scattering rates (see (6.1)); a more quantitative analysis will require a fully three-dimensional treatment of the transport problem (see below).

The evaluation of R_{N_b} in (7.1) is achieved via a proper rate-equation modeling of the electron dynamics in our device which allows us to access the charge population/density N_n of the generic nth level. Within such a simplified scheme, we also assume that the current density is proportional to the total charge density in the continuum and thus we may redefine R_{N_b} as

$$R_{N_b} = \frac{\sum_{n=N_b+1}^{N_b+N_c} N_n^{ph}}{\sum_{n=N_b+1}^{N_b+N_c} N_n^{dark}} , \tag{7.3}$$

where the superscripts again refer to the contribution due to the radiation to be detected (ph) and to the dark-current component (dark).

For the case of a monochromatic radiation resonant with our equally spaced levels, the light–matter dipole matrix elements corresponding to our parabolic-potential profile are known exactly, and analytical textbook formulas (see, e.g., [153]) are employed to model the effect of such incident classical radiation[1] on the first (bound-like) N_b levels; indeed, in the present model the N_c continuum levels do not interact with light as it would be for a real continuum.

---

[1] As discussed in Sect. 6.3, in the classical-quasiparticle regime ($N_{\boldsymbol{q}} \gg 1$) – corresponding in this case to a number of incident photons per mode $\boldsymbol{q}$ much greater than one – spontaneous-emission contributions may be safely disregarded, and the incident electromagnetic radiation may be treated as a classical-light field.

Following the phenomenological many-level picture introduced in Chap. 6 (see (6.1)), for a given level configuration the charge populations N_n in steady-state conditions may be obtained by solving a set of $N_b + N_c$ rate equations of the form:

$$\sum_{n'=1}^{N_b+N_c} (W_{nn'}N_{n'} - W_{n'n}N_n) = 0 , \tag{7.4}$$

where $W_{nn'}$ is the total, i.e., summed over all interaction mechanisms (electron–photon, electron–phonon, electron–electron, etc.), scattering rate from state n' to state n. More specifically, these total scattering rates may be written as

$$W_{nn'} = W^{\mathrm{ph}}_{nn'} + W^{\mathrm{dark}}_{nn'} , \tag{7.5}$$

where the first term describes carrier–photon interaction, while the second one accounts for all other non-radiative scattering processes.

As anticipated, for the present parabolic-potential model the carrier–photon scattering rates $W^{\mathrm{ph}}_{nn'}$ may be evaluated analytically via the standard Fermi's golden rule.[2] In contrast, the non-radiative scattering rates $W^{\mathrm{dark}}_{nn'}$ in (7.5), corresponding to various scattering mechanisms (like carrier–phonon, carrier–carrier, etc.), may be simply expressed in terms of a phenomenological life-time τ, independent from the level index n. More specifically, in the absence of external radiation (dark-operation regime), the scattering rates $W^{\mathrm{dark}}_{nn'}$ should maintain the carrier system at thermal equilibrium. Invoking once again the detailed-balance principle discussed in Sects. 4.4 and 6.3, it is possible to show that this requirement is fulfilled adopting phenomenological scattering rates given by

$$W^{\mathrm{dark}}_{nn'} = \frac{1}{N_{\mathrm{tot}}\tau} \tag{7.6}$$

for $n' > n$ (thermal decay) and

$$W^{\mathrm{dark}}_{nn'} = \frac{\mathrm{e}^{-\frac{\epsilon_n - \epsilon_{n'}}{k_B T}}}{N_{\mathrm{tot}}\tau} \tag{7.7}$$

for $n' < n$ (thermal excitation), where $N_{\mathrm{tot}} = N_b + N_c$ is the total number of states/levels. Employing a compact notation, the above phenomenological scattering rates may also be written as

$$W^{\mathrm{dark}}_{nn'} = \frac{\mathrm{e}^{-\frac{\Delta\epsilon_{nn'}}{k_B T}}}{N_{\mathrm{tot}}\tau} \tag{7.8}$$

with

$$\Delta\epsilon_{nn'} = \max(0, \epsilon_n - \epsilon_{n'}) . \tag{7.9}$$

---

[2] More precisely, in this simplified model the parabolic potential mimics an effective confinement potential within the usual envelope-function approximation; it follows that, according to the analysis presented in Sect. 1.2.2, the carrier–photon scattering rates $W^{\mathrm{ph}}_{nn'}$ are proportional to the squared modulus of the intraband dipole matrix element p^{env} in (1.56).

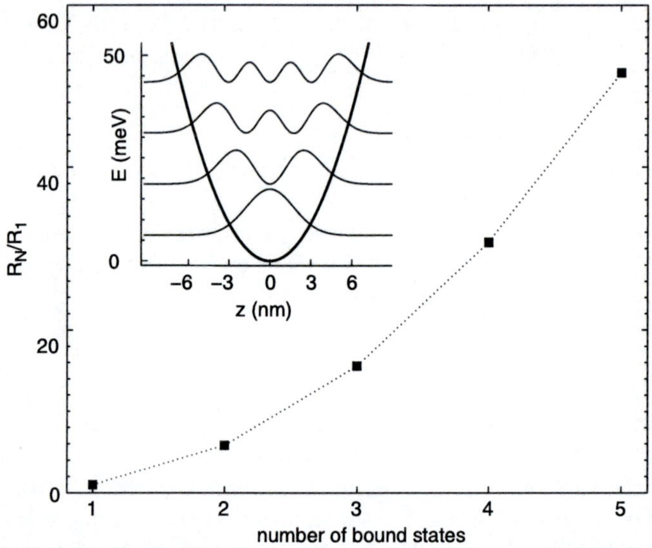

Fig. 7.2. Normalized signal-to-noise ratio R_N/R_1 in (7.2) for a simplified parabolic-potential design (shown in the inset for the case of a four-bound-level configuration) as a function of the number $N = N_\mathrm{b}$ of bound states. Here, the non-radiative electron life-time is $\tau = 1\,\mathrm{ps}$, the operation frequency is 3 THz, and the optical power is set in the linear response regime (see text). Reprinted from [309]

Figure 7.2 shows the simulated signal-to-noise ratio (7.2) as a function of the number of bound states, for the parabolic-potential profile sketched in the inset. Here, the non-radiative electron life-time is $\tau = 1\,\mathrm{ps}$, the operation frequency is 3 THz, and the optical power is set in the linear response regime of the device. The results demonstrate that R_{N_b}/R_1 increases with the number N_b of bound states, thus suggesting that the proposed detection scheme actually improves the response with respect to the conventional (single-bound-state) design.

Actually, a key feature of the parabolic-potential model considered so far is the fact that optical dipole matrix elements increase with the quantum number n, thus intrinsically improving the response as long as more bound states are considered. This is certainly not the case for a realistic potential profile, and the feasibility of the proposed architecture has to be tested on more concrete designs. To this end, the realistic potential profile employed in the fully three-dimensional analysis originally presented in [309] is a triple-quantum-well structure formed by a wide central well defining the number of bound states and by two thin lateral wells whose dimensions are tailored to properly tune the interlevel energy spacings. Such confinement profile is shown in the inset of Fig. 7.3.

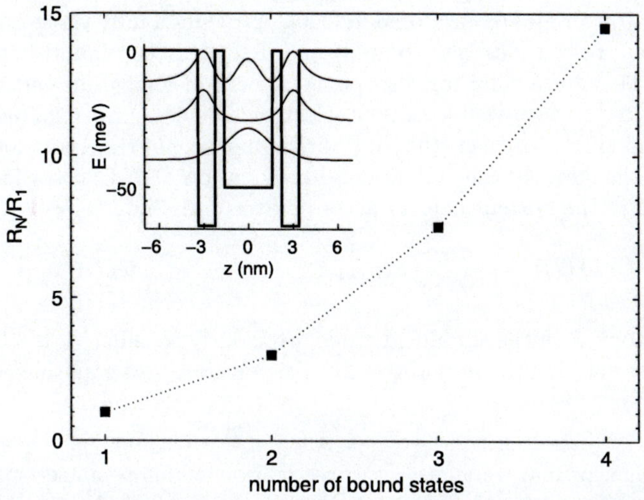

Fig. 7.3. Normalized signal-to-noise ratio R_{N}/R_1 for a realistic three-well potential profile (shown in the inset for the case of a three-bound-level configuration) as a function of the number $N = N_{\mathrm{b}}$ of bound states. Here, the prototypical device profile is tuned to operate at 3.3 THz, an electron mean life-time $\tau = 1\,\mathrm{ps}$ is assumed, and the incident power is kept within the linear response regime (see text). Reprinted from [309]

As discussed extensively in Chap. 6, in order to get a more refined modeling of the carrier dynamics in our nanodevice, the key step is to replace the previous phenomenological description based on a many-level system with the fully three-dimensional kinetic treatment in (6.2). In particular, in view of the periodic nature of quantum-well infrared photodetectors (obtained, e.g., by a periodic repetition of the basic unit/stage reported in the inset of Fig. 7.3) as well as of the low/moderate-field regime, vertical-transport phenomena across the proposed photodetector may be conveniently described via the conventional superlattice picture (see Fig. 6.2). More specifically, within the standard envelope-function treatment (see Sect. 1.2.2), the electronic-state label is specified via the three-dimensional carrier wavevector $\boldsymbol{k}$ plus the miniband index $\overline{\nu}$ (see (1.51)), i.e., $\overline{\alpha} = \boldsymbol{k}\overline{\nu}$. As shown in Sect. 6.1, the time evolution of the carrier distribution function $f_{\boldsymbol{k}\overline{\nu}}$ is described by the Boltzmann-like equation (6.9):

$$\left(\frac{\partial}{\partial t} - \frac{e\boldsymbol{E}}{\hbar} \cdot \boldsymbol{\nabla}_{\boldsymbol{k}}\right) f_{\boldsymbol{k}\overline{\nu}} = \sum_{\boldsymbol{k}'\overline{\nu}'} \left[P_{\boldsymbol{k}\overline{\nu},\boldsymbol{k}'\overline{\nu}'} f_{\boldsymbol{k}'\overline{\nu}'} - P_{\boldsymbol{k}'\overline{\nu}',\boldsymbol{k}\overline{\nu}} f_{\boldsymbol{k}\overline{\nu}}\right] . \tag{7.10}$$

As described in [311], the envelope-function electronic states $\boldsymbol{k}\overline{\nu}$ may be conveniently obtained via the plane-wave expansion presented in Sect. 5.1.2, which in turn allows for a microscopic evaluation of bound-to-bound as well as of bound-to-continuum carrier–photon scattering rates; indeed, the transport model (7.10) – originally proposed in [240] – provides a fully three-dimensional description of both localized (bound) and extended (continuum) states, crucial ingredient for a fully microscopic treatment of ionization versus capture processes in quantum-well-based optoelectronic devices (see, e.g., [64] and references therein). Moreover, the fully three-dimensional transport formulation based on the superlattice Boltzmann-like equation (7.10) allows for a direct evaluation of the current density according to (6.11), i.e.,

$$\boldsymbol{J}(\boldsymbol{E}) = -\frac{e}{\Omega}\sum_{\boldsymbol{k}\overline{\nu}} \boldsymbol{v}_{\boldsymbol{k}\overline{\nu}} f_{\boldsymbol{k}\overline{\nu}} \,, \qquad \boldsymbol{v}_{\boldsymbol{k}\overline{\nu}} = \frac{1}{\hbar}\boldsymbol{\nabla}_{\boldsymbol{k}}\epsilon_{\boldsymbol{k}\overline{\nu}} \,. \tag{7.11}$$

In analogy with the simplified model previously considered, the total scattering rates entering the fully three-dimensional transport equation (7.10) may be written as

$$P_{\boldsymbol{k}\overline{\nu},\boldsymbol{k}'\overline{\nu}'} = P^{\mathrm{ph}}_{\boldsymbol{k}\overline{\nu},\boldsymbol{k}'\overline{\nu}'} + P^{\mathrm{dark}}_{\boldsymbol{k}\overline{\nu},\boldsymbol{k}'\overline{\nu}'} \,. \tag{7.12}$$

Since our primary goal is the investigation of light–matter interaction, the carrier–photon scattering rates $P^{\mathrm{ph}}_{\boldsymbol{k}\overline{\nu},\boldsymbol{k}'\overline{\nu}'}$ are evaluated in a fully three-dimensional fashion. More specifically, according to the general treatment of carrier–quasiparticle interaction presented in Chap. 3 (see (3.137)), the corresponding scattering rates within the conventional Fermi's golden rule are given in (3.140). Within the present miniband picture, the latter – evaluated via the usual dipole approximation (see (1.26)) – are of the form

$$P^{\mathrm{ph}}_{\boldsymbol{k}\overline{\nu},\boldsymbol{k}'\overline{\nu}'} = P^{\mathrm{ph}}_{\boldsymbol{k},\overline{\nu}\overline{\nu}'}\delta_{\boldsymbol{k}\boldsymbol{k}'} \tag{7.13}$$

with

$$P^{\mathrm{ph}}_{\boldsymbol{k},\overline{\nu}\overline{\nu}'} = \frac{2\pi}{\hbar}\sum_{\boldsymbol{q}\pm}\left(N^{\circ}_{\boldsymbol{q}} + \frac{1}{2} \pm \frac{1}{2}\right)\left|g^{\pm}_{\boldsymbol{k},\overline{\nu}\overline{\nu}',\boldsymbol{q}}\right|^2 \delta\left(\epsilon_{\boldsymbol{k}\overline{\nu}} - \epsilon_{\boldsymbol{k}\overline{\nu}'} \pm \epsilon_{\boldsymbol{q}}\right) \,. \tag{7.14}$$

Moreover, it is possible to show that the carrier–photon matrix elements $g^{\pm}_{\boldsymbol{k},\overline{\nu}\overline{\nu}',\boldsymbol{q}}$ evaluated within the dipole approximation are proportional to the intraband dipole matrix elements p^{env} in (1.56).

At this point, it is imperative to stress that the carrier–photon scattering rates (7.13) correspond to a fully quantum-mechanical treatment of the electromagnetic radiation. The latter is indeed an essential prerequisite for the description of the black-body-radiation phenomena presented in Sect. 7.3; in contrast, as pointed out in Sect. 6.3, in the classical-quasiparticle limit corresponding to an average number of photons per mode much greater than one ($N_{\boldsymbol{q}} \gg 1$), spontaneous-emission contributions can be safely neglected, and the quantum-mechanical photon picture is replaced by a classical-field description.[3]

[3] As discussed in [311], the link between the average number of photons ($N_{\boldsymbol{q}}$) and the corresponding classical-light field is obtained imposing that the quantum-mechanical energy density ($\frac{N_{\boldsymbol{q}}\epsilon_{\boldsymbol{q}}}{\Omega}$) corresponding to a given mode $\boldsymbol{q}$ is equal to its classical counterpart (proportional to the square of the classical electromagnetic field).

Let us now come to the scattering rates $P^{\rm dark}_{\boldsymbol{k}\overline{\nu},\boldsymbol{k}'\overline{\nu}'}$; as for the simplified model previously considered, the latter – describing other (non-radiative) scattering mechanisms (e.g., carrier–phonon, carrier–carrier) – are still modeled via a phenomenological life-time. More specifically, denoting with

$$\Gamma^{\rm dark}_{\boldsymbol{k}\overline{\nu}} = \sum_{\boldsymbol{k}'\overline{\nu}'} P^{\rm dark}_{\boldsymbol{k}'\overline{\nu}',\boldsymbol{k}\overline{\nu}} \tag{7.15}$$

the total (i.e., summed over all possible final states $\boldsymbol{k}'\overline{\nu}'$) scattering rate due to non-radiative processes, the inverse of such a phenomenological electron life-time may be defined as the average of the total rate $\Gamma^{\rm dark}$ over all relevant microscopic states $\boldsymbol{k}\overline{\nu}$, i.e.,

$$\frac{1}{\tau} = \frac{1}{N_{\rm tot}} \sum_{\boldsymbol{k}\overline{\nu}} \Gamma^{\rm dark}_{\boldsymbol{k}\overline{\nu}} = \frac{1}{N_{\rm tot}} \sum_{\boldsymbol{k}\overline{\nu},\boldsymbol{k}'\overline{\nu}'} P^{\rm dark}_{\boldsymbol{k}'\overline{\nu}',\boldsymbol{k}\overline{\nu}} , \tag{7.16}$$

where $N_{\rm tot}$ now denotes the total number of microscopic states $\boldsymbol{k}\overline{\nu}$.

In the absence of applied fields ($\boldsymbol{E} = 0$) as well as of incident radiation ($P^{\rm ph}_{\boldsymbol{k}\overline{\nu},\boldsymbol{k}'\overline{\nu}'} = 0$), the superlattice Boltzmann equation (7.10) in steady-state conditions reduces to

$$\sum_{\boldsymbol{k}'\overline{\nu}'} \left[P^{\rm dark}_{\boldsymbol{k}\overline{\nu},\boldsymbol{k}'\overline{\nu}'} f_{\boldsymbol{k}'\overline{\nu}'} - P^{\rm dark}_{\boldsymbol{k}'\overline{\nu}',\boldsymbol{k}\overline{\nu}} f_{\boldsymbol{k}\overline{\nu}} \right] = 0 . \tag{7.17}$$

In this case the desired solution must coincide with the equilibrium distribution $f^\circ_{\boldsymbol{k}\overline{\nu}}$. By applying once again the detailed-balance principle to the above thermal-equilibrium equation in the low-density limit ($f_{\boldsymbol{k}\overline{\nu}} \ll 1$) we get

$$P^{\rm dark}_{\boldsymbol{k}\overline{\nu},\boldsymbol{k}'\overline{\nu}'} f^\circ_{\boldsymbol{k}'\overline{\nu}'} = P^{\rm dark}_{\boldsymbol{k}'\overline{\nu}',\boldsymbol{k}\overline{\nu}} f^\circ_{\boldsymbol{k}\overline{\nu}} , \tag{7.18}$$

which implies that

$$\frac{P^{\rm dark}_{\boldsymbol{k}'\overline{\nu}',\boldsymbol{k}\overline{\nu}}}{P^{\rm dark}_{\boldsymbol{k}\overline{\nu},\boldsymbol{k}'\overline{\nu}'}} = \frac{f^\circ_{\boldsymbol{k}'\overline{\nu}'}}{f^\circ_{\boldsymbol{k}\overline{\nu}}} = {\rm e}^{-\frac{\epsilon_{\boldsymbol{k}'\overline{\nu}'} - \epsilon_{\boldsymbol{k}\overline{\nu}}}{k_{\rm B}T}} . \tag{7.19}$$

In total analogy to the phenomenological scattering rates of the parabolic model (see (7.8)), this requirement is fulfilled choosing non-radiative scattering rates of the form

$$P^{\rm dark}_{\boldsymbol{k}\overline{\nu},\boldsymbol{k}'\overline{\nu}'} = \frac{\gamma^{\rm dark}}{N_{\rm tot}} {\rm e}^{-\frac{\Delta\epsilon_{\boldsymbol{k}\overline{\nu},\boldsymbol{k}'\overline{\nu}'}}{k_{\rm B}T}} \tag{7.20}$$

with

$$\Delta\epsilon_{\boldsymbol{k}\overline{\nu},\boldsymbol{k}'\overline{\nu}'} = \max(0, \epsilon_{\boldsymbol{k}\overline{\nu}} - \epsilon_{\boldsymbol{k}'\overline{\nu}'}) . \tag{7.21}$$

By inserting the form of the scattering rates (7.20) into the definition of the electronic life-time τ in (7.16), we get

$$\gamma^{\rm dark} = \frac{1}{\tau} \frac{N^2_{\rm tot}}{\sum_{\boldsymbol{k}\overline{\nu},\boldsymbol{k}'\overline{\nu}'} {\rm e}^{-\frac{\Delta\epsilon_{\boldsymbol{k}\overline{\nu},\boldsymbol{k}'\overline{\nu}'}}{k_{\rm B}T}}} , \tag{7.22}$$

which allows us to express the scattering rates (7.20) in terms of the phenomenological life-time τ according to

$$P^{\text{dark}}_{k\overline{\nu},k'\overline{\nu}'} = \frac{N_{\text{tot}}}{\tau} \frac{e^{-\frac{\Delta\epsilon_{k\overline{\nu},k'\overline{\nu}'}}{k_B T}}}{\sum_{k_1\overline{\nu}_1,k_2\overline{\nu}_2} e^{-\frac{\Delta\epsilon_{k_1\overline{\nu}_1,k_2\overline{\nu}_2}}{k_B T}}} . \tag{7.23}$$

As discussed extensively in [240], a standard Monte Carlo simulation (see Chap. 5) of the device current–voltage characteristics based on the Boltzmann transport equation (7.10) is often characterized by strong stochastic fluctuations; for this reason, it is useful to adopt more refined simulation schemes based, e.g., on the Weighted Monte Carlo method introduced in Sect. 5.2.5. As alternative strategy adopted in [309], in steady-state conditions the transport equation (7.10) may be easily solved via the finite-difference approach described in Sect. 5.1.1 (see (5.15)).

The fully three-dimensional finite-difference calculations presented in [309] are carried out by applying a static electric field $E = 500\,\text{V/cm}$ along the growth direction and then evaluating the current density with and without a monochromatic incident radiation resonant with the device interlevel splitting. The signal-to-noise ratio is then estimated according to the prescription (7.1) and its values are reported in Fig. 7.3 as a function of the number of bound states. In these simulations the prototypical device profile (sketched in the inset) is tuned to operate at 3.3 THz, an electron mean lifetime $\tau = 1\,\text{ps}$ is assumed, and the incident power is kept within the linear response regime. Figure 7.3 shows that R_{N_b}/R_1 grows super-linearly with the number N_b of bound states, thus confirming the results obtained with the simplified parabolic-potential model reported in Fig. 7.2.

This occurs in spite of the fact that the potential profile sketched in the inset of Fig. 7.3 is not the best choice for the optimization of the electron–photon coupling; indeed, the latter depends strongly on the values of the dipole matrix elements between the envelope wavefunctions of the initial and final states of the various interminiband transitions (see (1.55)) and thus a proper device design is crucial, in this respect, for the improvement of the overall response of the photodetector. However, the main focus of the investigation presented in [309] was to achieve an evenly spaced set of bound states, since primary goal was to verify whether a multi-photon absorption scheme could improve the detector responsivity over the conventional single-bound-state design.

7.3 Operational-Temperature Optimization of Terahertz Photodetectors

In the reminder of this chapter we shall quantitatively analyze the performances of the multi-photon scheme introduced so far, focusing on a specific figure of merit of infrared detection, namely the background-limited infrared-photodetection temperature T_{blip}; the latter is defined as the temperature at which the current due to thermal noise in the device (dark current) equals the photocurrent induced by a room-temperature (300 K) black-body radiation (background photocurrent). To properly evaluate this characteristic quantity, the fully three-dimensional analysis of the photocurrent response to a monochromatic infrared radiation presented in Fig. 7.3 has been extended to the case of a generic black-body source corresponding to the detection background. To this end, the fully quantum-mechanical carrier–photon scattering rates (7.13) are employed, assuming a black-body thermal occupation $N^{\circ}_{\boldsymbol{q}}$ and accounting for the specific field-of-view – i.e., the effective irradiation solid angle – of the device under examination. A detailed derivation of such black-body carrier-photon scattering rates can be found in [311].

The first and crucial step for a quantitative analysis is the design of our confinement-potential profile; the latter should satisfy several requirements. First of all, the main constraint is to have equally spaced bound levels; second, we want to be able to control the number of such levels as well as their radiative coupling. As discussed in Sect. 7.2, speaking of equally spaced levels the first solution seems to be that of a parabolic-potential profile considered in Fig. 7.2. The implementation of the latter, however, besides non-trivial growth issues, poses more fundamental problems: in order to have vertical carrier transport we need a continuum of states and thus this parabolic potential must be truncated at some point; such a truncated parabola would not support equally spaced levels anymore. For these reasons it is convenient to adopt a different strategy; the latter, originally employed in [309] (see Fig. 7.3), is based on periodically repeated multiple quantum-well structures. More specifically, a single quantum well is used to produce the single-bound level providing the bound-to-continuum transition exploited in conventional infrared photodetectors; two energetically equal transitions can still be obtained with a single quantum well of proper geometry. In contrast, the fine tuning of the energetic separations of three bound levels cannot be achieved via a potential characterized by two free parameters only (width and depth) and thus one has to switch to more complex structures.

The nested multiple quantum-well confinement profiles proposed in [311] are shown in Fig. 7.4. The introduction of additional geometrical parameters to the standard quantum-well design allows one to control the number and energy of the desired bound states. A detailed description of the optimization algorithm used to determine such confinement-potential profiles can be found in [311]. Figure 7.4 shows, in particular, the supercells of these prototypical structures, which are infinitely replicated along the growth direction. As stressed in Chap. 6, the use of many repetitions of the basic unit/stage is indeed the technological strategy exploited in this kind of unipolar devices to optimize detection efficiency; it follows that finite-size effects due to spatial boundaries (see Chap. 4) are of minor importance.

Starting from the prototypical confinement-potential profiles depicted in Fig. 7.4, we shall consider again the fully three-dimensional transport

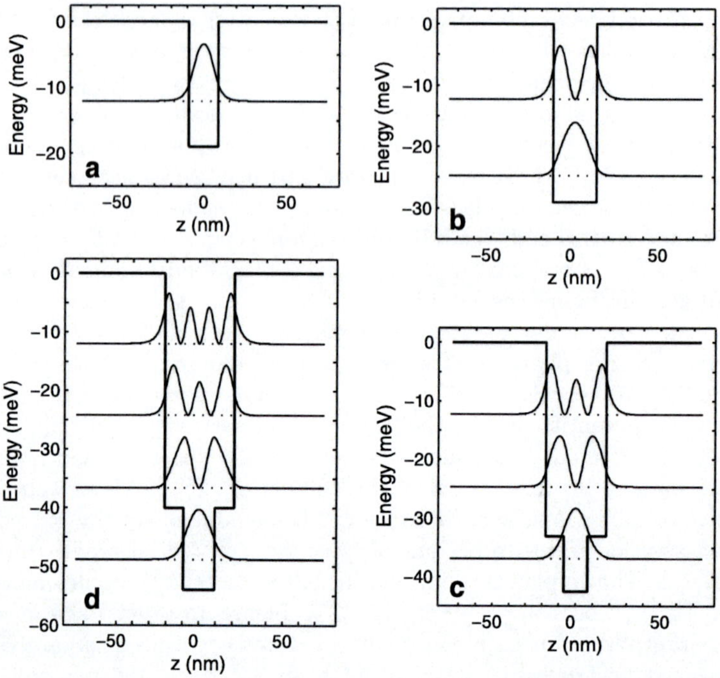

Fig. 7.4. Confinement-potential profiles along the growth direction of our prototypical devices, designed to operate at 3 THz, with a number of bound states varying from one (**a**) to four (**d**). The symmetric nested-quantum-well structures in (**c**) and (**d**) provide additional geometric parameters (with respect to the single-quantum-well design) that can be varied to properly tune the intersubband energy separation (see text). Reprinted from [311]

equation (7.10) replacing the carrier–photon scattering rates due to a monochromatic infrared radiation with those corresponding to a room-temperature black-body source. As anticipated, the proposed transport model contains a free parameter, the non-radiative life-time τ, whose value has to be adjusted in order to reproduce some experimental data; indeed, in principle, the latter is crucial in determining the T_{blip} of the simulated devices, since changing the electron life-time τ will unavoidably change the weight of thermal scattering with respect to light–matter coupling and thus will affect the point at which these two competing processes balance. More specifically, for the simulated experiments presented in [311] the choice is to adjust τ in order to reproduce the measured $T_{\text{blip}} = 12\,\text{K}$ of the bound-to-continuum infrared detector operating at 3.2 THz reported in [308].

Figure 7.5 shows the simulated current density along the growth direction versus device temperature for the single-bound-level design reported in Fig. 7.4 (operating at 3 THz) in the presence of a room-temperature black-body radiation. Here, the different symbols correspond to different values of the non-

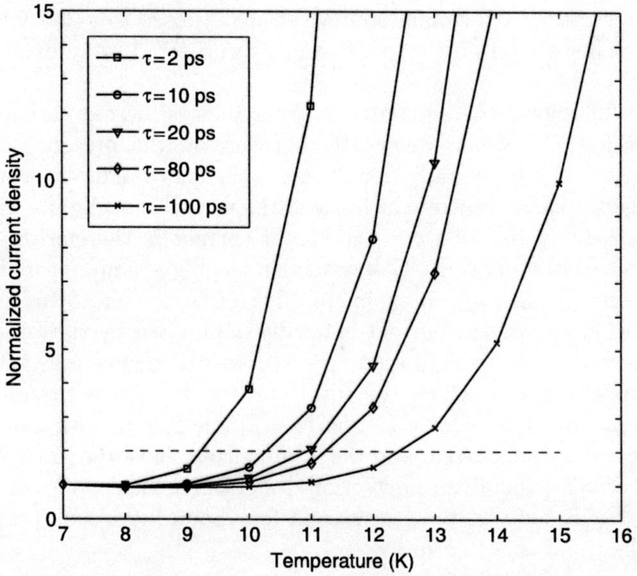

Fig. 7.5. Normalized current density along the growth direction versus device temperature for the single-bound-level design reported in Fig. 7.4 (operating at 3 THz) in the presence of a room-temperature black-body radiation. Here, the different symbols correspond to different values of the non-radiative life-time τ, the applied electric field is 50 V/cm, and the field-of-view of the device is 90°. The *dashed line* marks the current doubling (see text). Reprinted from [311]

radiative life-time τ, the applied electric field is 50 V/cm, and the field-of-view of the device is 90°. Although τ plays a key role in determining the value of T_{blip}, a closer inspection of Fig. 7.5 shows that within the interval $\tau \sim 50$–100 ps the latter shows little variation around a reference value of 12 K. In order to reproduce the experimental data, one can thus safely assume for the non-radiative life-time τ any value in this range, like, e.g., $\tau = 80$ ps. It is important to stress once more that this relatively large value derives from the fact that we are considering a simplified model for thermal scattering; indeed, this global/average fitting parameter τ does not reflect the actual (state-dependent) electronic scattering time in the nanodevice under examination.

On the basis of the above discussion, once a proper value of τ has been identified, the latter may be employed to analyze the current response of various detection designs operating at identical frequencies but employing the proposed bound-to-bound-to-continuum strategy, thus differing in the number of bound states. More specifically, let us now review the quantitative analysis originally presented in [311] corresponding to the four prototypical devices in Fig. 7.4, characterized by a number of bound levels ranging from one (standard configuration) to four and designed according to our bound-to-bound-to-continuum strategy. All simulated devices are exposed to a 300 K black-body radiation under a field-of-view of 90° and are subjected to an external bias of 50 V/cm.

Figure 7.6 shows the simulated current density across each of the four devices of Fig. 7.4 versus device temperature in the presence of a room-temperature black-body radiation. Here, each curve allows one to identify a low-temperature regime in which the dark current is negligible with respect to the black-body photocurrent; the total current is therefore independent from the device temperature. Conversely, in the high-temperature region, the dark current increases almost exponentially so that the photocurrent becomes quickly negligible and the current is totally dominated by the "dark" contribution. According to its definition, the T_{blip} of the device may be identified as the temperature at which the total current doubles with respect to its low-temperature value (see dashed horizontal line in Fig. 7.6); indeed, at this temperature the dark current and the photocurrent have the same magnitude.

As we can see, the diverse detection designs give T_{blip} values of 11.5, 19.5, 23.5, and 28.5 K for one, two, three, and four bound levels, respectively, evidencing the trend reported in Fig. 7.7.

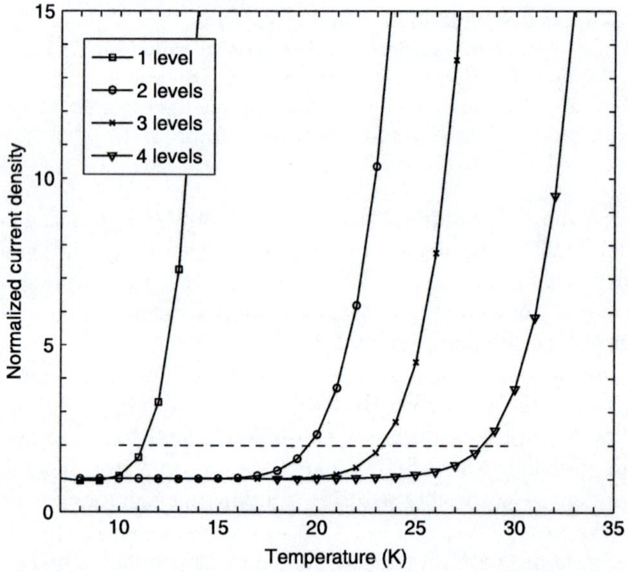

Fig. 7.6. Normalized current density along the growth direction versus device temperature corresponding to the four multi-level designs reported in Fig. 7.4 in the presence of a room-temperature black-body radiation. Here, the applied electric field is 50 V/cm, and the field-of-view of the device is 90° (see text). Reprinted from [311]

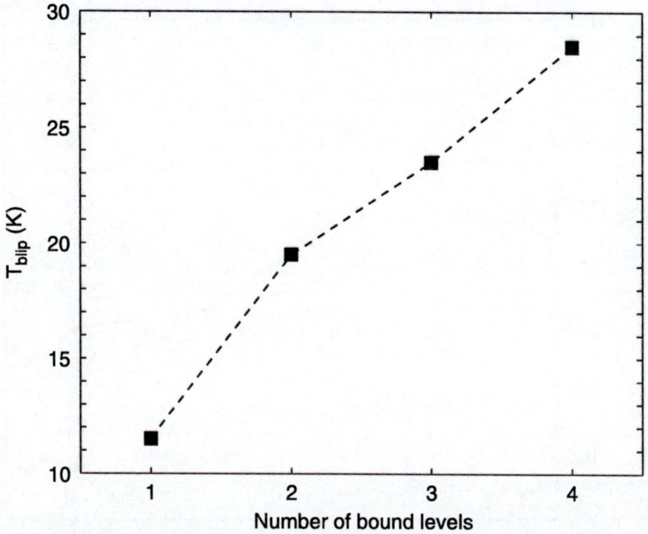

Fig. 7.7. Estimated values of T_{blip} for the four devices of Fig. 7.4, as deduced from the simulation data reported in Fig. 7.6. The *dashed line* is a guide to the eye (see text). Reprinted from [311]

This increase of T_{blip} may be better interpreted by looking at Fig. 7.8, where the black-body photocurrent and the dark current are plotted as a function of the number of bound levels. Here, both quantities decrease, but the dark current decreases faster than the photo-induced one; therefore, the temperature at which the two are equal moves toward higher values.

As can be seen from Fig. 7.8 there is a dramatic decrease in the photocurrent when switching from two to three bound states, which is mainly ascribed to the reduction of the photoconductive gain.[4] In fact, for the four-level design (see Fig. 7.4) the latter reduces to just the 0.2% of the single-bound-state value. Conversely, the quantum efficiency[5] is only lowered by 14% and thus its variation does not significantly affect the photocurrent. This behavior can be explained by looking at Fig. 7.4 and noticing that there is a remarkable geometrical difference between the two-level and the three-level designs. The presence of the nested quantum well introduces a new ground state whose envelope wavefunction has little overlap with those of higher energy states, thus reducing the corresponding intersubband oscillator strength. This conclusion

[4] The photoconductive gain (or photoconductive gain factor) is defined as the ratio of the number of electrons per second flowing through a circuit containing a given volume of semiconducting material to the number of photons per second absorbed in this volume (see, e.g., [64] and references therein).

[5] The quantum efficiency is defined as the probability that a photon is converted into a conducting electron (see, e.g., [64] and references therein).

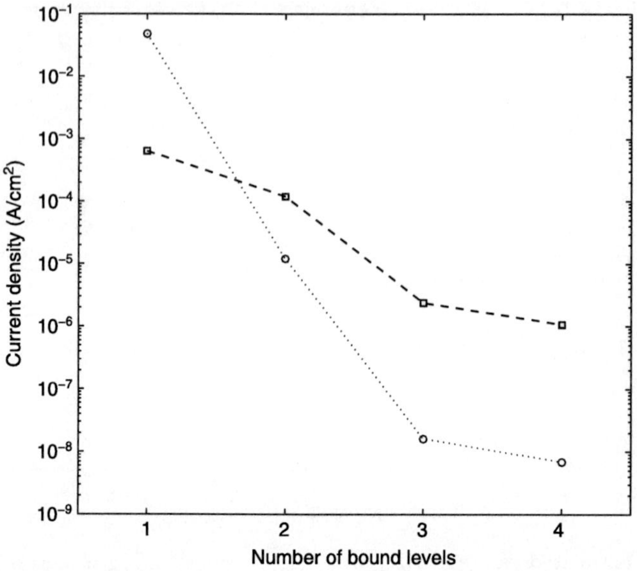

Fig. 7.8. Black-body photocurrent (*squares*) and dark current (*circles*) at 17 K as a function of the number of bound states in the device. The dark current decreases more rapidly than the photocurrent, and thus the $T_{\rm blip}$ increases. *Lines* are a guide to the eye (see text). Reprinted from [311]

is supported by the fact that the photocurrent reduction between the three-level and four-level designs – where no major structural change is introduced – is comparable to the decrease between the one-level and the two-level cases.

As anticipated, in the analysis reviewed so far the optimization of the device performance was not the central issue; in this respect, a more refined tuning of the device geometry – aimed at achieving higher oscillator strengths – would surely allow for better operational results and conditions.

8

Quantum-Cascade Lasers

In this chapter we shall discuss basic features of quantum-cascade coherent-light sources; to this aim we shall review a few simulated experiments focusing on the microscopic explanation of the gain regime, both in the mid-infrared and in the far-infrared spectral regions.

8.1 Fundamentals of Quantum-Cascade Devices

Based on the bandgap-engineering paradigm originally envisioned in 1970 by Leona Esaki and Raphael Tzu [312] as well as on the revolutionary intraband-lasing strategy originally proposed in 1971 by Rudolf F. Kazarinov and Robert A. Suris [313], unipolar coherent-light sources such as quantum-cascade lasers (see, e.g., [66, 314] and references therein) are complex nanodevices, whose core is a multiple quantum-well structure made up of a periodic repetition of identical stages of active regions sandwiched between carrier-injecting and carrier-collecting portions. When a proper bias is applied, a sort of electronic cascade along the subsequent quantized-level energy staircase takes place, the so-called quantum cascade originally demonstrated in 1994 by Federico Capasso and co-workers [315]. Thanks to the continuous progress in bandgap engineering, ultrafast spectroscopy, and semiconductor-based quantum optics, several successful quantum-cascade designs [315–350] have been proposed and further developed.

To better understand the basic physical processes governing the operation of such state-of-the-art electro-optical devices, it can be of some help just to recall the main features which are common to many of these quantum-cascade realizations, regardless of the specificities of each design. As a prototypical example, Fig. 8.1 shows the conduction-band profile along the growth direction of the GaAs-based quantum-cascade laser proposed by Carlo Sirtori and co-workers in [351], emitting in the mid-infrared portion of the electromagnetic spectrum. The figure presents the generic stage ($\ell = 0$) plus a portion of both the neighboring (left and right) stages. Also shown are the square

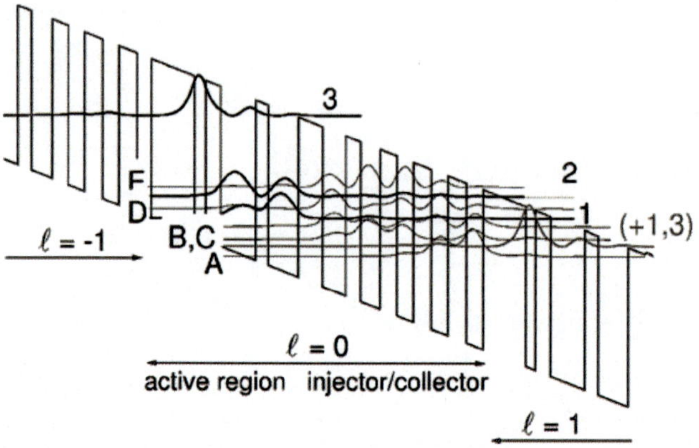

Fig. 8.1. Schematic representation of the conduction-band profile along the growth direction for the diagonal-configuration quantum-cascade laser proposed in [351]. The multiple quantum-well structure is biased by an electric field of 48 kV/cm. The levels $\tilde{n} = 1, 2, 3$ and $\tilde{n} = \mathrm{A, B, C, D, E}$ in the active and collector regions of the simulated stage ($\ell = 0$) are also plotted together with the corresponding charge distributions. The replica of level 3 in the following stage ($\ell = +1$) is also shown for clarity. The separation between subbands 1 and 2 matches the GaAs longitudinal optical (LO) phonon energy ($\hbar\omega_{\mathrm{LO}} \sim 35\,\mathrm{meV}$) and the stage length is about 450 Å (the figure is in scale). Reprinted from [352]

moduli of the carrier wavefunctions of the confined states in the active region ($\tilde{n} = 1, 2, 3$) and in the collector/injector ($\tilde{n} = \mathrm{A, B, C, D, E}$), evaluated within the usual envelope-function framework (see Sect. 1.2.2 and Appendix A). In the quantum-cascade device of Fig. 8.1, electrons are injected via resonant tunneling into subband 3 of the generic stage ($\ell = 0$) and can then relax into lower energy states of subbands 2 and 1. By a proper choice of the confinement geometry, optical matrix elements as well as carrier–phonon scattering rates can be optimized to establish such a population inversion between subbands 3 and 2, corresponding to the device lasing transition. Once a photon has been emitted in this stage, the electron is drifted off into the collector, whence it can be recycled into the downstream neighboring stage ($\ell = 1$) to emit a second photon, and so on so forth. It follows that these are intrinsically high-power devices (see also Fig. 8.2), which can be designed to emit over a wide range of frequencies, just by modifying the nanoscale geometrical parameters while keeping the same constituent materials; this is indeed the essence of the bandgap-engineering paradigm previously mentioned (see, e.g., [66]).

Since their appearance, quantum-cascade lasers have been subject of a rapid experimental development in emission wavelength, lasing threshold, output power, and operating temperature (see, e.g., [314] and references therein). This has stimulated considerable theoretical interest, mainly motivated by the desire to improve the nanodevice performances by optimizing the semiconductor-heterostructure design.

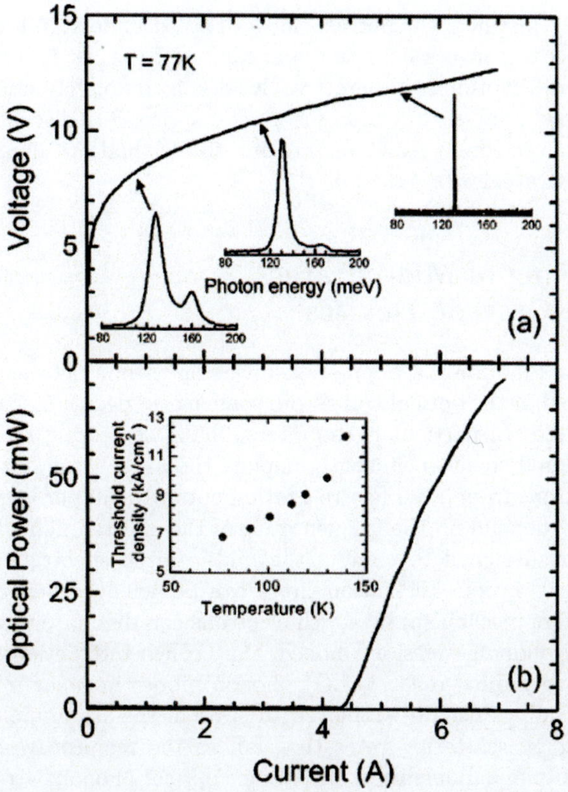

Fig. 8.2. Experimentally measured optoelectronic response of the GaAs-based quantum-cascade device proposed by Carlo Sirtori and co-workers, whose active region is depicted in Fig. 8.1. (**a**) Voltage versus current characteristics measured at 77 K. In the insets, three emission spectra at various drive currents: 0.8 and 2.8 A below threshold and 5.1 A above it. In the spectra below threshold, the high-energy peak corresponds to the $3 \to 1$ transition. The measured separation between the two peaks is 35 meV. Note the strong line-narrowing signature of laser action. The spontaneous emission and the laser radiation are polarized normal to the layer (TM polarization). (**b**) Collected peak optical power from a single facet versus drive current. The device is 24-μm wide and 2.5-mm long. The inset shows the threshold current density in pulsed operation as a function of the heat-sink temperature. Reprinted with permission from [351]

As discussed extensively in Chap. 6, various theoretical approaches have been proposed to describe the electro-optical response of unipolar nanodevices, including current–voltage characteristics and gain spectra of quantum-cascade lasers (see, e.g., [280] and references therein). However, as anticipated, these approaches generally resort on macroscopic rate-equation models to describe the carrier dynamics (see (6.1)), whose application in this context is not physically justified. In contrast, for a detailed understanding of the basic physical processes involved in the operation of these devices, a fully three-dimensional description is imperative. Besides, a completely microscopic (i.e., parameter-free) analysis based on the kinetic approaches discussed in Chap. 6 constitutes a precious predictive tool for the evaluation of new-generation designs and strategies (see Sect. 6.2).

8.2 Modeling of Mid-infrared Quantum-Cascade Devices

To achieve and maintain a population-inversion regime between the two subbands involved in the optical transition, some clever design of the lasing structure is required. In particular, the intersubband scattering rates have to be properly tailored so as to efficiently deplete the lower laser states, while preventing electrons from a too fast relaxation out of the upper laser states. This does strongly depend on the frequency $\omega_\circ$ of the emitted light. For quantum-cascade lasers designed to emit in the mid-infrared spectral region, such a tailoring of the various relaxation times can be achieved benefiting from an energy-selective mechanism by which electrons can dissipate their energy: the polar optical-phonon emission. Indeed, the Frölich interaction between electrons and longitudinal optical (LO) phonon modes in polar materials (such as the III–V alloys usually employed in state-of-the-art unipolar devices) is characterized by scattering rates that fall as the reciprocal of the squared transferred in-plane momentum $\Delta \boldsymbol{k}_{\parallel}$. Since optical phonons are basically dispersionless (see Fig. 1.8), an electronic intersubband relaxation process via optical-phonon emission comes out to be vertical in momentum space – and therefore highly probable – when the intersubband energy difference matches the phonon energy $\hbar\omega_{\mathrm{LO}}$. In contrast, when the subbands are more than one phonon far apart, the electronic transition comes out to be strongly diagonal in $\boldsymbol{k}$-space and thus less probable than in the previous (i.e., phonon-resonant) case.

8.2.1 Partially Phenomenological Approach

Let us start our analysis by reviewing a few simulated experiments of mid-infrared lasers originally presented in [353]. The latter are based on the partially phenomenological transport equations (6.2), which have been numerically solved via a proper combination of deterministic (finite-difference) and stochastic (Monte Carlo) strategies described in Sect. 5.3. In particular, the actual problem requires a Monte Carlo simulation where the total number of particles (in the active region of the device) is not constant.

To this end, let us consider once again the GaAs-based diagonal-configuration quantum-cascade laser proposed in [351], whose potential profile is schematically depicted in Fig. 8.1. More specifically, the analysis in [353] is mainly devoted to the hot-carrier dynamics within the active region of the device, which allows one to limit the kinetic treatment to the subset $\alpha = \boldsymbol{k}_{\|}, \ell = 0, \tilde{n} = 1, 2, 3$ (see Sect. 6.2). As anticipated, the kinetic transport model in (6.2) provides a fully three-dimensional description of energy relaxation based on microscopic intra- as well as intersubband scattering mechanisms within the bare active region. Contrarily, the injection/loss contributions (describing the coupling with injector and collector) are still treated on a partially phenomenological level. The only interaction mechanism considered in the simulated experiments reviewed below is carrier–optical phonon scattering: As previously recalled, this is indeed the key effect inspiring the device design at these frequencies. A more detailed discussion on the role of other scattering mechanisms and carrier–carrier interaction is provided in the following section.

Figure 8.3 shows the three global carrier populations (see (6.5)) of the active-region subbands as a function of time for an injection current of $10\,\mathrm{kA/cm^2}$ at $T = 77\,\mathrm{K}$. At the initial time ($t = 0$) the device active region is "empty"; then electrons are continuously injected into level 3, according to a Fermi–Dirac distribution corresponding to the lattice temperature and simulated current density, and the "cascade" – $3 \to 2$, $2 \to 1$ – occurs on time-scales corresponding to the different values of the interminiband scattering rates. Eventually, the steady-state condition is reached, leading to the desired $3-2$ population inversion. Figure 8.3a has been obtained using a typical value for the carrier escape time from level 1 of 1 ps. Here, the results of our Monte Carlo simulated experiments (solid curves) are compared to the corresponding results obtained by neglecting various Pauli-blocking factors $(1 - f)$ (dotted curves) entering the Boltzmann collision term in (6.3) (see also (1.74)). Since in this simulation carrier escape from level 1 is relatively fast (1 ps), we have $f_{\boldsymbol{k}_{\|},1} \ll 1$ and thus Pauli-blocking effects are not very important. In contrast, by artificially increasing the escape time from level 1, nonlinear effects due to Pauli blocking may play a crucial role. This can be clearly seen in Fig. 8.3b, where the same simulated experiments are shown for an escape time of 10 ps. In this case, the strong Pauli blocking leads to a significant reduction of the intersubband relaxation $2 \to 1$, thus giving rise to a phonon bottleneck and preventing the $3 \to 2$ population inversion.[1] Although a quantitative evaluation of carrier dynamics in this regime of slow depletion of subband 1 would require to take into account also the competing carrier–carrier scattering channels, Fig. 8.3b shows that, as a general prescription, it is important to limit the carrier population of subband 1, also if the latter is not directly involved in the lasing process. Such Pauli-blocking effects are intrinsically time dependent: during the first stage of the simulation the population N_1 is still small and the two results (solid and dotted curves) basically coincide; at later times N_1 increases and significant deviations come into

[1] Note the different time-scales in Fig. 8.3a,b, corresponding to the different values of the escape time from subband 1.

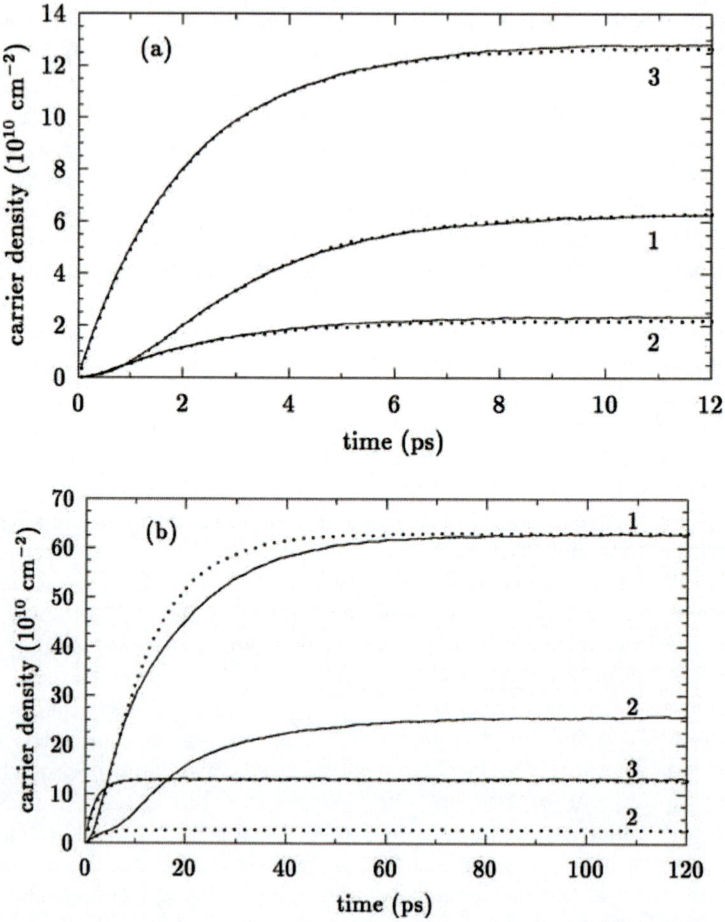

Fig. 8.3. Various subband carrier concentrations ($N_{\tilde{n}}, \tilde{n} = 1, 2, 3$) as a function of time with (*solid*) and without (*dotted*) Pauli-blocking effects. Simulated experiments in (**a**) and (**b**) correspond to a phenomenological escape/loss time from level 1 of 1 and 10 ps, respectively (see text). Reprinted from [353]

play. As anticipated, such a nonlinear behavior cannot be described by the rate-equation model in (6.1). The same is true for the so-called in-plane (or intrasubband) relaxation. This is a purely three-dimensional feature, whose analysis is straightforward within the present Monte Carlo simulation scheme.

Figure 8.4 presents a direct energy versus time two-dimensional map of the carrier distribution corresponding to the simulated experiment in Fig. 8.3a. We can clearly see that, in addition to the $3 \to 2 \to 1$ intersubband relaxation (which is the only one considered in a purely macroscopic model), the quantum-cascade process is also characterized by a significant intrasubband energy relaxation, resulting in the typical phonon-replica scenario. Since the

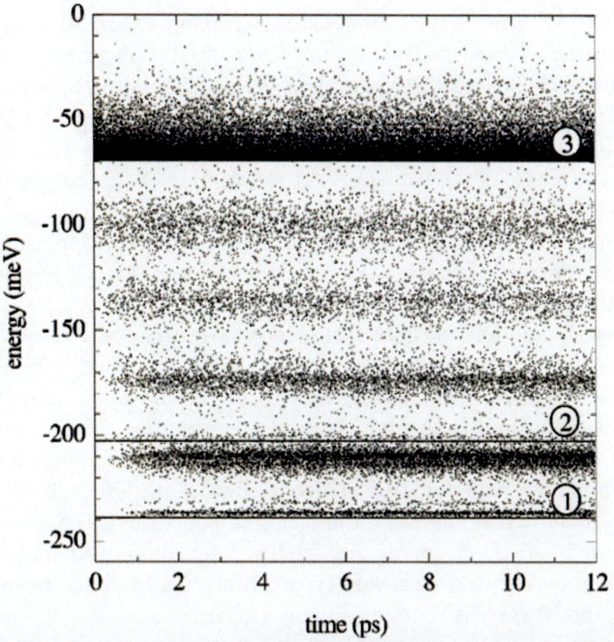

Fig. 8.4. Energy versus time two-dimensional map of the simulated carriers involved in the $3 \rightarrow 2 \rightarrow 1$ quantum-cascade process corresponding to the simulated experiment of Fig. 8.3a (see text). Reprinted from [353]

overall intersubband carrier transfer depends on the scattering rates $P_{\mathbf{k}_\parallel \tilde{n}, \mathbf{k}'_\parallel \tilde{n}'}$ as well as on the actual carrier distributions $f_{\mathbf{k}_\parallel \tilde{n}}$ in the various subbands, the introduction of phenomenological intersubband transition rates $W_{\tilde{n}\tilde{n}'}$ is meaningless, since the latter are unavoidably time-dependent quantities (see (6.8)), thus giving rise to nonlinear effects. In particular, for steady-state conditions the energy position of the lowest phonon replica in subband 2 is crucial in determining the $2 \rightarrow 1$ phonon-induced carrier transfer. Therefore, in order to improve the quantum-cascade performance it is essential to locate the phonon replica in phase-space regions characterized by a strong $2 \rightarrow 1$ phonon coupling.

8.2.2 Global-Simulation Scheme

The partially phenomenological description reviewed in the previous section has proven to be quite useful to address the microscopic nature of the hot-carrier relaxation within a portion of the quantum-cascade structure, namely the bare active region of the generic stage; however, due to the presence of free/fitting parameters – coupling the region of interest to the rest of the device – the latter does not allow to address the nature of the physical mechanisms governing charge transport through injector/active-region/collector interfaces. To this aim, the partially phenomenological model proposed in [353] has to be replaced by the fully microscopic description of the whole multiple quantum-well structure in (6.17). As shown in Sect. 6.2, in view of the spatial periodicity of our nanodevice (see Fig. 8.1), it is possible to employ the single-stage transport formulation in (6.19) originally proposed in [354].

Let us now review the key results of such a fully kinetic treatment – and of its particle-conserving Monte Carlo simulation (see Chap. 5) – applied again to the prototypical quantum-cascade device of Fig. 8.1.

To properly model phase-breaking hopping processes, all various intra- as well as intersubband carrier–optical phonon and carrier–carrier scattering mechanisms have been considered. As far as the first interaction is concerned, a conventional bulk phonon model is assumed, thus neglecting both quantized/localized phonon states and hot-phonon effects, which should give only minor corrections for the device design considered so far (see, e.g., [355]). As far as carrier–carrier interaction is concerned, the well-established time-dependent static-screening model has been employed; indeed, the latter is commonly adopted in a variety of Monte Carlo simulations of energy-relaxation and transport phenomena in two-dimensional systems (see, e.g., [356] and references therein). Other intrinsic scattering mechanisms, not included in the simulation, are expected to play a minor role. In particular, the interaction with acoustic phonons – in spite of its quasielastic nature – does not affect charge transport significantly due to its small coupling constant.

The role of extrinsic mechanisms, such as, e.g., interface roughness and carrier–impurity scattering, does considerably vary according to material growth and processing, with the result that the former is strongly material/device dependent. In this respect, a modeling of their effect would unavoidably require some "a posteriori" adjustment of free parameters to reproduce the diverse experimental data. For this reason, although they might be relevant to properly reproduce specific properties of realistic devices at the various operating conditions, extrinsic scattering mechanisms are not included in the microscopic simulations presented in [354].

As a starting point, let us discuss the relative weight of the carrier–carrier and carrier–phonon competing energy-relaxation channels. The time evolution of the carrier population in the various subbands as well as of the total current density in the presence of carrier–phonon scattering is depicted in Fig. 8.5. Here, parts (a) and (b) refer, respectively, to simulated experiments without and with two-body carrier–carrier scattering. In the "charge conserving" Monte Carlo scheme proposed in [354] (see Sect. 5.2) – based on the global single-stage formulation presented in Sect. 6.2 – the simulation starts assuming the total number of carriers to be, e.g., equally distributed among the different subbands; then the electron distribution function $f_{\mathbf{k}_\parallel \tilde{n}}$ evolves according to the Boltzmann-like equation (6.19) and a steady-state condition is eventually reached, leading to the desired $3 \rightarrow 2$ population inversion. As

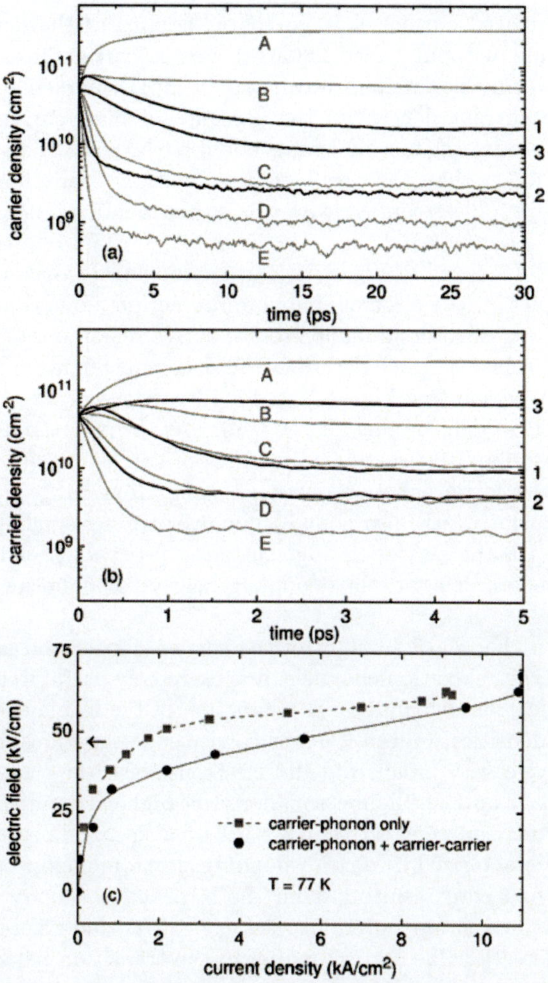

Fig. 8.5. Time evolution of simulated carrier densities in the various subbands of the quantum-cascade structure in Fig. 8.1, without (**a**) and with (**b**) carrier–carrier scattering. (**c**) Simulated applied-field versus current-density characteristics of the whole structure with carrier–phonon scattering only (*squares*) and with both carrier–carrier and carrier–phonon scattering (*discs*). *Lines* are a guide to the eye (see text). Reprinted from [352]

shown in Fig. 8.5, the inclusion of carrier–carrier scattering has significant effects: It strongly increases intersubband carrier redistribution, thus reducing the electron accumulation in the lowest energy level A and optimizing the coupling between active region and injector/collector (the populations of subbands 3 and B get comparable). This effect comes out to be crucial in determining the electron flux through the quantum-cascade structure: Fig. 8.5c shows the simulated current–voltage characteristics of the device, obtained with (discs) and without (filled squares) carrier–carrier interaction; at the threshold operating parameters estimated in [351] (marked by an arrow in Fig. 8.5c) the current density in the presence of both carrier–phonon and carrier–carrier scattering mechanisms is about 5.6 kA/cm^2. This value is about a factor of 3 higher than the result obtained with carrier–phonon scattering only and is in good agreement with the experimental findings reported in Fig. 8.2 (~ 7 kA/cm^2) [351].

The results displayed in Fig. 8.5 clearly demonstrate that within a purely semiclassical picture the electron–phonon interaction alone is not able to efficiently couple the injector subbands to the active region ones: While carrier–phonon energy relaxation well describes the electronic quantum cascade within the bare active region (see Figs. 8.3 and 8.4), carrier–carrier scattering plays an essential role in determining charge transport through the full core region. This can be ascribed to two typical features of carrier–carrier interaction as compared to the carrier–phonon one: (i) this is a long-range two-body interaction mechanism, which also couples non-overlapping single-particle states (see Fig. 8.1); (ii) the corresponding scattering process at relatively low carrier densities is quasielastic, thus coupling nearly resonant energy levels like states 3 and B.

To focus on the relative weight of the carrier–carrier and carrier–phonon competing energy-relaxation channels, it is extremely useful to investigate the steady-state electron distribution as a function of the in-plane energy for different carrier densities. Indeed, for low carrier concentrations, carrier–carrier scattering plays a very minor role and electrons relax their energy via a cascade of successive optical-phonon emissions; for high carrier densities, carrier–carrier scattering is more effective in setting up a so-called heated Maxwellian distribution characterized by a corresponding electronic temperature $T_{\tilde{n}}$. This is a well-known trend, resulting from the screened two-body nature of the carrier–carrier interaction potential (see, e.g., [278] and references therein). The specific density value corresponding to the transition between these two different regimes, however, depends on the heterostructure details and can be established only via corresponding fully three-dimensional simulations.

The results presented in [354] clearly show that, for typical operating conditions (corresponding to a sheet density of 3.9×10^{11} cm^{-2}), electrons tend to thermalize within each quantum state of the confinement-potential profile (see Fig. 8.1). Indeed, electronic temperatures corresponding to the diverse subbands vary in a quite narrow range ($T_{\tilde{n}} = 600$–750 K), as shown in Fig. 8.6. Again, the role of carrier–carrier interaction is crucial in setting up this behavior: This effect can be mainly ascribed to so-called bi-intrasubband two-body processes, i.e., carrier–carrier scattering processes coupling electrons in different subbands without changing their subband quantum number $\tilde{n}$. They provide, in fact, a very efficient way of redistributing excess kinetic energy in order to achieve a common effective temperature; the latter comes out to be about 700 K, to be compared with a lattice temperature of 77 K as-

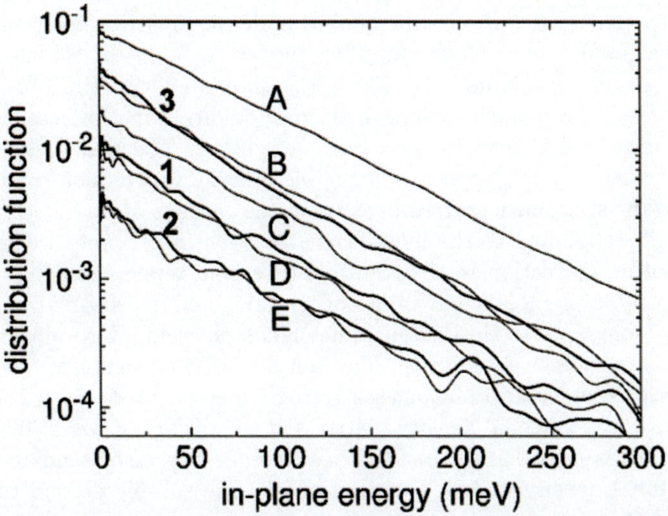

Fig. 8.6. Electron distribution function versus in-plane energy, for carriers in subbands $\tilde{n} = 1, 2, 3$ and $\tilde{n} = \mathrm{A, B, C, D, E}$ corresponding to the quantum-cascade structure of Fig. 8.1 (characterized by a sheet density of $3.9 \times 10^{11}\,\mathrm{cm}^{-2}$). Here, the slope of each curve provides an estimate of the effective carrier temperature $T_{\tilde{n}}$ in the various subbands, ranging from 600 to 750 K (see text). Reprinted from [354]

sumed in these simulated experiments. This heating is a clear fingerprint of a strong hot-carrier regime: the carrier system is not able to dissipate – via optical-phonon emission – the relatively large amount of energy provided by the applied bias. Recent experimental investigations (see, e.g., [357]) appear to confirm the above theoretical findings.

8.2.3 Quantum-Transport Phenomena

Due to the semiclassical character of the simulated experiments reviewed so far, no genuine quantum-mechanical effects have been considered. This remark addresses, in particular, the nature (coherent versus incoherent/sequential) of the resonant-tunneling injection of electrons into the active region of the device, which has been a long-standing open issue (see, e.g., [358] and references therein).

As originally proposed in [325] and discussed extensively in Sect. 6.2, in order to account for coherent phenomena, the single-stage semiclassical treatment in [354] (see (6.19)) needs to be replaced by its quantum-mechanical generalization given by a proper set of equations for the single-particle density matrix (6.26); more specifically, in the low-density regime the quantum formulation reduces to the linear transport equation (6.28), while in the presence of significant carrier concentrations one is forced to employ its nonlinear version in (6.31). Based on such nonlinear density-matrix equation – solved via a Weighted Monte Carlo sampling (see Sect. 5.2.5) of the in- and out-scattering superoperators in (6.31) – the simulated experiments presented in [325] provide a quantitative evaluation of quantum-mechanical corrections to the semiclassical scenario previously discussed. The main result of such density-matrix investigation is a negligible quantum correction (of a few percents) to the stationary current density: in the absence of carrier–carrier scattering (corresponding to the most favorable quantum regime) we get, e.g., a 2 % quantum correction to the result at threshold reported in Fig. 8.5. This is due to the extremely small values of non-diagonal density-matrix elements $\rho_{\mathbf{k}_\parallel,\tilde{n}\neq\tilde{n}'}$ (compared to the diagonal ones). The physical interpretation of such a behavior proceeds as follows: The non-diagonal elements of the generalized carrier–phonon in- and out-scattering superoperators in (6.31) tend to maintain a non-diagonal density matrix also in stationary conditions. On the other hand, energy-dissipation and decoherence processes tend to suppress non-diagonal density-matrix contributions on the sub-picosecond time-scale. Since the average transit time – the time needed by a carrier to travel through the device period and thus to be re-injected into the simulation region – is of the order of several picoseconds, the degree of carrier coherence (i.e., the weight of non-diagonal density-matrix elements) in stationary conditions is extremely small.

This steady-state analysis does not allow us to conclude that wavefunction-interference phenomena, like resonant-tunneling processes, are absent. Indeed, the latter – negligible in stationary conditions – are expected to strongly influence the ultrafast electro-optical response of our nanodevice. The first fully three-dimensional simulated experiments of coherent versus incoherent ultrafast carrier dynamics in quantum-cascade lasers, originally presented in [325], are shown in Fig. 8.7. Here, we focus on the picosecond time evolution of a properly tailored electron wavepacket within the core region of the quantum-cascade nanodevice sketched in Fig. 8.1. More specifically, the aim is to focus on the injector/active-region tunneling mechanisms. For this purpose, at the initial time ($t = 0$) the carrier system has been prepared as a coherent superposition of the two quasi-resonant states $\ell = 0, \tilde{n} = \mathrm{B}$ and $\ell = 1, \tilde{n} = 3$ shown in Fig. 8.1, which corresponds to a spatial charge distribution fully localized in the $\ell = 0$ injector region. As clearly shown, the transient dynamics is characterized by a strong interplay between quantum-mechanical phase coherence and dissipation/decoherence; only after several picoseconds the carrier system will eventually reach the stationary transport regime, where incoherent/sequential tunneling is the dominant interwell mechanism.

The results of Fig. 8.7 refer to the case where only carrier–phonon interaction is taken into account. When also carrier–carrier contributions are included the transient evolution (not reported here) gets much shorter and the carrier system reaches the steady-state regime on a sub-picosecond time-scale; this time is much shorter than the average transit time across one period,

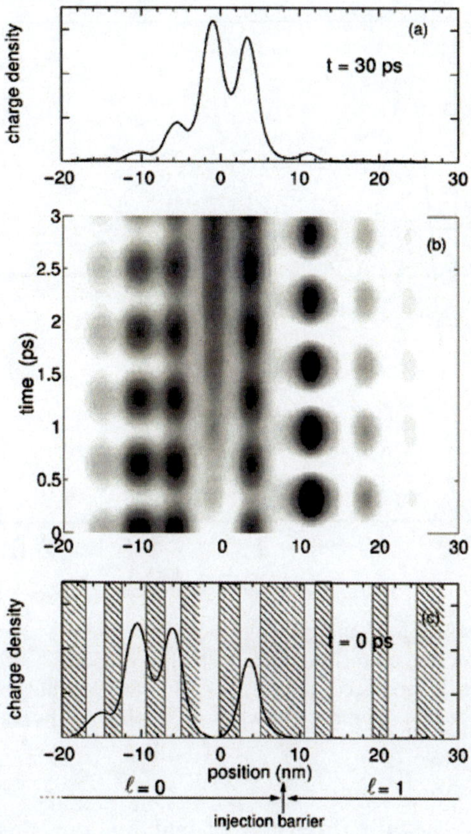

Fig. 8.7. Time evolution of the charge density for an electron wavepacket properly tailored to study the carrier tunneling dynamics across the injection barrier for the quantum-cascade design reported in Fig. 8.1. At $t = 0$ (**c**) the wavepacket is fully localized within the injector. The *shaded regions* correspond to the (Al,Ga)As barriers in the nanostructure design. The transient dynamics (**b**) is characterized by a strong interplay between phase-coherence and dissipation/decoherence processes. At much longer times (**a**) the system will eventually evolve into the stationary transport solution (see text). Reprinted from [325]

thus confirming the incoherent nature of steady-state transport. This sub-picosecond scenario has been originally described in [359]; in particular, during the initial transient, a clear gain overshoot appears, as shown in Fig. 8.8. These results suggested that ultrafast optical experiments – like pump-probe or four-wave-mixing measurements (see Sect. 2.5) – should provide a clear fingerprint of such interplay between coherent and incoherent carrier dynamics. This has been indeed confirmed by recent ultrafast experiments (see, e.g., [329]).

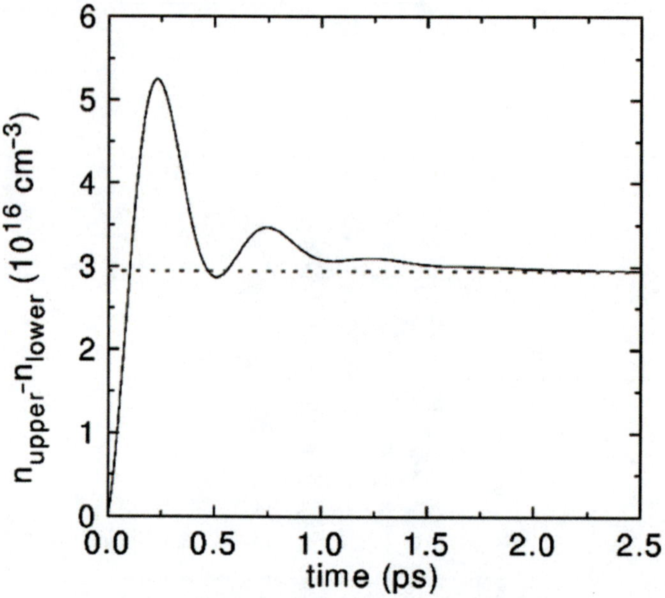

Fig. 8.8. Time evolution of the population difference between the upper and the lower lasing state for the quantum-cascade structure of Fig. 8.1. At the initial time ($t = 0$) the system is prepared so that the charge distribution is fully localized within the injector region. The *dotted line* is set to the steady-state value (see text). Reprinted from [359]

On the basis of the fully three-dimensional quantum simulations reviewed above as well as of recent experimental investigations, we are in the position to provide a definite answer to the long-standing controversial question on the nature – coherent versus incoherent – of charge transport in quantum-cascade lasers: For the typical quantum-cascade structure previously considered (see Fig. 8.1), energy-dissipation and decoherence processes are so strong to destroy any phase-coherence effect on a sub-picosecond time-scale; as a result, the usual semiclassical (or incoherent) description of stationary charge transport presented in Sect. 6.2 is found to be in excellent agreement with experimental results corresponding to typical steady-state operation conditions.

8.2.4 Active-Region/Cavity–Mode Coupling

The investigation of quantum-cascade lasers reviewed so far has mainly focused on the bare electron system, without considering any optical-cavity effect; in other words, we have modeled the properties of the so-called electroluminescent device and not of the laser (see, e.g., [66] and references therein). A global modeling of the coupled carrier–photon system in nonequilibrium conditions is a highly non-trivial task and may of course be performed at various levels of accuracy, ranging from purely phenomenological models to fully kinetic treatments (see, e.g., [360–362] and references therein).

In order to benefit from the fully kinetic description of the luminescent device employed so far (via the single-stage picture introduced in Sect. 6.2), it is desirable to adopt a fully kinetic treatment of the photonic system as well. As discussed extensively in Sect. 6.3, this requires to solve the coupled set of nonlinear equations (6.35), describing, in general, a coupled carrier–quasiparticle nonequilibrium dynamics. For the specific problem under examination, the nonequilibrium quasiparticle population $N_{\boldsymbol{q}}$ in (6.35) corresponds to one or more photonic modes of our laser cavity. More specifically, by adopting once again the single-stage description of the carrier system introduced in Sect. 6.2 (see (6.17)) and limiting our treatment to a single photonic cavity mode, the explicit form of the coupled (carrier plus photon) set (6.35) in the presence of additional (i.e., non-radiative (nr)) carrier scattering (e.g., carrier–phonon, carrier–carrier, etc.) is given by

$$\begin{aligned}
\frac{\mathrm{d}f_{\boldsymbol{k}_\parallel \tilde{n}}}{\mathrm{d}t} =& \sum_{\tilde{n}'\pm} \left(N + \frac{1}{2} \pm \frac{1}{2}\right) \left[\left(1 - f_{\boldsymbol{k}_\parallel \tilde{n}}\right) \tilde{P}^{\pm}_{\boldsymbol{k}_\parallel, \tilde{n}\tilde{n}'} f_{\boldsymbol{k}_\parallel \tilde{n}'} - \left(1 - f_{\boldsymbol{k}_\parallel \tilde{n}'}\right) \tilde{P}^{\pm}_{\boldsymbol{k}_\parallel, \tilde{n}'\tilde{n}} f_{\boldsymbol{k}_\parallel \tilde{n}}\right] \\
&+ \sum_{\boldsymbol{k}'_\parallel \tilde{n}'} \left[\tilde{P}^{\mathrm{nr}}_{\boldsymbol{k}_\parallel \tilde{n}, \boldsymbol{k}'_\parallel \tilde{n}'} f_{\boldsymbol{k}'_\parallel \tilde{n}'} - \tilde{P}^{\mathrm{nr}}_{\boldsymbol{k}'_\parallel \tilde{n}', \boldsymbol{k}_\parallel \tilde{n}} f_{\boldsymbol{k}_\parallel \tilde{n}}\right] \\
\frac{\mathrm{d}N}{\mathrm{d}t} =& \sum_{\boldsymbol{k}_\parallel, \tilde{n} \neq \tilde{n}', \pm} \pm \left(N + \frac{1}{2} \pm \frac{1}{2}\right) \left(1 - f_{\boldsymbol{k}_\parallel \tilde{n}}\right) \tilde{P}^{\pm}_{\boldsymbol{k}_\parallel, \tilde{n}\tilde{n}'} f_{\boldsymbol{k}_\parallel \tilde{n}'} - \frac{N - N^\circ}{\tau^{\mathrm{loss}}} \, .
\end{aligned} \tag{8.1}$$

Here, the first equation describes the effect on the carrier distribution $f_{\boldsymbol{k}_\parallel \tilde{n}}$ of cavity–mode absorption (−) and emission (+) processes (in terms of the corresponding single-stage carrier–photon rates $\tilde{P}^{\pm}$) as well as of all other non-radiative (nr) processes (in terms of their total single-stage rate $\tilde{P}^{\mathrm{nr}}$). The second equation, conversely, describes the effect on the cavity–mode average photon number N of carrier–photon absorption and emission as well as of cavity-loss processes; the latter are accounted for via a phenomenological escape/loss time τ^{loss}. Here, the carrier–photon interaction is treated within the well-known dipole approximation (see (1.26)), and the corresponding scattering rates – diagonal/vertical in $\boldsymbol{k}_\parallel$ – are dictated once again by the intraband dipole matrix elements p^{env} in (1.56) (see Sect. 1.2.2).

Employing the symmetry property of the carrier-photon transition rates $\tilde{P}^{\pm}_{\boldsymbol{k}_\parallel, \tilde{n}\tilde{n}'} = \tilde{P}^{\mp}_{\boldsymbol{k}_\parallel, \tilde{n}'\tilde{n}}$ (see also (6.36)), it is convenient to rewrite the photon equation in (8.1) as

$$\frac{\mathrm{d}N}{\mathrm{d}t} = \sum_{\boldsymbol{k}_\parallel, \tilde{n} \neq \tilde{n}'} \tilde{P}^{+}_{\boldsymbol{k}_\parallel, \tilde{n}\tilde{n}'} \left[N \left(f_{\boldsymbol{k}_\parallel \tilde{n}'} - f_{\boldsymbol{k}_\parallel \tilde{n}}\right) + \left(1 - f_{\boldsymbol{k}_\parallel \tilde{n}}\right) f_{\boldsymbol{k}_\parallel \tilde{n}'}\right] - \frac{N - N^\circ}{\tau^{\mathrm{loss}}} \, . \tag{8.2}$$

As we can see, apart from the cavity-loss term, carrier–photon coupling manifests itself via two distinct contributions: a standard stimulated-emission contribution (proportional to the photon occupation N as well as to the population-inversion factor $f_{\mathbf{k}_\parallel \tilde{n}'} - f_{\mathbf{k}_\parallel \tilde{n}}$) plus a spontaneous-emission term (proportional to $(1 - f_{\mathbf{k}_\parallel \tilde{n}})f_{\mathbf{k}_\parallel \tilde{n}'}$). The result in (8.2) confirms that, in order to have steady-state light amplification, the stimulated-emission term should be positive, which requires, in turn, at least one resonant intersubband transition $\tilde{n}' \rightarrow \tilde{n}$ characterized by population inversion, i.e., $f_{\mathbf{k}_\parallel \tilde{n}'} > f_{\mathbf{k}_\parallel \tilde{n}}$.

The distinguished advantage of treating the carrier–photon system via the coupled set of kinetic equations (8.1) is that – due to their partial Boltzmann-like structure – they can be easily sampled via Monte Carlo simulations. As discussed extensively in [280], two aspects, however, make this task not so straightforward. The first one is represented by the extremely different time-scales characterizing the carrier and photon dynamics during the transient response of the quantum-cascade laser: While the electronic ensemble reorganizes itself within picoseconds, the buildup of the photon population in an "empty device" might take several hundreds of picoseconds. The second reason is linked to the control of the statistical fluctuations, which are unavoidable when dealing with rare events; this applies, in particular, to the initial simulation transient characterized by a very small number of photons. These remarks suggest that the most convenient modeling strategy to simulate the transient evolution of the coupled electron-photon system is the Weighted Monte Carlo approach presented in Sect. 5.2.5.

The fully kinetic treatment of the carrier–photon coupling in a quantum-cascade device reviewed so far together with its Weighted Monte Carlo simulation has been originally proposed in [280]. In agreement with the general prescription of the Weighted Monte Carlo technique (see Sect. 5.2), the microscopic carrier–photon scattering probabilities $\tilde{P}^{\pm}$ entering the coupled set (8.1) are artificially enhanced to effectively limit/control the statistical fluctuations in the photon number. This simulative artifact is then exactly compensated by a proper "weighting" of the carrier and photon counters, so that the average system behavior does not depend on such weighting procedure. Due to the rare-event nature of the problem, however, the transient dynamics may show significant variations, especially when simulating the photon buildup starting from an initially empty cavity; this is shown in Fig. 8.9 where the photon population is reported as a function of the simulated time for the initial condition of 10 photons in the cavity at time $t = 0$ and for two different random-number sequences (see Sect. 5.2.2). It is imperative to stress that such transient instability is a well-known mathematical feature of the photon equation in (8.1) (see, e.g., [360]) and not an anomaly of such weighting procedure.

To reduce computing resources, the Weighted Monte Carlo simulations of Fig. 8.9 have been performed in the presence of carrier–phonon scattering only. The relevant cavity parameters entering the coupled set of equations (8.1) are taken from [351], while the spontaneous-emission intersubband line-width is

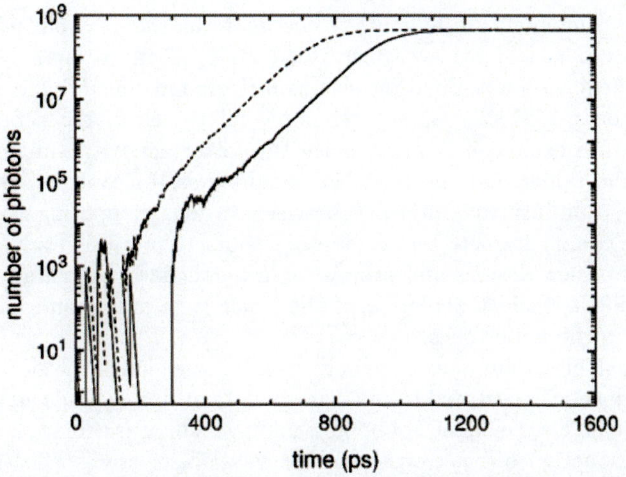

Fig. 8.9. Fully three-dimensional Weighted Monte Carlo simulation of the lasing buildup in the quantum-cascade device of Fig. 8.1. The two *curves* refer to simulated experiments differing in the random-number sequence (see Sect. 5.2.2): the initial transient shows the typical statistical fluctuations of a rare-event problem (see text). Reprinted from [280]

adjusted to reproduce the measured emitted power of the operating device.[2] Such a free parameter (varying in the range 9–12 meV in extremely good agreement with the experimental findings reported in [351]) includes the contributions of all intrinsic (carrier–carrier, carrier–phonon, etc.) as well as extrinsic (carrier–impurity, interface roughness, etc.) scattering mechanisms and is therefore strongly device dependent.

8.3 Toward Terahertz Laser Sources

Since their first demonstration in 1994 [315], the performances of quantum-cascade lasers have experienced tremendous improvements, and the range of emission wavelength has been continuously extended [315–350]. In principle, quantum-cascade devices can be designed to emit at any frequency over an extremely wide range, using the same combination of materials and varying the nanostructure design only. However, the translation of the conventional cascade scheme into devices operating at photon energies below the longitudinal

[2] Dealing with a single-mode cavity and thus with a discrete photonic spectrum, it is crucial to replace the energy-conserving Dirac delta function in (6.36) with a corresponding Lorentzian spectral function accounting for the intersubband linewidth (see (2.42)).

optical (LO) phonon threshold of the host material ($\hbar\omega_{\mathrm{LO}} \sim 35\,\mathrm{meV}$ in GaAs) is definitely not straightforward; in this configuration, in fact, the main difficulty in achieving a population-inversion regime in the active region is that optical-phonon emission cannot be employed to selectively depopulate the nanostructure excited subbands, since the latter act with similar efficiency on both the upper and the lower laser subbands. In contrast, the complex and highly non-intuitive interplay between various competing non-radiative relaxation channels has to be taken into account to properly evaluate the performances of new designs and proposals; in particular, two main targets have to be fulfilled: efficient depletion of the lower laser states and long electron life-times in the upper ones.

Although electroluminescence in the terahertz region of the electromagnetic spectrum has been detected from a variety of quantum-cascade structures, for a long time no clear evidence of population inversion was achieved, mainly due to slow extraction of electrons from the lower laser states. The parameter-free simulation strategy previously reviewed has proven to be an ideal tool to have some insight into the main limitations of these early structures and for the characterization and optimization of novel quantum-cascade prototypes. Indeed, the simulated experiments originally presented in [363] allowed to propose promising designs for which the predicted lasing action was later demonstrated [328].

More specifically, the quantum-cascade structure of the first heterostructure terahertz laser, proposed in [328] and schematically depicted in Figs. 8.10 and 8.11, has been designed to emit at 4.4 THz ($\lambda \sim 70\,\mu\mathrm{m}$) [363]. The latter is based on a vertical-transition configuration (see Fig. 8.10a), which is known to lead to larger dipole matrix elements as well as narrower line-widths, and employs a conventional chirped-superlattice[3] design. Its operating strategy exploits the same concepts successfully implemented for the shortest wavelength ($\lambda \sim 17 - 24\,\mu\mathrm{m}$) mid-infrared quantum-cascade lasers (see, e.g., [364–366]). To minimize the occupation of the lower laser subband, the key idea is to employ a dense miniband with seven subbands, which provides a large phase-space where electrons scattered either from the upper laser subband or directly from the injector can spread. The miniband dispersion is chosen as large as possible compatibly with the need of avoiding cross absorption. This suppresses thermal backfilling and provides a large operating range of currents and voltages. Energy relaxation within the first miniband would appear to be hindered by the lack of final states with appropriate energy to allow for optical-phonon emission. Nevertheless – as originally pointed out in [363] – carrier–carrier interaction does beneficially operate as an activation

[3] A so-called CHIRP (coherent hetero-interfaces for reflection and penetration) superlattice, also referred to as chirped superlattice, is a superlattice-like structure where the spatial periodicity is gradually changed.

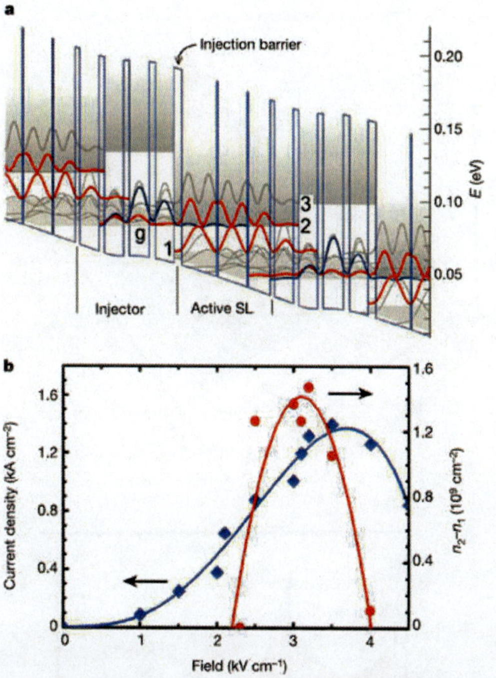

Fig. 8.10. Conduction-band structure of the laser active core and simulated population inversion and current density at several applied biases. (**a**) Self-consistent calculation of the conduction-band structure of a portion of the layer stack in the waveguide core under a field of 3.5 kV/cm. Injectors and superlattice (SL) active regions are alternating. The layer thicknesses, in nanometers, starting from the injection barrier are ***4.3***/*18.8*/***0.8***/*15.8*/***0.6***/*11.7*/***2.5***/10.3/**2.9**/10.2/**3.0**/10.8/**3.3**/9.9, where $Al_{0.15}Ga_{0.85}As$ layers are in bold face, the active region is in italic, and the 10.2 nm well is Si-doped at 4×10^{16} cm^{-3}. The moduli squared of the wavefunctions are shown, with miniband regions represented by *shaded areas*. The optical transition takes place across the 18-meV wide minigap between the second and the first miniband (states 2 and 1) and, being vertical in real space, presents a large dipole matrix element of 7.8 nm. Carriers are injected into state 2 via resonant tunneling from the injector ground state labeled g. (**b**) Simulated population inversion between states 2 and 1 of the active superlattice (*circles*), expressed in terms of electron sheet density (n_2 and n_1, respectively) and current density traversing the structure (*diamonds*). Both are evaluated as a function of bias field at 10 K. *Solid lines* are polynomial fits. From the simulation we also extract the level life-times: $\tau_2 = 0.8$ ps, $\tau_{21} = 8.3$ ps, $\tau_1 = 2.2$ ps with τ_2 and τ_{21} dominated by carrier–phonon scattering processes. Note that τ_{21} is much greater than τ_2 owing to the superlattice nature of the active region. This implies that $\tau_1 < \tau_{21}$, which is the condition required for population inversion. Reprinted from [328]

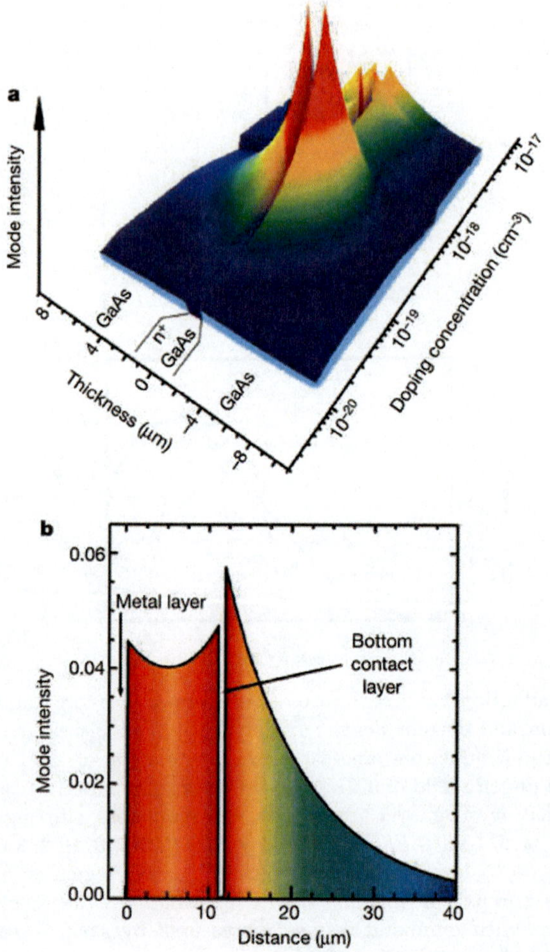

Fig. 8.11. Waveguide-design principle and radiation-mode confinement. (**a**) Evolution of the intensity profile of the optical mode bound to a 800-nm thick GaAs n-doped layer within undoped GaAs material calculated as a function of donor density. The dielectric constants were derived from a Drude–Lorentz model. (**b**) Calculated mode profile along the growth direction of the final device structure. The origin of the abscissa is at the top metal/semiconductor interface, and the waveguide core of 104 repetitions of the active region shown in Fig. 8.10 is between the bottom contact and top metal layer. Reprinted from [328]

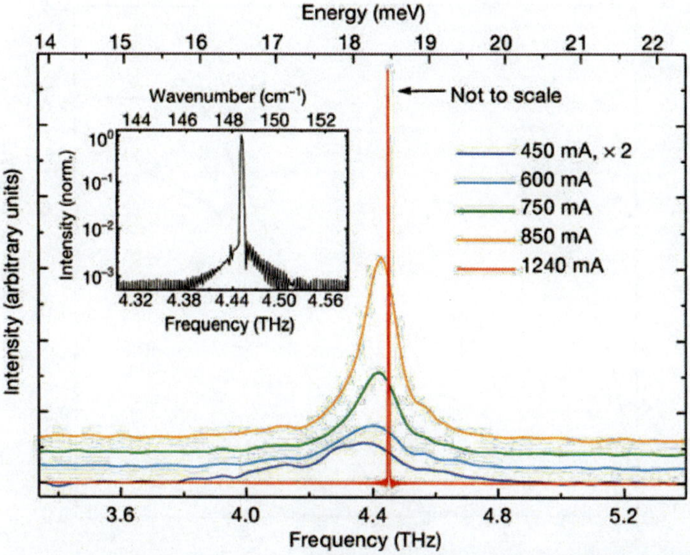

Fig. 8.12. Emission spectra from a 1.24-mm long and 180-μm wide laser device recorded at 8 K for different drive currents. The lowest curve is multiplied by a factor of 2 for clarity, while the 1, 240 mA laser spectrum is scaled down by several orders of magnitude. Current macro-pulses were applied at a repetition rate of 333 Hz and matched the frequency response of the detector; each macro-pulse comprised 750 micro-pulses of 200 ns width at intervals of 2 ms to avoid excess heating in the device. The characteristic narrowing of the emission line and nonlinear dependence of the intensity are clearly observed up to a current of about 880 mA, where laser threshold is reached. Lasing takes place at about 4.4 THz, on the high-energy side of the luminescence line, probably owing to the reduced waveguide losses at shorter wavelengths. Reprinted from [328]

mechanism since it provides sufficient in-plane momentum for the electrons to open a scattering path via optical-phonon emission to the lower subbands (see Fig. 8.14); indeed, in the presence of carrier–carrier interaction, electrons efficiently relax into the lower states of the injector, from where they are readily transferred into the upper laser state of the following period. A significant population inversion then results between the subbands of the laser transition (see Fig. 8.10b), as confirmed by the corresponding emission spectra in Fig. 8.12 as well as by the light–current characteristics in Fig. 8.13.

Apart from its strong technological implications, this revolutionary nanostructured device came out to be an ideal context to investigate the complex synergy between carrier–carrier thermalization and phonon-assisted energy relaxation previously mentioned; as shown in Fig. 8.14, the latter is indeed the key ingredient for a proper operation of these new light sources.

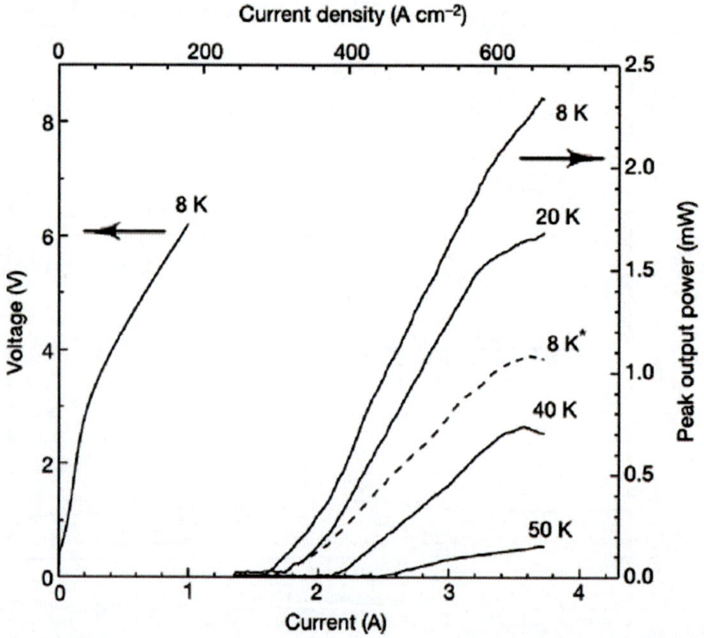

Fig. 8.13. Light–current characteristics of a 180 µm-wide and 3.1 mm-long laser ridge. Here, data were recorded in pulsed mode, applying 100 ns-long pulses at a repetition rate of 333 Hz. The peak power values represent what was collected from a single facet onto our detector after correction by a factor of 2 to account for retro-reflection effects. At the lowest temperature the device emits more than 2 mW. When the duty cycle was increased to 0.5%, the peak power reduced to 1 mW owing to the stronger heating of the device, as shown by the dashed curve labeled with an asterisk. At the maximum operating temperature of 50 K, we observed 120 µW of peak output power. The voltage–current curve at a temperature of 8 K is also displayed. Reprinted from [328]

As final issue related to far-infrared lasing sources, let us briefly mention the possible impact on the quantum-device performances of nonequilibrium phonon distributions, the so-called hot-phonon effects introduced in Sect. 6.3. While for mid-infrared systems the role of such nonequilibrium phonon dynamics is believed to be minor (see, e.g., [355]), for far-infrared ones the situation may be significatively different; indeed, according to recent experimental investigations (see, e.g., [369–371] and references therein), evidence of hot-phonon effects is found in the electro-optical response of terahertz quantum-cascade devices. As previously mentioned (see Sect. 6.3), such nonequilibrium-phonon regime may be properly modeled via the coupled set of Boltzmann-like equations (6.35), whose solution may be conveniently performed via a corresponding Monte Carlo sampling (see, e.g., [368]). Figure 8.15 shows the

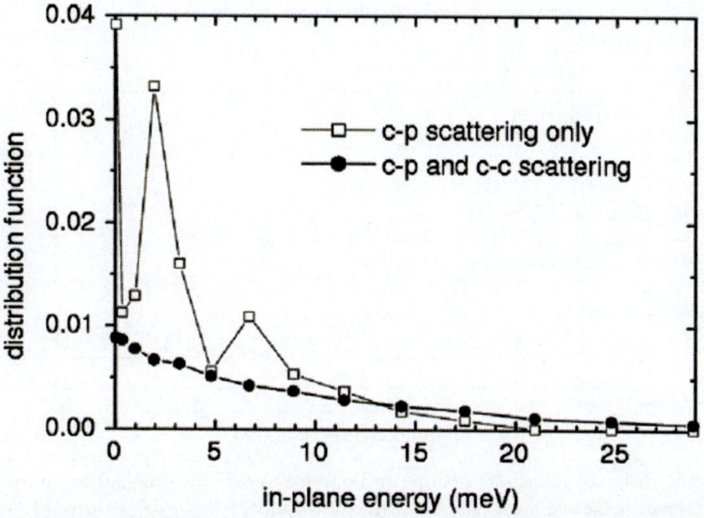

Fig. 8.14. Electronic distribution function versus in-plane energy for carriers in subband 1 (lower state of the lasing transition) of the terahertz quantum-cascade design proposed in [363]. When carrier–phonon scattering is the only active mechanism, electrons relax down in the subband and are mainly localized within less than 15 meV from its bottom; depletion of the lower laser states due to photon emission into the injector miniband, whose width amounts to about 18 meV, is therefore highly inhibited. In contrast, carrier–carrier interaction, on its side, acts as a thermalization mechanism within the subband; it spreads the distribution function over a wider energy range, therefore providing electrons with the necessary in-plane momentum to further relax, by phonon emission, out of the lower laser subband. Reprinted from [367]

hot-phonon distribution profile of a typical GaAs-based resonant-phonon THz quantum-cascade laser obtained via the fully three-dimensional Monte Carlo simulation strategy proposed in [368].

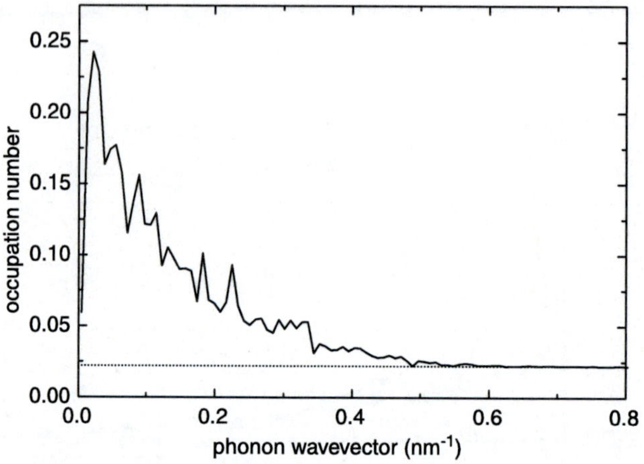

Fig. 8.15. Hot-phonon distribution profile in a typical GaAs-based resonant-phonon THz quantum-cascade laser (see text). Here, we show the nonequilibrium LO-phonon population as a function of the parallel (in-plane) wavevector $q_{\|}$ in the $q_{\perp} \rightarrow 0$ limit. In this simulated experiment the lattice temperature is $T = 110\,\mathrm{K}$, which corresponds to a thermal LO-phonon occupation of about 0.02 (*dotted line*). As we can see, due to the strong electron–phonon coupling, the LO-phonon population (*solid curve*) is driven out of equilibrium; this effect is particularly significant for $q_{\|} \sim 0.02\,\mathrm{nm}^{-1}$, where the nonequilibrium occupation is more than one order of magnitude larger than the thermal one. Reprinted from [368]

Part III

New-Generation Nanomaterials and Nanodevices

9

Few-Electron/Exciton Quantum Devices

In this chapter we shall discuss the basic properties and unique features of few-electron/exciton quantum systems, namely single and coupled semiconductor macroatoms, pointing out their potential role in designing a completely new class of optoelectronic quantum devices, like electron-state detectors and quantum logic gates.

9.1 Fundamentals of Semiconductor Macroatoms

As discussed extensively in Chap. 1, semiconductor quantum dots [42, 44–49] – also referred to as semiconductor macroatoms – represent the extreme evolution of low-dimensional semiconductor structures and constitute an ideal arena for the investigation of several fundamental phenomena related to solid-state quantum physics [372–377]. Opposite to quantum wells and wires, they exhibit a discrete – i.e., atomic-like – energy spectrum and their optical response is dominated by few-particle excitonic effects (see below).

During the 1990s, self-assembling techniques for the fabrication of quantum dots have been developed (see, e.g., [372] and references therein). The latter allow for a control of the areal density (between 10^9 and $10^{12}\,\mathrm{cm}^{-2}$) by varying the total deposition of the material for the quantum-dot layer. The in-plane average size of these dots usually varies between 10 and 50 nm, with a size fluctuation in the same sample which can be reduced to few percents depending on the growing conditions. Such a spontaneous size fluctuation is responsible for the inhomogeneous broadening of absorption as well as photoluminescence spectra, but it can also be employed to implement energy-selective addressing schemes for quantum-information processing (see Chap. 10).

Self-assembled quantum dots (see Figs. 1.11 and 1.12) show high optical efficiency due both to the strong carrier confinement (as well as to the resulting enhanced coupling to electromagnetic radiation) and to the absence of material degradation, which can instead be induced when semiconductor quantum dots are produced via nanolithographic techniques. Such self-organized zero-dimensional structures are often characterized by a symmetric in-plane confinement potential (which paves the road to an easier engineering of their electronic structure) and by a stronger confinement along the main growth direction. As we shall see, the possibility of optically driving their excitations allows for ultrafast state manipulation and coherent control of the few-carrier dynamics, a basic request for the quantum-computing devices discussed in Chap. 10. As discussed in [51], due to strain propagation through the barrier which separates different quantum-dot layers, similar-size quantum dots tend to vertically align (see Fig. 1.11), enhancing the vertical coupling between different dots. Such structures, which have been proposed as semiconductor-based quantum-bit (qubit) arrays for quantum information/computation [378–380], will be the prototypical quantum-dot systems investigated in the present chapter.

Self-organized quantum dots have been successfully fabricated using a wide range of semiconductor materials; as discussed in [51], they include III–V quantum-dot structures based on GaAs as well as on GaN compounds. The characterization of GaN-based quantum dots is still in progress, but interesting properties such as their long spin decoherence time (see Chap. 10) or the presence of a built-in electric field as strong as few MeV/cm [381–393] have already been demonstrated. GaN systems have wider bandgaps (of the order of 3.5 eV) compared to GaAs-based ones (of the order of 1.5 eV). Moreover, whereas GaAs and many of the other III–V compounds have a cubic (zinc-blend) structure, GaN (as well as other nitrides) exhibits a hexagonal (wurtzite) structure which leads to the strong built-in piezoelectric fields previously mentioned (of the order of MV/cm). As a consequence, in these nanostructures excitonic transitions are red shifted, and the corresponding interband emission is fractions of electronvolt below the bulk GaN bandgap (see below).

As discussed in Sect. 9.3, the strong built-in electric fields which characterize wurtzite GaN heterostructures are the sum of the spontaneous polarization and of the piezoelectric field. Spontaneous-polarization charge accumulates at the GaN/AlN interfaces as a consequence of a slight distortion of GaN and AlN unit cells, compared to those of an ideal hexagonal crystal. Piezoelectric fields are caused by uniform strain along the $[0\,0\,1]$ direction. Contrary to GaN/AlGaN quantum wells – where the spontaneous-polarization contribution is dominant (see, e.g., [394]) – in quantum dots the strain-induced piezoelectric field and the spontaneous-polarization potential are of similar magnitude and sign, both oriented along the growth direction. The strength of the intrinsic field along such direction is almost the same inside and outside the dot, but it comes out to be antiparallel to the growth direction inside the nanostructure, inverting its sign outside. Its importance in connection to single-electron devices lays on the possibility of creating strong couplings between neighboring quantum dots which can be switched on and off optically. Indeed, as described in [395–397], under the action of such internal field, electrons and holes are driven in opposite directions, thus creating electric dipoles which interact each other with coupling energies of the order of few millielectronvolt (see Fig. 9.3); this effect is, for example, the key ingredient of the GaN-based quantum-information/computation processing device described in Sect. 10.2.2 [380].

Primary goal of this chapter is twofold: on the one hand, we shall review efficient simulation tools for the estimation of few-particle Coulomb interactions, and we shall study how Coulomb-correlated few-electron/exciton states are affected by the presence of external [398] or built-in [396] electric fields (see Sect. 9.3). On the other hand, we shall review potential all-optical storage [399] and read-out devices [396] based on coupled quantum-dot structures (see Sects. 9.4 and 9.5). More specifically, the partially analytical model originally proposed in [398] and reviewed in Sect. 9.3 constitutes an ideal tool to scan the nanostructure parameter space – which usually depends non-trivially on many variables – and to quickly obtain the optimal parameters for the design of quantum-dot-based single-electron/exciton devices. Indeed, such an analytical approach has allowed (i) to single out systems with large exciton–exciton interaction energies – crucial ingredient for the realization of the ultrafast all-optical conditional quantum gates presented in Chap. 10 – and (ii) to investigate electron–exciton interactions in view of device applications, like those reviewed in Sects. 9.4 and 9.5.

9.2 Coulomb-Correlation Effects in Few-Carrier Systems

In what follows, the physical system under investigation is a gas of electron–hole pairs spatially confined within a zero-dimensional semiconductor structure (see Sect. 1.2.2), namely a single as well as a multiple quantum dot. According to the general formulation presented in Chap. 3, the total Hamiltonian of our semiconductor nanostructure (see (3.1)) can be regarded as the sum of a noninteracting contribution $\hat{H}_\circ$ plus a perturbation term $\hat{H}'$ (see (3.2)); in turn, the noninteracting Hamiltonian $\hat{H}_\circ$ may be regarded as the sum of two contributions, depending on relevant (ξ) and non-relevant (Ξ) degrees of freedom (see (3.62)). In particular, for few-electron systems it is imperative to treat carrier–carrier interaction exactly; this imposes to include carrier–carrier coupling terms within the noninteracting Hamiltonian, i.e.,

$$\hat{H}_\xi = \hat{H}^{\text{ic}} = \hat{H}^{\text{c}} + \hat{H}^{\text{cc}} . \tag{9.1}$$

In this case, the relevant-coordinate Hamiltonian $\hat{H}_\xi$ coincides with the interacting-carrier (ic) Hamiltonian $\hat{H}^{\text{ic}}$, namely free carriers plus confinement potential plus carrier–carrier Coulomb interaction, while the unperturbed Hamiltonian $\hat{H}_\Xi$ corresponding to non-relevant coordinates describes various quasiparticle excitations, namely phonons, photons, plasmons.

In contrast, the perturbation Hamiltonian in (3.2) may describe the interaction of the Coulomb-correlated carrier subsystem with light sources and environmental degrees of freedom, namely carrier–light plus carrier–phonon interactions.

9.2.1 Single-Particle Description

Let us first consider the gas of noninteracting carriers, electrons (e), and holes (h), spatially confined within the zero-dimensional semiconductor structure. As discussed in Sect. 1.2.2, the quantum confinement can be described in terms of an effective potential whose height is dictated by the conduction/valence band discontinuities. Since the energy region of interest is relatively close to the bandgap of the semiconductor materials forming our heterostructure, it is possible to describe the bulk band structure in terms of the usual effective-mass approximation (see (1.14)); in addition, since the confinement potential is a slowly varying function on the scale of the lattice periodicity, one is allowed to work within the envelope-function picture introduced in Sect. 1.2.2 (see also Appendix A). Within such approximation scheme, the noninteracting carriers in our low-dimensional structure are then described by single-particle wavefunctions of the form (see (1.35))

$$\psi_{\alpha^{e/h}}(\boldsymbol{r}) = \sqrt{\Omega}\overline{\psi}_{\overline{\alpha}^{e/h}}(\boldsymbol{r})\phi_{e/h}(\boldsymbol{r}) \;, \tag{9.2}$$

where $\phi_{e/h}$ are electron/hole bulk Bloch states, while the envelope wavefunctions $\overline{\psi}_{\overline{\alpha}^{e/h}}$ are the solutions of the envelope-function Schrödinger equation (1.41), i.e.,

$$\left[-\frac{\hbar^2\nabla_{\boldsymbol{r}}^2}{2m^*_{e/h}} + \overline{V}_{e/h}(\boldsymbol{r})\right]\overline{\psi}_{\overline{\alpha}^{e/h}}(\boldsymbol{r}) = \epsilon_{\overline{\alpha}^{e/h}}\overline{\psi}_{\overline{\alpha}^{e/h}}(\boldsymbol{r}) \;. \tag{9.3}$$

Here, $m^*_{e/h}$ is the bulk effective mass for electrons/holes while $\alpha^{e/h}$ denotes the set of single-particle quantum numbers, including charge as well as spin degrees of freedom. According to the energy region of interest, the different approaches commonly employed for the solution of the above Schrödinger equation range from direct three-dimensional plane-wave expansions (see Sect. 5.1.2) to approximated factorized-state solutions or to simplified bidimensional-parabolic-potential models (see, e.g., [379]).

9.2.2 Coulomb-Correlated Carrier System

Given the above single-particle representation $\{\alpha^e\}$ ($\{\alpha^h\}$) for electrons (holes) – i.e., the set of three-dimensional eigenfunctions $\psi_{\alpha^e}(\boldsymbol{r}) = \langle \boldsymbol{r}|\alpha^e\rangle$ ($\psi_{\alpha^h}(\boldsymbol{r}) = \langle \boldsymbol{r}|\alpha^h\rangle$) and the corresponding energy levels ϵ_{α^e} (ϵ_{α^h}) – let us now introduce the following creation and destruction operators for electrons and holes:

$$|\alpha^e\rangle = \hat{c}^\dagger_{\alpha^e}|0\rangle \to |0\rangle = \hat{c}_{\alpha^e}|\alpha^e\rangle \;, \quad |\alpha^h\rangle = \hat{d}^\dagger_{\alpha^h}|0\rangle \to |0\rangle = \hat{d}_{\alpha^h}|\alpha^h\rangle \;, \tag{9.4}$$

where $|0\rangle$ denotes the electron–hole vacuum state.

Within such second-quantization description based on the electron–hole picture introduced in Chap. 1, the single-particle Hamiltonian (3.110), i.e., the Hamiltonian describing the noninteracting carriers within our zero-dimensional confinement potential, can be written as

$$\hat{H}^{\mathrm{c}} = \hat{H}^{\mathrm{e}} + \hat{H}^{\mathrm{h}} = \sum_{\alpha^{\mathrm{e}}} \epsilon_{\alpha^{\mathrm{e}}} \hat{c}^{\dagger}_{\alpha^{\mathrm{e}}} \hat{c}_{\alpha^{\mathrm{e}}} + \sum_{\alpha^{\mathrm{h}}} \epsilon_{\alpha^{\mathrm{h}}} \hat{d}^{\dagger}_{\alpha^{\mathrm{h}}} \hat{d}_{\alpha^{\mathrm{h}}} \; . \tag{9.5}$$

The carriers (electrons and holes) within our low-dimensional nanostructure, however, interact via a two-body Coulomb potential screened by the crystal lattice. Due to such interaction, several correlation effects take place. Here, only processes conserving the total number of carriers are considered, thus Auger recombination and impact ionization are neglected. Such processes are known to become important only at very high carrier densities and at energies high up in the band (see, e.g., [110]). Within such approximation scheme, the carrier–carrier interaction Hamiltonian (3.131) rewritten in our electron–hole picture comes out to be

$$\begin{aligned} \hat{H}^{\mathrm{cc}} &= \hat{H}^{\mathrm{ee}} + \hat{H}^{\mathrm{hh}} + \hat{H}^{\mathrm{eh}} \\ &= \frac{1}{2} \sum_{\alpha^{\mathrm{e}}_1 \alpha^{\mathrm{e}}_2, \alpha^{\mathrm{e}\prime}_1 \alpha^{\mathrm{e}\prime}_2} V^{\mathrm{cc}}_{\alpha^{\mathrm{e}}_1 \alpha^{\mathrm{e}}_2, \alpha^{\mathrm{e}\prime}_1 \alpha^{\mathrm{e}\prime}_2} \hat{c}^{\dagger}_{\alpha^{\mathrm{e}}_1} \hat{c}^{\dagger}_{\alpha^{\mathrm{e}}_2} \hat{c}_{\alpha^{\mathrm{e}\prime}_2} \hat{c}_{\alpha^{\mathrm{e}\prime}_1} \\ &+ \frac{1}{2} \sum_{\alpha^{\mathrm{h}}_1 \alpha^{\mathrm{h}}_2, \alpha^{\mathrm{h}\prime}_1 \alpha^{\mathrm{h}\prime}_2} V^{\mathrm{cc}}_{\alpha^{\mathrm{h}}_1 \alpha^{\mathrm{h}}_2, \alpha^{\mathrm{h}\prime}_1 \alpha^{\mathrm{h}\prime}_2} \hat{d}^{\dagger}_{\alpha^{\mathrm{h}}_1} \hat{d}^{\dagger}_{\alpha^{\mathrm{h}}_2} \hat{d}_{\alpha^{\mathrm{h}\prime}_2} \hat{d}_{\alpha^{\mathrm{h}\prime}_1} \\ &- \sum_{\alpha^{\mathrm{e}} \alpha^{\mathrm{h}}, \alpha^{\mathrm{e}\prime} \alpha^{\mathrm{h}\prime}} V^{\mathrm{cc}}_{\alpha^{\mathrm{e}} \alpha^{\mathrm{h}}, \alpha^{\mathrm{e}\prime} \alpha^{\mathrm{h}\prime}} \hat{c}^{\dagger}_{\alpha^{\mathrm{e}}} \hat{d}^{\dagger}_{\alpha^{\mathrm{h}}} \hat{d}_{\alpha^{\mathrm{h}\prime}} \hat{c}_{\alpha^{\mathrm{e}\prime}} \; , \end{aligned} \tag{9.6}$$

where

$$V^{\mathrm{cc}}_{\alpha_1 \alpha_2, \alpha'_1 \alpha'_2} = \int \mathrm{d}^3 r \int \mathrm{d}^3 r' \overline{\psi}^*_{\overline{\alpha}_1}(\boldsymbol{r}) \overline{\psi}^*_{\overline{\alpha}_2}(\boldsymbol{r}') V^{\mathrm{cc}}(\boldsymbol{r} - \boldsymbol{r}') \overline{\psi}_{\overline{\alpha}'_2}(\boldsymbol{r}') \overline{\psi}_{\overline{\alpha}'_1}(\boldsymbol{r}) \tag{9.7}$$

are the matrix elements of the two-body Coulomb potential $V^{\mathrm{cc}}(\boldsymbol{r} - \boldsymbol{r}')$ for the generic two-particle transition $\alpha'_1 \alpha'_2 \to \alpha_1 \alpha_2$ written in our envelope-function picture. The first two terms in (9.6) describe the repulsive electron–electron and hole–hole interactions while the third one describes the attractive interaction between electrons and holes.

We stress the fully three-dimensional nature of the present approach based on the detailed knowledge of the three-dimensional single-particle wavefunctions. In particular, the explicit evaluation of the Coulomb matrix elements in (9.7) for a generic confinement-potential profile may be conveniently performed via a plane-wave-expansion approach, as described in Sect. 5.1.2.

Combining the single-particle Hamiltonian (9.5) with the Coulomb-interaction term (9.6), we get the following many-body Schrödinger equation for our Coulomb-correlated system:

$$\hat{H}^{\mathrm{ic}}|\xi\rangle = \left(\hat{H}^{\mathrm{c}} + \hat{H}^{\mathrm{cc}}\right)|\xi\rangle = \mathcal{E}|\xi\rangle \ . \tag{9.8}$$

Here, $|\xi\rangle$ denotes the interacting many-carrier state within our Fock space and $\mathcal{E}$ the corresponding total energy.

Let us now introduce the total-number operators for electrons and holes:

$$\hat{N}_{\mathrm{e}} = \sum_{\alpha^{\mathrm{e}}} \hat{c}^{\dagger}_{\alpha^{\mathrm{e}}}\hat{c}_{\alpha^{\mathrm{e}}} \ , \qquad \hat{N}_{\mathrm{h}} = \sum_{\alpha^{\mathrm{h}}} \hat{d}^{\dagger}_{\alpha^{\mathrm{h}}}\hat{d}_{\alpha^{\mathrm{h}}} \ . \tag{9.9}$$

It is easy to show that the above global operators commute with the carrier Hamiltonian $\hat{H}^{\mathrm{ic}}$ in (9.8). One can therefore look for many-body states $|\xi\rangle$ corresponding to a given number of electrons (N_{e}) and holes (N_{h}). In particular, we shall focus on intrinsic semiconductor materials, for which $N_{\mathrm{e}} = N_{\mathrm{h}}$; in this case the "good" quantum number is the total number of electron–hole pairs ($N = N_{\mathrm{e}} = N_{\mathrm{h}}$) and the Schrödinger equation (9.8) can be rewritten as

$$\hat{H}^{\mathrm{ic}}|\xi_N\rangle = \mathcal{E}_{\xi_N}|\xi_N\rangle \ , \tag{9.10}$$

where $|\xi_N\rangle$ and $\mathcal{E}_{\xi_N}$ denote, respectively, the ξth many-body state and energy level corresponding to N electron–hole pairs.

For any given number N of electron–hole pairs we thus identify a subspace of the original Fock space, for which there exists another natural basis $\{|l_N\rangle\}$, given by the eigenstates of the single-particle Hamiltonian in (9.5):

$$\hat{H}^{\mathrm{c}}|l_N\rangle = \epsilon_{l_N}|l_N\rangle \ . \tag{9.11}$$

Here, $l_N \equiv \alpha^{\mathrm{e}}_1\alpha^{\mathrm{h}}_1, \alpha^{\mathrm{e}}_2\alpha^{\mathrm{h}}_2, \ldots, \alpha^{\mathrm{e}}_i\alpha^{\mathrm{h}}_i, \ldots, \alpha^{\mathrm{e}}_N\alpha^{\mathrm{h}}_N$ is a compact notation for the set of noninteracting electron and hole single-particle quantum numbers corresponding to our N electron–hole pairs. Indeed, we have

$$|l_N\rangle = |\{\alpha^{\mathrm{e}}_i\alpha^{\mathrm{h}}_i\}\rangle = \prod_{i=1}^{N} \hat{c}^{\dagger}_{\alpha^{\mathrm{e}}_i}\hat{d}^{\dagger}_{\alpha^{\mathrm{h}}_i}|0\rangle \tag{9.12}$$

and $\epsilon_{l_N} = \sum_{i=1}^{N}\left(\epsilon_{\alpha^{\mathrm{e}}_i} + \epsilon_{\alpha^{\mathrm{h}}_i}\right)$.

The noninteracting basis set (9.12) constitutes the starting point of the direct-diagonalization approach employed for the solution of the many-body Schrödinger equation (9.10). Indeed, one can expand the unknown many-body state $|\xi_N\rangle$ over our single-particle basis:

$$|\xi_N\rangle = \sum_{l_N} U^{\xi_N}_{l_N}|l_N\rangle \ . \tag{9.13}$$

By inserting the above expansion into the many-body Schrödinger equation (9.10), the latter is transformed into the following eigenvalue problem:

$$\sum_{l'_N}\left(H^{\mathrm{ic}}_{l_N l'_N} - \mathcal{E}_{\xi_N}\delta_{l_N l'_N}\right)U^{\xi_N}_{l'_N} = 0 \ , \tag{9.14}$$

where

$$H^{\mathrm{ic}}_{l_N l'_N} = \langle l_N | \hat{H}^{\mathrm{ic}} | l'_N \rangle = \epsilon_{l_N} \delta_{l_N l'_N} + \mathcal{V}_{l_N l'_N} \tag{9.15}$$

are the matrix elements of the interacting-carrier Hamiltonian $\hat{H}^{\mathrm{ic}}$ in our single-particle basis. They are given by a diagonal – i.e., noninteracting – contribution plus a non-diagonal term given by the matrix elements of the Coulomb-interaction Hamiltonian (9.6):

$$\mathcal{V}_{l_N l'_N} = \langle l_N | \hat{H}^{\mathrm{cc}} | l'_N \rangle \ . \tag{9.16}$$

Their explicit form – involving the various two-body Coulomb matrix elements in (9.7) – is discussed in [379] for the excitonic ($N = 1$) and biexcitonic ($N = 2$) cases.

In the presence of Coulomb interaction, the Hamiltonian matrix (9.15) is non-diagonal; therefore, the interacting many-body states $|\xi_N\rangle$ are, in general, a linear superposition of all the single-particle states $|l_N\rangle$ (see (9.13)), whose coefficients $U^{\xi_N}_{l_N}$ can be regarded as elements of the unitary transformation connecting the single-particle to the interacting basis: $U^{\xi_N}_{l_N} = \langle l_N | \xi_N \rangle$.

The numerical evaluation of the Coulomb-correlated states is thus performed via a direct diagonalization of the Hamiltonian matrix (9.15), employing a large – but finite – single-particle basis set defined in terms of a suitable energy cutoff (see Sect. 5.1).

9.2.3 Interaction with External Light Sources

The Coulomb-correlated carrier system described so far will interact strongly with electromagnetic fields in the optical range. The light–matter interaction Hamiltonian in our second-quantization picture can be written as

$$\hat{H}' = \sum_{\alpha^{\mathrm{e}} \alpha^{\mathrm{h}}} \left[H'_{\alpha^{\mathrm{e}} \alpha^{\mathrm{h}}} \hat{c}^{\dagger}_{\alpha^{\mathrm{e}}} \hat{d}^{\dagger}_{\alpha^{\mathrm{h}}} + H'_{\alpha^{\mathrm{h}} \alpha^{\mathrm{e}}} \hat{d}_{\alpha^{\mathrm{h}}} \hat{c}_{\alpha^{\mathrm{e}}} \right] , \tag{9.17}$$

where $H'_{\alpha^{\mathrm{e}} \alpha^{\mathrm{h}}} \equiv H'_{e\overline{\alpha}^{\mathrm{e}}, h\overline{\alpha}^{\mathrm{h}}}$ are the single-particle matrix elements in (1.55) due to a classical-light field.[1]

In the presence of a time-dependent coherent optical excitation, the quantum-mechanical evolution of our electron–hole system will be described by the following time-dependent Schrödinger equation:

[1] It is imperative to stress that the light–matter Hamiltonian (9.17) neglects any quantum-mechanical feature of the electromagnetic field. Indeed, in order to treat our light source in quantum terms, one should employ an electron–photon coupling model according to the general carrier–quasiparticle interaction Hamiltonian (3.137). More specifically, the classical-light Hamiltonian (9.17) may be regarded as the average of the carrier–photon Hamiltonian (3.137) over the photonic quantum state; in particular, for the case of a so-called coherent state the photonic average of the destruction operator is proportional to the amplitude of the corresponding classical field (see, e.g., [110] and references therein).

$$i\hbar \frac{\mathrm{d}|\Psi(t)\rangle}{\mathrm{d}t} = \left(\hat{H}^{\mathrm{ic}} + \hat{H}'\right)|\Psi(t)\rangle . \tag{9.18}$$

Contrary to the interacting-carrier Hamiltonian $\hat{H}^{\mathrm{ic}}$ in (9.8), the carrier–light term $\hat{H}'$ does not commute with the global number operators in (9.9). Indeed, the two terms in (9.17) describe, respectively, the light-induced creation and destruction of an electron–hole pair. Therefore, N is no more a good quantum number and the many-body state at time t is, in general, a linear superposition of all the correlated N-pair basis states:

$$|\Psi(t)\rangle = \sum_N \sum_{\xi_N} a_{\xi_N}(t)|\xi_N\rangle . \tag{9.19}$$

By inserting the above expansion into the time-dependent Schrödinger (9.18) we get

$$i\hbar \frac{\mathrm{d}a_{\xi_N}(t)}{\mathrm{d}t} = \mathcal{E}_{\xi_N} a_{\xi_N}(t) + \sum_{N'} \sum_{\xi'_{N'}} H'_{\xi_N \xi'_{N'}} a_{\xi'_{N'}}(t) , \tag{9.20}$$

where

$$H'_{\xi_N \xi'_{N'}} = \langle \xi_N | \hat{H}' | \xi'_{N'} \rangle \tag{9.21}$$

are the matrix elements of the light–matter Hamiltonian (9.17) within our interacting N-pair basis $\{\xi_N\}$.

It can be easily shown that the above matrix elements are different from zero for $N' = N \pm 1$ only; this confirms that the only possible transitions are $N \to N+1$ or $N+1 \to N$ which correspond, respectively, to the creation (generation) and destruction (recombination) of Coulomb-correlated electron–hole pairs (excitons) discussed above. Moreover, we deal with well-precise spin selection rules: the only matrix elements in (9.21) different from zero are those conserving the total spin of the carrier–light system. Indeed, the possible final states $|\xi_N\rangle$ depend on the spin configuration of the initial many-body state $|\xi'_{N'}\rangle$ as well as on the polarization of the electromagnetic field. In particular, we are allowed to create two excitons with opposite spin orientation (i.e., antiparallel-spin configuration) in the same orbital quantum state. In contrast, due to the Pauli exclusion principle, two excitons with the same spin orientation (i.e., parallel-spin configuration) can never occupy the same orbital state.

By treating the Schrödinger equation (9.20) within the standard time-dependent perturbation-theory approach and assuming a monochromatic light source of frequency ω, we can define the absorption probability per time unit corresponding to the $\xi_{N-1} \to \xi_N$ transition:

$$P_{\xi_{N-1} \to \xi_N}(\omega) = \frac{2\pi}{\hbar} \left| H'_{\xi_N \xi_{N-1}} \right|^2 \delta(\mathcal{E}_{\xi_N} - \mathcal{E}_{\xi_{N-1}} - \hbar\omega) . \tag{9.22}$$

The latter describes the many-exciton optical response of our low-dimensional structure, i.e., the probability of creating a new exciton in the presence of $N-1$ Coulomb-correlated electron–hole pairs.

Excitonic Absorption

As a starting point, let us consider the so-called excitonic response, i.e., the optical response of our carrier system for the $0 \to 1$ transition. In this case, the initial ($N = 0$) state is the electron–hole vacuum state $|0\rangle$, while the final ($N = 1$) state $|\xi_1\rangle$ corresponds to a single Coulomb-correlated electron–hole pair, i.e., an exciton. Combining the results in (9.12) and (9.13), for $N = 1$, we have

$$|\xi_1\rangle = \sum_{l_1} U_{l_1}^{\xi_1} \hat{c}^{\dagger}_{\alpha_1^{\mathrm{e}}} \hat{d}^{\dagger}_{\alpha_1^{\mathrm{h}}} |0\rangle \ , \tag{9.23}$$

where $l_1 = \alpha_1^{\mathrm{e}}, \alpha_1^{\mathrm{h}}$ denotes the single-particle electron–hole basis for $N = 1$.

The excitonic-absorption probability is then given by (9.22) with $N = 1$:

$$P_{\xi_1}^{\mathrm{ex}}(\omega) = \frac{2\pi}{\hbar} \left| H'_{\xi_1 0} \right|^2 \delta(\mathcal{E}_{\xi_1} - \hbar\omega) \ , \tag{9.24}$$

where

$$H'_{\xi_1 0} = \langle \xi_1 | \hat{H}' | 0 \rangle \tag{9.25}$$

is the matrix element of the light–matter Hamiltonian (9.17) for the $0 \to 1$ optical transition. Its explicit form can be found in [379].

The excitonic spectrum is finally obtained by summing the absorption probability (9.24) over all possible final states $|\xi_1\rangle$:

$$\mathrm{A}^{\mathrm{ex}}(\omega) = \sum_{\xi_1} P_{\xi_1}^{\mathrm{ex}}(\omega) \ . \tag{9.26}$$

Biexcitonic Absorption

Let us now come to the so-called biexcitonic response, i.e., the optical response corresponding to the $1 \to 2$ transition. In this case, the initial ($N = 1$) state coincides with the excitonic state $|\xi_1\rangle$ in (9.23), while the final ($N = 2$) state $|\xi_2\rangle$ corresponds to two Coulomb-correlated electron–hole pairs, i.e., a so-called biexciton (see, e.g., [47]). Combining again the results in (9.12) and (9.13), for $N = 2$, we get

$$|\xi_2\rangle = \sum_{l_2} U_{l_2}^{\xi_2} \hat{c}^{\dagger}_{\alpha_1^{\mathrm{e}}} \hat{d}^{\dagger}_{\alpha_1^{\mathrm{h}}} \hat{c}^{\dagger}_{\alpha_2^{\mathrm{e}}} \hat{d}^{\dagger}_{\alpha_2^{\mathrm{h}}} |0\rangle \ , \tag{9.27}$$

where $l_2 \equiv \alpha_1^{\mathrm{e}}\alpha_1^{\mathrm{h}}, \alpha_2^{\mathrm{e}}\alpha_2^{\mathrm{h}}$ denotes the single-particle electron–hole basis for $N = 2$.

The biexcitonic-absorption probability is then given by (9.22) with $N = 2$:

$$P_{\xi_1 \to \xi_2}^{\mathrm{biex}}(\omega) = \frac{2\pi}{\hbar} \left| H'_{\xi_2 \xi_1} \right|^2 \delta(\mathcal{E}_{\xi_2} - \mathcal{E}_{\xi_1} - \hbar\omega) \ , \tag{9.28}$$

where

$$H'_{\xi_2\xi_1} = \langle \xi_2 | \hat{H}' | \xi_1 \rangle \tag{9.29}$$

is the matrix element of the light–matter Hamiltonian (9.17) for the $1 \to 2$ optical transition. Its explicit form can be found in [379].

The biexcitonic spectrum is finally obtained by summing the absorption probability (9.28) over all possible final states $|\xi_2\rangle$:

$$\mathrm{A}^{\mathrm{biex}}_{\xi_1}(\omega) = \sum_{\xi_2} P^{\mathrm{biex}}_{\xi_1\to\xi_2}(\omega) \ . \tag{9.30}$$

We stress that – opposite to the excitonic spectrum in (9.26) – the biexcitonic spectrum A^{biex} is a function of the initial excitonic state ξ_1.

For a better interpretation of the biexcitonic optical response of our few-carrier system, it is convenient to introduce the so-called biexcitonic shift (see, e.g., [47]). To this end, let us consider two generic excitonic states $|\xi_1^a\rangle$ and $|\xi_1^b\rangle$, together with the corresponding[2] biexcitonic state $|\xi_2^{ab}\rangle$; the biexcitonic shift associated with these two excitonic states is defined as

$$\Delta\mathcal{E} = \mathcal{E}_{\xi_2^{ab}} - \left(\mathcal{E}_{\xi_1^a} + \mathcal{E}_{\xi_1^b} \right) \ . \tag{9.31}$$

The latter can be regarded as the extra-energy (with respect to its single-exciton transition) required for the generation of a second electron–hole pair in the presence of a previously created exciton.

The absorption probabilities (9.24) and (9.28) will be employed in the reminder of this chapter as well as in Chap. 10 to investigate the electro-optical response of single as well as coupled quantum-dot structures. However, for the case of ultrafast optical excitations and strong light–matter coupling, the above perturbation-theory picture can no longer be applied, and the investigation of the time evolution of our many-body state $|\Psi(t)\rangle$ requires a direct solution of the time-dependent Schrödinger equation (9.18);[3] this applies, in particular, to the quantum-information processing schemes based on excitonic degrees of freedom presented in Sect. 10.2.

9.2.4 The Excitonic Picture

As discussed in Sect. 9.2.2, the generic Coulomb-correlated N-pair state $|\xi_N\rangle$ can be written as a linear combination (see (9.13)) of the noninteracting

[2] A natural correspondence between excitonic and biexcitonic states is unambiguously established in the limit of vanishing Coulomb interaction, where any biexcitonic state reduces to the product of two Coulomb-free excitonic states, i.e., two electron–hole pairs.

[3] We stress that opposite to the many-exciton absorption probability (9.22), the number of excitons, i.e.,

$$N(t) = \langle \Psi(t) | \hat{N}_{\mathrm{e}} | \Psi(t) \rangle = \langle \Psi(t) | \hat{N}_{\mathrm{h}} | \Psi(t) \rangle \ , \tag{9.32}$$

is a continuous function of time and changes according to the specific ultrafast laser-pulse sequence considered.

electron–hole basis states (9.12). Such single-particle picture is used to compute Coulomb-correlated N-pair states and energy levels via the exact-diagonalization approach previously mentioned (see, e.g., [379]). However, it is often convenient to adopt – instead of a single-particle description – an excitonic-like picture, i.e., a quasiparticle number representation based on Coulomb-coupled electron–hole pairs. The aim of this section is (i) to show that, in general, such an excitonic description is not possible and (ii) to identify the basic requirements needed for such a quasiparticle number representation in connection with quantum-information processing based on excitonic degrees of freedom (see Chap. 10).

To this aim, let us introduce the following set of excitonic creation operators:

$$|\xi_1\rangle = \hat{X}^\dagger_{\xi_1}|0\rangle \ , \tag{9.33}$$

where, as usual, $|0\rangle$ denotes the electron–hole vacuum state and ξ_1 is the label for the generic excitonic ($N = 1$) state. By comparing the result in (9.23) with the above definition, we can write these excitonic operators in terms of the electron and hole operators, i.e.,

$$\hat{X}^\dagger_{\xi_1} = \sum_{\alpha^e\alpha^h} U^{\xi_1}_{\alpha^e\alpha^h} \hat{c}^\dagger_{\alpha^e} \hat{d}^\dagger_{\alpha^h} \ . \tag{9.34}$$

Moreover, in view of the unitary character of the transformation U, one gets

$$\hat{c}^\dagger_{\alpha^e} \hat{d}^\dagger_{\alpha^h} = \sum_{\xi_1} U^{\xi_1 *}_{\alpha^e\alpha^h} \hat{X}^\dagger_{\xi_1} \ . \tag{9.35}$$

If we now consider the explicit form of the noninteracting basis states in (9.12), the generic N-pair many-body state (9.13) can be formally written as

$$|\xi_N\rangle = \sum_{\{\xi_1\}} C^{\xi_N}_{\{\xi_1\}} |\{\xi_1\}\rangle \tag{9.36}$$

with

$$|\{\xi_1\}\rangle = \prod_{\xi_1} \hat{X}^\dagger_{\xi_1}|0\rangle \ . \tag{9.37}$$

The expansion (9.36) would suggest to define a sort of excitonic number representation in terms of the N-pair states $|\{\xi_1\}\rangle$. We stress that, in general, this is not possible. The point is that, generally speaking, the set of states in (9.37) does not constitute a basis for our N-pair Hilbert subspace. This is intimately related to the fact that – contrary to electron and hole creation and destruction operators – the excitonic operators introduced in (9.34) do not obey canonical commutation relations, i.e., the commutator

$$\mathcal{C}_{\xi_1,\xi_1'} = \left[\hat{X}_{\xi_1}, \hat{X}^\dagger_{\xi_1'}\right] \tag{9.38}$$

is itself an operator. This clearly prevents the introduction of number operators and, therefore, of a genuine quasiparticle number representation.

As discussed in more detail in Chap. 10, two basic requirements are needed to perform quantum-information processing: (i) the tensor-product structure of the "computational space" considered and (ii) the SU(2)[4] character of the raising/lowering operators acting on our elementary computational units, known as qubits. The main question is thus to identify if – and under which conditions – the Coulomb-correlated electron–hole system discussed so far may act as "quantum hardware," i.e., may be employed to perform quantum-information processing. This requires to identify a set of independent degrees of freedom (qubits) with a SU(2) character, the one of spin-$\frac{1}{2}$ systems.

As a starting point, one should then check if there exists a set of independent excitonic degrees of freedom; this corresponds to verifying that for any pair of excitonic states ξ_1 and ξ_1' the commutator (9.38) is equal to zero.

Let us now discuss the tensor-product structure of our computational subspace. To this end, let us consider again the case of two qubits a and b. Generally speaking, we know that the Hilbert space of a bipartite system is $\mathcal{H}_a \otimes \mathcal{H}_b$, where $\mathcal{H}_{a/b}$ are the Hilbert spaces of the individual qubits. This means that if $\{|l_a\rangle\}$ is an orthonormal basis set for $\mathcal{H}_a$ and $\{|l_b\rangle\}$ is an orthonormal basis set for $\mathcal{H}_b$, then $\{|l_a\rangle \otimes |l_b\rangle\}$ is a basis set for the whole computational space. What one needs to test is the possibility of writing the many-body ground state – corresponding in this case to a biexcitonic state ξ_2 – as the product of two independent excitonic states ξ_1^a and ξ_1^b. This corresponds to verifying that

$$\langle \xi_2 | \left(|\xi_1^a\rangle |\xi_1^b\rangle \right) = 1 \, . \tag{9.39}$$

Provided that the above requirements are fulfilled, let us now focus on the single qubit, i.e., on one of the independent excitonic states ξ_1. In this case, one needs to check that the exciton creation/annihilation operators introduced in (9.33) obey the usual SU(2) commutation relations. More specifically, we are interested in defining the z-component pseudo-spin operator $S^z_{\xi_1}$ as

$$S^z_{\xi_1} = \frac{1}{2} \mathcal{C}_{\xi_1,\xi_1} \, . \tag{9.40}$$

In order to check that this is really a z-component spin operator, one should verify that its average value over our many-body state is either plus or minus one. Deviations from this ideal scenario can be regarded as a measure of the "leakage" from our computational space due to the presence of external, i.e., non-computational, excitonic states.

In Chap. 10 we shall show that for prototypical GaAs- and GaN-based quantum-dot molecules all the above requirements are well fulfilled and our excitonic subsystem can indeed be used as semiconductor-based quantum hardware.

[4] The special unitary group of degree n, denoted SU(n), is defined as the group of $n \times n$ unitary matrices with determinant 1.

9.3 Field-Induced Exciton–Exciton Dipole Coupling

We shall start our discussion by considering how an applied electric field acts on an exciton (i.e., a Coulomb-correlated electron–hole pair (see Sect. 1.2.2)) within a single quantum dot.

If the field is absent, charge neutrality is observed. Indeed, assuming that electron and hole experience nearly the same single-particle quantum confinement, it is possible to show that there is almost no net Coulomb interaction among excitons belonging to the same degenerate "shell" (see, e.g., [47] and references therein).

In contrast, an external electric field pulls apart charges opposite in sign, i.e., electrons and holes, as can be seen from panel A of Fig. 9.1: each exciton is turned into an electric dipole and the transition energies are red shifted according to the well-known Stark-shift effect. The net Coulomb interaction among excitons is now different from zero even if we consider two excitons in the same shell and with the same single-particle carrier confinement for electrons and holes.

Figure 9.1 shows the value of the biexcitonic shift $\Delta\mathcal{E}$ as a function of the applied electric field. As anticipated, the latter is defined as the extra-energy (with respect to its single-exciton transition) required for the generation of a second electron–hole pair in the presence of a previously created exciton (see (9.31)). The results reported in Fig. 9.1 have been obtained by a direct diagonalization of the three-dimensional Hamiltonian (9.1) describing the Coulomb-correlated few-carrier system quantum-confined within a realistic GaAs-based dot structure (see, e.g., [378, 379]). It is worth pointing out that in the single-dot case (panel B) the biexcitonic shift can be not only tuned at will but its sign can even be reversed. In fact, for small values of the field E the shift is mainly due to the exchange term (see, e.g., [47]) and is negative; when the field is increased, the latter tends to align the two dipoles so that $\Delta\mathcal{E}$ first vanishes and then becomes positive: the applied electric field – removing part of the system symmetries – allows one to turn on at will and to tune exciton–exciton interactions.

Let us now consider the interaction between excitons localized in two neighboring quantum dots stacked along the growth direction. In this case, even for different electron and hole charge distributions, the interaction between excitons sitting in different and far enough dots in the absence of applied fields is negligible (see Fig. 9.1c). In contrast, when a constant in-plane electric field is applied, we create again electron–hole dipoles inside each quantum dot, which are now parallel and pointing toward the same direction. Therefore, such polarized excitons will now interact via a positive $\Delta\mathcal{E}$, roughly proportional to the square of the field-induced excitonic dipole and inversely proportional to the third power of the interdot distance. This implies that, in order to exploit this interaction, the coupled dots must be close enough to allow for a non-vanishing dipole–dipole interaction energy, but far enough to prevent single-particle tunneling phenomena. Once more, the presence of

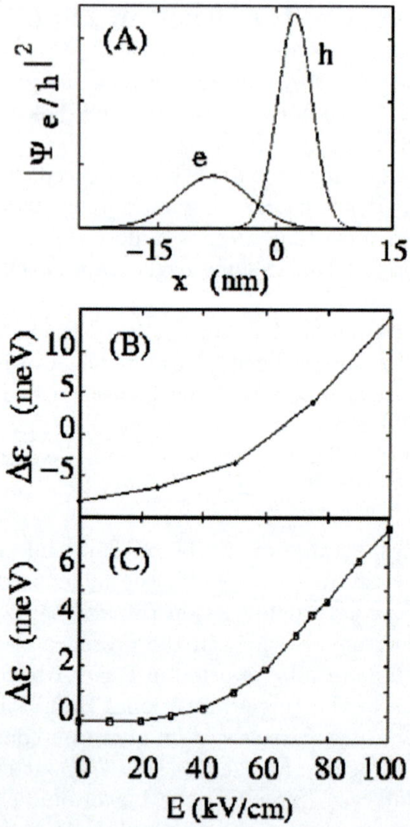

Fig. 9.1. (**a**) Electron and hole charge densities corresponding to the excitonic ground state along the direction of the applied electric field. (**b**) Biexcitonic shift $\Delta\mathcal{E}$ as a function of the applied field for the two-exciton ground state in a single dot. (**c**) Biexcitonic shift $\Delta\mathcal{E}$ as a function of the applied field for a two-exciton ground state in a double-dot structure. These results have been obtained via the exact-diagonalization approach proposed in [378, 379] employing realistic GaAs-based quantum-dot parameters (see text). Reprinted from [398]

an in-plane applied field is found to turn on exciton–exciton coupling, thus allowing for the formation of tunable interactions between different dots, i.e., for the creation of so-called artificial macromolecules.

As originally proposed in [398], the field-induced effects previously discussed can be described in terms of a simple analytical model; the latter constitutes an efficient tool for the quantitative evaluation of few-carrier effects and, more important, it allows one to identify key quantum-dot parameters and corresponding operation regimes. Within such simplified model, the quantum-dot single-particle carrier confinement along the growth direction (z) can be modeled as a narrow parabolic potential (or as a square well) $V(z)$, while the confinement in the dot plane ($\boldsymbol{r}$) is described via a two-dimensional parabolic potential. By denoting with $\boldsymbol{E}$ the in-plane electric field, our simplified single-exciton Hamiltonian is then given by

$$\begin{aligned}\hat{H}^{\text{ex}} &= \sum_{\nu=\text{e,h}} \left[-\frac{\hbar^2}{2m_\nu^*} \frac{\partial^2}{\partial^2 z_\nu} + V_\nu(z_\nu) \right] \\ &+ \sum_{\nu=\text{e,h}} \left[-\frac{\hbar^2 \nabla_{\boldsymbol{r}_\nu}}{2m_\nu^*} + \frac{1}{2} m_\nu^* \omega_\nu^2 \left| \boldsymbol{r}_\nu \pm \boldsymbol{d}_\nu \right|^2 \right] \\ &- \frac{e^2}{\varepsilon \sqrt{|\boldsymbol{r}_\text{e} - \boldsymbol{r}_\text{h}|^2 + (z_\text{e} - z_\text{h})^2}} , \end{aligned} \tag{9.41}$$

where the $\pm$ sign and the subscripts e and h refer, respectively, to electron and hole. Here, ε is the static dielectric constant and $\boldsymbol{d}_\nu = \frac{e\boldsymbol{E}}{m_\nu^* \omega_\nu^2}$ denotes the single-particle charge displacement induced by the in-plane applied field. As shown in [398], under suitable conditions the two-body Schrödinger equation corresponding to the excitonic Hamiltonian (9.41) can be analytically solved and all the relevant quantities can be easily estimated with a good degree of accuracy. More specifically, in view of the strong single-particle confinement along the z-direction, one is allowed to approximate/replace $(z_\text{e} - z_\text{h})^2$ with its average value l^2. We may choose l to be, e.g., twice the average length related to the ground state of an infinite-height square well of width a. It is then possible to split the excitonic Hamiltonian (9.41) into an in-plane-direction contribution and a growth-direction contribution. To this aim, it is convenient to introduce the following center-of-mass and relative in-plane coordinates:

$$\boldsymbol{R} = \frac{m_\text{e}^*(\boldsymbol{r}_\text{e} + \boldsymbol{d}_\text{e}) + m_\text{h}^*(\boldsymbol{r}_\text{h} - \boldsymbol{d}_\text{h})}{M} , \qquad \boldsymbol{r} = \boldsymbol{r}_\text{h} - \boldsymbol{r}_\text{e} , \tag{9.42}$$

where $M = m_\text{e}^* + m_\text{h}^*$. In terms of these new coordinates we get

$$\hat{H}^{\text{ex}} = \hat{H}^{\text{in-plane}}(\boldsymbol{R}, \boldsymbol{r}) + \sum_{\nu=\text{e,h}} \hat{H}_\nu^{\text{growth}}(z_\nu) \tag{9.43}$$

with

$$\begin{aligned}\hat{H}^{\text{in-plane}}(\boldsymbol{R}, \boldsymbol{r}) &= -\frac{\hbar^2 \nabla_{\boldsymbol{R}}^2}{2M} + \frac{1}{2} M \omega_\text{R}^2 R^2 - \frac{\hbar^2 \nabla_{\boldsymbol{r}}^2}{2\mu} + \frac{1}{2} \mu \omega_r^2 |\boldsymbol{r} - \boldsymbol{d}|^2 \\ &- \mu(\omega_\text{e}^2 - \omega_\text{h}^2) \boldsymbol{R} \cdot (\boldsymbol{r} - \boldsymbol{d}) - \frac{e^2}{\varepsilon \sqrt{r^2 + l^2}} \end{aligned} \tag{9.44}$$

and

$$\hat{H}_\nu^{\text{growth}}(z_\nu) = -\frac{\hbar^2}{2m_\nu^*} \frac{\partial^2}{\partial^2 z_\nu} + V_\nu(z_\nu) , \tag{9.45}$$

where $\mu = \frac{m_\text{e}^* m_\text{h}^*}{M}$ is the reduced mass, $\omega_\text{R}^2 = \frac{1}{2}(\omega_\text{e}^2 + \omega_\text{h}^2)(1 + \Delta)$, $\omega_r^2 = \frac{1}{2}(\omega_\text{e}^2 + \omega_\text{h}^2)(1 - \Delta)$, $\Delta = \frac{m_\text{e}^* - m_\text{h}^*}{M} \frac{\omega_\text{e}^2 - \omega_\text{h}^2}{\omega_\text{e}^2 + \omega_\text{h}^2}$, and $\boldsymbol{d} = \boldsymbol{d}_\text{e} + \boldsymbol{d}_\text{h}$ denotes the total electron–hole in-plane displacement. We stress that the growth-direction contribution $\hat{H}_\nu^{\text{growth}}$ corresponds to the single-particle Hamiltonian along the growth direction, exactly solvable for the case of a parabolic potential as well as of an infinite-height square well.

In the limit $\left|\frac{\omega_{\text{e}}^2-\omega_{\text{h}}^2}{\omega_{\text{e}}^2+\omega_{\text{h}}^2}\right| \ll 1$ the two in-plane coordinates $\boldsymbol{R}$ and $\boldsymbol{r}$ are only weakly coupled, and the Schrödinger equation associated with the center-of-mass coordinate $\boldsymbol{R}$ is exactly solvable. In contrast, for the general case the investigation is focused on the ground state. More specifically, adopting the standard variational-principle approach (see, e.g., [400] and references therein), it is convenient to approximate the ground-state wavefunction of $\hat{H}^{\text{in-plane}}$ as

$$\Psi(\boldsymbol{R},\boldsymbol{r}) = \Psi_x(x)\chi(\boldsymbol{R},y) \; , \tag{9.46}$$

with

$$\chi(\boldsymbol{R},y) = \frac{1}{(\pi\lambda_r^2)^{\frac{1}{4}}} \mathrm{e}^{-\frac{y^2}{2\lambda_r^2}} \frac{1}{(\pi\lambda_R^2)^{\frac{1}{2}}} \mathrm{e}^{-\frac{R^2}{2\lambda_R^2}} \; , \tag{9.47}$$

where x and y denote, respectively, the components of $\boldsymbol{r}$ parallel and perpendicular to the field $\boldsymbol{E}$, $\lambda_r = \sqrt{\frac{\hbar}{\mu\omega_r}}$, and $\lambda_R = \sqrt{\frac{\hbar}{M\omega_{\text{R}}}}$.

By averaging the in-plane Hamiltonian $\hat{H}^{\text{in-plane}}$ over $\boldsymbol{R}$ and y according to the probability density $|\chi(\boldsymbol{R},y)|^2$, we get an effective one-dimensional Hamiltonian of the form

$$\hat{H}^{\text{eff}} = \frac{1}{2}\hbar\omega_r + \hbar\omega_{\text{R}} - \frac{\hbar^2}{2\mu}\frac{\partial^2}{\partial^2 x} + V^{\text{eff}}(x) \tag{9.48}$$

characterized by the effective potential

$$V^{\text{eff}}(x) = \frac{1}{2}\mu\omega_r^2(x-d)^2 + V_C\left(\frac{x^2+l^2}{2\lambda_r^2}\right) \tag{9.49}$$

with

$$V_C(u) = -\frac{e^2}{\varepsilon\sqrt{\pi}\lambda_r}\mathrm{e}^u K_0(u) \; , \tag{9.50}$$

K_0 being the zero-order Bessel function.

If, as usual, one is interested in the low-energy states, it is possible to approximate V^{eff} around its minimum $V_\circ = V^{\text{eff}}(x_\circ)$ via a parabolic potential, i.e.,

$$V^{\text{eff}}(x) \approx V_\circ + \frac{1}{2}\mu\tilde{\omega}^2(x-x_\circ)^2 \tag{9.51}$$

where with

$$\mu\tilde{\omega}^2 = \left.\frac{\partial^2 V^{\text{eff}}}{\partial^2 x}\right|_{\text{x}_\circ} . \tag{9.52}$$

Within the approximation scheme reviewed so far, the eigenvalues and eigenfunctions of $\hat{H}^{\text{eff}}$ can be evaluated analytically. In particular, for the case of strong enough fields, the main effect of the attractive Coulombic interaction is to reduce the electron–hole displacement from d to $x_\circ$; for intermediate/strong fields, it is then convenient to write the effective displacement as $x_\circ = d - \Delta x$, with $\Delta x \ll d$. As discussed in [398], in this regime an analytical expression for Δx can be readily obtained and compared to the Coulomb-free displacement d.

The ratio $\frac{\Delta x}{d}$ as a function of the applied field is shown in the inset of Fig. 9.2. It is important to point out that the analytical expression of Δx previously mentioned [398] contains as a multiplying factor the ratio between λ_r and the bulk excitonic Bohr radius $a^* = \frac{\hbar^2 \varepsilon}{\mu e^2}$, which can be regarded as a measure of the single-particle carrier confinement.

In a similar way, writing $\tilde{\omega}$ as $\omega_r + \Delta\omega$, it is possible to evaluate the effect of the electron–hole Coulombic attraction on the curvature of the effective potential V^{eff}; as for Δx, also for the curvature correction $\Delta\omega$ an analytical expression can be readily obtained [398]. In particular, in the limit $\frac{\lambda_r^2}{d^2} \ll 1$ (strong-field limit), we have

$$\frac{\Delta\omega}{\omega_r} = -\frac{\Delta x}{d} \propto \frac{\lambda_r}{a^*} \frac{\lambda_r^3}{d^3} , \tag{9.53}$$

showing that in this limit Coulomb-correlation corrections decrease very fast for increasing fields. It is easy to show that – for the typical regimes of

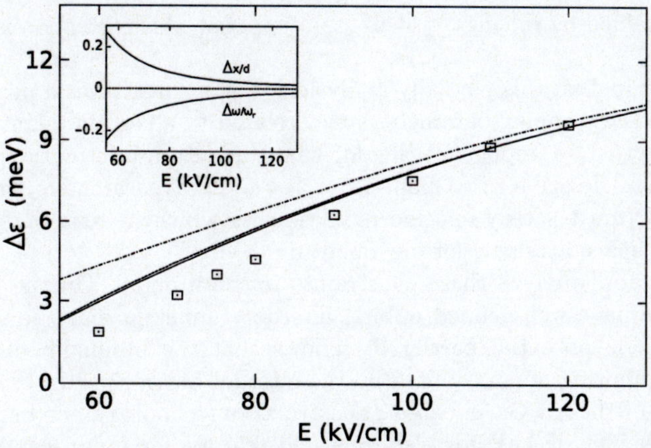

Fig. 9.2. Biexcitonic shift $\Delta\mathcal{E}$ versus applied field E for a quantum-dot system characterized by the following parameters: $m_e^* = 0.067m$, $m_h^* = 0.34m$ (m being the free-electron mass), $\hbar\omega_e = 30\,\text{meV}$, and $\hbar\omega_h = 24\,\text{meV}$. Here, the dipole–dipole coupling energy, i.e., the biexcitonic shift, is reported as a function of the in-plane field E. The *squares* indicate the results of the fully three-dimensional calculation, the *solid curve* represents the result of the simplified model, the *dotted curve* is the result obtained in the Coulomb-free case, and the *dashed curve* corresponds to setting $\Delta\omega = 0$ in the model. The *inset* displays the behavior of the two key quantities $\Delta\omega/\omega_r$ and $\Delta x/d$ as a function of the external field (see text). Reprinted from [398]

interest – the correction on the excitonic wavefunction due to $\frac{\Delta\omega}{\omega_r}$ is negligible with respect to the correction given by the shift $\frac{\Delta x}{d}$.

As anticipated, the most important quantity for the design and realization of the quantum-dot-based logic gates presented in Chap. 10 is the biexcitonic shift (9.31). For the case under examination this is the energy shift due to the Coulomb interaction between two excitons sitting in neighboring dots; we can approximate the corresponding biexcitonic ground state as the product of two excitonic wavefunctions sitting in different dots and built according to the simplified model just described. In particular, the wavefunction in the z-direction may be approximated by a Gaussian of width $l/2$, and the two dots are assumed to have the same width a; this is reasonable since, by construction, the two dots have almost the same width, while single-particle tunneling between the two dots is highly suppressed. The desired biexcitonic shift $\Delta\mathcal{E}$ is then obtained by averaging the corresponding two-exciton Hamiltonian over such factorized ground state.

By adopting this approximation scheme, the biexcitonic shift $\Delta\mathcal{E}$ becomes an easy-to-calculate sum of at most two-dimensional integrals. Within the corresponding validity region the analytical model reviewed so far provides a good estimate for $\Delta\mathcal{E}$; indeed, Fig. 9.2 shows a comparison between the exact fully three-dimensional calculations (squares), the results of the simplified model (solid curve), and the latter obtained employing Coulomb-free wavefunctions (dotted curve). The dashed curve shows the analytical-model results obtained by setting $\frac{\Delta\omega}{\omega_r} = 0$: as anticipated, this correction is generally negligible.

As anticipated, this partially analytical model constitutes a powerful tool to analyze the complex parameter space related to a certain quantum-device design. In order to implement the dot-based quantum-information/computation schemes described in Chap. 10, a few system parameters, such as $\omega_{\rm e}$, $\omega_{\rm h}$, and E, must satisfy specific requirements which in turn determine the parameter space available for our quantum device.

Let us now analyze these constraints in more detail. On the one hand, in order to have well-defined qubits, interdot tunneling must be suppressed via a suitable potential barrier; it follows that the minimum distance between neighboring dots is limited by the interdot barrier width. In agreement with state-of-the-art GaAs-based nanostructure technology, one may consider a barrier height of 1 eV for electrons and 0.58 eV for holes and a distance $D = 100$ Å between the centers of two neighboring dots. On the other hand, in order to implement the dot-based quantum-computing schemes presented in Chap. 10, Coulomb interaction between neighboring dots must be strong enough to produce a biexcitonic shift of the order of a few millielectronvolt (at least 3–4 meV). This can be realized either by tailoring the interdot distance or by varying the strength of the applied electric field E, since, as a rough approximation,

$$\Delta\mathcal{E} \propto \frac{d^2}{D^3} \,. \tag{9.54}$$

Unfortunately, a side effect induced by the applied field is a significant reduction of the interband oscillator strength and, accordingly, of the carrier-system response to ultrafast driving laser pulses (see Sect. 10.2): the applied field, in fact, induces a spatial separation of electron and hole single-particle wavefunctions, thus decreasing their overlap (see Fig. 9.1). Concerning the carrier confinement induced by the in-plane parabolic potential, in order to have well-defined quantum dots, the system must be in a strong-confinement regime, i.e., the characteristic length λ_r associated with the parabolic potential in (9.44) must be smaller than the bulk excitonic Bohr radius a^*. On the other side, according to the general trend in (9.54), a too strong confinement would in turn heavily decrease the biexcitonic shift $\Delta\mathcal{E}$. Last but not least, as discussed extensively in Chap. 10, in order to perform quantum-information processing one must be able to address specific excitations of the quantum-dot system unambiguously. This means that the peaks of interest in the optical spectrum (e.g., lowest single-exciton state in dot a, lowest single-exciton state in dot b, and lowest biexcitonic state) must be well defined and well isolated from peaks due to other high-energy excitations. This imposes additional constraints on the value of $\hbar\omega_\mathrm{e}$ and $\hbar\omega_\mathrm{h}$, since in the regime of interest single-particle excitations are well defined. In particular, for $\hbar\omega_\mathrm{e} > \hbar\omega_\mathrm{h}$ (typical situation for semiconductor materials with $m^*_\mathrm{h} > m^*_\mathrm{e}$), the closest additional peak in the spectrum corresponds to one electron in its ground state plus one hole in its first excited state and both in dot a. Defining Δ_E as the energy difference between the two excitonic ground states in dot a and dot b, one must then try to maximize $\hbar\omega_\mathrm{h} - \Delta_E$.

The present qualitative analysis shows that, in order to simultaneously satisfy all the requirements discussed so far, the system parameters must be fine tuned; it follows that a reliable and efficient way to scan the whole quantum-dot parameter space is imperative.

Figure 9.3 shows the available parameter space for a reasonable field of $E = 75\,\mathrm{kV/cm}$; the latter has been evaluated via the simplified analytical model presented so far. The typical error in the calculated values of $\Delta\mathcal{E}$ is of the order of 10%. The constraints corresponding to the results shown in Fig. 9.3 are $\Delta\mathcal{E} \geq 3.5\,\mathrm{meV}$, relative oscillator strength greater than 0.15, $\hbar\omega_\mathrm{e} > \hbar\omega_\mathrm{h}$, $\hbar\omega_\mathrm{h} - \Delta_E \geq 10\,\mathrm{meV}$, $\frac{\lambda_\mathrm{r}}{a^*} \leq 0.6$. As anticipated, the scenario presented in Fig. 9.3 has played a crucial role in the design and optimization of the GaAs-based quantum hardware discussed in Sect. 10.2.1.

The analysis of field-induced exciton–exciton coupling presented so far applies to GaAs-based quantum-dot systems, where the creation of excitonic dipoles may be realized via the application of an external field oriented along the quantum-dot plane. In the reminder of this section we shall discuss excitonic as well as biexcitonic effects in a different class of materials, namely GaN-based quantum-dot structures. Indeed, opposite to GaAs-based systems, the latter are characterized by intense built-in electric fields, giving rise to spontaneous excitonic dipoles.

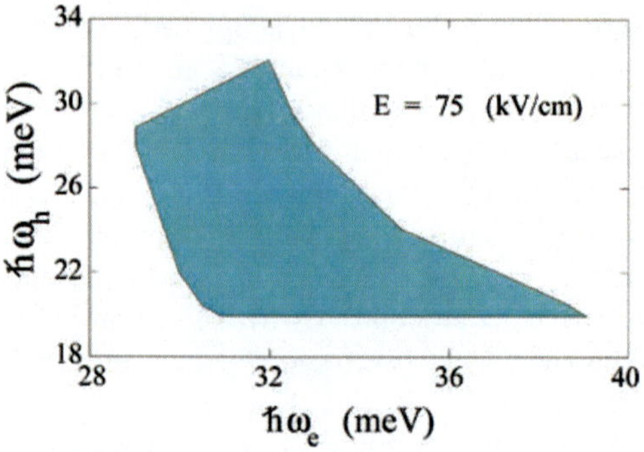

Fig. 9.3. Plot of the parameter space available for constructing a prototypical two-qubit gate, evaluated employing the simplified analytical model reviewed in this section. Here, the physical constraints imposed are $\Delta\mathcal{E} \geq 3.5\,\mathrm{meV}$, relative oscillator strength greater than 0.15, $\hbar\omega_{\mathrm{e}} > \hbar\omega_{\mathrm{h}}$, $\hbar\omega_{\mathrm{h}} - \Delta_E \geq 10\,\mathrm{meV}$ (Δ_E being the energy difference between the lowest excitonic transitions of the two neighboring dots), $\frac{\lambda_r}{a^*} \leq 0.6$ (see text). Reprinted from [398]

More specifically, we shall start our analysis by reviewing fully three-dimensional investigations of the many-exciton optical response of vertically stacked quantum dots via the direct-diagonalization approach previously mentioned. In particular, we shall focus on the interdot exciton–exciton coupling – key ingredient for the realization of the all-optical GaN-based quantum processor proposed in Sect. 10.2.2 – showing that there is a large window of realistic parameters for which both biexcitonic shift and oscillator strength are compatible with the GaN-based implementation scheme originally proposed in [380]. Also for this second class of nanomaterials, it is convenient to adopt a semi-analytical model to estimate few-carrier interactions between stacked quantum dots; indeed, the latter provides a high degree of accuracy in evaluating both exciton–exciton and single particle–exciton coupling strengths and has allowed for an easy modeling of the prototypical GaN-based quantum devices originally proposed in [396] and reviewed in Sect. 9.5.

As already underlined, the main peculiarity of wurtzite GaN heterostructures is a strong built-in electric field; the latter is the sum of the spontaneous polarization and of the piezoelectric field. Such built-in electric field in GaN quantum dots and AlN barriers may be calculated according to [394]

$$E_{\mathrm{d}} = \frac{L_{\mathrm{b}}(P_{\mathrm{b}}^{\mathrm{tot}} - P_{\mathrm{d}}^{\mathrm{tot}})}{L_{\mathrm{d}}\varepsilon_{\mathrm{b}} + L_{\mathrm{b}}\varepsilon_{\mathrm{d}}}\,, \tag{9.55}$$

where $\varepsilon_{\mathrm{d/b}}$ and $P_{\mathrm{d/b}}^{\mathrm{tot}}$ are, respectively, the relative dielectric constant and the total polarization of the dot/barrier, while $L_{\mathrm{b/d}}$ is the width of the barrier/height of the dot. The value of the electric field in the barrier (E_{b}) is obtained by simply interchanging the barrier (b) and dot (d) indices. The result in (9.55) is derived assuming an alternating sequence of quantum wells and barriers, but the latter is also a good approximation for the case of an array of similar quantum dots along the growth (z) direction. As for the case of the GaAs-based structures previously considered, the spatial confinement induced by the lateral shape of the quantum dot is simply approximated by a two-dimensional parabolic potential, which mimics the strong in-plane carrier confinement caused by the built-in electric field and preserves the spherical symmetry of the single-particle ground state [381–393]. As anticipated, the intrinsic field is the sum of the spontaneous-polarization charge – accumulating at GaN/AlN interfaces – and of the piezoelectric one.

According to the analysis originally presented in [396], we shall focus on the interplay between single-particle carrier confinement and two-body Coulomb correlation; in particular, we shall discuss exciton–exciton dipole coupling versus oscillator strength. The relevance of this analysis is twofold: (i) we shall address a distinguished few-particle phenomenon typical of nitride quantum dots, i.e., the presence of an intrinsic exciton–exciton dipole coupling induced by built-in polarization fields and; (ii) we shall provide detailed information on the set of parameters needed for the experimental realization of the GaN-based quantum-information processing strategy [380] presented in Sect. 10.2.2.

In the analysis reviewed below the dot height is varied from 2 to 4 nm and the dot-base diameter from 10 to 17 nm, assuming a linear dependence between these two parameters, in agreement with experimental and theoretical findings. The difference between the well width of two neighboring quantum dots is assumed to be 8% in order to allow for energy-selective generation of ground-state excitons in neighboring dots via ultrafast laser pulses (see Chap. 10). The interdot barrier width is such to prevent single-particle tunneling and to realize at the same time a significant dipole–dipole Coulomb coupling: the giant internal field, in fact, induces a separation of electron and hole charge distributions, thus creating intrinsic electric dipoles. If we consider two stacked dots occupied by one exciton each, the resulting charge distribution can be regarded as a system of two dipoles aligned along the growth direction and thus favorably coupled (See also Fig. 10.3).

As for the GaAs-based quantum-dot structures previously considered, let us start our analysis by investigating the optical response of our GaN system via the fully three-dimensional exact-diagonalization approach previously mentioned (see, e.g., [379]). More specifically, we consider electrons (e) and holes (h) quantum-confined within stacked GaN-based dots as depicted in Fig. 10.3; the confinement potential is assumed to be parabolic within the quantum-dot plane and is described by a square-well profile modified by the built-in electric field along the growth (z) direction.

We shall focus once again on the excitonic and biexcitonic optical spectra in the presence of the built-in electric field. For all the structures considered in [396] the two lowest optical transitions correspond to the formation of direct ground-state excitons in dots a and b, respectively. Here, we shall consider parallel-spin configurations only.

Let us focus once again on the biexcitonic shift corresponding to the energy difference between the ground-state biexcitonic transition (given a ground-state exciton in dot a) and the ground-state excitonic transition of dot b. This quantity – which constitutes a direct measure of the ground-state exciton–exciton coupling – plays a crucial role in the design and optimization of the all-optical quantum hardware presented in Sect. 10.2.2.

Figure 9.4 shows how the biexcitonic shift increases with the height of the dot; here, the barrier width is kept constant and is equal to 2.5 nm. In (A) both the height L_d and the diameter D of the dots are varied according to the relation $D = 3.5L_\mathrm{d} + 3\,\mathrm{nm}$, while in (B) only the height of the dot is changed. We notice that, for realistic parameters, it is possible to achieve biexcitonic shifts up to 9.1 meV. Opposite to what happens for the stacked GaAs-based dots previously considered, in GaN-based structures characterized by this parameter range the lowest states preserve their atomic-like shape even for barriers as low as 2.5 nm. In these structures, in fact, both electron and hole effective masses and band discontinuities are much higher than in GaAs-like materials, thus decreasing the macroatomic-like wavefunction overlap responsible for the so-called macromolecular bonding. Moreover, the excitonic dipole

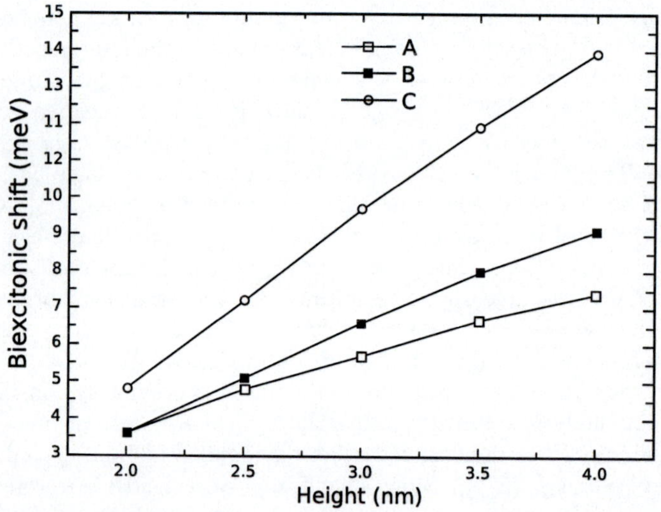

Fig. 9.4. Biexcitonic shift (*upper panel*) of the ground-state transition in dot b versus quantum-dot height for two coupled GaN dots separated by a barrier of 2.5 nm. In (*B*) only the height of the dots is changed ($D = 10\,\mathrm{nm}$), while in (*A*) D is varied proportionally to the height from 10 to 17 nm. The result in (*C*) shows the biexcitonic shift in the point-like charge approximation. Here, the parameters used are as follows: effective masses $m_\mathrm{e}^* = 0.2m$ and $m_\mathrm{h}^* = m$; in-plane parabolic-confinement energies $\hbar\omega_\mathrm{e} = 74\,\mathrm{meV}$ and $\hbar\omega_\mathrm{h} = 33\,\mathrm{meV}$ for curve (*B*); $\hbar\omega_\mathrm{e} = 74$–$290\,\mathrm{meV}$, $\hbar\omega_\mathrm{h} = 33$–$130\,\mathrm{meV}$ for curve (*A*) (see text). Reprinted from [396]

length is roughly proportional to the height of the dot because of the strong built-in electric field; therefore, it is crucial to evaluate the dependence of the exciton–exciton interaction on the height of the quantum dot. In this respect, the spreading of the carrier wavefunction affects the biexcitonic shift, as one can notice by comparing curves A and B in Fig. 9.4. More specifically, the biexcitonic shift is larger (up to 20% for the parameters considered here) when the wavefunction is more localized, since the system is closer to the point-like charge limit (see curve C in the same figure). These results demonstrate that there exists a wide range of parameters for which the biexcitonic shift is at least a few millielectronvolt. This is a central prerequisite for realizing energy-selective addressing with sub-picosecond laser pulses, as required, e.g., by the all-optical quantum-information processing devices described in Sect. 10.2.

In view of the analysis presented so far, the best strategy to achieve large values of the biexcitonic shift is to grow "high-" and "small-diameter" dots. The drawback is that the oscillator strength of the interband ground-state transition – proportional to the overlap of electron and hole wavefunctions – strongly decreases with the dot height. A small value of the oscillator strength, in fact, implies all the well-known difficulties of single-dot signal detection (see, e.g., [51] and references therein).

In the range of height values considered in Fig. 9.4, the oscillator strength varies over three orders of magnitude, so care must be taken in a future quantum-information processing experiment in order to optimize at the same time biexcitonic shift and oscillator strength. As originally suggested in [396], a reasonable figure of merit is the product between the biexcitonic shift and

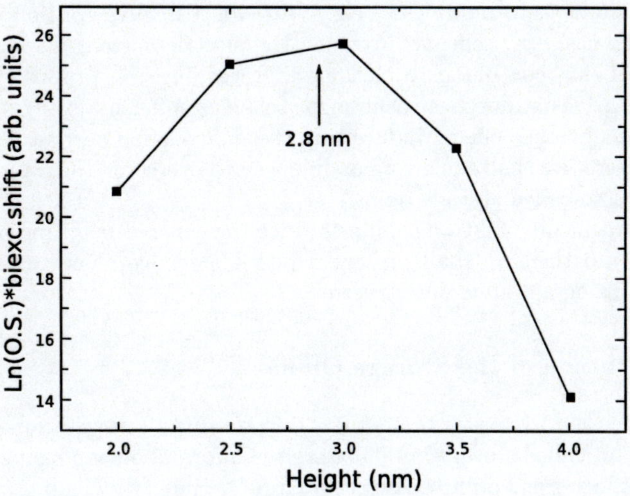

Fig. 9.5. Figure of merit (biexcitonic shift times logarithm of oscillator strength) versus quantum-dot height. Here, the *arrow* indicates the maximum obtained via a parabolic fit (see text). Reprinted from [396]

the logarithm of the oscillator strength. Such quantity is reported in Fig. 9.5, displaying its maximum for a dot height of 2.5–3 nm.

As a final remark, we stress that in order to perform a fast and efficient scanning of the quantum-dot parameter space – and thus to identify optimal operational conditions for prototypical quantum devices – also for the case of the present GaN-based structures a simplified analytical model has been developed; the latter, not reviewed here, can be found in [396].

9.4 Semiconductor Double Quantum Dots as "Storage Qubits"

The last part of this chapter is devoted to the description of measuring devices, useful both for quantum-computing purposes (see Chap. 10) and for detecting charge variations in nanoscopic structures.

As discussed extensively in Sect. 10.1, one of the main problems with solid-state quantum-computing schemes based on charge degrees of freedom is to find a way to overcome "fast" decoherence times. A major step in facing this problem has been the idea of using ultrafast gating operations driven by laser pulses which address energy-selected interband optical transitions. In the original proposal [378, 379] reviewed in Sect. 10.2.1 the qubit is implemented using excitonic degrees of freedom; in what follows, we review a possible scheme – originally proposed in [399] – to measure the state of such quantum-dot-based qubit.

Generally speaking, many of the proposals for measuring the quantum state of a solid-state qubit (see, e.g., [401–404]) involve continuous-measurement schemes, e.g., schemes in which the electric current through a point contact is being continuously measured. Regarding the proposal described in Sect. 10.2.1, continuous-measurement schemes suffer major drawbacks because of decoherence effects induced by the unavoidable current fluctuations involved. Here, we shall review a possible way to overcome these problems via the use of a so-called storage qubit.

More specifically, first we shall introduce the general definition of the storage qubit and then we shall present a possible semiconductor-based implementation using a double-dot structure.

9.4.1 Definition of the Storage Qubit

The key idea is to transfer the information from the computational qubit to another qubit called storage qubit, where the information can reside for a long time, i.e., the storage qubit possesses a large T_1 time (see Chap. 2) compared to that of the computational qubit. Indeed, the latter is characterized by a relatively short T_1 time (due in this case to the finite excitonic life-time), which limits the available time for the measurement process, even when no gating operations are being performed. In addition, the use of a storage qubit can increase the spatial distance between the qubit and the measurement device, thus decreasing the decoherence rate when no measurement is taking place. Due to its relatively large T_1 time, the information inside the storage qubit can be extracted, e.g., via continuous-measurement schemes, without affecting the state of the computational qubits.

The storage qubit must measure the generic computational qubit in a time much shorter than the decoherence time and store the information. The generic way to describe this measurement is through the so-called controlled not (CNOT) gate (see Chap. 10), also referred to as the measurement gate.

Let us now describe a possible realization of a storage qubit which could be employed for measuring the state of the computational qubit corresponding to the quantum-dot implementation reviewed in Sect. 10.2.1. In this case the qubit is implemented through the excitonic degrees of freedom of a quantum dot: the two basis states $|0\rangle$ and $|1\rangle$ of the computational qubit correspond, respectively, to the absence and presence of a ground-state exciton in the quantum dot.

The storage qubit, designed to measure the excitonic state of the computational qubit, consists of two coupled semiconductor quantum dots. Through the application of an external gate voltage, a surplus hole occupies the double-dot system. The storage-qubit states are thus defined as an excess hole in the right dot, $|R\rangle$, and an excess hole in the left dot, $|L\rangle$. The original symmetry between the two states is lifted through the application of an electric field along the growth direction (see, e.g., [405]). Due to the applied field, the energy levels are lowered in the left dot with respect to the right one. For measuring the computational qubit one may exploit coherent population transfer in coupled semiconductor quantum dots, as originally proposed in [405]. Such coherent population transfer (in this case the transfer of an excess hole from the left to the right dot) is achieved through a stimulated Raman adiabatic passage (STIRAP) (see, e.g., [406] and references therein). The idea is to use the Coulomb interaction – between the exciton in the computational qubit and the surplus hole in the double dot – to detune the coherent population transfer in the double dot (see Fig. 9.6).

In order to consider a double-dot system as a concrete implementation of a storage qubit, one should check first that the measured information about the state of the quantum dot stored inside the double dot is actually long lived, and, second, that the measurement of the computational qubit via the storage qubit is fast and reliable.

It can be shown that the states $|R\rangle$ and $|L\rangle$ are indeed long lasting, i.e., the tunneling between them is highly improbable on a time-scale set by the decoherence time T_2 or by the exciton recombination time T_1 of the quantum dot. This is due to the relatively large distance (of the order of 100 Å) and high-energy barrier (of the order of 200 meV) between the two dots. Additionally, the application of an external electric field along the growth direction

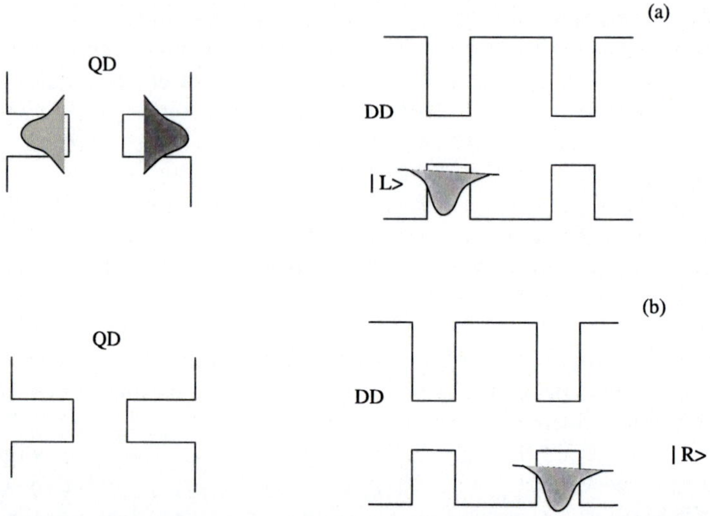

Fig. 9.6. Schematic illustration of a storage qubit based on a double-dot structure. The double-dot states are labeled $|L\rangle$ and $|R\rangle$ and correspond, respectively, to a hole in the left dot or in the right dot. The degeneracy between these two states is lifted by an external electric field. (**a**) State of the double dot corresponding to an exciton in the quantum dot: the STIRAP is detuned and the hole remains in state $|L\rangle$. (**b**) State of the double dot corresponding to no exciton in the quantum dot: STIRAP is not detuned and the hole is transferred to state $|R\rangle$ (see text). Reprinted from [399]

introduces an energy-level mismatch of tenths of millielectronvolt between the two dots, which strongly inhibits tunneling from $|L\rangle$ to $|R\rangle$.

For the measurement to be fast, the typical time for extracting information on the excitonic state of the computational qubit should be much shorter than T_1 as well as T_2; for the measurement to be reliable, the energy shift of the double-dot states due to the presence of an exciton in the computational qubit should be larger than the energy uncertainty of the laser pulses (see Chap. 2) and larger than the typical width of the energy levels due to interaction with the environment.

The following estimates are based on the same physical parameters employed in [405], except for the distance between the two dots which has been extended to 100 Å. This particular choice of distance will be discussed extensively in the description of the measurement scheme.

9.4.2 State Measurement via a STIRAP Process

As discussed extensively in [406], the STIRAP process involves three quantum states, two of which are the long-lived lower energy states $|L\rangle$ and $|R\rangle$, while the third one is a high-energy state, in our case a charged exciton labeled $|X^+\rangle$. It is crucial to stress that between $|L\rangle$ and $|R\rangle$ there are no dipole-allowed transitions, while both levels are dipole-coupled to $|X^+\rangle$. Employing, e.g., two delayed laser pulses, coherent population transfer can be achieved between $|L\rangle$ and $|R\rangle$ without ever occupying $|X^+\rangle$. The first pulse ("pump") is tuned to the L–X^+ resonance and the second pulse ("Stokes") is tuned to the R–X^+ resonance.

For the above STIRAP process to be effective, the coupling of the excited state $|X^+\rangle$ to the two long-lived states should be of the same order. Moreover, the two long-lived states should be non-degenerate. According to the scheme proposed in [405], such conditions may be realized applying an electric field in the growth direction (to lift the original degeneracy between states $|L\rangle$ and $|R\rangle$) and choosing an excited state $|X^+\rangle$ with an electronic wavefunction split between the two quantum dots. In this way the coupling between $|X^+\rangle$ and the two localized states $|L\rangle$ and $|R\rangle$ can be of the same order. In the storage-qubit scheme proposed in [399] the spatial separation between the two dots has been increased to 100 Å; this localizes the ground state and the first excited states of the electron in one of the dots. To have an electron wavefunction[5] equally spread over the two quantum dots – as needed for an effective STIRAP process – two options may be considered. The first one is to use a charged exciton in which the holes are in the ground states of both dots, and the electron occupies a level with energy comparable to the confining potential. In this case $|X^+\rangle$ is composed of two holes localized in the two quantum dots and an electron wavefunction which is split between the quantum-dot wells, as sketched in Fig. 9.7.

A second possibility is to keep the holes in their ground states while the electron is excited to a continuum level above the quantum-dot confining potential. In this case the charged exciton state is a hybrid state formed by a confined exciton state for the hole and a bulk exciton state for the electron. The typical length-scale for the hole wavefunction is given in this case by the confining-potential width, $l = 50$ Å, and for the electron by the bulk excitonic Bohr radius, $a^* \sim 100$ Å.

It is imperative to stress that both choices for $|X^+\rangle$ are very susceptible to decoherence, especially the hybrid state where the electron, in a continuum level, is bound to the hole by Coulomb interaction only: outside the dot the electron is not shielded by the quantum-dot confining potential from interacting with phonons or other decohering mechanisms.

The measurement of the computational-qubit state (i.e., presence or absence of the ground-state exciton in the dot) exploits the Coulomb interaction between the exciton in the computational qubit and the charged states in the double dot. Indeed, such an interaction shifts the energy levels in the double-dot system and may detune coherent population transfer: the presence of an

[5] Here, we are in the strong-confinement regime, i.e., the confining-potential width is much smaller than the bulk excitonic Bohr radius and therefore it is meaningful to describe the state in terms of a product of electron and hole wavefunctions rather than using an excitonic wavefunction.

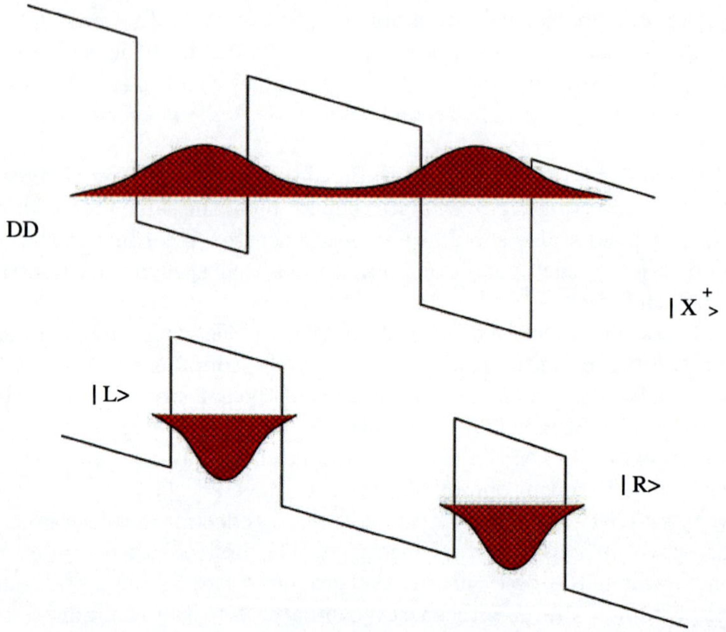

Fig. 9.7. Schematic illustration of the charged exciton state $|X^+\rangle$ in the double-dot structure. Here, the two holes are in their quantum-dot ground states, while the electron is in an excited state characterized by a wavefunction split between the two quantum dots (see text). Reprinted from [399]

exciton in the computational qubit manifests itself by detuning the STIRAP process, thus preventing the coherent transfer of the excess hole from the left to the right side of the double dot (see Fig. 9.6).

Concerning decoherence, a basic requirement for the storage qubit is not to induce decoherence on the computational qubit when no measurement is taking place. This requirement is fulfilled, since the presence of the hole in the double-dot apparatus does not disturb the quantum-dot states but rather causes a constant (time independent on the computation time-scale) shift of the energy levels. Thus, the measuring device will not affect the quantum computer when the measurement is not taking place.

Regarding the typical time on which the measurement takes place, the detection of the state of the computational qubit by the double-dot system occurs on a time-scale which is dictated by the duration of the laser pulses ("pump" and "Stokes"), which induce the coherent population transfer. The duration of the laser pulses is typically of the order of 10 ps (see, e.g., [405]), and thus the typical time for extracting information on the state of the computational qubit is fast compared to the electron–hole recombination times as well as to the excitonic decoherence (the latter being typically of the order of 100 ps).

The measurement process of the computational qubit by the double-dot system is performed by detuning the STIRAP process: the idea is that the STIRAP will take place only when there is no exciton in the quantum dot. For the STIRAP to occur (i) the adiabatic condition $\Omega_R \tau \gg 1$ must be fulfilled (τ denoting the typical overlap time between the pulses and Ω_R the typical Rabi frequency associated with the STIRAP process) and (ii) energy conservation during the transfer must be preserved (see, e.g., [407, 408]), i.e., the initial and final states must be into resonance.

We shall now discuss how the presence of an exciton in the computational qubit affects the constraints just mentioned, destroying the probability for a STIRAP process to occur by detuning it. Let us consider a detuning Δ_p of the pump laser from resonance with the L–X^+ transition and a detuning Δ_s of the Stokes laser from the R–X^+ transition. The effective Hamiltonian for such a three-level system within the usual rotating-wave approximation is of the form (see, e.g., [406])

$$\begin{aligned} \hat{H}^{\mathrm{STIRAP}} &= \frac{\hbar}{2}\left[\left(\Omega_p |X^+\rangle\langle L| + \Omega_s |X^+\rangle\langle R| + \mathrm{H.c.} \right) + 2\Delta_p |X^+\rangle\langle X^+| \right. \\ &\quad \left. + 2(\Delta_p - \Delta_s)|R\rangle\langle R| \right] , \end{aligned} \tag{9.56}$$

where Ω_p and Ω_s are the coupling Rabi frequencies corresponding to the pump and Stokes, respectively.

Failure of the Adiabatic Condition

If the two-photon resonance condition applies, i.e., $\Delta_p = \Delta_s$, the instantaneous eigenstates and eigenfunction of the Hamiltonian (9.56) are given by

$$\begin{aligned} |a_0\rangle &= \cos\theta |L\rangle - \sin\theta |R\rangle, & \omega_0 &= 0 \\ |a_\pm\rangle &\propto \sin\theta |L\rangle \pm \cot^{\pm 1}\phi |X^+\rangle + \cos\theta |R\rangle, & \omega_\pm &= \Delta_p \pm \sqrt{\Delta_p^2 + \Omega_p^2 + \Omega_s^2}, \end{aligned} \tag{9.57}$$

where θ is the mixing angle defined by $\tan\theta = \frac{\Omega_p}{\Omega_s}$, ϕ depends on detuning as well as Rabi frequencies and is of no importance in the following discussion, and $|a_0\rangle$ is usually referred to as the "dark state" since it includes no contributions from the "leaky state" $|X^+\rangle$.

The adiabatic-transfer condition (constraint (i)) is fulfilled when $|\omega^\pm - \omega_0|\tau \gg 1$. For the parameters employed in [405] ($\hbar\Omega_{s/p} = 1\,\mathrm{meV}$ and $\tau = 10\,\mathrm{ps}$), when the laser detuning Δ_p becomes of the order of the effective Rabi frequency, $\Omega_{\mathrm{eff}} = \sqrt{\Omega_p^2 + \Omega_s^2}$, this condition is no longer satisfied. In this case the STIRAP process is detuned when the levels in the double-dot structure are perturbed such that the energy difference used in the transition L–X^+ is shifted by more than 1 meV.

When the adiabatic condition is no longer fulfilled, there is a non-vanishing occupation probability of the leaky state; once the latter is occupied, there exists a significant probability to observe transitions to a state different from

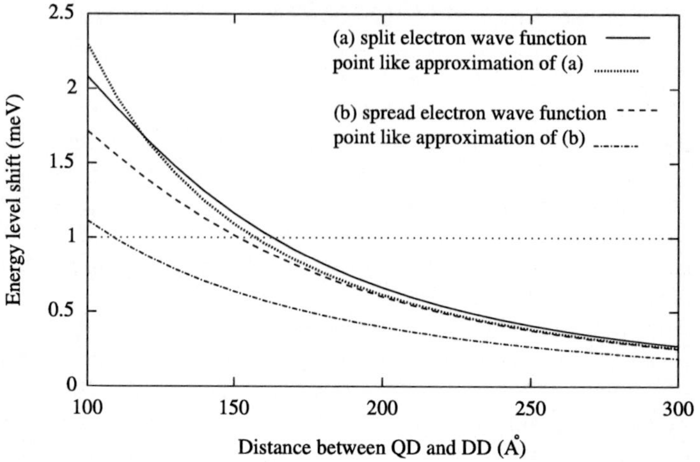

Fig. 9.8. Energy-level shift in the double-dot structure as a function of the distance from the quantum dot. Here, the distance is measured from the center of the electron wavefunction in the quantum dot to the center of the left (*closest*) unit of the double-dot structure. For the case of an excited electron state spread over the dots we took the typical length-scale for the wavefunction (along the *growth direction*) to be 100 Å. The results correspond to Gaussian and point-like wavefunctions for two cases: electron wavefunction split between the two quantum dots and electron wavefunction spread over the double-dot structure (see text). Reprinted from [399]

the three states involved in the STIRAP process, and in this case the transfer of the hole from the left to the right dot does not take place.

As described in [399], by modeling the involved (excitonic and hole) wavefunctions as products of Gaussian functions – whose parameters have been carefully selected based on the structure and material features – it is possible to estimate both the shift of the energy levels in the double dot (see Fig. 9.8) and the difference in the energy shifts of the $|R\rangle$ and $|L\rangle$ states (see Fig. 9.9) due to the presence of an exciton in the computational qubit. This analysis allows us to conclude that, either when the electron wavefunction of state $|X^+\rangle$ is split between the two dots or when its energy is above the confining potential, it is possible to obtain an energy-level shift in the double dot bigger than 1 meV for distances between double dot and computational qubit up to 150 Å. It is therefore not crucial to have a specific configuration for the excited electron wavefunction (see Fig. 9.8).

Failure of the Energy-Conservation Requirement

When $\Delta_\mathrm{p} \neq \Delta_\mathrm{s}$, the STIRAP process is destroyed much sooner. For $\Delta_\mathrm{ps} \equiv \Delta_\mathrm{p} - \Delta_\mathrm{s} \neq 0$ the zero eigenvalue moves toward a value $\tilde{\omega}_0$ of the order of $\frac{2\Omega_\mathrm{p}^2 \Delta_\mathrm{ps}}{\Omega_\mathrm{eff}^2}$, and the (ex) dark state, $|a_0\rangle$, now includes contributions from the

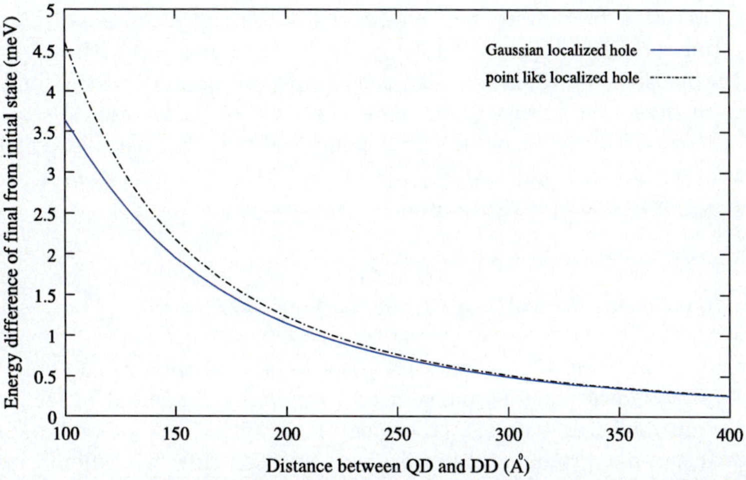

Fig. 9.9. Difference in the shifts of the initial ($|L\rangle$) and final state ($|R\rangle$) of the STIRAP process in the presence of an exciton in the quantum dot as a function of the distance from the quantum dot (the latter is measured in the same way as described for Fig. 9.8). Here, results are presented for Gaussian and point-like wavefunctions (see text). Reprinted from [399]

leaky state, $|X^+\rangle$, which are of the same order of $\tilde{\omega}_0$. Since in this case the energy-conservation requirement is not fulfilled, i.e., initial and final levels are not into resonance, in order to establish if the STIRAP process takes place, one needs to compare the energy uncertainty of the pulse with the difference in the energy shift of the initial and final states. Therefore, the condition for the STIRAP to take place is given by $\Delta_{\mathrm{ps}}\tau \lesssim 1$. Since τ is of the order of 10 ps, an energy shift larger than 0.5 meV is sufficient.

Figure 9.9 shows the energy difference between the initial state $|L\rangle$ and the final state $|R\rangle$ of the double-dot system in the presence of an exciton in the computational qubit. This energy difference is much more sensitive to the presence of an exciton in the quantum dot and is shown to be greater than 0.5 meV up to distances of 300 Å between the quantum dot and the double-dot structure.

The reason why it is much easier to detune the transition with respect to the initial and final states (i.e., hole in $|L\rangle$ and hole in $|R\rangle$) rather than the initial (or final) state with respect to the intermediate state is due to the different charge configurations: the intermediate leaky state ($|X^+\rangle$) couples weakly to an exciton in the quantum dot, since the detuning is basically due to a dipole–dipole interaction, whereas for the initial and final states the detuning is due to a monopole–dipole interaction.

As anticipated, the concept of storage qubit may greatly help practical implementations of some of the quantum-computing devices presented in Chap. 10. Indeed, apart from measuring the qubit state, a storage qubit stores the information for a time longer than the decoherence times which characterize the computing qubits; additionally, by incrementing with its physical size the spatial separation between the computing qubits and the final measuring devices, the storage qubit may allow to use conventional strategies, e.g., a single-electron transistor or a point contact (see, e.g., [401–404]), to extract the required information, without affecting the quantum-mechanical coherence of the computing qubits.

9.5 Potential All-Optical Read-Out Devices

In this section we shall review a proposal for an all-optical read-out device based on a semiconductor quantum dot, originally presented in [396]. Such a quantum device is based on the idea of exploiting the energetic shift on the quantum-dot ground-state excitonic transition due to Coulomb interaction with external charges. Indeed, similar to the biexcitonic shift discussed in Sect. 9.3, Coulomb interaction modifies the transition energy in the optical-absorption spectrum corresponding to the creation of an exciton in a certain dot in the presence, e.g., of an electron (or a hole) trapped in the vicinity. However, it is important to notice that the dipole–monopole interaction decreases as $\frac{d}{D^2}$, d denoting again the dipole length and D the distance between the dipole and the charge, i.e., much slower than the dipole–dipole interaction characterizing the biexcitonic shift (see (9.54)). This implies that it should be possible to detect such an energy shift (and thus the electron (hole) presence) even if the dot and the trapped charge are relatively far apart.

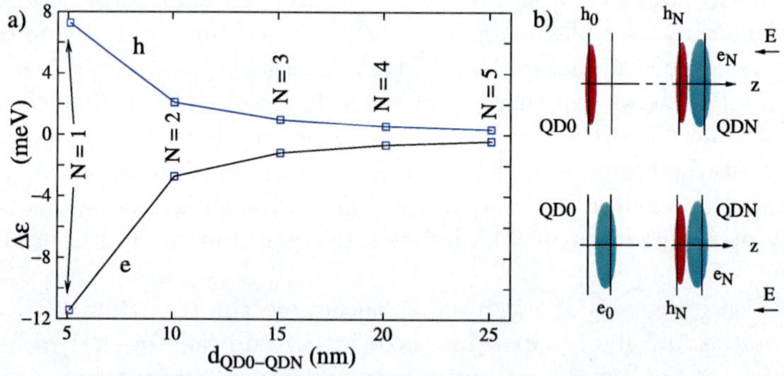

Fig. 9.10. (**a**) Charge–exciton interaction energy $\Delta\epsilon$ versus distance between "written dot" $N = 0$ and "reading dot" N. Here, e (h) corresponds to an electron (*hole*) in dot $N = 0$. (**b**) Schematic view of the stacked-quantum-dot array corresponding to the occupation of dot $N = 0$ by an electron (*bottom panel*) or by a hole (*upper panel*). The position along the growth direction occupied by each dot is marked by couples of *parallel lines*. The electronic (*light gray*, e_i) and hole (*dark gray*, h_i) clouds are sketched as well (see text). Reprinted from [396]

Let us now focus on a specific quantum-dot system (though the following discussion can be easily generalized to structures with different geometry or materials) considering the response of an array of slightly different stacked GaN dots (see Fig. 9.10b), whose height is about 2.5 nm and which are separated by barriers 2.5 nm wide.

As shown in Fig. 9.10, let us assume that the electron (hole) to be detected is trapped in dot $N = 0$; by employing the theoretical modeling proposed in [396], it is possible to evaluate the energy shift connected to the creation of an exciton in the Nth dot ($N = 1, 2, 3, \ldots$). The calculated energy shift $\Delta\epsilon$ is plotted in Fig. 9.10a as a function of the distance between dot 0 and dot N. Here, the curve labeled by e (h) corresponds to the presence of an electron (hole) in quantum dot 0. For $N = 1$, $\Delta\epsilon$ is of the order of 10 meV, but even considering N as high as 5, the energy shift is still of the order of 0.5 meV, i.e., it could be resolved via laser excitations as short as 2–3 ps. The asymmetry between the e and the h curves reflects the corresponding asymmetry between electron and hole wavefunctions.

Based on the results reported in Fig. 9.10, a non-invasive optical read-out device has been originally proposed in [396]; the latter is able to detect the state of a quantum-dot-based memory corresponding to the presence (state 1) or absence (state 0) of a charge in each quantum dot (see, e.g., [409]); indeed, state 1 (0) will correspond to the absence (presence) of the charge-free excitonic peak in the absorption spectrum.

Such a scheme may play a key role in the measuring process of the quantum-computing devices presented in Sect. 10.2 and could in general be an alternative strategy to more conventional reading schemes performed, e.g., via point-contact devices, since it avoids charge fluctuations due to the presence of electric currents in the system. Moreover, this new class of quantum devices may be employed also to monitor the net charge status of a certain mesoscopic system/memory S. For far enough distances between the "written" memory S and the "reading" device, the interaction becomes in fact proportional to the total net charge inside S; therefore, in principle, the corresponding excitonic shift is given by $\delta \approx n\Delta\epsilon$, where n is the number of elementary charges inside S. A measure of δ could be then used to "count" the electrons (holes) that have been injected/written in S.

A schematics of the proposed quantum device is reported in Fig. 9.11: for simplicity we suppose that one electron (upper panel) or a negative net charge (lower panel) may be stored in S. In the first case we send on quantum dot N a polarized laser pulse centered around its excitonic-transition energy $\epsilon_N + \Delta\epsilon$, which include the shift due to the interaction with the possible electronic charge in S (upper panel). The pulse must be long enough to energy-discriminate/select between ϵ_N and $\epsilon_N + \Delta\epsilon$. An exciton will then be generated in quantum dot N if and only if there is a single electron in S. For measuring an unknown net charge instead (lower panel), one may think of illuminating quantum dot N with a shorter pulse, corresponding to an energy spreading of about $n\Delta\epsilon$, n being of the order of the maximum net charge we can expect in

S. In this case an exciton will be always generated in quantum dot N, but the shift δ of such a transition with respect to the benchmark ε_N (corresponding to no charge in S) will be a measure of the net charge in S. We stress that this kind of measurement may be repeated at will by means of laser-pulse trains, as long as the charge to be measured remains in S. This possibility enhances considerably the efficiency of the read-out scheme presented so far.

The excitonic transitions corresponding to the different dots can be generated by energy-selective schemes (see, e.g., [378, 379]) (if quantum dots N and S are too close to be spatially resolved by the laser pulse) or via near-field techniques. A plus of this kind of quantum devices is that, by exploiting a long-range interaction, the presence of the exciton in the "reading" dot would not perturb significantly the carrier system in the "written" one.

Generally speaking, an important advantage of the present read-out scheme is that the latter does not perturb the system during the time in which the process leading to the charge storage (e.g., a computation) is carried out, since the exciton in quantum dot N is created during the measurement process only. Moreover, we underline that this scheme does not require S to be subjected to

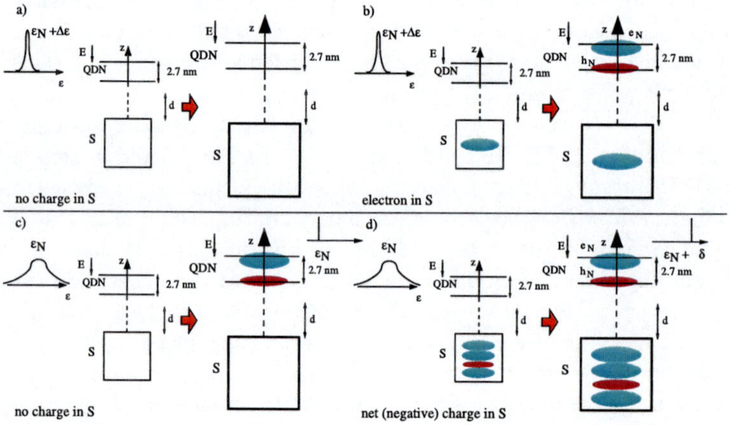

Fig. 9.11. *Upper panel*: detection of a known number of charges (a single electron in the figure); the laser pulse is a peak narrow in energy around the transition energy expected when including Coulomb interaction between quantum dots N and S. (**a**) No charge in S: the pulse does not generate any exciton in quantum dot N. (**b**) One electron in S: the pulse generates the corresponding exciton in quantum dot N (conditional dynamics). *Lower panel*: detection of an unknown amount of charge in S; the laser pulse is a peak relatively broad in energy centered around the excitonic transition energy expected when Coulomb interaction between quantum dots N and S is not included. (**c**) No charge in S: the pulse generates an exciton in quantum dot N at the absorption energy ϵ_N (*upper right* corner of the panel). (**d**) Net charge in S: the pulse generates an exciton in quantum dot N at the absorption energy $\epsilon_N + \delta$, where $\delta \approx n\Delta\epsilon$, n being the net number of elementary charges in S (*upper right corner of the panel*) (see text). Reprinted from [396]

an electric field (either intrinsic or applied). In principle, it is indeed sufficient to grow a single GaN quantum dot in the vicinity of the memory which needs to be measured, or – if we consider, for example, a stack of GaAs quantum dots – the required applied field can be applied to the "reading" GaAs quantum dot only. The "written" system, in turn, can be a generic nanoscopic charge-based memory (e.g., a gate-defined quantum dot), while the geometry of the device can be different from the one depicted in Fig. 9.11 as long as the distances between the "written" and "reading" units are of the same order.

We stress once again that the main technological advantage of using GaN quantum dots is the presence of strong built-in electric fields, which on the one hand simplify the setup and on the other hand will never let the trapped exciton ionize. If, in particular, the device geometry is the one depicted in Fig. 9.11, Coulomb interaction between quantum dot N and S comes out to be maximized, since (i) in GaN-based dots dipoles are spontaneously aligned along the growth direction and (ii) due to the strength of the built-in polarization field, the wavefunction spreading in the growth direction is significantly reduced.

As a final remark, we stress that all physical parameters employed in the design and optimization of the prototypical quantum devices reviewed in this chapter are compatible with present or near-future nanomaterial technology.

10

Semiconductor-Based Quantum Logic Gates

In this chapter we shall review a few potential implementation strategies for the concrete realization of quantum information processing using specifically designed semiconductor nanostructures, namely quantum dots and wires.

10.1 Fundamentals of Quantum Information Processing

It has been recently recognized that processing information in a quantum-mechanical fashion can provide, in some cases, a huge computational speedup with respect to "any" classical device [139–152]. Just to mention the two best-known quantum algorithms, a quantum computer would allow for ultra-fast factoring of large integer numbers or would efficiently perform database search (see, e.g., [140] and references therein). From the fundamental point of view, the novel computational power owned by such devices relies on two basic quantum features: (i) superposition of states, i.e., linear structure of the state-space, and (ii) quantum entanglement, i.e., tensor-product structure of a multipartite state-space. These two ingredients, along with the unitary character of quantum dynamics, lie at the heart of the additional capability provided by quantum information/computation processing. The first one endows a quantum computer with a sort of massive built-in parallelism, while the latter makes it possible to use uniquely quantum correlations, i.e., entanglement, for designing efficient algorithms.

Even though from a theoretical point of view quantum information processing (QIP) can be already considered as a well-established field, a major challenge arises when one starts to move toward the experimental side, addressing the problem of concrete implementations. Indeed, from the practical point of view this goal is exceptionally demanding: once a system with a well-defined state-space has been identified, one should be able to perform precise state preparation (input), any – arbitrarily long – coherent state manipulation (data processing), and state measurement (output) as well. In analogy to classical computation, such manipulations are thought of as a sequence (network) of elementary logic quantum gates, e.g., universal sets of single plus two quantum-bit (qubit) operations. Such quantum gates, in turn, have to be realized exploiting physically affordable coupling mechanisms, in terms of, e.g., single- and two-body interactions.

The primary obstacle toward a straightforward implementation of this idealized scheme is the destruction of the unitary nature of quantum evolution due to the unavoidable coupling of the computational degrees of freedom to their environment. Indeed, any quantum computer is an "open system"; it follows that, typically on a very short time-scale, a system prepared in a pure quantum state (coherent superposition) decays into a statistical (incoherent) mixture. This process is known as "decoherence": it represents a kind of noise peculiar to quantum mechanics; it spoils quantum computation making it basically a classical process.

Mostly due to the need for low decoherence rates, the first proposals for the experimental realization of quantum information processing devices originated from specialties in atomic physics [410–413], in quantum optics [414, 415], and in magnetic resonance spectroscopy [416]. On the other hand, practically relevant QIP requires a large number of quantum-hardware units (qubits), a condition hardly achievable in terms of such systems.

In contrast, in spite of the serious difficulties related to the "fast" decoherence times, a solid-state implementation of QIP (see, e.g., [51, 417]) seems to be the only way to benefit synergistically from the continuous progress in ultrafast optics [90–92] as well as in superconducting-device [418–420] and semiconductor-nanostructure fabrication and characterization [37–40]. Among the potential solid-state implementations not treated in this book, it is imperative to mention those in superconducting-device physics (see, e.g., [421, 422] and references therein) as well as in nuclear- (see, e.g., [423–425] and references therein) and electronic-spin physics (see, e.g., [426] and references therein).

10.2 All-Optical QIP with Semiconductor Macroatoms

In this section we shall address the problem of a semiconductor-based realization of QIP using charge and/or spin degrees of freedom. Along this line, we shall present an all-optical implementation [378–380] of QIP with semiconductor quantum dots, where the computational degrees of freedom are interband optical transitions (i.e., excitonic states) manipulated/controlled by ultrafast sequences of multi-color laser-pulse trains; this kind of implementation scheme has indeed stimulated a significant experimental activity aimed at realizing basic QIP with semiconductor quantum dots [427–432]. Moreover, we shall review a recently proposed scheme [433] based on a proper combination of charge and spin degrees of freedom.

The first semiconductor-based implementation of QIP has been proposed by Daniel Loss and David P. DiVincenzo in 1998 [434]; the latter relies on electronic-spin dynamics in semiconductor quantum dots, thus exploiting the low decoherence of spin degrees of freedom, in comparison to the one of charge excitations (see Sect. 10.2.3). On the other hand, as originally envisioned in [435, 436], gating of charge excitations could be performed by exploiting present ultrafast laser technology that allows to generate and manipulate electron–hole quantum states on a sub-picosecond time-scale (see, e.g., [90]). In this respect, decoherence times on nano/microsecond scales can be regarded as "long" ones. Following this spirit, the first all-optical implementation schemes with semiconductor macroatoms/molecules [51] have been put forward [378–380, 433, 437–442].

The scheme proposed by Troiani and co-workers [437] is based on ultrafast optical manipulations of ground as well as excited excitonic transitions in a single quantum-dot structure. Indeed, the implementation of quantum gates in a single dot is expected to provide a first proof-of-principle of QIP schemes based on the use of interband excitations as computational degrees of freedom; however, one should keep in mind that – as discussed extensively in Sect. 9.2.4 – single quantum dots are not suited for the QIP implementation beyond the 2–3 qubit limit. This is essentially due to the tensorial structure of the Hilbert space, which is required by the most efficient quantum algorithms and calls for the quantum hardware to consist in a collection of spatially separated subsystems.

In contrast, in the scheme proposed by Biolatti and co-workers [378, 379] the quantum information is encoded in the ground-state excitons of quantum-dot molecules. More specifically, the original proposal [378, 379], discussed in Sect. 10.2.1, deals with Coulomb-coupled GaAs-based quantum-dot structures; as we shall see, in this scheme an applied external field is needed. In order to overcome such requirement/limitation, De Rinaldis and co-workers have then proposed a modified version [380] of the original scheme; the latter, based on GaN quantum-dot structures, is described in Sect. 10.2.2.

10.2.1 GaAs-Based Quantum Hardware

As multiple-dot quantum hardware, we consider a vertically staked array of GaAs-based self-assembled quantum-dot structures, whose unit cell $a + b$ is schematically depicted in Fig. 10.1a. Here, the in-plane carrier confinement is described by a two-dimensional (xy) parabolic potential, while along the growth (z) direction we deal with a square-like potential profile. The latter is tailored in such a way to allow for an energy-selective creation/destruction of bound excitons in dots a and b. The interdot barrier width (of about 5 nm) is to prevent such single-particle tunneling and at the same time to allow for significant interdot Coulomb coupling. Moreover, as discussed extensively in Sect. 9.3, in order to induce a significant exciton–exciton dipole coupling, an in-plane static electric field F is applied. As a starting point, let us discuss the optical response of the semiconductor macromolecule $(a + b)$ in Fig. 10.1a to a single- as well as a double-pulse sequence. The excitonic spectrum in

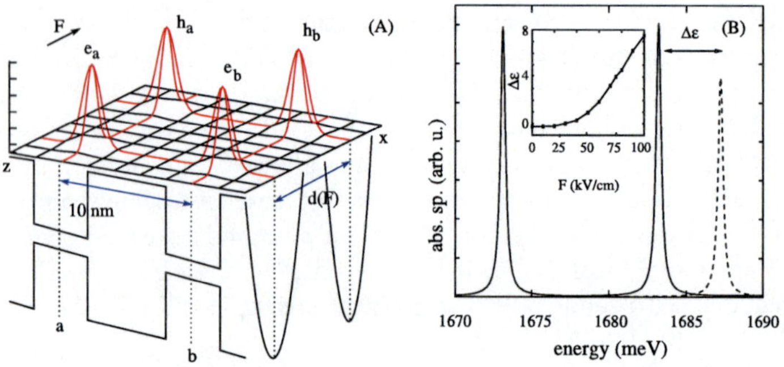

Fig. 10.1. (**a**) Schematic representation of the electron and hole charge distribution as well as of the confinement potential profile in our Coulomb-coupled GaAs-based quantum-dot structure. (**b**) Excitonic (*solid curve*) and biexcitonic response (*dashed curve*) of the GaAs-based coupled quantum-dot structure in (**a**) for an in-plane field $F = 75\,\mathrm{kV/cm}$. The biexcitonic shift $\Delta\mathcal{E}$ as a function of the in-plane field F is also reported in the inset (see text). Reprinted from [379]

the presence of an in-plane electric field $F = 75\,\mathrm{k\,V/cm}$ is shown in Fig. 10.1b (solid curve) and compared to the corresponding biexcitonic spectrum (dashed curve); the latter describes the generation of a second electron–hole pair in the presence of a previously created exciton. The crucial feature in Fig. 10.1b is again the magnitude of the biexcitonic shift $\Delta\mathcal{E}$ introduced in Chap. 9 (see (9.31)), i.e., the energy difference between the excitonic and the biexcitonic transitions (see solid and dashed curves). As discussed in [398], this is due to the in-plane field F, which induces an electron–hole charge separation along the field direction. This, in turn, gives rise to significant dipole–dipole coupling between adjacent excitonic states (see Fig. 10.1a).

The central idea in this quantum-dot-based QIP proposal is to exploit such few-exciton effect to design conditional operations: the presence of an exciton in a quantum dot can prevent the generation of a second exciton by varying the detuning and thus controlling the resonance condition in the photogeneration process. In order to define the computational space for the present QIP scheme, let us introduce the excitonic occupation number operators $\hat{n}_l$, where l denotes the generic quantum dot in our array. The two states with eigenvalues $n_l = 0$ and $n_l = 1$ correspond, respectively, to the absence (no conduction-band electrons) and to the presence of a ground-state exciton (a Coulomb-correlated electron–hole pair) in dot l (see also Sect. 9.2.4). Each individual quantum dot in the array can be regarded as a qubit, whose generic quantum state is a vector in a two-dimensional subspace defined by the computational basis $|0_l\rangle$ and $|1_l\rangle$, i.e.,

$$|\psi_l\rangle = a_l|0_l\rangle + b_l|1_l\rangle \ . \tag{10.1}$$

The whole computational state-space $\mathcal{H}$ is then spanned by the global basis $|\{n_l\}\rangle = \otimes_l|n_l\rangle$, $(n_l = 0, 1)$. As discussed in Chap. 9, the full many-body

Hamiltonian describing our interacting electron–hole system restricted to the above computational space $\mathcal{H}$ reduces to

$$\hat{H}^{\mathrm{exc}} = \sum_l \mathcal{E}_l \, \hat{n}_l + \frac{1}{2} \sum_{l \neq l'} \Delta\mathcal{E}_{ll'} \, \hat{n}_l \, \hat{n}_{l'} \; . \tag{10.2}$$

Here, $\mathcal{E}_l$ denotes the energy of the ground-state exciton in dot l while $\Delta\mathcal{E}_{ll'}$ is the biexcitonic shift in (9.31) due to the Coulomb interaction between dots l and l', previously discussed (see Fig. 10.1b). The effective Hamiltonian (10.2) has exactly the same structure of the one proposed in 1993 by Seth Lloyd in his pioneering paper on quantum cellular automata [443], and it is the model Hamiltonian currently used in many of the nuclear magnetic resonance (NMR) quantum-computing schemes.

As can be readily derived from (10.2), the single-exciton energy $\mathcal{E}_l$ is simply renormalized by the biexcitonic shift $\Delta\mathcal{E}_{ll'}$, induced by the presence of a second exciton in dot l'. The dependence of the single-qubit energy on the occupation of the neighboring sites,

$$\tilde{\mathcal{E}}_l = \mathcal{E}_l + \sum_{l' \neq l} \Delta\mathcal{E}_{ll'} \, n_{l'} \; , \tag{10.3}$$

is the crucial ingredient allowing for the implementation of a conditional dynamics required for universal QIP. In order to illustrate this idea, let us focus again on the two-dot structure of Fig. 10.1a and point our attention on one of the two dots, say dot b. The effective energy gap between $|0_b\rangle$ and $|1_b\rangle$ depends now on the occupation of dot a. This elementary remark suggests to design properly tailored laser-pulse sequences to implement controlled-not logic gates among the two quantum dots as well as single-qubit rotations. Indeed, by sending an ultrafast laser π pulse with central energy $\hbar\omega_b^{n_a} = \mathcal{E}_b + \Delta\mathcal{E}_{ba} n_a$, the transition $|n_b\rangle \to |1 - n_b\rangle$ (π rotation) of the "target" qubit (dot b) is obtained if and only if the "control" qubit (dot a) is in the state n_a. In the very same way, by using a laser pulse with central energy $\hbar\omega_a^{n_b}$, the role of the target and control qubit is interchanged. By denoting with $\hat{U}_l^n$ the unitary transformation induced by the laser π pulse of central frequency ω_l^n, the above conditional gate corresponds to the transformation $\hat{U}_b^{n_a}$. Moreover, one can easily check that the two-color pulse sequence $\hat{U}_l^0 \, \hat{U}_l^1$ achieves the unconditional π rotation of the lth qubit.

In order to test the viability of the proposed QIP strategy, Biolatti and co-workers have performed a few simulated experiments of basic quantum information processing [378, 379]. They are based on a numerical solution of the Liouville–von Neumann equation describing the exact quantum-mechanical evolution of the many-exciton system (10.2) within the computational subspace $\mathcal{H}$, in the presence of environment-induced decoherence processes. The latter are accounted for via the density-matrix formalism introduced in Chap. 3 by means of a standard T_1/T_2 model (see (2.40)): a band-to-band recombination time $T_1 = 1\,\mathrm{ns}$ has been employed, and phonon-induced decoherence processes have been described in terms of an exciton–phonon decoherence time $T_2 = 50\,\text{ps}$, which is fully compatible with typical experimental findings (see, e.g., [444]). A microscopic analysis of phonon-induced decoherence in optically excited quantum-dot structures has been performed by Gerald Bastard and co-workers [445–456] as well as by Tilmann Kuhn and co-workers [457–474].

The above simulation scheme has been applied to the coupled-dot structure of Fig. 10.1a in the presence of an in-plane static field $F = 75\,\text{kV/cm}$: $\mathcal{E}_a = 1.673\,\text{eV}$, $\mathcal{E}_b = 1.683\,\text{eV}$, and $\Delta\mathcal{E} = 4\,\text{meV}$ (see inset in Fig. 10.1b).

We shall start our analysis by considering simulated experiments of a basic conditional two-qubit operation, the so-called controlled not (CNOT) gate. The first simulated experiment is shown in Fig. 10.2a. Here, the multi-color laser-pulse train (see central panel) is able to perform first a π rotation of the qubit a ($|0_a\rangle \to |1_a\rangle$); then, the second pulse is tuned to the frequency $\mathcal{E}_b + \Delta\mathcal{E}$, thus performing a π rotation of the qubit b ($|0_b\rangle \to |1_b\rangle$), since this corresponds to its renormalized transition energy when the neighboring qubit a is in state $|1_a\rangle$. The scenario described so far is confirmed by the time evolution of the exciton occupation numbers n_a and n_b (upper panel) as well as of the diagonal elements of the density matrix in the corresponding four-dimensional computational basis (lower panel).

This realizes the first part of the well-known CNOT gate: the target qubit b is rotated if the control qubit a is in state $|1_a\rangle$. To complete it, one has to show that the state of the target qubit b remains unchanged if the control qubit a is in state $|0_a\rangle$. This has been checked by a second simulated

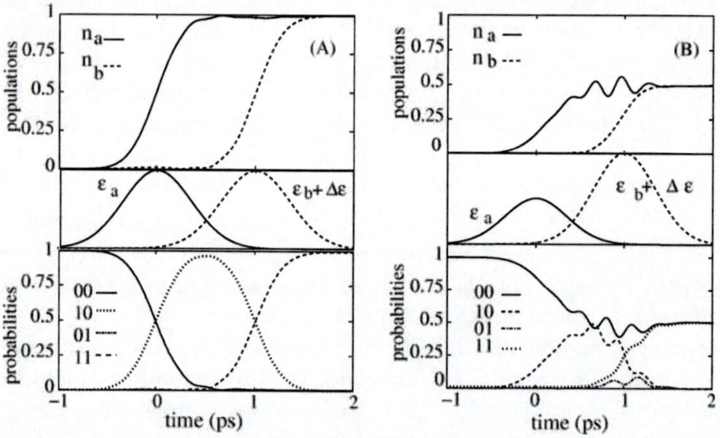

Fig. 10.2. Time-dependent simulation of a CNOT quantum gate (**a**) realizing the prescription ($|1,0\rangle \to |1,1\rangle$) and (**b**) transforming the factorized state $|0,0\rangle + |1,0\rangle$ into a maximally entangled state $|0,0\rangle + |1,1\rangle$ (see text). Exciton populations n_a and n_b (*upper panels*) and diagonal density-matrix elements (*lower panels*) as a function of time. The laser-pulse sequences are also sketched. Reprinted from [379]

experiment (not reported here) where the first pulse, being now off-resonant (with respect to dot a), does not change the computational state of the system. As a consequence, the second pulse is no more into resonance with the excitonic-transition energy of dot b, since the latter is no more renormalized by the excitonic occupation of dot a. Therefore, the initial state of the system is $|0,0\rangle$ and the final one is again $|0,0\rangle$.

The simulated experiments discussed so far clearly show the potential realization of the CNOT gate, thus confirming the validity of the proposed semiconductor-based QIP strategy. However, the analysis presented so far deals with factorized states only. It is well known [139–152] that the key ingredient in any QIP is entanglement. Generally speaking, this corresponds to a non-trivial linear combination of our basis states.

We shall now show that the CNOT gate previously discussed is able to transform a factorized state into a maximally entangled one. Figure 10.2b shows a simulated two-qubit operation driven again by a two-color laser-pulse sequence (see central panel). Initially, the system is in the state $|0,0\rangle$. The first laser pulse (at $t = 0$) is tailored in such a way to induce now a $\frac{\pi}{2}$ rotation of the qubit a: $|0,0\rangle \rightarrow (|0,0\rangle + |1,0\rangle)/\sqrt{2}$. At time $t = 1$ ps a second pulse induces a conditional π rotation of the qubit b: $|0,0\rangle + |1,0\rangle \rightarrow |0,0\rangle + |1,1\rangle$. This last operation plays a central role in any QIP, since it transforms a factorized state $((|0_a\rangle + |1_a\rangle) \otimes |0_b\rangle)$ into a maximally entangled state $(|0_a\rangle \otimes |0_b\rangle + |1_a\rangle \otimes |1_b\rangle)$.

The above simulated experiments (see Figs. 10.2a and 10.2b) clearly show that the gating time is fully compatible with the typical decoherence time-scale of quantum-dot structures; indeed, this is also confirmed by a number of recent experiments demonstrating Rabi oscillations in quantum dots [427–431]. Moreover, our simulated experiments show that the energy scale of the biexcitonic splitting $\Delta\mathcal{E}$ in our quantum-dot molecule (see inset in Fig. 10.1b) is compatible with the sub-picosecond operation time-scale of modern ultrafast laser technology.

10.2.2 GaN-Based Quantum Hardware

As anticipated in Chap. 9, one of the major technological problems in implementing the solid-state quantum gate previously discussed [378, 379] is the need for an external electric field to drive two-qubit conditional operations. Indeed, from the technological point of view, this requires the presence of electric contacts which limit the time response of the system and strongly complicate the physical interconnection of the device; the ideal scheme would thus be a quantum-dot structure with built-in electric fields. GaN-based quantum dots seem to match these requirements quite well [381–393], since they are known to exhibit strong built-in polarization and piezoelectric fields (of the order of MV/cm).

To overcome this limitation, De Rinaldis and co-workers have proposed an all-optical quantum gate based on GaN quantum-dot molecules [380]. The central idea is to properly tailor the internal field such to induce built-in excitonic dipoles. This gives rise to an intrinsic exciton–exciton coupling, i.e., without the need for additional external fields. In Sect. 9.3 it has been shown that the biexcitonic shift due to such dipole–dipole interaction is again of the order of a few milli electronvolt, thus allowing for sub-picosecond quantum state manipulations via ultrafast laser-pulse sequences.

As already mentioned, one of the most interesting features of GaN heterostructures is the strong built-in electric field: this is induced both by the spontaneous polarization and by the piezoelectric field. Contrary to GaN quantum wells – where the major role is played by the spontaneous polarization charge accumulated at the GaN/AlGaN interfaces while strain-induced piezoelectric fields have a minor importance (see, e.g., [394]) – in GaN quantum dots piezoelectric effects become relevant because of the non-central symmetry of the wurtzite structure. Moreover, since the dimensions of GaN and AlN unit cells differ slightly from those of an ideal hexagonal crystal, a spontaneous electrostatic polarization is present as well [381–393]. More specifically, the strain-induced piezoelectric potential and the spontaneous-polarization contribution are of similar magnitude and equal sign, both oriented along the growth direction. The magnitude of the intrinsic electric field along the growth direction is almost the same inside and outside the dot, but it is opposite in sign.

The resulting internal field can be tailored by varying the size of the dot: The models usually considered for its estimation neglect the actual in-plane shape of the dot; the latter will be primarily responsible for the relatively strong in-plane carrier confinement, which – as for the case of GaAs-based structures – can be well described in terms of a two-dimensional parabolic potential. Such built-in electric field induces electron–hole charge separation, which results in a charge dipole – also called excitonic dipole – in each quantum dot. This, in turn, will generate an effective interdot dipole–dipole coupling. For the case of two staked quantum dots a and b along the growth (z) direction (see Fig. 10.3) – for which such intrinsic excitonic dipole $\hat{p}$ is directed along z as well – we get

$$\Delta\mathcal{E} = -2\frac{p_a p_b}{Z^3} , \tag{10.4}$$

where Z is the interdipole distance.

The prototypical scheme of our two-qubit quantum gate consists of two stacked dots of slightly different size. For the sake of clarity, here we discuss the case of two coupled dots with identical bases, stacked along the growth axis, and having heights of 2.5 and 2.7 nm, respectively (analogous results would be obtained by two dots with same height and different base radius). The resulting difference in the potential well width (of about 0.2 nm) allows us to shift the excitonic transitions (see below). Moreover, as for the case of the GaAs-based nanostructures previously considered, the barrier width is chosen such as to prevent single-particle tunneling and to allow at the same time for significant dipole–dipole Coulomb coupling. Such internal field strongly modifies the valence- and conduction-band edges along the growth direction

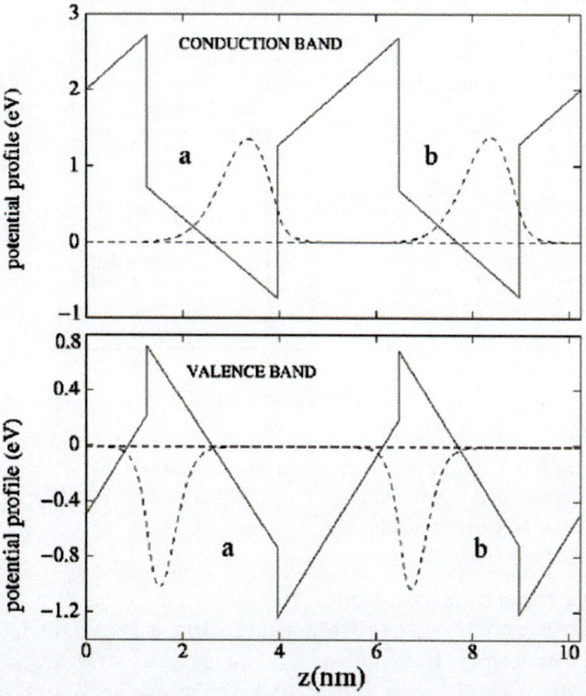

Fig. 10.3. Effects of the intrinsic electric field in a GaN-based quantum-dot molecule: schematic representation of the square-well potential profile for electrons and holes along the growth direction modified by the internal field (*solid lines*) and single-particle spatial charge distributions for electrons (e_a and e_b) and holes (h_a and h_b) (*dashed curves*) (see text). Reprinted from [380]

(see Fig. 10.3); this, in turn, induces the spatial separation of the electronic states and the creation of the desired intrinsic dipoles, as can be seen in Fig. 10.3. Here, the internal field shifts in opposite directions electron and hole charge distributions: electrons are forced to move toward the top of the dot (right) while holes are driven toward its bottom (left). Therefore, if we consider two dots (a and b) along the growth direction and with one exciton each, we are left with two aligned electric dipoles which are both pointing in the same direction. As recalled before, this results in a negative dipole–dipole coupling term (see (10.4)). The energy shift $\Delta\mathcal{E}$ is opposite and its absolute value is doubled with respect to the configuration discussed in Sect. 10.2.1, in which the two excitonic dipoles are still parallel but lying side by side.

Let us now discuss the optical response of the GaN-based macromolecule $(a + b)$ depicted in Fig. 10.3. This analysis is based again on the fully three-dimensional approach proposed in [379]. The excitonic absorption spectrum in the presence of the intrinsic field is shown in Fig. 10.4 (solid curve). The two

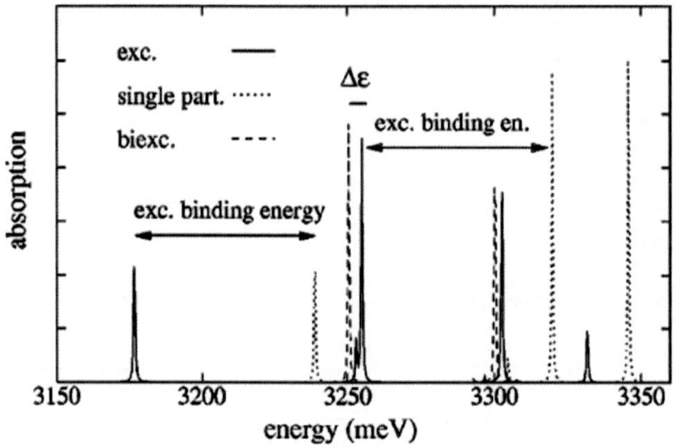

Fig. 10.4. Optical response of the GaN-based macromolecule in Fig. 10.3: excitonic (*solid curve*) and biexcitonic absorption (*dashed curve*); here, the single-particle spectrum is also reported (*dotted curve*) as well as the corresponding exciton binding energy (see text). Reprinted from [380]

lowest optical transitions correspond to the formation of direct ground-state excitons in dots a and b, respectively. Due to the strong built-in electric field in such heterostructures, it is known that the excitonic transitions energies are red shifted and, as can be seen in Fig. 10.4, the interband emission is fractions of electronvolt below the GaN bulk energy gap (3.5 eV) (see also Fig. 10.3). It is worth noticing that the difference in the ground-state excitonic transitions of the two dots is one order of magnitude bigger than the one employed in [378, 379], and this is due to the presence of the strong intrinsic field along the growth direction. Moreover, the two ground-state transitions have different amplitudes, due to the non-symmetric nanostructure considered: indeed, the two dots have different heights along the growth direction, which implies that the internal electric field experienced by the carriers in each dot is not the same. As a result, the oscillator strength – i.e., the overlap of electron and hole single-particle wavefunctions – changes according to the specific dot. The peaks occurring at higher energies correspond to optical transitions involving excited states of the in-plane parabolic potential in each quantum dot.

It is important to stress that, since the electric field is oriented along the growth direction, the in-plane bidimensional confinement potentials for electrons and holes exhibit the same symmetry axis. It follows that, due to parity arguments, only a few interband optical transitions are allowed, which makes the interpretation as well as the energy-selective addressing of specific lines in the spectrum much easier. This is a distinguished advantage of this GaN-based scheme, compared to the GaAs-based implementation previously discussed, where the presence of the in-plane field breaks all interband selection rules related to the bidimensional cylindrical symmetry of single-particle wavefunctions. This feature may be of crucial importance when considering energy-selective addressing schemes, since it allows one to employ shorter light pulses, i.e., pulses with a larger spectral width.

Let us now discuss the biexcitonic spectrum (dashed curve in Fig. 10.4); the latter describes the generation of a second electron–hole pair in the presence of a previously created exciton. Once again, the crucial feature in Fig. 10.4 is the magnitude of the biexcitonic shift, i.e., the energy difference between the excitonic and the biexcitonic transition (see solid and dashed curves). For the quantum-dot structure under investigation we get an energy splitting of 4.4 meV, which is quite comparable to the one in [378] and can still be resolved by sub-picosecond optical excitations.

In order to test the viability of the present GaN-based quantum hardware in the framework of the quantum-computing scheme described in Sect. 10.2.1, we shall review a few simulated experiments of basic QIP. The latter have been performed starting from the realistic state-of-the-art multiple-dot structure depicted in Fig. 10.3, for which $\mathcal{E}_a = 3.177\,\mathrm{eV}$, $\mathcal{E}_b = 3.255\,\mathrm{eV}$, and $\Delta\mathcal{E} = -4.4\,\mathrm{meV}$ (see Fig. 10.4). These simulated experiments are based again on a numerical solution of the corresponding effective Liouville–von Neumann equation; in this case interband recombination is described in terms of a $T_1 = 1\,\mathrm{ns}$ while exciton–phonon decoherence is accounted for via a time $T_2 = 30\,\mathrm{ps}$. More specifically, we shall now reconsider for the present GaN-based nanostructure the same time-dependent analysis performed in Sect. 10.2.1, i.e., a simulated experiment of a CNOT quantum gate.

Figure 10.5a shows a simulated sequence of two-qubit operations driven by a two-color laser-pulse train (dashed curve in the upper panel). This realizes again the CNOT previously discussed (see Fig. 10.2): the state of the second quantum dot is changed if and only if the first dot is in the state $|1\rangle$. Indeed, the multi-color laser-pulse train is able to perform first a π rotation of the qubit a

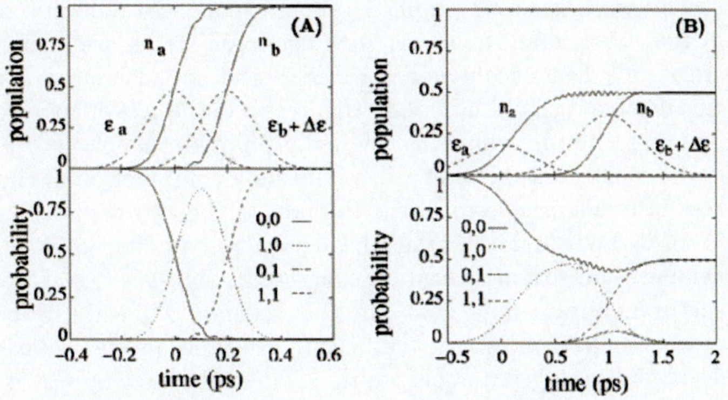

Fig. 10.5. Same as in Fig. 10.2 but for the GaN-based molecule depicted in Fig. 10.3 (see text). Excitonic populations n_a and n_b (*upper panel*) and diagonal density-matrix elements (*lower panel*) as a function of time. The laser-pulse sequences are also sketched (*dashed curves in the upper panel*). Reprinted from [380]

($|0_a\rangle \rightarrow |1_a\rangle$); then, the second pulse is tuned to the frequency $\mathcal{E}_b + \Delta\mathcal{E}$, thus performing a π rotation of the qubit b ($|0_b\rangle \rightarrow |1_b\rangle$), since this corresponds to its renormalized transition energy when the neighboring qubit a is in state $|1_a\rangle$. The scenario described so far is confirmed by the time evolution of the exciton occupation numbers n_a and n_b (upper panel) as well as of the diagonal elements of the density matrix in our four-dimensional computational basis (lower panel). We stress that, again, the operation is performed on a sub-picosecond time-scale; in particular, the time delay between the pulses is only $\delta t = 0.2\,\mathrm{ps}$.

Let us now come again to the ultrafast generation of entangled states shown in Fig. 10.5b. As for the CNOT-gate simulation reported in Fig. 10.5a, initially the system is in the state $|0,0\rangle \equiv |0_a\rangle \otimes |0_b\rangle$. The first laser pulse (at $t = 0$) is now tailored in such a way to induce a $\frac{\pi}{2}$ rotation of the qubit a: $|0,0\rangle \rightarrow (|0,0\rangle + |1,0\rangle)/\sqrt{2}$. At time $t = 0.4$ ps a second pulse induces a conditional π rotation of the qubit b: $|0,0\rangle + |1,0\rangle \rightarrow |0,0\rangle + |1,1\rangle$. We stress once again that the conditional operation simulated in Fig. 10.5b plays a central role in any quantum information processing.

The simulated experiments reviewed above (see Fig. 10.5) clearly show that the energy scale of the biexcitonic splitting $\Delta\mathcal{E}$ of the present GaN-based quantum-dot molecule – which is now totally driven/controlled by the built-in polarization field – is compatible with the sub-picosecond operation time-scale of modern ultrafast laser technology.

10.2.3 Combination of Charge and Spin Degrees of Freedom

The implementation schemes discussed so far are both based on charge degrees of freedom corresponding to interband optical excitations in semiconductors. As such, on the one hand, they can be easily manipulated/controlled on a sub-picosecond time-scale via properly tailored sequences of multi-color laser pulses; on the other hand, however, such excitonic states are known to suffer from relatively fast decoherence processes and are ultimately limited by interband radiative recombination. In this respect, the use of spin degrees of freedom seems to be preferable, as originally pointed out in [434]. It is then clear that the most desirable scenario would be a combination of charge and spin degrees of freedom able to merge the best of the two worlds.

In this spirit, Pazy and co-workers have proposed an alternative all-optical QIP implementation scheme based on charge-plus-spin degrees of freedom in semiconductor quantum dots [433]. In this scheme, while the qubit is the spin of an excess electron in a dot, quantum information/computation operations rely on swapping spin superpositions via short laser pulses to charged excitonic states. This strategy merges ideas from both the fields of spintronics and optoelectronics: using spin as quantum memory, and charge for the interaction between qubits, we can benefit (i) from the relatively low spin decoherence rates of conduction electrons (see, e.g., [475]) and (ii) from ultrafast (sub-picosecond) optical gating of charge excitations. Indeed, coherent optical control of electronic spins (see, e.g., [476] and references therein) as well as of excitonic states [427–432] together with a proper tailoring of exciton–exciton Coulomb coupling may allow for the implementation of single- as well as two-qubit gates, i.e., the full set of basic operations to implement quantum computing.

More specifically, in the implementation scheme proposed by Pazy and co-workers [433] one considers an excess electron in a semiconductor quantum dot, its spin degree of freedom (up or down) defining the two qubit states $|0\rangle$ and $|1\rangle$. The corresponding quantum register consists of an array of quantum dots, each dot containing one excess conduction electron. Primary goal of the present scheme is the implementation of a two-qubit phase gate. The dynamics required to perform the gating operation exploits a "Pauli blocking" mechanism, as observed experimentally in quantum dots (see, e.g., [428]). Within this scheme, one assumes that the quantum dots can be individually addressed via ultrafast laser excitations, by means of energy-selective gating enabled by the size fluctuations of self-assembled quantum dots (see Sect. 1.2.2). As schematically depicted in Figs 10.6 and 10.7, the control of the quantum-mechanical phase accumulated by Coulomb interactions is obtained by shining a σ^+-polarized laser pulse on the dot: due to the Pauli exclusion principle, a $|M_J^{\mathrm{e}} = -1/2, M_J^{\mathrm{h}} = +3/2\rangle$ electron–heavy hole pair is created in the s-shell only if the excess electron – already present in the dot – has a spin projection 1/2 (see Fig. 10.6). Thus with a π pulse one creates an exciton conditional to the spin state (qubit) of the original excess electron. In this way one obtains precise spin control of the switching on (and off) of further Coulomb

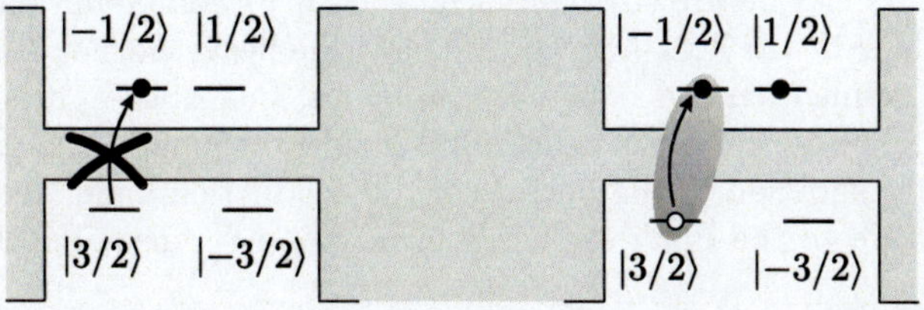

Fig. 10.6. Quantum dots and energy-level scheme. *Left*: the excess electron is in state $|-1/2\rangle \equiv |0\rangle$ and the transition induced by a σ^+-polarized light is blocked. *Right*: the excess electron is in $|+1/2\rangle \equiv |1\rangle$ and the exciton can be created (see text). Reprinted from [433]

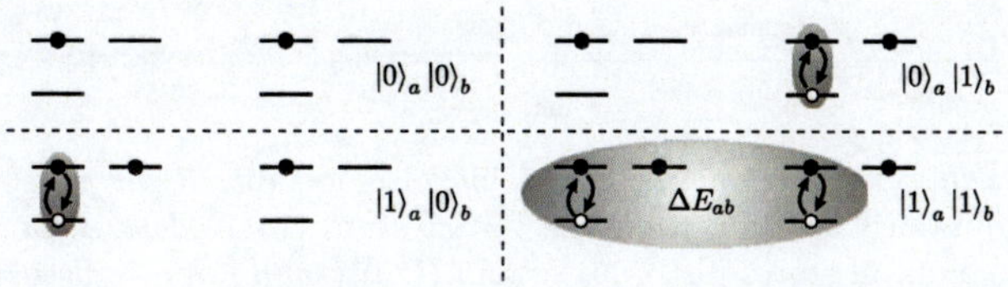

Fig. 10.7. Dynamics of the two-qubit gate for the computational basis states $|\alpha_a\rangle \otimes |\beta_b\rangle$ ($\alpha, \beta \in \{0,1\}$). In the ideal case of perfect Pauli blocking, the effective dipole–dipole interaction is present only for the $|1,1\rangle \equiv |1_a\rangle \otimes |1_b\rangle$ component (see text). Reprinted from [433]

interactions. In particular, this allows one to switch on exciton–exciton dipole coupling between neighboring quantum dots previously discussed conditional to the spins (qubits) being in state $|1_a\rangle \otimes |1_b\rangle$ (see Fig. 10.7). It is important to emphasize that the presence of the photogenerated electron–hole pair is only required during the gating, after which the latter will be annihilated via a second laser pulse. Thus excitonic interactions can be switched on and off, in contrast to the proposals based on charge excitations only (see Sects. 10.2.1 and 10.2.2). We stress that the two-pulse sequence previously mentioned may also be replaced by a properly tailored single-pulse sequence, as shown in Fig. 10.8 and discussed in [433].

The simulated experiments reported in Fig. 10.8 clearly show that the proposed implementation scheme benefits from the vast time-scale separation between excitonic and spin decoherence processes: whereas the proposed qubit is given by the spin of a conduction electron and thus decoheres on a microsecond time-scale, the proposed conditional two-qubit phase gate is driven/controlled by Coulomb interaction on a picosecond time-scale. The ratio between gating operation time and the coherence time of the quantum memory is therefore of the order of 10^5.

The above charge-plus-spin implementation strategy can be extended to single-qubit operations (see Fig. 10.9), thus leading to a fully optical spin-based QIP scheme. More specifically, what Pazy and co-workers have proposed is to implement the single-qubit gate via a Raman process involving the lowest light-hole states ($M_J^{\rm h} = \pm 1/2$) which however, being excited hole states, suffer from significant decoherence. On the other hand, semiconductor quantum dots can be grown where strain in II–VI compound materials can energetically shift the light-hole states below the heavy-hole ones, thus avoiding the above-mentioned decoherence problems. In this case, it is possible to apply the following pulse sequence: first a linearly polarized laser, coupling the light-hole subband to the bottom of the conduction band (see Fig. 10.9); due once again to the Pauli exclusion principle, a π pulse of such a linearly

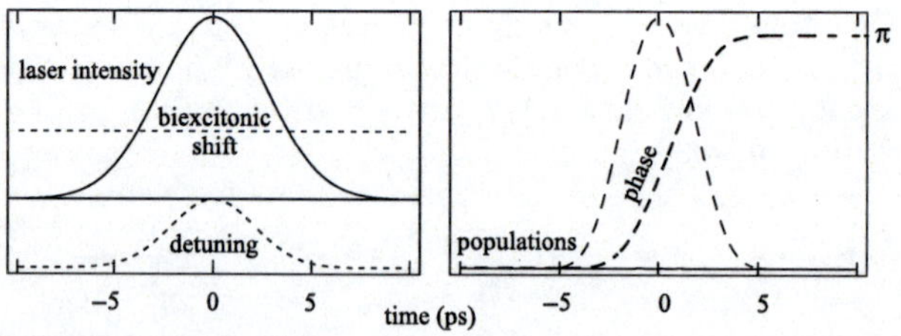

Fig. 10.8. *Left panel*: Pulse shapes and biexcitonic shift. *Right panel*: Biexcitonic population obtained starting from the initial state $|1,1\rangle$ (*dashed line*) as well as from the three initial states $|0,0\rangle$, $|0,1\rangle$, and $|1,0\rangle$ (*solid line*) (indistinguishable on this *plot scale*), and accumulated phase. We obtain the desired phase value $\theta = \pi$, while the population eventually *left* in the unwanted single- and two-exciton states remains at most of the order of 10^{-6} (see text). Reprinted from [433]

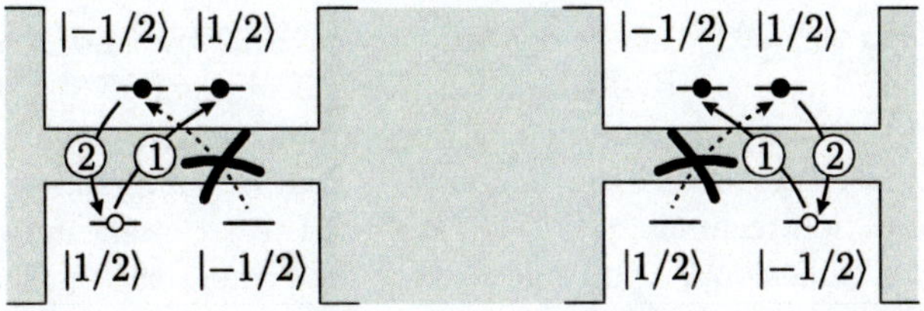

Fig. 10.9. Prototypical scheme for all-optical single-qubit rotations involving light-hole states (see text). Reprinted from [433]

polarized light will promote an electron only into the unoccupied qubit state (processes (1) in Fig. 10.9). The hole created in this first photoexcitation process is then recombined with the original excess electron via a further π pulse of σ^+-polarized light propagating in the plane orthogonal to the dot symmetry axis (processes (2) in Fig. 10.9). By adjusting the duration and phase of the pulses, the net effect of the above laser-pulse sequence is a coherent rotation between $|0\rangle$ and $|1\rangle$ on the fast picosecond time-scale. We also notice that the two-qubit gate described above can also be adapted to the inverted heavy-light hole configuration previously discussed. As said before, a major advancement toward a quantum-dot-based implementation of QIP would result from combining the most favorable features of the exciton-based schemes (ultrafast optical gating and efficient Coulomb coupling between qubits) with the long-lived memory of the electron spins. This is also the ultimate goal of an implementation scheme recently proposed by Troiani and co-workers [438], which relies on an efficient interplay between spin and charge degrees of freedom of quantum-confined carriers: the former store the information, the latter provide the auxiliary states required for their optical manipulation. As an essential step within this quantum-gate implementation, they propose to map literally the spin state of each electron onto its charge state by means of optically induced (Raman) transitions (see, e.g., [406]): more specifically, the two computational basis states ($|0\rangle \equiv |\downarrow\rangle$ and $|1\rangle \equiv |\uparrow\rangle$) are mapped onto spatially separated orbitals and therefore produce different electrostatic fields, which are felt by the neighboring electrons (qubits). If simultaneously applied to two neighboring qubits, the overall process thus results in an effective spin–spin coupling, mediated by a Coulomb (dipole–dipole) interaction, that carries the information and thus allows to implement the conditional dynamics.

10.3 QIP with Ballistic Electrons in Semiconductor Nanowires

In this section we shall review and discuss a second QIP implementation scheme based on charge degrees of freedom in semiconductor quantum wires, proposed simultaneously and independently in 1999 by Carlo Jacoboni and co-workers [477] and by Radu Ionicioiu and co-workers [478]. The key idea is to employ ballistic electrons as "flying qubits" (see below) in quantum-wire structures. In view of the nanometric spatial confinement reached by current fabrication technology [37–40], state-of-the-art quantum wires behave as quasi one-dimensional electronic waveguides, where the carrier coherence length may reach values of a few microns; it follows that on the nanometric scale electrons are in the so-called ballistic regime (see Sect. 1.4) and their quantum-mechanical phase coherence is preserved. This regime has been the natural arena for a number of interferometric experiments (see, e.g., [479] and references therein), and more recently such coherent-transport phenomena have been further investigated in the presence of surface acoustic waves (SAW) [480–483].

10.3.1 Quantum Hardware and Basic Logic Operations

As schematically depicted in Fig. 10.10, the building block of the present quantum hardware is a pair of adjacent quantum wires. The qubit state is defined according to the electron spatial location: within the so-called dual-rail representation [477, 478], one defines the basis state $|0\rangle$ by the presence of the electron in one of the wires (called the 0-rail) and the basis state $|1\rangle$ by the presence of the electron in the other one (the 1-rail).

As already mentioned, any QIP device can be built using single- and two-qubit gates only [139–152]. In the present implementation scheme a possible universal set of quantum logic gates is given by $\{H,\ P_\varphi,\ P_\pi^{\mathrm{C}}\}$, where

$$H = \frac{1}{\sqrt{2}} \begin{pmatrix} 1 & 1 \\ 1 & -1 \end{pmatrix} \tag{10.5}$$

is the well-known Hadamard gate,

$$P_\varphi = \begin{pmatrix} 1 & 0 \\ 0 & \mathrm{e}^{\mathrm{i}\varphi} \end{pmatrix} \tag{10.6}$$

is a single-qubit phase-shift gate, and P_π^{C} is a so-called controlled sign flip, whose action may be expressed via the two-qubit controlled phase-shift gate

$$P_\varphi^{\mathrm{C}} = \begin{pmatrix} 1 & 0 & 0 & 0 \\ 0 & 1 & 0 & 0 \\ 0 & 0 & 1 & 0 \\ 0 & 0 & 0 & \mathrm{e}^{\mathrm{i}\varphi} \end{pmatrix} . \tag{10.7}$$

The physical implementation of the above elementary gates in terms of the dual-rail representation of Fig. 10.10 is discussed extensively in [477, 478].

The single-qubit Hadamard gate (10.5) can be implemented using a sort of electronic beam splitter, also referred to as "waveguide coupler" (see, e.g.,

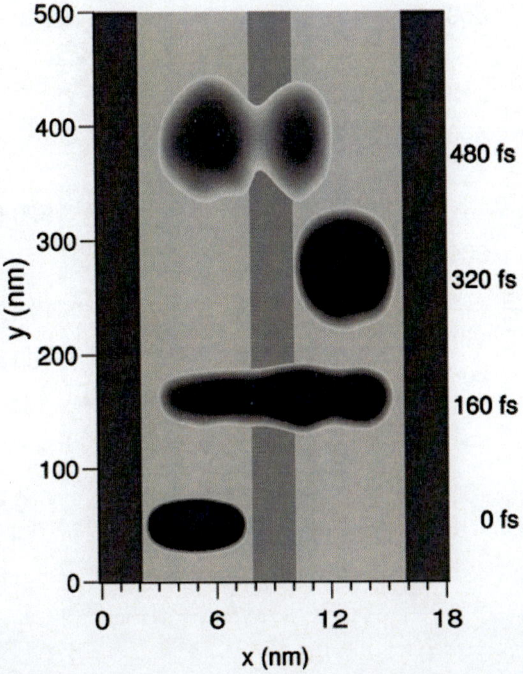

Fig. 10.10. Continuous single-qubit rotation corresponding to the oscillation of the electron density between two coupled quantum wires at different times. Here, the lateral extension of the single wire is 6 nm, the potential in the outer zones is assumed to be infinite, and the interwire barrier height is 0.1 eV. It is assumed that at $t = 0$ the electron is located in the left wire with a kinetic energy of 0.1 eV. Note the different scales of the x and y axes (see text). Reprinted with permission from [477]

[484] and references therein). As schematically shown in Figs. 10.11 and 10.12, the idea is to design the two-wire nanostructure in such a way to spatially control the interwire electron tunneling. For a given interwire distance, a proper modulation (along the quantum-wire direction) of the interwire potential barrier can produce any linear superposition

$$|\psi\rangle = a|0\rangle + b|1\rangle \tag{10.8}$$

(see Fig. 10.11) of our computational basis states.

The single-qubit phase-shift gate P_φ in (10.6) can be implemented using either a potential step or a potential well along the wire direction; the well is preferable, since the induced quantum-mechanical phase shift is more stable under voltage fluctuations.

Finally, the controlled phase-shift gate P_φ^{C} in (10.7) is implemented using a so-called Coulomb coupler (see, e.g., [485] and references therein). Within the present implementation scheme, such two-qubit quantum gate exploits the

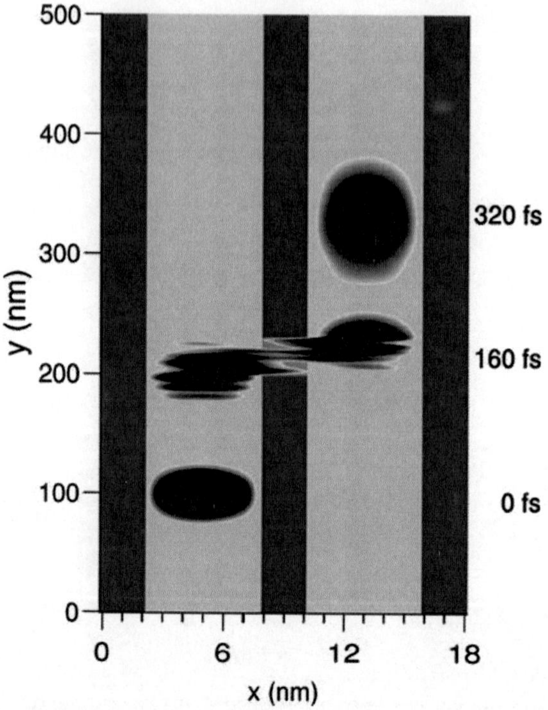

Fig. 10.11. Electron density in the two-wire nanostructure of Fig. 10.10 at different times, where the interwire coupling region is reduced to a suitable window in order to realize a complete transfer of the electron from the left to the right wire (NOT gate). Again, it is assumed that at $t = 0$ the electron is confined in the left wire with the same kinetic energy of Fig. 10.10 (see text). Reprinted with permission from [477]

Coulomb interaction between two single electrons in different quantum-wire pairs (representing the two qubits). The gate is similar in construction to the beam splitter previously mentioned.

An appealing feature of the present scheme is the mobile character of the qubits: using so-called flying qubits one can transfer entanglement from one place to another, without the need to interconvert stationary into mobile qubits. One important requirement of such quantum hardware is that electrons within different wires need to be synchronized at all times in order to properly perform two-qubit gating (the two electron wavepackets should reach simultaneously the Coulomb-coupling window). It is thus essential to have highly monoenergetic electrons launched simultaneously. This can be accomplished by properly tailored energy filters and synchronized single-electron injectors at the preparation stage.

Starting from the original quantum hardware (based on charge degrees of freedom) presented so far, a spin-based version has been also put forward

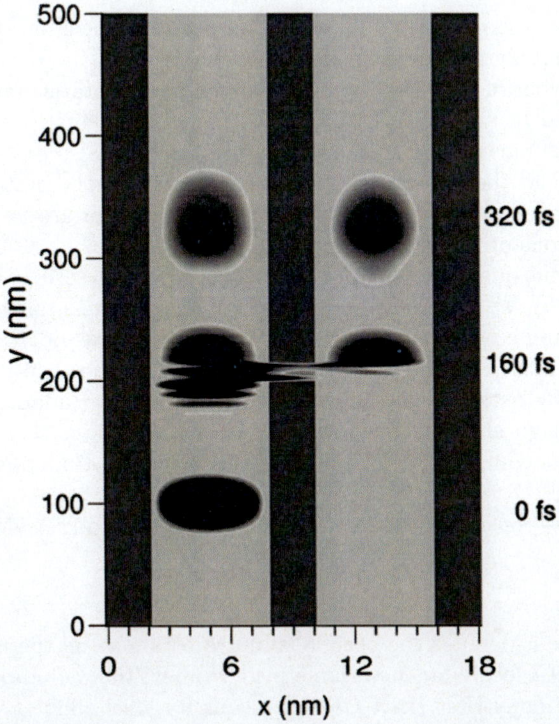

Fig. 10.12. Electron density in the two-wire nanostructure of Fig. 10.10 at different times, where the interwire coupling region is reduced to a suitable window in order to realize an equal splitting of the wavefunction between the two wires. Again, it is assumed that at $t = 0$ the electron is confined in the left wire with the same kinetic energy of Fig. 10.10 (see text). Reprinted with permission from [477]

[486]; moreover, a GaAs/AlGaAs-based two-qubit quantum device for the controlled generation and straightforward detection of entanglement by measuring a stationary current–voltage characteristics has been recently proposed [487].

It is finally imperative to mention that, in view of the excellent room-temperature coherent-transport properties of carbon nanotubes (see, e.g., [488] and references therein), the latter are expected to replace conventional semiconductor quantum wires in many classical- as well as quantum-device architectures (see Chap. 11), including the present QIP implementation scheme.

10.3.2 Testing Bell's Inequality Violations in Semiconductors

As already mentioned, the key ingredient for computational speed-up in QIP is entanglement. Einstein–Podolsky–Rosen (EPR) pairs [489] and three-particle Greenberger–Horne–Zeilinger (GHZ) states [490] are at the heart of quantum cryptography, teleportation, dense coding, entanglement swapping, and of many quantum algorithms [139–152].

On the experimental side, two-particle entangled states have been originally prepared using photons [491] and trapped ions [492]; more recently, a photonic three-particle GHZ entangled state has been also measured [493]. A few proposals for the generation of entangled states in solid-state physics have been also put forward (see, e.g., [494]), but so far there are no experimental implementations of robust entanglement protocols.

Based on the quantum-wire prototypical hardware described in Sect. 10.3.1 (see Fig. 10.10), we shall now review a possible experimental setup proposed by Ionicioiu and co-workers [478] to test Bell's inequality violations in a solid-state environment. Within this implementation scheme, two-particle entangled states (Bell states) can be generated using three Hadamard gates plus a controlled sign shift, as schematically depicted in Fig. 10.13 (see dashed box); here, the controlled sign shift plus the lower two Hadamard gates form a CNOT gate.

Let us now consider the correlation function for two (pseudo)spins

$$P(\boldsymbol{a}, \boldsymbol{b}) = \langle \Psi | \hat{\sigma}_{\boldsymbol{a}}^{(1)} \hat{\sigma}_{\boldsymbol{b}}^{(2)} | \Psi \rangle \; , \tag{10.9}$$

where $\hat{\sigma}_{\boldsymbol{a}} = \hat{\sigma}_i a_i$ denotes the pseudo-spin projection along the unit vector $\boldsymbol{a}$.

Any realistic local hidden-variable theory[1] obeys the well-known Bell/Clauser-Horne-Shimony-Holt (Bell–CHSH) inequality [495, 496]:

$$|P(\boldsymbol{a}, \boldsymbol{b}) + P(\boldsymbol{a}', \boldsymbol{b}) + P(\boldsymbol{a}', \boldsymbol{b}') - P(\boldsymbol{a}, \boldsymbol{b}')| \leq 2 \; . \tag{10.10}$$

However, the latter can be violated in quantum mechanics. Indeed, for the case of a singlet state

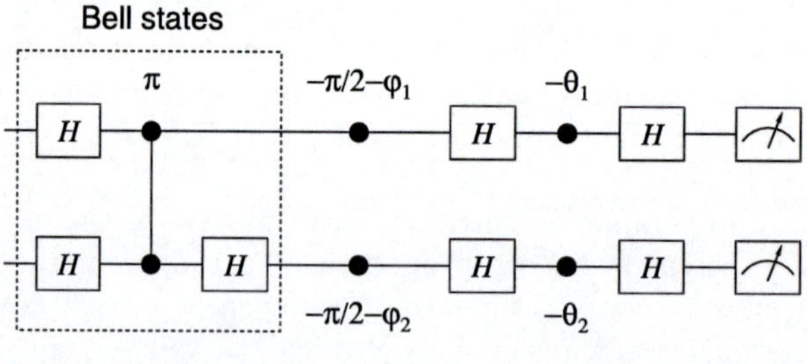

Fig. 10.13. Prototypical quantum network for testing Bell's inequality: entangled Bell states are prepared in the *dashed box*; then the first qubit is measured along the direction $\boldsymbol{a} = (\theta_1, \varphi_1)$ and the second qubit along the direction $\boldsymbol{b} = (\theta_2, \varphi_2)$ (see text). Reprinted from [478]

[1] In quantum mechanics a local hidden-variable theory is one in which distant events are assumed to have no instantaneous (or at least faster-than-light) effect on local ones.

$$|\Psi^-\rangle = |1,0\rangle - |0,1\rangle \tag{10.11}$$

a straightforward calculation gives

$$P(\boldsymbol{a},\boldsymbol{b}) = -\boldsymbol{a}\cdot\boldsymbol{b}\ ; \tag{10.12}$$

by choosing $\boldsymbol{a}\cdot\boldsymbol{b} = \sqrt{2}/2$, it is easy to verify that the Bell–CHSH inequality (10.10) is maximally violated.

Let us now focus on the correlation function (10.9). In the conventional EPR–Bohm "Gedankenexperiment" one needs to measure the spin component of one particle along a direction $\boldsymbol{n}$; however, in our semiconductor-based setup this is not possible directly, since we can measure only σ_z, i.e., whether the electron is in the 0- or in the 1-rail. As discussed in [478], the solution is to do a unitary transformation $|\psi\rangle \to |\psi'\rangle = \hat{U}|\psi\rangle$, such that the operator $\hat{\sigma}_{\boldsymbol{n}}$ is diagonalized to $\hat{\sigma}_z$, i.e., $\langle\psi|\hat{\sigma}_{\boldsymbol{n}}|\psi\rangle = \langle\psi'|\hat{\sigma}_z|\psi'\rangle$. We are then looking for a unitary transformation $\hat{U}$ which satisfies $\hat{U}^\dagger\hat{\sigma}_z\hat{U} = \hat{\sigma}_{\boldsymbol{n}}$, with $\boldsymbol{n} = (\sin\theta\cos\varphi, \sin\theta\sin\varphi, \cos\theta)$ denoting the generic unit vector. More specifically, in terms of our elementary gates one obtains

$$\hat{U}(\theta,\varphi) = H\,P_{-\theta}\,H\,P_{-\varphi-\frac{\pi}{2}}\ . \tag{10.13}$$

Thus, measuring the spin (in the EPR–Bohm setup) along a direction $\boldsymbol{n}$ is equivalent to performing the above unitary transformation $\hat{U}(\theta,\varphi)$ followed by a measurement of σ_z.

Going back to the entangled pair in Fig. 10.13, we now apply locally on each qubit the two transformations $\hat{U}(\theta_1,\varphi_1)$ and $\hat{U}(\theta_2,\varphi_2)$, respectively. Here, $\boldsymbol{a} = (\theta_1,\varphi_1)$ and $\boldsymbol{b} = (\theta_2,\varphi_2)$ are the two directions discussed above. At the very end, one measures σ_z, i.e., the presence of the electron in the 0- or 1-rail (see Fig. 10.13).

For the singlet state $|\Psi^-\rangle$ in (10.11) the correlation function depends on the scalar product only (see (10.12)), and hence only on the relative angle. Without loss of generality we can then choose $\varphi_1 = \varphi_2 = -\pi/2$, $\theta_1 = 0$, and

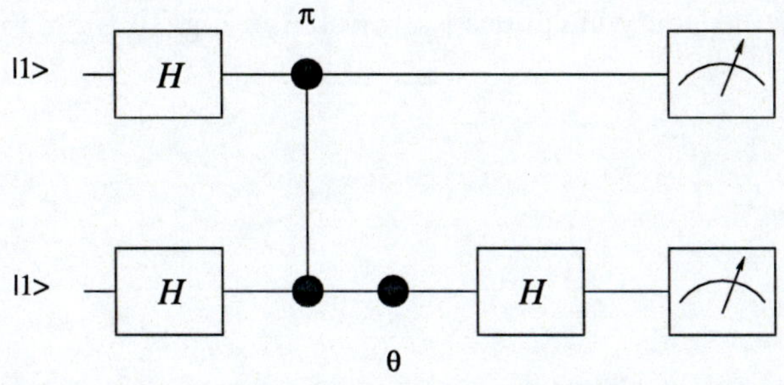

Fig. 10.14. Testing Bell's inequality for the singlet state $|\Psi^-\rangle$ in (10.11): the present quantum network is obtained from the one reported in Fig. 10.13 by setting $\varphi_1 = \varphi_2 = -\pi/2$, $\theta_1 = 0$ and relabeling θ_2 with $-\theta$ (see text). Reprinted from [478]

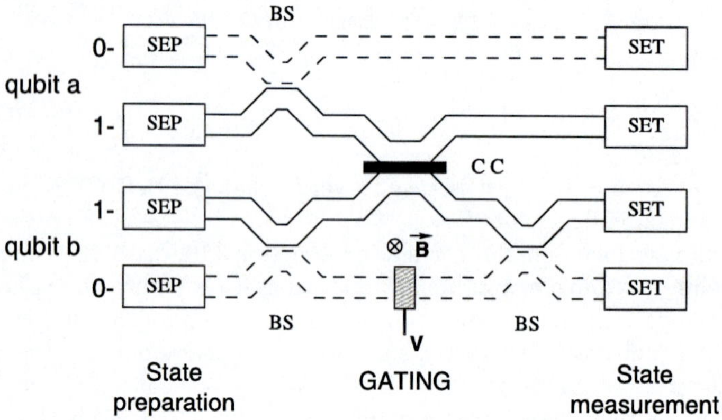

Fig. 10.15. Potential experimental setup to test Bell–CHSH inequality violations for a singlet state: the 0-rails of each qubit are dashed for clarity; an electric potential V applied on top of the 0-rail (*dashed box* in the figure) is used to produce a phase shift $P_{-\theta}$ on the second qubit; alternatively, the same effect can be achieved via a magnetic field $\boldsymbol{B}$ (see text). Reprinted from [478]

relabel θ_2 with $-\theta$. Recalling that $H^2 = 1$, the quantum network (i.e., the gating sequence) in Fig. 10.13 simplifies to five gates only, as schematically shown in Fig. 10.14.

Employing this relatively simple network, in principle one can measure the correlation function (10.9) and test Bell's inequality violations. In practice, however, the situation is more complex. The essential ingredient for producing entanglement is the controlled phase-shift gate (10.7) which involves an interaction between the two qubits. As anticipated, experimentally this requires a good "timing" of the two electrons (they should reach simultaneously the two-qubit gating region). In this respect, the detailed analysis of the two-qubit gate imperfections given in [478] shows that the scheme is robust against the potential synchronization problems previously mentioned.

A schematic representation of the proposed experimental setup for measuring Bell's inequality violations is presented in Fig. 10.15. More details can be found in [478].

11

New Frontiers of Electronic and Optoelectronic Device Physics and Technology

In this chapter we shall finally address two extremely active and stimulating research topics, namely molecular and spin-transport electronics, whose development may lead to completely new paradigms in semiconductor-based electronics and optoelectronics physics and technology. A detailed treatment of these rapidly developing fields is out of the scope of the present book; aim of the following pages is to provide a brief historical account and a concise description of such strategic research areas.

11.1 Molecular Electronics (Moletronics)

Molecular electronics – also referred to as moletronics – is an interdisciplinary research area that spans physics, chemistry, and material science (see, e.g., [497, 498] and references therein); the unifying feature of this area is the use of molecular building blocks for the fabrication of electronic components, both passive (e.g., resistive wires) and active (e.g., transistors). The concept of molecular electronics has aroused much excitement both in science fiction and among researchers due to the prospect of size reduction in device technology offered by such molecular-level control; indeed, molecular electronics provides means to extend Moore's law beyond the foreseen limits of small-scale conventional silicon-based integrated circuits (see Sect. 1.1). Due to the broad use of the term, molecular electronics can be split into two related but separate subdisciplines: molecular materials for electronics utilize the properties of the molecules to affect the bulk properties of a material, while molecular-scale electronics focuses on single-molecule applications.

Study of charge transfer in molecules was advanced in the 1940s by Robert Mulliken and Albert Szent-Györgyi by investigating so-called donor-acceptor systems and developing the control of charge and energy transfer in molecules (see, e.g., [500] and references therein). Likewise, in 1974 Mark Ratner and Ari Aviram [501] proposed a possible scheme for a molecular rectifier; later, in 1988, Aviram described in detail a theoretical single-molecule field-effect transistor; further concepts were also proposed, including single-molecule logic gates. These were all theoretical constructs and not concrete devices. The direct measurement of the electronic characteristics of individual molecules awaited the development of methods for making molecular-scale electric contacts. This was no easy task; indeed, the first unambiguous experiment measuring the conductance of a single molecule (see Fig. 11.1) was only reported in 1997 by Mark Reed and co-workers [499]. Since then, this branch of the field has progressed rapidly. Likewise, as it has become possible to measure such properties directly, the theoretical predictions of the early proposals have been substantially confirmed.

It is worth stressing that while mostly operating in the quantum realm of less than 100 nm, molecular electronics processes often collectively manifest themselves at the macroscale. Examples include quantum tunneling, negative resistance, phonon-assisted hopping, polarons. Thus, macroscale active organic electronic devices were described decades before molecular-scale ones. For example, in 1974 John E. McGinness and co-workers described the putative "first experimental demonstration of an operating molecular electronic device" [502]: this was a voltage-controlled switch; as its active element, this device used DOPA melanin, an oxidized mixed polymer of polyacetylene, polypyrrole, and polyaniline, and its "ON" state exhibited almost metallic conductivity. Since the 1970s, scientists have developed an entire panoply of new materials and devices. These findings have opened the door to plastic electronics and optoelectronics, which are beginning to find significant commercial application.

As anticipated, the first highly conductive organic compounds were the charge-transfer complexes. In 1954, researchers at Bell Labs and elsewhere reported charge-transfer complexes with resistivities as low as $8\,\Omega/\text{cm}$ (see, e.g., [503]). In the early 1970s, salts of tetrathiafulvalene were shown to exhibit almost metallic conductivity, while superconductivity was demonstrated in 1980. Broad research on charge-transfer salts continues today.

The so-called linear-backbone "polymer blacks" (polyacetylene, polypyrrole, and polyaniline) and their copolymers are the main class of conductive polymers. Historically, these are known as melanins. In 1963 Weiss and co-workers reported iodine-doped oxidized polypyrrole blacks with resistivities as low as $1\,\Omega/\text{cm}$ [504]. Subsequent papers reported resistances as low as $0.03\,\Omega/\text{cm}$ (see, e.g., [500] and references therein). With the notable exception of charge-transfer complexes (some of which are even superconductors), organic molecules had previously been considered insulators or at best weakly conducting semiconductors. Over a decade later, in 1977 Alan J. Heeger, Alan G. MacDiarmid, and Hideki Shirakawa reported equivalent high conductivity in rather similarly oxidized and iodine-doped polyacetylene; they later received the 2000 Nobel prize in chemistry "for the discovery and development of conductive polymers."

In polymers, classical organic molecules are composed of both carbon and hydrogen (and sometimes of additional compounds such as nitrogen, chlorine,

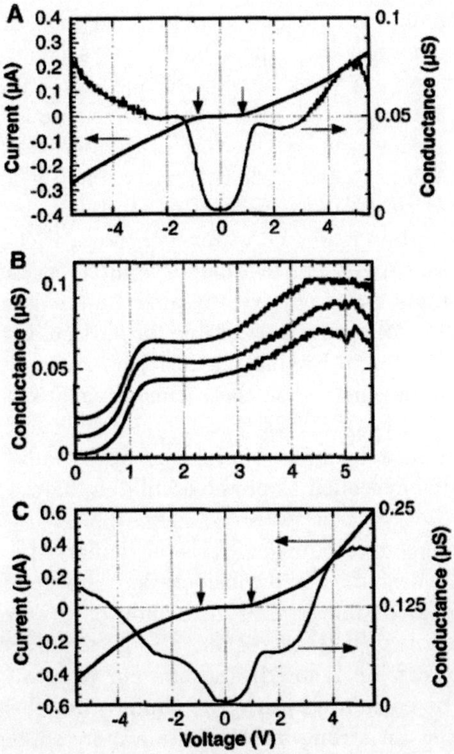

Fig. 11.1. Experimental results by Mark Reed and co-workers on the conductance of a molecular junction (see text). (**a**) Typical current–voltage characteristics $I(V)$ illustrating a gap of 0.7 V, and the corresponding , conductance (first derivative) $G(V)$ showing a step-like structure. (**b**) Three independent measurements of $G(V)$ illustrating the reproducibility of the conductance values. The measurements were performed with the same mechanically controllable break junction but for different retractions/contacts and thus different contact configurations. Offsets of 0.01 μS for the middle curve and 0.02 μS for the top curve are used for clarity. The first step for these three measurements gives values of 22.2, 22.2, and 22.7 MΩ (*top to bottom*); the next step gives values of 12.5, 13.3, and 14.3 MΩ. The middle curve corresponds to the same data as in (**a**). (**c**) A particular $I(V)$ and corresponding $G(V)$ measurement illustrating conductance values approximately twice the observed minimum conductance values. Resistances of 14 MΩ for the first step and 7.1 MΩ (negative bias) as well as 5 MΩ (positive bias) for the second step were measured. Reprinted with permission from [499]

or sulphur). They are obtained from petrol and can often be synthesized in large amounts. Most of these molecules are insulating systems when their length exceeds a few nanometers. However, naturally occurring carbon is an electric conductor. In particular, graphite (recovered from coal or encountered naturally) is a conducting material. From a theoretical point of view, graphite is a semimetal, a category in between metals and semiconductors. It has a layered structure, each sheet being one atom thick. Between each sheet, the interactions are weak enough to allow an easy manual cleavage. Tailoring the graphite sheet to obtain well-defined nanometer-sized objects remains a challenge. However, by the close of the twentieth century, chemists were exploring methods to fabricate extremely small graphitic objects that could be considered single molecules. After studying the interstellar conditions under which carbon is known to form clusters, Richard E. Smalley and co-workers set up an experiment in which graphite was vaporized using laser irradiation [505]; mass spectrometry revealed that clusters containing specific "magic numbers" of atoms were stable, in particular those clusters of 60 atoms. Harold W. Kroto, an English chemist who assisted in the experiment, suggested a possible geometry for these clustering atoms covalently bound with the exact symmetry of a soccer ball, and coined for these new molecular complexes names as buckminsterfullerenes, buckyballs, or C60 [506]; these objects were rapidly envisioned as possible building blocks for molecular electronics.

The theory of single-molecule devices is particularly interesting since, generally speaking, the system under consideration is a genuine open quantum system (see Chap. 4) in nonequilibrium conditions (driven by the applied voltage). In the low-bias/voltage regime, the nonequilibrium nature of the molecular junction can be ignored, and the current–voltage characteristics of the device can be calculated using the equilibrium electronic structure of the system. However, for strong-bias regimes a more sophisticated treatment is required. In the elastic – i.e., dissipation-free – tunneling case (where the passing electron does not exchange energy with the system), the well-known Landauer–Büttiker formalism (see, e.g., [101] and references therein) can be used to calculate the transmission through the system as a function of bias voltage, and hence the current. For nonelastic-tunneling regimes, an elegant formalism based on the nonequilibrium Green's-function formalism [157, 158] (see Chap. 3) has been put forward by Yigal Meir and Ned Wingreen [507]. This nonequilibrium formulation has been successfully employed by the molecular electronics community to examine the more difficult and interesting cases where the transient electron exchanges energy with the molecular system (e.g., through electron–phonon coupling or electronic excitations).

Recent progress in nanoscience and nanotechnology has facilitated both experimental and theoretical investigations of molecular electronics. In particular, the development of the scanning tunneling microscopy (STM) and later the atomic force microscopy (AFM) have facilitated significantly the manipulation of single-molecule systems.

The first measurement of the conductance of a single molecule was published in 1995 [508]; this was the conclusion of 10 years of research using the scanning tunneling microscope tip apex to switch a single molecule, as already explored by Aviram, Joachim, and Pomerantz at the end of the 1980s [509]. This first experiment was followed by the important results by Mark Reed and James Tour [499] obtained using a mechanical-break-junction approach to connect two gold electrodes to a sulfur-terminated molecular wire (see Fig. 11.1).

A single-molecule amplifier was then implemented by Joachim and Gimzewski [510]; this experiment involving a single C60 molecule demonstrated that the latter can provide gain in a circuit just by playing with C60 intramolecular quantum interference effects. Relevant work is also being done on the use of single-wall carbon nanotubes (see, e.g., [511] and references therein) as field-effect transistors.

The previously mentioned Aviram–Ratner model for a molecular rectifier [501] – which until recently was entirely theoretical – has been unambiguously confirmed via a number of experiments by Geoffrey Ashwell and co-workers (see, e.g., [512]). Many rectifying molecules have so far been identified, and the number and efficiency of these molecular systems is expanding rapidly.

Generally speaking, the microscopic modeling of single-molecule quantum devices is a highly non-trivial task. In this context, many of the issues addressed in this book – e.g., the proper description of nonequilibrium regimes in the presence of open spatial boundaries – play a crucial role, and their treatment is still subject of intense investigation.

11.2 Spin-Transport Electronics (Spintronics)

Spintronics – a neologism meaning "spin-transport electronics" [513–516] (also known as magnetoelectronics) – is an emerging solid-state technology aimed at exploiting the intrinsic spin of the electron and its associated magnetic moment (in addition to its fundamental electronic charge). The research field of Spintronics originated from experiments on spin-dependent electron-transport phenomena in solid-state devices performed in the 1980s, including the observation of spin-polarized electron injection from a ferromagnetic metal to a normal metal by Johnson and Silsbee [517], and the discovery of giant magnetoresistance (see Fig. 11.3) independently by Fert and co-workers [518] and by Grünberg and co-workers [519]. The use of semiconductors for spintronics can be traced back to the theoretical proposal of a spin-based field-effect transistor published in 1990 by Datta and Das [520] and schematically depicted in Fig. 11.2.

As recalled in Chap. 10, electrons are spin-1/2 fermions and therefore constitute a two-state quantum system with spin "up" and spin "down." According to Fig. 11.2, to realize a spintronic device, the primary requirements are a system able to generate a current of spin-polarized electrons comprising more of one spin species – up or down – than the other (called a spin injector), and a separate system that is sensitive to the spin polarization of the electrons (spin detector). Manipulation of the electron spin during transport between injector and detector (especially in semiconductors) via spin precession can be accomplished using real external magnetic fields or effective fields caused by spin–orbit interaction. Spin polarization in non-magnetic materials can be

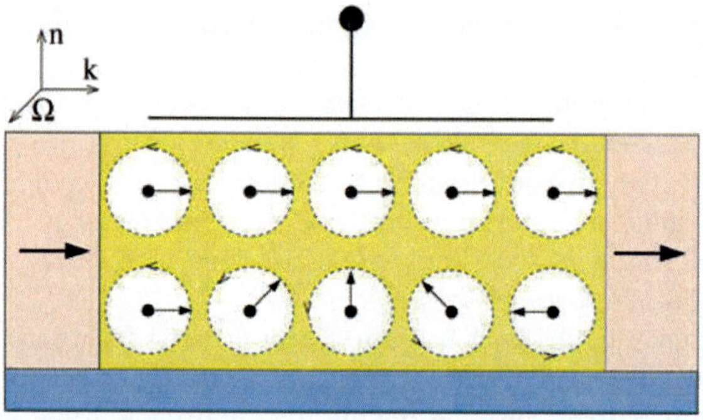

Fig. 11.2. Schematic illustration of the Datta–Das spin-based field-effect transistor [520]. Here, the source (spin injector) and the drain (spin detector) are ferromagnetic metals or semiconductors, with parallel magnetic moments. The injected spin-polarized electrons with wavevector $\boldsymbol{k}$ move ballistically along a quasi-one-dimensional channel formed by, e.g., an InGaAs/InAlAs heterojunction in a plane normal to $\boldsymbol{n}$. Electron spins process around the precession vector $\boldsymbol{\Omega}$, which arises from spin–orbit coupling and which is defined by the structure and material properties of the channel. The magnitude of $\boldsymbol{\Omega}$ is tunable by the gate voltage V_G at the top of the channel. The current is large if the electron spin at the drain points in the initial direction (*top row*) – e.g., if the precession period is much larger than the time-of-flight – and small if the direction is reversed (*bottom*). Reprinted with permission from [513]

achieved either through the Zeeman effect in large magnetic fields and low temperatures or by nonequilibrium methods. In the latter case, the nonequilibrium polarization will decay over a time-scale called "spin life-time"; for the case of conduction electrons in metals the spin life-time is relatively short (typically less than 1 ns) but in semiconductors the latter can be very long (microseconds at low temperatures), especially when the electrons are isolated in local trapping potentials (e.g., at impurities, where life-times can be milliseconds).

The simplest method of generating a spin-polarized current in a metal is to pass the current through a ferromagnetic material. The most common application of this effect is a giant magnetoresistance (GMR) device; such a device may consist of at least two layers of ferromagnetic materials separated by a spacer layer. When the two magnetization vectors of the ferromagnetic layers are aligned, the electric resistance will be lower (so a higher current flows at constant voltage) than if the ferromagnetic layers are anti-aligned. This constitutes a magnetic-field sensor. As schematically depicted in Fig. 11.4, two variants of GMR have been implemented: (i) current-in-plane (CIP), where the electric current flows parallel to the layers, and (ii) current-perpendicular-to-plane (CPP), where the electric current flows in a direction perpendicular to the layers. Other metal-based spintronic devices include Tunnel Magnetoresistance (TMR), where CPP transport is achieved by using quantum-mechanical tunneling of electrons through a thin insulator separating ferromagnetic layers

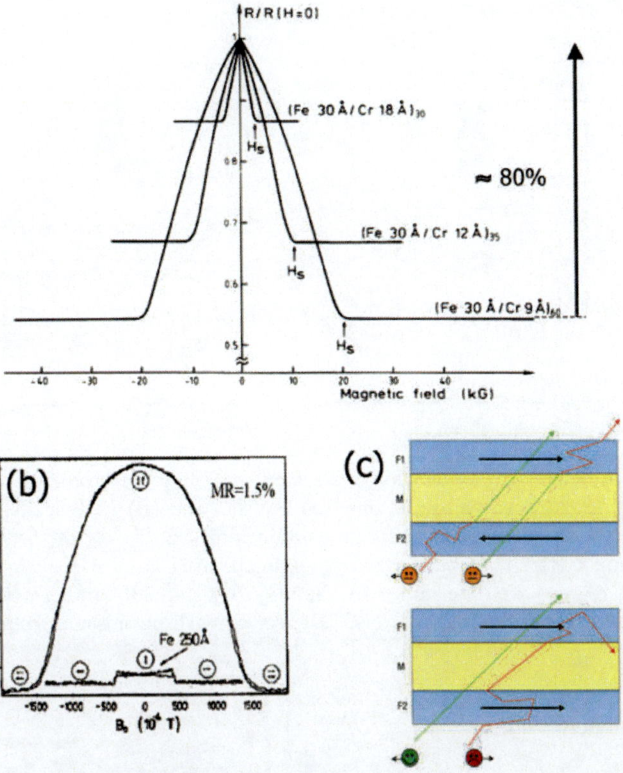

Fig. 11.3. First observations of giant magnetoresistance (GMR). (**a**) Case of Fe/Cr(001)multilayers [518]. (**b**) Case of Fe/Cr/Fe trilayers [519]. (**c**) Schematics of the mechanism of the GMR. In the parallel magnetic configuration (*bottom*), the electrons of one of the spin directions can go easily through all the magnetic layers and the short circuit through this channel leads to a small resistance. In the antiparallel configuration (*top*), the electrons of each channel are slowed down every second magnetic layer and the resistance is high. Reprinted with permission from [515]

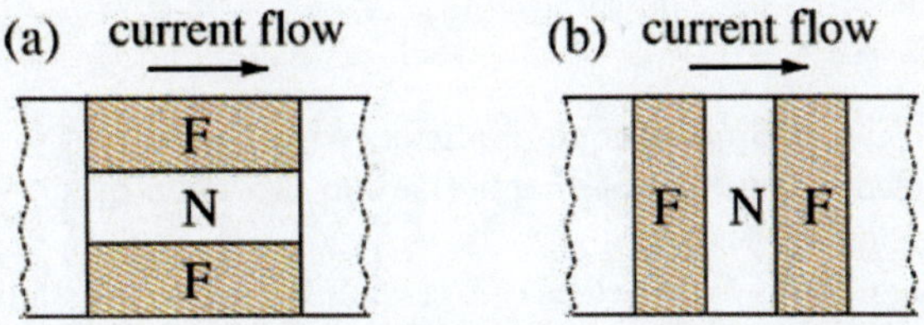

Fig. 11.4. Schematic illustration of (**a**) the current-in-plane (CIP) and (**b**) the current-perpendicular-to-plane (CPP) giant magnetoresistance geometry (see text). Reprinted with permission from [513]

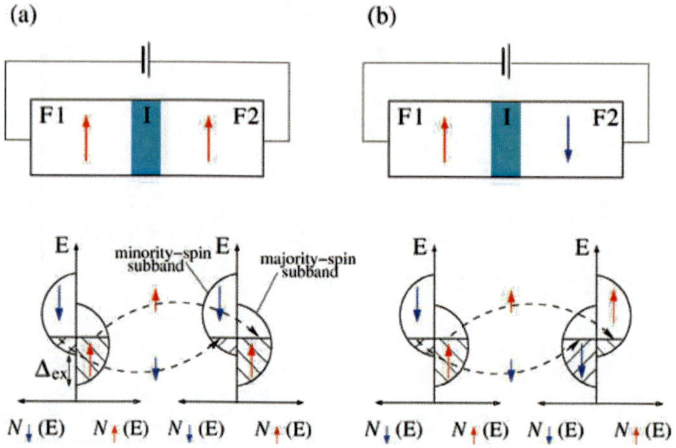

Fig. 11.5. Schematic illustration of electron tunneling in ferromagnet/insulator/ferromagnet (F/I/F) tunnel junctions. (**a**) Parallel and (**b**) antiparallel orientation of magnetizations with the corresponding spin-resolved density of the *d* states in ferromagnetic metals that have exchange spin splitting Δ_{ex}. *Arrows* in the two ferromagnetic regions are determined by the majority-spin subband. *Dashed lines* depict spin-conserving tunneling (see text). Reprinted with permission from [513]

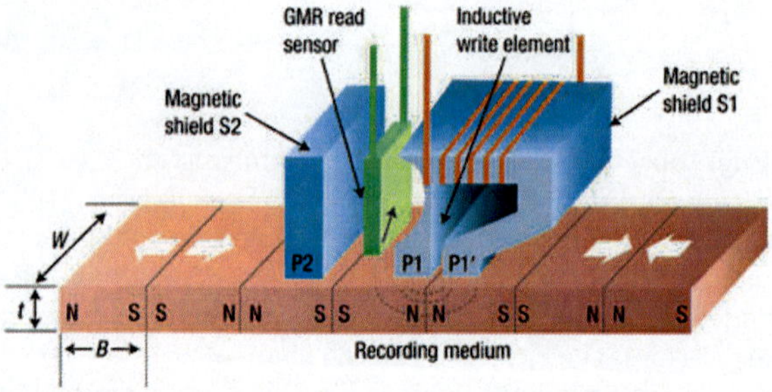

Fig. 11.6. Typical GMR head for hard disks (see text). Reprinted with permission from [515]

(see Fig. 11.5), and Spin Torque Transfer, where a current of spin-polarized electrons is used to control the magnetization direction of ferromagnetic electrodes in the device.

The storage density of hard drives is rapidly increasing along an exponential growth curve, in part because spintronics-enabled devices like GMR and TMR sensors have increased the sensitivity of the read heads which measure the magnetic state of small magnetic domains (bits) on the spinning platter (see Fig. 11.6). The doubling period for the areal density of information storage is 12 months, much shorter than Moore's law, stating that the number

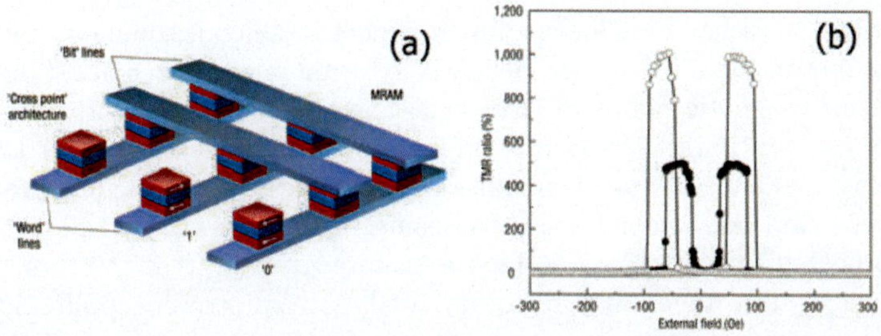

Fig. 11.7. (**a**) Principle of the magnetic random access memory (MRAM) in the basic "cross point" architecture. The binary information "0" and "1" is recorded on the two opposite magnetization orientations of the free layer of magnetic tunnel junctions (MTJs), which are connected to the crossing points of two perpendicular arrays of parallel conducting lines. For writing, current pulses are sent through one line of each array, and only at the crossing point of these lines the resulting magnetic field is high enough to orient the magnetization of the free layer. For reading, one measures the resistance between the two lines connecting the addressed cell. (**b**) High magnetoresistance measured by Lee and co-workers [521] for the magnetic stack $(Co_{25}Fe_{75})_{80}B_{20}$(4 nm)/MgO(2.1 nm)/$(Co_{25}Fe_{75})_{80}B_{20}$(4.3 nm) annealed at 475° C after growth, measured at room temperature (*closed circles*) and low temperature (*open circles*). Reprinted with permission from [515]

of transistors that can cheaply be incorporated in an integrated circuit doubles every 2 years (see Sect. 1.1). A magnetic random access memory (MRAM), schematically shown in Fig. 11.7, uses a grid of magnetic storage elements called magnetic tunnel junctions (MTJs). It is important to stress that MRAM is nonvolatile (unlike charge-based DRAM in today's computers), so information is stored even when power is turned off, potentially providing "instant-on" computing.

In early efforts, spin-polarized electrons were generated via optical orientation induced by circularly polarized photons incident on semiconductors with appreciable spin–orbit interaction (like GaAs and ZnSe). Although electric spin injection can be achieved in metallic systems by simply passing a current through a ferromagnet, the large impedance mismatch between ferromagnetic metals and semiconductors prevents efficient injection across metal/semiconductor interfaces. A solution to this problem is to employ ferromagnetic semiconductor sources, like manganese-doped gallium arsenide (GaMnAs) [522], increasing the interface resistance with a tunnel barrier [523] or using hot-electron injection [524].

Spin detection in semiconductors is another key challenge, which has been met with the following techniques: Faraday/Kerr rotation of transmitted/reflected photons [525], circular-polarization analysis of electroluminescence [526], non-local spin valve (adapted from the work of Johnson and Silsbee with metals) [527], and ballistic spin filtering [528]. The latter technique has been successfully employed to overcome the lack of spin–orbit interaction and other material issues to achieve spin transport in silicon, the most important semiconductor for electronics. Because external magnetic fields (and stray fields from magnetic contacts) can cause large Hall effects and magnetoresistance in semiconductors (which mimic spin-valve effects), so far the only conclusive evidence of spin transport in semiconductors is demonstration of spin precession and decoherence/dephasing in a magnetic field non-collinear to the injected spin orientation. This is usually referred to as the Hanle effect.

Primary advantages of semiconductor-based spintronic applications are potentially lower power use and a smaller footprint than electrical devices used for information processing [529]. In addition, applications such as semiconductor lasers using spin-polarized electric injection have shown threshold current reduction and controllable circularly polarized coherent-light output [530]. Future applications may include spin-based transistors having advantages over MOSFET devices such as steeper sub-threshold slope. Moreover, as pointed out in Chap. 10, the reduced decoherence times of spin degrees of freedom – compared to the charge ones – may play a crucial role for the realization of semiconductor-based quantum information processing.

Part IV

Appendices

A

The Envelope-Function Approximation

In order to provide a general derivation of the envelope-function theory, let us consider a total electronic Hamiltonian given by the sum of the single-electron Hamiltonian $\hat{H}_{1e}$ in (1.6) plus a generic non-periodic potential $\overline{V}(\boldsymbol{r})$. The corresponding energy spectrum is obtained by solving the following eigenvalue equation:

$$\left[\hat{H}_{1e} + \overline{V}(\boldsymbol{r})\right] \psi(\boldsymbol{r}) = \epsilon \psi(\boldsymbol{r}) \; . \tag{A.1}$$

To this aim, it is convenient to employ the usual interaction picture of quantum mechanics (see, e.g., [86]). In particular, for the case of a three-dimensional semiconductor crystal it is convenient to write the unknown total wavefunction $\psi(\boldsymbol{r})$ as a linear combination of the conventional (i.e., three-dimensional) Bloch states in (1.8), i.e.,

$$\psi(\boldsymbol{r}) = \sum_{\boldsymbol{k}\nu} c_{\boldsymbol{k}\nu} \phi_{\boldsymbol{k}\nu}(\boldsymbol{r}) \; . \tag{A.2}$$

Inserting the above expansion into the eigenvalue equation (A.1) and taking into account that the Bloch states $\phi_{\boldsymbol{k}\nu}(\boldsymbol{r})$ are eigenstates of the single-electron Hamiltonian $\hat{H}_{1e}$ (corresponding to the eigenvalues $\epsilon_{\boldsymbol{k}\nu}$), it is easy to derive the following set of coupled linear equations for the unknown coefficients $c_{\boldsymbol{k}\nu}$:

$$\epsilon_{\boldsymbol{k}\nu} c_{\boldsymbol{k}\nu} + \sum_{\boldsymbol{k}'\nu'} \overline{V}_{\boldsymbol{k}\nu,\boldsymbol{k}'\nu'} c_{\boldsymbol{k}'\nu'} = \epsilon c_{\boldsymbol{k}\nu} \; , \tag{A.3}$$

where

$$\overline{V}_{\boldsymbol{k}\nu,\boldsymbol{k}'\nu'} = \int \phi^*_{\boldsymbol{k}\nu}(\boldsymbol{r}) \overline{V}(\boldsymbol{r}) \phi_{\boldsymbol{k}'\nu'}(\boldsymbol{r}) \mathrm{d}^3 r \tag{A.4}$$

are the matrix elements of the potential $\overline{V}$ within our perfect-crystal Bloch basis $\phi_{\boldsymbol{k}\nu}$. By inserting into (A.4) the explicit form of the Bloch states (1.8) and using the spatial lattice periodicity of the functions $u_{\boldsymbol{k}\nu}(\boldsymbol{r})$, the previous matrix elements may also be written as

$$\overline{V}_{\boldsymbol{k}\nu,\boldsymbol{k}'\nu'} = \sum_{\boldsymbol{G}} F_{\boldsymbol{k}\nu,\boldsymbol{k}'\nu'}(\boldsymbol{G})\tilde{V}_{\boldsymbol{k}-\boldsymbol{k}'-\boldsymbol{G}} \tag{A.5}$$

with

$$F_{\boldsymbol{k}\nu,\boldsymbol{k}'\nu'}(\boldsymbol{G}) = \int u^*_{\boldsymbol{k}\nu}(\boldsymbol{r})u_{\boldsymbol{k}'\nu'}(\boldsymbol{r})\mathrm{e}^{-\mathrm{i}\boldsymbol{G}\cdot\boldsymbol{r}}\mathrm{d}^3r \tag{A.6}$$

and

$$\tilde{V}_{\boldsymbol{q}} = \frac{1}{\Omega}\int \overline{V}(\boldsymbol{r})\mathrm{e}^{-\mathrm{i}\boldsymbol{q}\cdot\boldsymbol{r}}\mathrm{d}^3r \ . \tag{A.7}$$

Here, Ω denotes the crystal volume, $\boldsymbol{G}$ are reciprocal-lattice vectors, and the function in (A.7) is simply the Fourier transform of the potential $\overline{V}(\boldsymbol{r})$.

We stress that, so far, no approximation has been introduced; therefore, the eigenvalue equation (A.3) together with the definition of the potential matrix elements in (A.5) is fully equivalent to the original Schrödinger equation (A.1).

As anticipated in Sect. 1.2.2, the envelope-function approximation is a spatial decoupling scheme, valid when the space-scale of the perturbation is much larger than the lattice period. More specifically, this amounts to assuming that the potential $\overline{V}(\boldsymbol{r})$ is a slowly varying function on the atomic scale, which in turn implies that its Fourier transform $\tilde{V}$ in (A.7) will contribute for vanishing q values only, i.e., for $q \ll \frac{\pi}{a_0}$. This distinguished feature allows for a strong simplification of the result in (A.5). Indeed, in this limit the only matrix elements significantly different from 0 correspond to $\boldsymbol{q} = \boldsymbol{k} - \boldsymbol{k}' - \boldsymbol{G} \to 0$, but in this limit, in view of the orthonormality properties of the Bloch states, it is possible to show that

$$F_{\boldsymbol{k}\nu,\boldsymbol{k}'\nu'}(\boldsymbol{G}) \approx \delta_{\nu\nu'}\delta(\boldsymbol{G}) \ . \tag{A.8}$$

By inserting the above approximated result into (A.5), we finally get

$$\overline{V}_{\boldsymbol{k}\nu,\boldsymbol{k}'\nu'} \approx \tilde{V}_{\boldsymbol{k}-\boldsymbol{k}'}\delta_{\nu\nu'} \ . \tag{A.9}$$

This constitutes the central result of the envelope-function approximation: in the presence of a slowly varying potential, its matrix elements within the Bloch basis (i) are diagonal with respect to the band index ν and (ii) are simply given by its Fourier transform $\tilde{V}$ evaluated in $\boldsymbol{k} - \boldsymbol{k}'$. This implies that within this limit (i) all interband coupling terms ($\nu \neq \nu'$) in (A.3) vanish and (ii) for intraband transitions ($\nu = \nu'$) the form of the Bloch functions $u_{\boldsymbol{k}\nu}$ is totally irrelevant; indeed, the same result is obtained by replacing our Bloch basis with conventional plane waves (corresponding to free electrons). From a different perspective, we may also say that the potential variation is so slow that it "feels" just a macroscopic average of the crystalline lattice.

By inserting the approximated version of the potential matrix elements (A.9) into the original set of coupled linear equations (A.3), we have

$$\epsilon_{\boldsymbol{k}\nu}c_{\boldsymbol{k}\nu} + \sum_{\boldsymbol{k}'} \tilde{V}_{\boldsymbol{k}-\boldsymbol{k}'}c_{\boldsymbol{k}'\nu} = \epsilon c_{\boldsymbol{k}\nu} \ . \tag{A.10}$$

As anticipated, within the present approximation scheme all interband coupling terms vanish; thus, from now on, the band index ν plays simply the role of a parameter, and the linear combination of Bloch states in (A.2) reduces to

$$\psi_\nu(\boldsymbol{r}) = \sum_{\boldsymbol{k}} c_{\boldsymbol{k}\nu}\phi_{\boldsymbol{k}\nu}(\boldsymbol{r}) \ . \tag{A.11}$$

For a better understanding of the result in (A.10), let us consider first the vanishing-potential limit ($\tilde{V}_{\boldsymbol{k}-\boldsymbol{k}'} \to 0$), for which the solutions of the linear set of equations (A.10) are simply given by

$$c_{\boldsymbol{k}\nu} = \delta_{\boldsymbol{k}-\boldsymbol{k}_\circ} \ , \qquad \epsilon = \epsilon_{\boldsymbol{k}_\circ\nu} \ . \tag{A.12}$$

Indeed, by inserting the above potential-free solutions into the linear combination (A.11), the usual Bloch state (corresponding to a given wavevector $\boldsymbol{k}_\circ$) is readily recovered. In contrast, in the presence of the potential $\overline{V}$, the unknown wavefunction ψ_ν in (A.11) will also involve $\boldsymbol{k}$ values different from the reference wavevector $\boldsymbol{k}_\circ$. However, due to the slowly varying nature of the perturbation – corresponding to a sharp-peak structure of its Fourier transform $\tilde{V}$ in (A.10) – the corresponding solutions $c_{\boldsymbol{k}\nu}$ will be strongly peaked around the reference wavevector $\boldsymbol{k}_\circ$. It follows that only $\boldsymbol{k}$ values close to $\boldsymbol{k}_\circ$ will contribute significantly to the linear combination (A.11).

A closer inspection of the result in (A.10) shows also the presence of a so-called convolution sum (over $\boldsymbol{k}'$), which suggests to Fourier transform it back to the original real-space coordinates $\boldsymbol{r}$. More specifically, recalling that the coefficients $c_{\boldsymbol{k}\nu}$ will be strongly peaked around the reference wavevector $\boldsymbol{k}_\circ$, it is convenient to perform the inverse Fourier transform of the eigenvalue equation (A.10) with respect to the relative wavevector $\Delta\boldsymbol{k} = \boldsymbol{k} - \boldsymbol{k}_\circ$. By denoting with

$$\overline{\psi}_{\boldsymbol{k}_\circ\nu}(\boldsymbol{r}) = \frac{1}{\sqrt{\Omega}} \sum_{\boldsymbol{k}} c_{\boldsymbol{k}\nu} \mathrm{e}^{\mathrm{i}\Delta\boldsymbol{k}\cdot\boldsymbol{r}} \tag{A.13}$$

the inverse Fourier transform of the coefficients $c_{\boldsymbol{k}\nu}$ with respect to $\Delta\boldsymbol{k}$, and applying the same inverse Fourier transform to (A.10), we get the following Schrödinger-like equation for the effective wavefunction in (A.13):

$$\epsilon_\nu(\boldsymbol{k}_\circ - \mathrm{i}\boldsymbol{\nabla}_{\boldsymbol{r}})\overline{\psi}_{\boldsymbol{k}_\circ\nu}(\boldsymbol{r}) + \overline{V}(\boldsymbol{r})\overline{\psi}_{\boldsymbol{k}_\circ\nu}(\boldsymbol{r}) = \epsilon\overline{\psi}_{\boldsymbol{k}_\circ\nu}(\boldsymbol{r}) \ , \tag{A.14}$$

where $\epsilon_\nu(\boldsymbol{k}_\circ - \mathrm{i}\boldsymbol{\nabla}_{\boldsymbol{r}})$ denotes the operatorial version ($\Delta\boldsymbol{k} \to -\mathrm{i}\boldsymbol{\nabla}_{\boldsymbol{r}}$) of the crystal band structure $\epsilon_{\boldsymbol{k}\nu} \equiv \epsilon_\nu(\boldsymbol{k}_\circ + \Delta\boldsymbol{k})$.

Compared to the original Schrödinger equation (A.1), the effective equation (A.14) involves the potential $\overline{V}$ only. Indeed, the effect of the crystalline periodic potential V in (1.6) is fully expressed via the band structure operator $\epsilon_\nu(\boldsymbol{k}_\circ - \mathrm{i}\boldsymbol{\nabla}_{\boldsymbol{r}})$. As a result, it is possible to solve the original problem in (A.1) starting directly from the knowledge of the perfect-crystal band structure $\epsilon_{\boldsymbol{k}\nu}$ – or from an approximated version of it (see also Sect. 1.2.2) – and solving an extremely simplified Schrödinger-like equation involving the potential $\overline{V}$ only.

Let us now come to the link between the original wavefunction ψ and the effective wavefunction $\overline{\psi}$ in (A.13). More specifically, by inserting the explicit form of the Bloch states (1.8) into the linear combination (A.11) and taking into account that $u_{\boldsymbol{k}\nu}(\boldsymbol{r})$ is a slowly varying function of the wavevector $\boldsymbol{k}$ (compared to $c_{\boldsymbol{k}\nu}$), the latter may be taken out of the sum and evaluated in $\boldsymbol{k}_\circ$:

$$\psi_{\boldsymbol{k}_\circ\nu}(\boldsymbol{r}) = \sum_{\boldsymbol{k}} c_{\boldsymbol{k}\nu} \mathrm{e}^{\mathrm{i}\boldsymbol{k}\cdot\boldsymbol{r}} u_{\boldsymbol{k}\nu}(\boldsymbol{r}) \approx \left(\sum_{\boldsymbol{k}} c_{\boldsymbol{k}\nu} \mathrm{e}^{\mathrm{i}\Delta\boldsymbol{k}\cdot\boldsymbol{r}} \right) \phi_{\boldsymbol{k}_\circ\nu}(\boldsymbol{r}) \ . \tag{A.15}$$

Taking into account that the quantity in parenthesis is just proportional to the effective wavefunction $\overline{\psi}$ in (A.13), we finally get

$$\psi_{\boldsymbol{k}_\circ\nu}(\boldsymbol{r}) \approx \sqrt{\Omega}\,\overline{\psi}_{\boldsymbol{k}_\circ\nu}(\boldsymbol{r}) \phi_{\boldsymbol{k}_\circ\nu}(\boldsymbol{r}) \ . \tag{A.16}$$

This is the fundamental result of the envelope-function approximation: for any given band ν and reference wavevector $\boldsymbol{k}_\circ$, the corresponding global wavefunction ψ is simply given by the original (i.e., unperturbed) Bloch function ϕ multiplied by the slowly varying effective wavefunction $\overline{\psi}$, called "envelope function;" the latter is obtained by solving the effective Schrödinger equation (A.14), also referred to as "envelope-function equation."

B

The U Boundary-Condition Scheme

Aim of this appendix is to provide a general formulation – applicable to Boltzmann as well as Wigner equations – of the U boundary-condition scheme, depicted in Fig. 1.22 for the one-dimensional case. To this end, let us consider a linear equation of the form

$$\frac{\partial F_{\boldsymbol{k}\nu}(\boldsymbol{r})}{\partial t} = \sum_{\boldsymbol{k}'\nu'} \int d^3 r' \mathcal{L}_{\boldsymbol{k}\nu,\boldsymbol{k}'\nu'}(\boldsymbol{r},\boldsymbol{r}') F_{\boldsymbol{k}'\nu'}(\boldsymbol{r}') , \tag{B.1}$$

where $\mathcal{L}$ is a first-order differential operator in $\boldsymbol{r}$ given by a negative diffusion-like term plus a generic local contribution, i.e.,

$$\begin{aligned} \int d^3 r' \mathcal{L}_{\boldsymbol{k}\nu,\boldsymbol{k}'\nu'}(\boldsymbol{r},\boldsymbol{r}') F_{\boldsymbol{k}'\nu'}(\boldsymbol{r}') &= -\delta_{\boldsymbol{k}\nu,\boldsymbol{k}'\nu'} \boldsymbol{v}_{\boldsymbol{k}\nu} \cdot \nabla_{\boldsymbol{r}} F_{\boldsymbol{k}\nu}(\boldsymbol{r}) \\ &+ \mathcal{A}_{\boldsymbol{k}\nu,\boldsymbol{k}'\nu'}(\boldsymbol{r}) F_{\boldsymbol{k}'\nu'}(\boldsymbol{r}) . \end{aligned} \tag{B.2}$$

We stress that the generic linear equation (B.1) – defined over the whole coordinate space $\boldsymbol{r}$ – has exactly the structure of the multiband semiconductor Boltzmann equation (1.67), and in the single-band limit plus effective-mass approximation the latter reduces to the closed-system Wigner equation (4.10).

In order to properly impose to the solution $F_{\boldsymbol{k}\nu}$ the desired value $F^b_{\boldsymbol{k}\nu}$ on the boundaries $\boldsymbol{r}^b$ of the simulated region Ω_d, the crucial step is to add to the generic closed-system equation (B.1) a delta-like boundary contribution of the form

$$\mathcal{B}_{\boldsymbol{k}\nu}(\boldsymbol{r}) = \delta(\boldsymbol{r}-\boldsymbol{r}^b) v^\mathrm{in}_{\boldsymbol{k}\nu} \left(F^b_{\boldsymbol{k}\nu} - F_{\boldsymbol{k}\nu}(\boldsymbol{r}^b) \right) , \tag{B.3}$$

where $v^\mathrm{in}_{\boldsymbol{k}\nu}$ denotes the incoming part of the carrier group velocity $\boldsymbol{v}_{\boldsymbol{k}\nu}$ normal to the boundary surface.[1] The resulting open-system equation is then given by

[1] More precisely, we have $v^\mathrm{in}_{\boldsymbol{k}\nu} = \max(0, -\boldsymbol{v}_{\boldsymbol{k}\nu} \cdot \boldsymbol{n}^b)$, where $\boldsymbol{n}^b$ denotes the unit vector normal to the boundary surface in $\boldsymbol{r}^b$.

$$\frac{\partial F_{\boldsymbol{k}\nu}(\boldsymbol{r})}{\partial t} = \sum_{\boldsymbol{k}'\nu'} \int d^3r' \mathcal{L}_{\boldsymbol{k}\nu,\boldsymbol{k}'\nu'}(\boldsymbol{r},\boldsymbol{r}') F_{\boldsymbol{k}'\nu'}(\boldsymbol{r}') + \mathcal{B}_{\boldsymbol{k}\nu}(\boldsymbol{r}) \; . \tag{B.4}$$

In view of the strictly local action of the boundary term (B.3), within the simulated region Ω_{d} the transport equations (B.1) and (B.4) are fully equivalent. Moreover, assuming a regular behavior of the unknown solution $F_{\boldsymbol{k}\nu}$ as well as of the local term $\mathcal{A}$ in (B.2), in order to fulfill the open-system equation (B.4) on the boundary region $\boldsymbol{r}^b$, the singular (i.e., delta-like) boundary term $\mathcal{B}$ in (B.3) needs to be always equal to zero; this, in turn, translates into the following boundary condition:

$$v^{\mathrm{in}}_{\boldsymbol{k}\nu} \left(F^b_{\boldsymbol{k}\nu} - F_{\boldsymbol{k}\nu}(\boldsymbol{r}^b) \right) = 0 \; . \tag{B.5}$$

At this point, it is crucial to notice that for a carrier exiting the simulated region Ω_{d} the incoming part of its group velocity ($v^{\mathrm{in}}_{\boldsymbol{k}\nu}$) is always equal to zero, and thus the boundary condition (B.5) is automatically fulfilled. In contrast, for incoming carriers, $v^{\mathrm{in}}_{\boldsymbol{k}\nu}$ is always different from zero, and the above boundary condition requires to impose $F_{\boldsymbol{k}\nu}(\boldsymbol{r}^b) = F^b_{\boldsymbol{k}\nu}$. As anticipated in Sect. 1.4, the present first-order boundary-condition scheme corresponds to fixing the value of the carrier distribution $F_{\boldsymbol{k}\nu}$ entering the device region Ω_{d}, which in turn will fix the flux of incoming particles; in contrast, no condition is imposed on the distribution of outgoing carriers. This is fully consistent with a first-order boundary problem, where – for each value of $\boldsymbol{k}\nu$ – we are allowed to specify the value of the unknown solution $F_{\boldsymbol{k}\nu}$ at a given spatial coordinate only.[2]

Compared to the original (closed-system) equation in (B.1), the new equation is intrinsically nonhomogeneous, due to the presence of a boundary source term. In particular, the boundary term (B.3) may be regarded as the sum of a source/injection term

$$S_{\boldsymbol{k}\nu}(\boldsymbol{r}) = \delta(\boldsymbol{r} - \boldsymbol{r}^b) v^{\mathrm{in}}_{\boldsymbol{k}\nu} F^b_{\boldsymbol{k}\nu} \tag{B.6}$$

and of a corresponding (negative) boundary loss term

$$-\Lambda_{\boldsymbol{k}\nu}(\boldsymbol{r}) = -\delta(\boldsymbol{r} - \boldsymbol{r}^b) v^{\mathrm{in}}_{\boldsymbol{k}\nu} F_{\boldsymbol{k}\nu}(\boldsymbol{r}^b) = -\delta(\boldsymbol{r} - \boldsymbol{r}^b) v^{\mathrm{in}}_{\boldsymbol{k}\nu} F_{\boldsymbol{k}\nu}(\boldsymbol{r}) \; . \tag{B.7}$$

As we can see, the incorporation of spatial boundary conditions manifests itself via a source or injection term plus a sort of loss contribution (i.e., proportional to the current value of the carrier distribution function):

$$\mathcal{B}_{\boldsymbol{k}\nu}(\boldsymbol{r}) = S_{\boldsymbol{k}\nu}(\boldsymbol{r}) - \Lambda_{\boldsymbol{k}\nu}(\boldsymbol{r}) \; . \tag{B.8}$$

---

[2] We stress that, in spite of its natural physical interpretation, the choice of fixing the incoming distribution is not unique; indeed, according to the nature of our first-order problem, one could alternatively fix, e.g., the value of the outgoing carrier distribution.

As a final step, it is crucial to realize that the loss contribution Λ in (B.7) may also be expressed as a renormalization of the original (closed-system) linear operator in (B.1) according to

$$-\Lambda_{\boldsymbol{k}\nu}(\boldsymbol{r}) = \sum_{\boldsymbol{k}'\nu'} \int \mathrm{d}^3r' \Delta\mathcal{L}_{\boldsymbol{k}\nu,\boldsymbol{k}'\nu'}(\boldsymbol{r},\boldsymbol{r}') F_{\boldsymbol{k}'\nu'}(\boldsymbol{r}') \tag{B.9}$$

with

$$\Delta\mathcal{L}_{\boldsymbol{k}\nu,\boldsymbol{k}'\nu'}(\boldsymbol{r},\boldsymbol{r}') = -\delta_{\boldsymbol{k}\nu,\boldsymbol{k}'\nu'}\delta(\boldsymbol{r}-\boldsymbol{r}')\Lambda_{\boldsymbol{k}'\nu'}(\boldsymbol{r}') \ . \tag{B.10}$$

Combining the explicit form of the boundary term in (B.8) with the renormalization version of the loss contribution in (B.9), the open-system equation (B.4) may also be formulated as

$$\frac{\partial F_{\boldsymbol{k}\nu}(\boldsymbol{r})}{\partial t} = \sum_{\boldsymbol{k}'\nu'} \int \mathrm{d}^3r' \mathcal{L}^{\mathrm{open}}_{\boldsymbol{k}\nu,\boldsymbol{k}'\nu'}(\boldsymbol{r},\boldsymbol{r}') F_{\boldsymbol{k}'\nu'}(\boldsymbol{r}') + S_{\boldsymbol{k}\nu}(\boldsymbol{r}) \tag{B.11}$$

with

$$\mathcal{L}^{\mathrm{open}}_{\boldsymbol{k}\nu,\boldsymbol{k}'\nu'}(\boldsymbol{r},\boldsymbol{r}') = \mathcal{L}_{\boldsymbol{k}\nu,\boldsymbol{k}'\nu'}(\boldsymbol{r},\boldsymbol{r}') + \Delta\mathcal{L}_{\boldsymbol{k}\nu,\boldsymbol{k}'\nu'}(\boldsymbol{r},\boldsymbol{r}') \ . \tag{B.12}$$

As already pointed out in Chaps. 1 and 4, both within a semiclassical picture (see Sect. 1.4) and within a quantum-mechanical treatment (see Sect. 4.1), the presence of spatial boundary conditions may induce energy dissipation/decoherence phenomena, described by the renormalization $\Delta\mathcal{L}$ to the original (closed-system) operator $\mathcal{L}$ (see (B.12)).

C

Evaluation of the Carrier–Quasiparticle Scattering Superoperator

In this appendix we shall discuss how to derive the explicit form of the conventional Markov superoperator within the global treatment of Sect. 3.3.1 as well as within the reduced or electronic description of Sect. 3.3.2.

Let us start our analysis by evaluating the global $(\lambda\lambda')$ matrix element of the double commutator in (3.32):

$$\begin{aligned}
\Gamma(\hat{\rho})_{\lambda_1\lambda_2} &= -\left[\hat{\mathcal{H}}, \left[\hat{\mathcal{K}}, \hat{\rho}\right]\right]_{\lambda_1\lambda_2} \\
&= -\left(\hat{\mathcal{H}}\hat{\mathcal{K}}\hat{\rho}\right)_{\lambda_1\lambda_2} + \left(\hat{\mathcal{H}}\hat{\rho}\hat{\mathcal{K}}\right)_{\lambda_1\lambda_2} \\
&\quad - \left(\hat{\rho}\hat{\mathcal{K}}\hat{\mathcal{H}}\right)_{\lambda_1\lambda_2} + \left(\hat{\mathcal{K}}\hat{\rho}\hat{\mathcal{H}}\right)_{\lambda_1\lambda_2} \\
&= \left\{\left(\hat{\mathcal{H}}\hat{\rho}\hat{\mathcal{K}}\right)_{\lambda_1\lambda_2} - \left(\hat{\mathcal{H}}\hat{\mathcal{K}}\hat{\rho}\right)_{\lambda_1\lambda_2}\right\} + \mathrm{H.c.}\,, \qquad \text{(C.1)}
\end{aligned}$$

where "H.c." denotes the Hermitian conjugate. As we can see, the Markov superoperator Γ can be regarded as the sum of a positive-like ("in-scattering") and a negative-like ("out-scattering") term; more specifically, the result in (C.1) may also be expressed as

$$\Gamma(\hat{\rho})_{\lambda_1\lambda_2} = \frac{1}{2}\sum_{\lambda_1'\lambda_2'}\left[\mathcal{P}_{\lambda_1\lambda_2,\lambda_1'\lambda_2'}\rho_{\lambda_1'\lambda_2'} - \mathcal{P}^*_{\lambda_1'\lambda_1',\lambda_1\lambda_2'}\rho_{\lambda_2'\lambda_2}\right] + \mathrm{H.c.} \qquad \text{(C.2)}$$

with generalized scattering rates given by

$$\mathcal{P}_{\lambda_1\lambda_2,\lambda_1'\lambda_2'} = 2\mathcal{H}_{\lambda_1\lambda_1'}\mathcal{K}^*_{\lambda_2\lambda_2'}\,. \qquad \text{(C.3)}$$

Let us now derive the reduced or electronic version of the conventional Markov superoperator in (3.88) due to carrier-quasiparticle interaction. As a starting point, we shall expand the double-commutator structure in (3.87); in analogy with the result in (C.1) we have

$$\Gamma^{\mathrm{c}}\left(\hat{\rho}^{\mathrm{c}}\right)=\left\{-\operatorname{tr}\left\{\hat{\mathcal{H}}\hat{\mathcal{K}}\hat{\rho}^{\mathrm{c}}\hat{\rho}^{\circ}\right\}_{\{n_{\boldsymbol{q}}\}}\right.$$
$$\left.+\operatorname{tr}\left\{\hat{\mathcal{H}}\hat{\rho}^{\mathrm{c}}\hat{\rho}^{\circ}\hat{\mathcal{K}}\right\}_{\{n_{\boldsymbol{q}}\}}\right\}+\text{H.c.}\,. \tag{C.4}$$

Let us now focus, e.g., on the first of the two terms. In order to evaluate the trace over the $\boldsymbol{q}$ coordinates, we shall derive the explicit form of the operator $\hat{\mathcal{K}}$. To this end, let us insert into (3.42) the explicit form of the interaction Hamiltonian (3.86):

$$\hat{\mathcal{K}}=\int_{0}^{t-t_0}\mathrm{d}\tau\sum_{\boldsymbol{q}}\mathrm{e}^{\frac{\hat{H}_{\circ}\tau}{\mathrm{i}\hbar}}\left(\hat{\mathcal{H}}_{\boldsymbol{q}}^{-}\hat{b}_{\boldsymbol{q}}+\hat{\mathcal{H}}_{\boldsymbol{q}}^{+}\hat{b}_{\boldsymbol{q}}^{\dagger}\right)\mathrm{e}^{-\frac{\hat{H}_{\circ}\tau}{\mathrm{i}\hbar}}\,. \tag{C.5}$$

Combining the above result with the explicit form of the interaction Hamiltonian in (3.86) we obtain

$$\operatorname{tr}\left\{\hat{\mathcal{H}}\hat{\mathcal{K}}\hat{\rho}^{\mathrm{c}}\hat{\rho}^{\circ}\right\}_{\{n_{\boldsymbol{q}}\}}=\int_{0}^{t-t_0}\mathrm{d}\tau\sum_{\boldsymbol{q}\boldsymbol{q}'}\operatorname{tr}\left\{\left(\hat{\mathcal{H}}_{\boldsymbol{q}}^{-}\hat{b}_{\boldsymbol{q}}+\hat{\mathcal{H}}_{\boldsymbol{q}}^{+}\hat{b}_{\boldsymbol{q}}^{\dagger}\right)\cdot\right.$$
$$\left.\cdot\,\mathrm{e}^{\frac{\hat{H}_{\circ}\tau}{\mathrm{i}\hbar}}\left(\hat{\mathcal{H}}_{\boldsymbol{q}'}^{-}\hat{b}_{\boldsymbol{q}'}+\hat{\mathcal{H}}_{\boldsymbol{q}'}^{+}\hat{b}_{\boldsymbol{q}'}^{\dagger}\right)\mathrm{e}^{-\frac{\hat{H}_{\circ}\tau}{\mathrm{i}\hbar}}\hat{\rho}^{\mathrm{c}}\hat{\rho}^{\circ}\right\}_{\{n_{\boldsymbol{q}}\}}\,. \tag{C.6}$$

By employing the cyclic property of the trace

$$\operatorname{tr}\left\{\hat{A}\hat{B}\hat{C}\right\}=\operatorname{tr}\left\{\hat{C}\hat{A}\hat{B}\right\}=\operatorname{tr}\left\{\hat{B}\hat{C}\hat{A}\right\} \tag{C.7}$$

together with various commutation relations for the bosonic operators, in particular

$$\mathrm{e}^{\frac{\hat{H}_{\circ}^{\mathrm{qp}}\tau}{\mathrm{i}\hbar}}\hat{b}_{\boldsymbol{q}}\mathrm{e}^{-\frac{\hat{H}_{\circ}^{\mathrm{qp}}\tau}{\mathrm{i}\hbar}}=\hat{b}_{\boldsymbol{q}}\mathrm{e}^{-\frac{\epsilon_{\boldsymbol{q}}\tau}{\mathrm{i}\hbar}} \tag{C.8}$$

and

$$\mathrm{e}^{\frac{\hat{H}_{\circ}^{\mathrm{qp}}\tau}{\mathrm{i}\hbar}}\hat{b}_{\boldsymbol{q}}^{\dagger}\mathrm{e}^{-\frac{\hat{H}_{\circ}^{\mathrm{qp}}\tau}{\mathrm{i}\hbar}}=\hat{b}_{\boldsymbol{q}}^{\dagger}\mathrm{e}^{+\frac{\epsilon_{\boldsymbol{q}}\tau}{\mathrm{i}\hbar}} \tag{C.9}$$

as well as

$$\operatorname{tr}\left\{\hat{b}_{\boldsymbol{q}}^{\dagger}\hat{b}_{\boldsymbol{q}'}\hat{\rho}^{\circ}\right\}=N_{\boldsymbol{q}}^{\circ}\delta_{\boldsymbol{q}\boldsymbol{q}'} \tag{C.10}$$

and

$$\operatorname{tr}\left\{\hat{b}_{\boldsymbol{q}}\hat{b}_{\boldsymbol{q}'}^{\dagger}\hat{\rho}^{\circ}\right\}=\left(N_{\boldsymbol{q}}^{\circ}+1\right)\delta_{\boldsymbol{q}\boldsymbol{q}'}\,, \tag{C.11}$$

we finally get

$$\operatorname{tr}\left\{\hat{\mathcal{H}}\hat{\mathcal{K}}\hat{\rho}^{\mathrm{c}}\hat{\rho}^{\circ}\right\}_{\{n_{\boldsymbol{q}}\}}=\sum_{\boldsymbol{q}\pm}\left(N_{\boldsymbol{q}}^{\circ}+\frac{1}{2}\pm\frac{1}{2}\right)\hat{\mathcal{H}}_{\boldsymbol{q}}^{\pm\dagger}\hat{\mathcal{K}}_{\boldsymbol{q}}^{\pm}\hat{\rho}^{\mathrm{c}} \tag{C.12}$$

with

$$\hat{\mathcal{K}}_{\boldsymbol{q}}^{\pm}=\int_{0}^{t-t_0}\mathrm{d}\tau\,\mathrm{e}^{\frac{\hat{H}_{\circ}^{\mathrm{c}}\tau}{\mathrm{i}\hbar}}\hat{\mathcal{H}}_{\boldsymbol{q}}^{\pm}\mathrm{e}^{-\frac{\hat{H}_{\circ}^{\mathrm{c}}\tau}{\mathrm{i}\hbar}}\mathrm{e}^{\pm\frac{\epsilon_{\boldsymbol{q}}\tau}{\mathrm{i}\hbar}}=\hat{\mathcal{K}}_{\boldsymbol{q}}^{\mp\dagger}\,. \tag{C.13}$$

In a similar way, for the second term in (C.4) one gets

$$\mathrm{tr}\left\{\hat{\mathcal{H}}\hat{\rho}^{\mathrm{c}}\hat{\rho}^{\circ}\hat{\mathcal{K}}\right\}_{\{n_{\boldsymbol{q}}\}} = \sum_{\boldsymbol{q}\pm}\left(N^{\circ}_{\boldsymbol{q}} + \frac{1}{2} \pm \frac{1}{2}\right)\hat{\mathcal{H}}^{\pm}_{\boldsymbol{q}}\hat{\rho}^{\mathrm{c}}\hat{\mathcal{K}}^{\pm\dagger}_{\boldsymbol{q}} \,. \tag{C.14}$$

Combining the original definition in (C.4) with the two results in (C.12) and (C.14), it is easy to get the desired electronic superoperator (3.88):

$$\Gamma^{\mathrm{c}}\left(\hat{\rho}^{\mathrm{c}}\right) = -\sum_{\boldsymbol{q}\pm}\left(N^{\circ}_{\boldsymbol{q}} + \frac{1}{2} \pm \frac{1}{2}\right)\left(\hat{\mathcal{H}}^{\pm\dagger}_{\boldsymbol{q}}\hat{\mathcal{K}}^{\pm}_{\boldsymbol{q}}\hat{\rho}^{\mathrm{c}} - \hat{\mathcal{H}}^{\pm}_{\boldsymbol{q}}\hat{\rho}^{\mathrm{c}}\hat{\mathcal{K}}^{\pm\dagger}_{\boldsymbol{q}}\right) + \mathrm{H.c.}\,. \tag{C.15}$$

D

Derivation of the Wigner Transport Equation

In what follows we shall recall the derivation of the well-known Wigner transport equation (4.27) (see, e.g., [85] and references therein). To this aim, we shall consider a fully coherent one-electron system described by the following one-dimensional Hamiltonian:

$$H = -\frac{\hbar^2}{2m^*}\frac{\partial^2}{\partial^2 z} + V(z) \,, \tag{D.1}$$

where the electron band structure is described via the effective mass m^*, while $V(z)$ denotes the device potential profile. For this particular case, the Liouville–von Neumann equation (3.9) written in our one-dimensional real-space representation z comes out to be

$$\frac{\partial \rho(z_1,z_2)}{\partial t} = \frac{1}{\mathrm{i}\hbar}\left\{-\frac{\hbar^2}{2m^*}\left[\frac{\partial^2}{\partial^2 z_1} - \frac{\partial^2}{\partial^2 z_2}\right]\rho(z_1,z_2) + [V(z_1) - V(z_2)]\rho(z_1,z_2)\right\} \,. \tag{D.2}$$

Since our goal is to derive a transport equation within the Wigner formalism, let us move to the usual Weyl–Wigner coordinates:

$$z = \frac{z_1 + z_2}{2} \,, \qquad z' = z_1 - z_2 \,. \tag{D.3}$$

Recalling that

$$\frac{\partial^2}{\partial^2 z_1} - \frac{\partial^2}{\partial^2 z_2} = 2\frac{\partial}{\partial z}\frac{\partial}{\partial z'} \,, \tag{D.4}$$

the Liouville–von Neumann equation (D.2) written in the new coordinates comes out to be

$$\begin{aligned}\frac{\partial}{\partial t}\rho\left(z+\frac{z'}{2}, z-\frac{z'}{2}\right) &= \frac{\mathrm{i}\hbar}{m^*}\frac{\partial}{\partial z}\frac{\partial}{\partial z'}\rho\left(z+\frac{z'}{2}, z-\frac{z'}{2}\right) \\ &+ \frac{1}{\mathrm{i}\hbar}\left[V\left(z+\frac{z'}{2}\right) - V\left(z-\frac{z'}{2}\right)\right]\rho\left(z+\frac{z'}{2}, z-\frac{z'}{2}\right) \,.\end{aligned} \tag{D.5}$$

In order to get the desired Wigner transport equation, the key step is to apply to the above Liouville–von Neumann equation the Weyl–Wigner transform linking the real-space single-particle density matrix to the corresponding Wigner function (see, e.g., [85]):

$$f_{\mathrm{W}}(z,k) = \int \mathrm{d}z' \frac{\mathrm{e}^{-ikz'}}{\sqrt{2\pi}} \rho\left(z+\frac{z'}{2}, z-\frac{z'}{2}\right) . \tag{D.6}$$

The result of such a phase-space transformation is

$$\frac{\partial f_{\mathrm{W}}(z,k)}{\partial t} = \frac{\mathrm{i}\hbar}{m^*}\frac{\partial}{\partial z} \int \mathrm{d}z' \frac{\mathrm{e}^{-ikz'}}{\sqrt{2\pi}} \frac{\partial}{\partial z'} \rho\left(z+\frac{z'}{2}, z-\frac{z'}{2}\right)$$

$$+\frac{1}{\mathrm{i}\hbar}\int \mathrm{d}z' \frac{\mathrm{e}^{-ikz'}}{\sqrt{2\pi}} \left[V\left(z+\frac{z'}{2}\right) - V\left(z-\frac{z'}{2}\right)\right] \rho\left(z+\frac{z'}{2}, z-\frac{z'}{2}\right) . \tag{D.7}$$

As far as the first (kinetic) term is concerned, by performing the integration over z' by parts we have

$$\int \mathrm{d}z' \frac{\mathrm{e}^{-ikz'}}{\sqrt{2\pi}} \frac{\partial}{\partial z'} \rho\left(z+\frac{z'}{2}, z-\frac{z'}{2}\right) = \left[\frac{\mathrm{e}^{-ikz'}}{\sqrt{2\pi}} \rho\left(z+\frac{z'}{2}, z-\frac{z'}{2}\right)\right]_{-\infty}^{+\infty}$$
$$+\, \mathrm{i}k \int \mathrm{d}z' \frac{\mathrm{e}^{-ikz'}}{\sqrt{2\pi}} \rho\left(z+\frac{z'}{2}, z-\frac{z'}{2}\right) . \tag{D.8}$$

At this point, the crucial assumption is to neglect the first contribution, stating that

$$\lim_{z'\to\pm\infty} \rho\left(z+\frac{z'}{2}, z-\frac{z'}{2}\right) = 0 . \tag{D.9}$$

This assumption is usually justified arguing that, due to possible phase-breaking interactions, non-diagonal density-matrix elements $\rho(z_1, z_2)$ corresponding to values of $z' = z_1 - z_2$ much larger than the carrier coherence length may be safely neglected. Employing such approximation scheme and recalling the definition of the Wigner function in (D.6), the integral in (D.8) comes out to be

$$\int \mathrm{d}z' \frac{\mathrm{e}^{-ikz'}}{\sqrt{2\pi}} \frac{\partial}{\partial z'} \rho\left(z+\frac{z'}{2}, z-\frac{z'}{2}\right) = \mathrm{i}k f_{\mathrm{W}}(z,k) . \tag{D.10}$$

Let us now come to the second (potential) term in (D.7). By employing the inverse of the Weyl–Wigner transform (D.6), i.e.,

$$\rho\left(z+\frac{z'}{2}, z-\frac{z'}{2}\right) = \int \mathrm{d}k' \frac{\mathrm{e}^{\mathrm{i}k'z'}}{\sqrt{2\pi}} f_{\mathrm{W}}(z,k') , \tag{D.11}$$

it is possible to rewrite our potential contribution as

$$\frac{1}{\mathrm{i}\hbar}\int \mathrm{d}z' \frac{\mathrm{e}^{-ikz'}}{\sqrt{2\pi}} \left[V\left(z+\frac{z'}{2}\right) - V\left(z-\frac{z'}{2}\right)\right] \rho\left(z+\frac{z'}{2}, z-\frac{z'}{2}\right)$$

$$= -\int \mathrm{d}k' \mathcal{V}(z, k-k') f_{\mathrm{W}}(z, k') \qquad \text{(D.12)}$$

with

$$\mathcal{V}(z, k-k') = \frac{\mathrm{i}}{\hbar}\int \mathrm{d}z' \frac{\mathrm{e}^{-\mathrm{i}(k-k')z'}}{2\pi} \left[V\left(z+\frac{z'}{2}\right) - V\left(z-\frac{z'}{2}\right)\right] . \qquad \text{(D.13)}$$

By inserting the two results (D.10) and (D.12) into the equation of motion (D.7), we finally get the desired Wigner transport equation (4.27)

$$\frac{\partial f_{\mathrm{W}}(z,k)}{\partial t} + v(k)\frac{\partial f_{\mathrm{W}}(z,k)}{\partial z} + \int \mathrm{d}k' \mathcal{V}(z, k-k') f_{\mathrm{W}}(z, k') = 0 \, , \qquad \text{(D.14)}$$

where $v(k) = \frac{\hbar k}{m^*}$ denotes the electron group velocity within the usual effective-mass approximation.

References

1. M. Stutzmann, An introduction to semiconductors (2008), http://www.wsi.tum.de
2. N.W. Ashcroft, N.D. Mermin, *Solid State Physics* (Cengage Learning, London, 1976)
3. J. Ziman, *Principles of the Theory of Solids*, 2nd edn. (Cambridge University Press, Cambridge, 1979)
4. W.A. Harrison, *Solid State Theory* (Dover Publications, New York, NY, 1980)
5. J.S. Blakemore, *Solid State Physics*, 2nd edn. (Cambridge University Press, Cambridge, 1985)
6. C. Kittel, *Quantum Theory of Solids*, 2nd edn. (Wiley, New York, NY, 1987)
7. H.M. Rosenberg, *The Solid State: Introduction to the Physics of Crystals for Students of Physics, Materials Science and Engineering*, 3rd edn. (Oxford University Press, Oxford, 1988)
8. J.R. Hook, H.E. Hall, *Solid State Physics*, 2nd edn. (Wiley, Chichester, 1991)
9. M.N. Rudden, J. Wilson, *Elements of Solid State Physics*, 2nd edn. (Wiley, Chichester, 1993)
10. M.A. Omar, *Elementary Solid State Physics: Principles and Applications* (Addison Wesley, London, 1993)
11. H.P. Myers, *Introductory Solid State Physics*, 2nd edn. (Taylor & Francis, London, 1997)
12. R.J. Turton, *The Physics of Solids* (Oxford University Press, Oxford, 2000)
13. R.E. Peierls, *Quantum Theory of Solids* (Oxford University Press, Oxford, 2001)
14. E. O'Reilly, *Quantum Theory of Solids* (Taylor & Francis, London, 2002)
15. C. Kittel, *Introduction to Solid State Physics*, 8th edn. (Wiley, New York, NY, 2004)
16. L. Kantorovich, *Quantum Theory of the Solid State: An Introduction* (Springer Publishing Map, Dordrecht, 2004)
17. H. Ibach, H. Lüth, *Solid-State Physics*, 3rd edn. (Springer, Berlin Heidelberg, 2006)
18. P. Hofmann, *Solid State Physics* (Wiley-VCH Verlag GmbH, Weinheim, 2008)
19. H.F. Wolf, *Semiconductors* (Wiley, New York, NY, 1971)
20. M.L. Cohen, *Electronic Structure and Optical Properties of Semiconductors*, 2nd edn. (Springer, Berlin, 1989)

21. D.K. Ferry, *Semiconductors* (Prentice Hall, Englewood Cliffs, NJ, 1991)
22. P.Y. Yu, M. Cardona, *Fundamentals of Semiconductors*, 2nd edn. (Springer, Berlin, 1999)
23. C. Hamaguchi, *Basic Semiconductor Physics* (Springer, Berlin Heidelberg, 2004)
24. M. Grundmann, *The Physics of Semiconductors: An Introduction Including Devices and Nanophysics* (Springer, Berlin Heidelberg, 2006)
25. M. Stutzmann, M. Brandt, *Basic Semiconductor Physics* (Wiley-VCH Verlag GmbH, Weinheim, 2009)
26. C. Jacoboni, *Theory of Electron Transport in Semiconductor Materials and Structures* (Springer, Berlin Heidelberg, 2010)
27. J. Singleton, *Band Theory and Electronic Properties of Solids* (Oxford University Press, Oxford, 2001)
28. C. Jacoboni, P. Lugli, *The Monte Carlo Method for Semiconductor Device Simulation* (Springer, Vienna, 1989)
29. G. Nilsson, G. Nelin, Phys. Rev. B **6**, 3777 (1972)
30. W. Weber, Phys. Rev. B **15**, 4789 (1977)
31. D. Strauch, B. Dorner, J. Phys. Condens. Matter **2**, 1457 (1990)
32. K. Leo, *High-Field Transport in Semiconductor Superlattices* (Springer, Berlin Heidelberg, 2003)
33. H. Haug, S.W. Koch, *Quantum Theory of the Optical and Electronic Properties of Semiconductors*, 5th edn. (World Scientific Publishing, Singapore, 2009)
34. G. Bastard, *Wave Mechanics Applied to Semiconductor Heterostructures* (Wiley, New York, NY, 1990)
35. M.J. Kelly, *Low-dimensional Semiconductors: Materials, Physics, Technology, Devices* (Oxford University Press, Oxford, 1995)
36. J.H. Davies, *The Physics of Low-dimensional Semiconductors: An Introduction* (Cambridge University Press, Cambridge, 1997)
37. V.A. Shchukin, N.N. Ledentsov, D. Bimberg, *Epitaxy of Nanostructures* (Springer, Berlin Heidelberg, 2003)
38. G. Cao, *Nanostructures and Nanomaterials: Synthesis, Properties and Applications* (Imperial College Press, London, 2004)
39. O. Manasreh, *Semiconductor Heterojunctions and Nanostructures* (McGraw-Hill Education – Europe, Maidenhead, UK, 2005)
40. C.C. Koch, *Nanostructured Materials: Processing, Properties and Applications*, 2nd edn. (William Andrew Publishing, Norwich, 2006)
41. J. Hermanson, J.C. Phillips, Phys. Rev. **150**, 652 (1966)
42. P. Harrison, *Quantum Wells, Wires and Dots: Theoretical and Computational Physics of Semiconductor Nanostructures*, 2nd edn. (Wiley, Chichester, 2005)
43. A. Shik, *Quantum Wells: Physics and Electronics of Two-dimensional Systems* (World Scientific Publishing, Singapore, 1998)
44. L. Banyai, S.W. Koch, *Semiconductor Quantum Dots* (World Scientific Publishing, Singapore, 1993)
45. R. Turton, *The Quantum Dot: A Journey Into the Future of Microelectronics* (Oxford University Press, Oxford, 1996)
46. D. Bimberg, M. Grundmann, N.N. Ledentskov, *Quantum Dot Heterostructures* (Wiley, Chichester, 1998)
47. L. Jacak, P. Hawrylak, A. Wojs, *Quantum Dots* (Springer, Berlin, 1998)

48. T. Chakraborty, *Quantum Dots: A Survey of the Properties of Artificial Atoms* (Elsevier Science & Technology, Oxford, 1999)
49. V.I. Klimov, *Nanocrystal Quantum Dots* (Taylor & Francis, London, 2009)
50. W. Wegscheider et al., Phys. Rev. Lett. **71**, 4071 (1993)
51. *Semiconductor Macroatoms: Basic Physics and Quantum-Device Applications*, ed. by F. Rossi (Imperial College Press, London, 2005)
52. Y.-C. Chang, L.L. Chang, L. Esaki, Appl. Phys. Lett. **47**, 1324 (1985)
53. L. Pfeiffer et al., Appl. Phys. Lett. **56**, 1697 (1990)
54. M.A. Kastner, Phys. Today **46**, 24 (1993)
55. S. Glutsch, *Excitons in Low-Dimensional Semiconductors: Theory, Numerical Methods, Applications* (Springer, Berlin Heidelberg, 2004)
56. *Handbook of Semiconductor Nanostructures and Nanodevices*, vols. 1–5, ed. by A.A. Balandin, K.L. Wang (American Scientific Publishers, Los Angeles, CA, 2005)
57. *Heterojunction Band Discontinuities: Physics and Device Applications*, ed. by F. Capasso, G. Margaritondo (Elsevier Science & Technology, Oxford, 1989)
58. F. Capasso, *Physics of Quantum Electron Devices* (Springer, Berlin Heidelberg, 1990)
59. *Semiconductor Quantum Wells and Superlattices for Long-wavelength Infrared Detectors*, ed. by M.O. Manasreh (Artech House, London, 1993)
60. K.K. Choi, *Physics of Quantum Well Infrared Photodetectors* (World Scientific Publishing, Singapore, 1997)
61. *Intersubband Transitions in Quantum Wells: Physics and Devices*, ed. by S.S. Li, Y.-K. Su (Springer, Boston, MA, 1998)
62. *Intersubband Transitions in Quantum Wells: Physics and Device Applications I*, ed. by H.C. Liu, F. Capasso (Academic Press, London, 1999)
63. *Intersubband Transitions in Quantum Wells: Physics and Device Applications II*, ed. by G. Mueller (Academic Press, London, 1999)
64. H. Schneider, H.C. Liu, *Quantum Well Infrared Photodetectors* (Springer, Berlin Heidelberg, 2006)
65. R. Paiella, *Intersubband Transitions In Quantum Structures* (McGraw-Hill Education – Europe, Maidenhead, UK, 2008)
66. F. Capasso, *Bandgap Engineering: the Physics of Heterostructure Semiconductor Devices* (Wiley, Chichester, 2009)
67. P.K. Bhattacharya, *Semiconductor Optoelectronic Devices*, 2nd edn. (Pearson Higher Education, Upper Saddle River, NJ, 1996)
68. B.R. Nag, *Physics of Quantum Well Devices* (Springer Publishing Map, Dordrecht, 2001)
69. *Semiconductor Quantum Bits*, ed. by O. Benson, F. Henneberger (Pan Stanford Publishing, Singapore, 2007)
70. F. Rossi, E. Molinari, Phys. Rev. Lett. **76**, 3642 (1996)
71. F. Rossi, E. Molinari, Phys. Rev. B **53**, 16462 (1996)
72. C. Cercignani, *The Boltzmann Equation and Its Applications* (Springer, New York, NY, 1987)
73. D. Zubarev, V. Morozov, G. Ropke, *Statistical Mechanics of Nonequilibrium Processes: Basic Concepts, Kinetic Theory*, vol. 1 (Wiley, Weinheim, 1996)
74. A.I. Burshtein, *Introduction to Thermodynamics and Kinetic Theory of Matter*, 2nd edn. (Wiley-VCH Verlag GmbH, Weinheim, 2005)

75. C. Cercignani, *Ludwig Boltzmann: The Man Who Trusted Atoms* (Oxford University Press, Oxford, 2006)
76. L. Boltzmann, *Lectures on Gas Theory* (Dover Publications, New York, NY, 1995)
77. D.K. Ferry, *Semiconductor Transport* (Taylor & Francis, London, 2000)
78. A. Jüngel, *Transport Equations for Semiconductors* (Springer, Berlin Heidelberg, 2009)
79. S. Selberherr, *Analysis and Simulation of Semiconductor Devices* (Springer, New York, NY, 1984)
80. R. Proietti Zaccaria, F. Rossi, Phys. Rev. B **67**, 113311 (2003)
81. D.A. Neamen, *Semiconductor Physics and Devices*, 3rd edn. (McGraw-Hill, London, 2002)
82. K. Seeger, *Semiconductor Physics: An Introduction*, 9th edn. (Springer, Berlin Heidelberg, 2004)
83. J.-P. Colinge, *Physics of Semiconductor Devices* (Springer, Boston, MA, 2005)
84. S.M. Sze, *Physics of Semiconductor Devices*, 3rd edn. (Wiley, New York, NY, 2006)
85. W.R. Frensley, Rev. Mod. Phys. **62**, 745 (1990)
86. J.J. Sakurai, *Modern Quantum Mechanics*, 2nd edn. (Prentice Hall, Englewood Cliffs, NJ, 1993)
87. E. Ciancio, R.C. Iotti, F. Rossi, Europhys. Lett. **65**, 242 (2004)
88. E. Ciancio, R.C. Iotti, F. Rossi, Phys. Rev. B **69**, 165319 (2004)
89. W. Magnus, W. Schoenmaker, *Quantum Transport in Submicron Devices: A Theoretical Introduction* (Springer, Berlin Heidelberg, 2002)
90. J. Shah, *Ultrafast Spectroscopy of Semiconductors and Semiconductor Nanostructures*, 2nd edn. (Springer, Berlin Heidelberg, 1999)
91. A.M. Weiner, *Ultrafast Optics* (Wiley, Chichester, 2008)
92. E. Glezer, *Ultrafast Optics: Principles and Techniques* (Wiley, Chichester, 2009)
93. *Quantum Transport in Semiconductors*, ed. by D.K. Ferry, C. Jacoboni (Springer, Boston, MA, 1992)
94. *Hot Carriers in Semiconductor Nanostructures: Physics and Applications*, ed. by J. Shah (Academic Press, Boston, MA, 1992)
95. *Quantum Transport in Ultrasmall Devices*, ed. by D.K. Ferry, H.L. Grubin, C. Jacoboni, A.-P. Jauho (Springer, Boston, MA, 1995)
96. S. Datta, *Electronic Transport in Mesoscopic Systems* (Cambridge University Press, Cambridge, 1997)
97. Y. Fu, M. Willander, *Physical Models of Semiconductor Quantum Devices* (Springer, Boston, MA, 1999)
98. K. Goser, *Nanoelectronics and Nanosystems: From Transistors to Molecular and Quantum Devices* (Springer, Berlin Heidelberg, 2003)
99. T. Ihn, *Electronic Quantum Transport in Mesoscopic Semiconductor Structures* (Springer, New York, NY, 2004)
100. J. Rammer, *Quantum Transport Theory* (The Perseus Books Group, Boulder, CO, 2004)
101. S. Datta, *Quantum Transport: Atom to Transistor* (Cambridge University Press, Cambridge, 2005)

102. H. Haug, A.-P. Jauho, *Quantum Kinetics in Transport and Optics of Semiconductors*, 2nd edn. (Springer, Berlin Heidelberg, 2007)
103. F.A. Buot, *Quantum Transport Physics in Mesoscopic Systems and Nanodevices: Foundation of Computational Nonequilibrium Physics in Nanoscience and Nanotechnology* (World Scientific Publishing, Singapore, 2007)
104. T. Ouisse, *Electron Transport in Nanostructures and Mesoscopic Devices* (ISTE and Wiley, London, 2008)
105. D.K. Ferry, S.M. Goodnick, J. Bird, *Transport in Nanostructures*, 2nd edn. (Cambridge University Press, Cambridge, 2009)
106. L. Allen, J.H. Eberly, *Optical Resonance and Two-level Atoms* (Dover Publications, New York, NY, 1988)
107. F. Rossi, Topical review on: coherent phenomena in semiconductors. Semicond. Sci. Technol. **13**, 147 (1998)
108. R. Brunetti, C. Jacoboni, F. Rossi, Phys. Rev. B **39**, 10781 (1989)
109. C. Fürst, A. Leitenstorfer, A. Laubereau, R. Zimmermann, Phys. Rev. Lett. **78**, 3733 (1997)
110. F. Rossi, T. Kuhn, Rev. Mod. Phys. **74**, 895 (2002)
111. F. Bloch, Phys. Rev. **70**, 460 (1946)
112. M. Flanagan, *Density Matrix Theory and Applications* (Kluwer Academic Publishers Group, Dordrecht, 1981)
113. K. Blum, *Density Matrix Theory and Applications*, 2nd edn. (Springer, Boston, MA, 1996)
114. P. Meystre, M. Sargent III, *Elements of Quantum Optics*, 4th edn. (Springer, Berlin Heidelberg, 2007)
115. T. Meier, P. Thomas, S.W. Koch, *Coherent Semiconductor Optics: From Basic Concepts to Nanostructure Applications* (Springer, Berlin Heidelberg, 2006)
116. E. Wigner, Phys. Rev. **40**, 749 (1932)
117. O. Hess, T. Kuhn, Phys. Rev. A **54**, 3347 (1996)
118. J. Shah, R.C. Leite, Phys. Rev. Lett. **22**, 1304 (1969)
119. R.G. Ulbrich, Solid State Electron. **21**, 51 (1978)
120. C.V. Shank, R.L. Fork, R.F. Leheny, J. Shah, Phys. Rev. Lett. **42**, 112 (1979)
121. P. Lugli, S.M. Goodnick, Phys. Rev. Lett. **59**, 716 (1987)
122. D. von der Linde, J. Kuhl, H. Klingenberg, Phys. Rev. Lett. **44**, 1505 (1980)
123. J. Shah, Phys. Rev. Lett. **59**, 2222 (1987)
124. S. Schmitt-Rink, D.S. Chemla, Phys. Rev. Lett. **57**, 2752 (1986)
125. E.L. Hahn, Phys. Rev. **80**, 580 (1950)
126. N.A. Kurnit, I.D. Abella, S.R. Hartmann, Phys. Rev. Lett. **13**, 567 (1964)
127. L. Schultheis, M.D. Sturge, Appl. Phys. Lett. **47**, 995 (1985)
128. P.C. Becker et al., Phys. Rev. Lett. **61**, 1647 (1988)
129. E.O. Göbel et al., Phys. Rev. Lett. **64**, 1801 (1990)
130. R.W. Schoenlein et al., Phys. Rev. Lett. **70**, 1014 (1993)
131. L. Bányai et al., Phys. Rev. Lett. **75**, 2188 (1995)
132. M. Joschko et al., Phys. Rev. Lett. **78**, 737 (1997)
133. H.G. Roskos et al., Phys. Rev. Lett. **68**, 2216 (1992)
134. A. Leitenstorfer et al., Phys. Rev. Lett. **82**, 5140 (1999)
135. A.P. Heberle, J.J. Baumberg, K. Köhler, Phys. Rev. Lett. **75**, 2598 (1995)
136. R. Atanasov et al., Phys. Rev. Lett. **76**, 1703 (1996)

137. *Coherent Control in Atoms, Molecules, and Semiconductors*, ed. by W. Pötz, W.A. Schroeder (Springer Publishing Map, Dordrecht, 1999)
138. G.C. Cho, W. Kütt, H. Kurz, Phys. Rev. Lett. **65**, 764 (1990)
139. T. Spiller, H.-K. Lo, S. Popescu, *Introduction to Quantum Computation and Information* (World Scientific Publishing, Singapore, 1998)
140. M.A. Nielsen, I.L. Chuang, *Quantum Computation and Quantum Information* (Cambridge University Press, Cambridge, 2000)
141. G. Benenti, G. Casati, G. Strini, *Principles of Quantum Computation and Information: Basic Concepts* (World Scientific Publishing, Singapore, 2004)
142. I. Burda, *Introduction to Quantum Computation* (Universal Publishers, Boca Raton, FL, 2005)
143. M. Hayashi, *Quantum Information: An Introduction* (Springer, Berlin Heidelberg, 2006)
144. M. Bellac, *A Short Introduction to Quantum Information and Quantum Computation* (Cambridge University Press, Cambridge, 2006)
145. G. Jaeger, *Quantum Information: An Overview* (Springer, New York, NY, 2006)
146. L. Diósi, *A Short Course in Quantum Information Theory: An Approach From Theoretical Physics* (Springer, Berlin Heidelberg, 2006)
147. V. Vedral, *Introduction to Quantum Information Science* (Oxford University Press, Oxford, 2006)
148. P. Kaye, R. Laflamme, M. Mosca, *An Introduction to Quantum Computing* (Oxford University Press, Oxford, 2006)
149. N.D. Mermin, *Quantum Computer Science: An Introduction* (Cambridge University Press, Cambridge, 2007)
150. T. Spiller, B. Munro, *Introduction to Quantum Information Processing (QIP)* (William Andrew Publishing, Norwich, 2009)
151. C. Bennett, D.P. DiVincenzo, W.K. Wootters, *Quantum Information Theory* (Springer, New York, NY, 2009)
152. S.M. Barnett, *Quantum Information* (Oxford University Press, Oxford, 2009)
153. A. Messiah, *Quantum Mechanics*, vol. 1 (Dover Publications, New York, NY, 2003)
154. J.C. Inkson, *Many-Body Theory of Solids: An Introduction* (Plenum Publishing, New York, NY, 1984)
155. M. Toda, R. Kubo, N. Saito, *Statistical Physics I: Equilibrium Statistical Mechanics*, 2nd edn. (Springer, Berlin Heidelberg, 1998)
156. R. Kubo, M. Toda, N. Hashitsume, *Statistical Physics II: Nonequilibrium Statistical Mechanics*, 2nd edn. (Springer, Berlin Heidelberg, 1998)
157. L.P. Kadanoff, G. Baym, *Quantum Statistical Mechanics* (Benjamin, New York, NY, 1962)
158. L.V. Keldysh, Sov. Phys. JETP **20**, 1018 (1965)
159. N.N. Bogoliubov, *Lectures on Quantum Statistics* (Gordon and Breach, New York, NY, 1967)
160. V.M. Axt, S. Mukamel, Rev. Mod. Phys. **70**, 145 (1998)
161. W. Quade, E. Schöll, F. Rossi, C. Jacoboni, Phys. Rev. B **50**, 7398 (1994)
162. R. Brunetti, C. Jacoboni, P.J. Price, Phys. Rev. B **50**, 11872 (1994)
163. R. Brunetti, C. Jacoboni, Phys. Rev. B **57**, 1723 (1998)
164. M. Pascoli, P. Bordone, R. Brunetti, C. Jacoboni, Phys. Rev. B **58**, 3503 (1998)

165. P. Bordone et al., Phys. Rev. B **59**, 3060 (1999)
166. C. Jacoboni, R. Brunetti, S. Monastra, Phys. Rev. B **68**, 125205 (2003)
167. E. Cancellieri, P. Bordone, C. Jacoboni, Phys. Rev. B **76**, 214301 (2007)
168. R.C. Iotti, E. Ciancio, F. Rossi, Phys. Rev. B **72**, 125347 (2005)
169. G. Lindblad, Commun. Math. Phys. **48**, 119 (1976)
170. M.O. Scully, M.S. Zubairy, *Quantum Optics* (Cambridge University Press, Cambridge, 1997)
171. E.B. Davies, *Quantum Theory of Open Systems* (Academic Press, London, 1976)
172. D. Taj, F. Rossi, Phys. Rev. A **78**, 052113 (2008)
173. D. Taj, R.C. Iotti, F. Rossi, Semicond. Sci. Technol. **24**, 065004 (2009)
174. D. Taj, R.C. Iotti, F. Rossi, Eur. Phys. J. B **72**, 305 (2009)
175. F. Bloch, Z. Phys. **52**, 555 (1928)
176. G. Nenciu, Rev. Mod. Phys. **63**, 91 (1991)
177. W.V. Houston, Phys. Rev. **57**, 184 (1940)
178. G.H. Wannier, Rev. Mod. Phys. **34**, 645 (1962)
179. F. Rossi, in *Theory of Transport Properties of Semiconductor Nanostructures*, ed. by E. Schöll (Springer, Boston, 1997)
180. T. Meier, F. Rossi, P. Thomas, S.W. Koch, Phys. Rev. Lett. **75**, 2558 (1995)
181. B. Rosam et al., Phys. Rev. Lett. **86**, 1307 (2001)
182. I.B. Levinson, Y. Yasevichyute, Sov. Phys. JETP **35**, 991 (1972)
183. J.R. Barker, D.K. Ferry, Phys. Rev. Lett. **42**, 1779 (1979)
184. R. Bertoncini, A.-P. Jauho, Phys. Rev. B **44**, 3655 (1991)
185. J. Hader et al., Phys. Rev. B **55**, 13799 (1997)
186. L. Reggiani, P. Lugli, A.-P. Jauho, Phys. Rev. B **36**, 6602 (1987)
187. R. Dümcke, H. Spohn, Z. Phys. B **34**, 419 (1979)
188. E.B. Davies, Comm. Math. Phys. **39**, 91 (1974)
189. R. Alicki, Int. J. Theor. Phys. **16**, 351 (1977)
190. H. Spohn, Rev. Mod. Phys. **52**, 569 (1980)
191. Y.S. Kim, M.E. Noz, *Phase Space Picture of Quantum Mechanics* (World Scientific Publishing, Singapore, 1991)
192. *Open Quantum Systems*, ed. by S. Attal, A. Joye, C.A. Pillet (Springer, Berlin Heidelberg, 2006)
193. E.B. Davies, Math. Ann. **219**, 147 (1976)
194. K. Kraus, Ann. Phys. **64**, 311 (1971)
195. F. Rossi, A. Di Carlo, P. Lugli, Phys. Rev. Lett. **80**, 3348 (1998)
196. M.V. Fischetti, Phys. Rev. B **59**, 4901 (1999)
197. A. Di Carlo, P. Vogl, W. Pötz, Phys. Rev. B **50**, 8358 (1994)
198. W.R. Frensley, Phys. Rev. Lett. **57**, 2853 (1986)
199. W.R. Frensley, Phys. Rev. B **36**, 1570 (1987)
200. A.M. Kriman, N.C. Kluksdahl, D.K. Ferry, Phys. Rev. B **36**, 5953 (1987)
201. N.C. Kluksdahl, A.M. Kriman, D.K. Ferry, C. Ringhofer, Phys. Rev. B **39**, 7720 (1989)
202. F.A. Buot, K.L. Jensen, Phys. Rev. B **42**, 9429 (1990)
203. D.R. Miller, D.P. Neikirk, Appl. Phys. Lett. **58**, 2803 (1991)
204. M.J. McLennan, Y. Lee, S. Datta, Phys. Rev. B **43**, 13846 (1991)
205. H.C. Tso, N.J. Horing, Phys. Rev. B **44**, 11358 (1991)
206. D.K. Ferry, J.-R. Zhou, Phys. Rev. B **48**, 7944 (1993)
207. K.K. Gullapalli, D.R. Miller, D.P. Neikirk, Phys. Rev. B **49**, 2622 (1994)

208. C.L. Fernando, W.R. Frensley, Phys. Rev. B **52**, 5092 (1995)
209. K.-Y. Kim, B. Lee, Phys. Rev. B **64**, 115304 (2001)
210. M. Nedjalkov et al., Phys. Rev. B **70**, 115319 (2004)
211. M. Nedjalkov et al., Phys. Rev. B **74**, 035311 (2006)
212. D. Querlioz, J. Saint-Martin, A. Bournel, P. Dollfus, Phys. Rev. B **78**, 165306 (2008)
213. D. Taj, L. Genovese, F. Rossi, Europhys. Lett. **74**, 1060 (2006)
214. R.P. Feynman, A.R. Hibbs, *Quantum Mechanics and Path Integrals* (McGraw-Hill, New York, NY, 1965)
215. D. Taj, F. Rossi, Phys. Stat. Sol. C **5**, 66 (2008)
216. R. Proietti Zaccaria, R.C. Iotti, F. Rossi, Appl. Phys. Lett. **84**, 139 (2004)
217. R. Proietti Zaccaria, E. Ciancio, R.C. Iotti, F. Rossi, Phys. Rev. B **70**, 195311 (2004)
218. O.C. Zienkiewicz, *Introductory Lectures on the Finite Element Method* (Springer, Vienna, 1972)
219. C.S. Desai, *Elementary Finite Element Method* (Prentice Hall, Englewood Cliffs, NJ, 1979)
220. D.W. Pepper, J.C. Heinrich, *Finite Element Method: Basic Concepts and Applications* (Taylor & Francis, Washington, DC, 1992)
221. O.C. Zienkiewicz, *The Finite Element Method*, 5th edn. (Wiley, New York, NY, 2001)
222. I. Babuska, T. Strouboulis, *The Finite Element Method and its Reliability* (Oxford University Press, Oxford, 2001)
223. E.G. Thompson, *Introduction to the Finite Element Method: Theory, Programming and Applications* (Wiley, New York, NY, 2004)
224. M.R. Gosz, *Finite Element Method* (Taylor & Francis, Boca Raton, FL, 2005)
225. P. Solin, *Partial Differential Equations and the Finite Element Method* (Wiley, New York, NY, 2005)
226. W. Kuntjoro, *An Introduction to the Finite Element Method* (McGraw-Hill Education – Europe, Maidenhead, UK, 2006)
227. M.S. Gockenbach, *Understanding and Implementing the Finite Element Method* (Society for Industrial & Applied Mathematics, New York, NY, 2008)
228. J.O. Dow, *A Unified Approach to the Finite Element Method and Error Analysis Procedures* (Academic Press, London, 2009)
229. P. Tong, J.N. Rossettos, *Finite Element Method: Basic Technique and Implementation* (Dover Publications, New York, NY, 2009)
230. C. Johnson, *Numerical Solution of Partial Differential Equations by the Finite Element Method* (Dover Publications, New York, NY, 2009)
231. J.N. Reddy, *An Introduction to the Finite Element Method*, 3rd edn. (McGraw-Hill Education – Europe, Maidenhead, UK, 2009)
232. A.R. Mitchell, D.F. Griffiths, *The Finite Difference Method in Partial Differential Equations* (Wiley, Chichester, 1980)
233. *Handbook of Numerical Analysis: Finite Difference Methods*, vol. 1, ed. by P.G. Ciarlet, J.L. Lions (Elsevier Science & Technology, Oxford, 1990)
234. R.E. Mickens, *Nonstandard Finite Difference Models of Differential Equations* (World Scientific Publishing, Singapore, 1993)
235. E. Suli, *Analysis of Finite Difference Schemes* (Kluwer Academic Publishers Group, Dordrecht, 1996)

236. J.W. Thomas, *Numerical Partial Differential Equations: Finite Difference Methods* (Springer Publishing Map, New York, NY, 1998)
237. R.E. Mickens, *Applications of Nonstandard Finite Difference Schemes* (World Scientific Publishing, Singapore, 2000)
238. R.J. LeVeque, *Finite Difference Methods for Ordinary and Partial Differential Equations: Steady-state and Time-dependent Problems* (Society for Industrial & Applied Mathematics, New York, NY, 2007)
239. J.C. Strikwerda, *Finite Difference Schemes and Partial Differential Equations*, 2nd edn. (Society for Industrial & Applied Mathematics, New York, NY, 2007)
240. S. Barbieri, F. Beltram, F. Rossi, Phys. Rev. B **60**, 1953 (1999)
241. J.C. Butcher, *Numerical Methods for Ordinary Differential Equations*, 2nd edn. (Wiley, Chicester, 2008)
242. J.M. Hammersley, D.C. Handscomb, *Monte Carlo Methods* (Methuen, London, 1964)
243. I.M. Sobol, *The Monte Carlo Method* (Mir, Moscow, 1975)
244. K.K. Sabelfeld, *Monte Carlo Methods in Boundary Value Problems* (Springer, Berlin Heidelberg, 1991)
245. I.M. Sobol, *A Primer for the Monte Carlo Method* (Taylor & Francis, Boca Raton, FL, 1994)
246. C.P. Robert, G.C. Casella, *Monte Carlo Statistical Methods* (Springer, New York, NY, 1999)
247. G.S. Fishman, *Monte Carlo: Concepts, Algorithms, and Applications* (Springer, New York, NY, 2003)
248. J.E. Gentle, *Random Number Generation and Monte Carlo Methods*, 2nd edn. (Springer Publishing Map, New York, NY, 2004)
249. G.S. Fishman, *First Course in Monte Carlo Simulation* (Cengage Learning, New York, NY, 2005)
250. J.S. Liu, *Monte Carlo Strategies in Scientific Computing* (Springer Publishing Map, New York, NY, 2008)
251. R.Y. Rubinstein, *Simulation and the Monte Carlo Method*, 2nd edn. (Wiley, Chicester, 2008)
252. M.H. Kalos, P.A. Whitlock, *Monte Carlo Methods*, 2nd edn. (Wiley-VCH Verlag GmbH, Weinheim, 2008)
253. C. Lemieux, *Monte Carlo and Quasi-Monte Carlo Sampling* (Springer Publishing Map, New York, NY, 2009)
254. C. Rebbi, *Lattice Gauge Theories and Monte Carlo Simulations* (World Scientific Publishing, Singapore, 1983)
255. W.A. Lester, B.L. Hammond, P.J. Reynolds, *Monte Carlo Methods in Ab Initio Quantum Chemistry* (World Scientific Publishing, Singapore, 1991)
256. R. Billinton, W. Li, *Reliability Assessment of Electrical Power Systems Using Monte Carlo Methods* (Springer, Boston, MA, 1994)
257. G. Winkler, *Image Analysis, Random Fields and Dynamic Monte Carlo Methods: A Mathematical Introduction* (Springer, Berlin Heidelberg, 1994)
258. D.C. Joy, *Monte Carlo Modeling for Electron Microscopy and Microanalysis* (Oxford University Press, New York, NY, 1995)
259. S.A. Dupree, S.K. Fraley, *A Monte Carlo Primer: A Practical Approach to Radiation Transport* (Springer Publishing Map, Boston, MA, 2001)
260. H. Zaidi, G. Sgouros, *Therapeutic Applications of Monte Carlo Calculations in Nuclear Medicine* (Taylor & Francis, London, 2002)

261. D.L. McLeish, *Monte Carlo Simulation and Finance* (Wiley, New York, NY, 2005)
262. D. Gamerman, *Markov Chain Monte Carlo: Stochastic Simulation for Bayesian Inference*, 2nd edn. (Taylor & Francis, London, 2006)
263. B.F.J. Manly, *Randomization, Bootstrap and Monte Carlo Methods in Biology*, 3rd edn. (Taylor & Francis, London, 2006)
264. C. Lim, J. Nebus, *Vorticity, Statistical Mechanics, and Monte Carlo Simulation* (Springer Publishing Map, New York, NY, 2006)
265. J.S. Dagpunar, *Simulation and Monte Carlo: With Applications in Finance and MCMC* (Wiley, Chichester, 2007)
266. J.B. Anderson, *Quantum Monte Carlo: Origins, Development, Applications* (Oxford University Press, New York, NY, 2007)
267. J.J. Buckley, L.J. Jowers, *Monte Carlo Methods in Fuzzy Optimization* (Springer Publishing Map, Berlin Heidelberg, 2008)
268. M.N.O. Sadiku, *Monte Carlo Methods for Electromagnetics* (Taylor & Francis, Boca Raton, FL, 2009)
269. D.P. Landau, K. Binder, *A Guide to Monte Carlo Simulations in Statistical Physics*, 3rd edn. (Cambridge University Press, Cambridge, 2009)
270. R. Korn, E. Korn, G. Kroisandt, *Monte Carlo Methods and Models in Finance and Insurance* (Taylor & Francis, Boca Raton, FL, 2009)
271. K. Binder, D.W. Heermann, *Monte Carlo Simulation in Statistical Physics*, 5th edn. (Springer, Berlin Heidelberg, 2010)
272. H.D. Rees, Phys. Lett. A **26**, 416 (1968)
273. M. Reed, B. Simon, *Functional Analysis* (Elsevier Science, Oxford, UK, 1981)
274. F. Rossi, P. Poli, C. Jacoboni, Topical review on: Weighted Monte Carlo approach to electron transport in semiconductors. Semicond. Sci. Technol. **7**, 1017 (1992)
275. R.G. Chambers, Proc. Phys. Soc. A **65**, 458 (1952)
276. T. Kuhn, F. Rossi, Phys. Rev. Lett. **69**, 977 (1992)
277. T. Kuhn, F. Rossi, Phys. Rev. B **46**, 7496 (1992)
278. F. Rossi, S. Haas, T. Kuhn, Phys. Rev. Lett. **72**, 152 (1994)
279. S. Haas, F. Rossi, T. Kuhn, Phys. Rev. B **53**, 12855 (1996)
280. R.C. Iotti, F. Rossi, Rep. Prog. Phys. **68**, 2533 (2005)
281. A. Lohner et al., Phys. Rev. Lett. **71**, 77 (1993)
282. A. Leitenstorfer et al., Phys. Rev. B **49**, 16372 (1994)
283. A. Leitenstorfer et al., Phys. Rev. Lett. **73**, 1687 (1994)
284. A. Leitenstorfer et al., Phys. Rev. B **53**, 9876 (1996)
285. F. Rossi et al., Phys. Rev. B **51**, 16943 (1995)
286. S. Portolan et al., Phys. Rev. A **73**, 020101 (2006)
287. S. Portolan et al., Phys. Rev. B **77**, 035433 (2008)
288. S. Portolan et al., Phys. Rev. B **77**, 195305 (2008)
289. Y.-S. Lee, *Principles of Terahertz Science and Technology* (Springer, Boston, MA, 2008)
290. E. Bründermann, H.W. Hübers, M. Kimmitt, *Terahertz Techniques* (Springer, Berlin Heidelberg, 2009)
291. W. Knap et al., Appl. Phys. Lett. **80**, 3433 (2002)
292. W. Knap et al., Appl. Phys. Lett. **81**, 4637 (2002)
293. E.A. Shaner et al., Appl. Phys. Lett. **87**, 193507 (2005)
294. C. Winnewisser et al., Appl. Phys. Lett. **70**, 3069 (1997)

295. A. Nahata, J.T. Yardley, T.F. Heinz, Appl. Phys. Lett. **75**, 2524 (1999)
296. D. Hofstetter, M. Beck, J. Faist, Appl. Phys. Lett. **81**, 2683 (2002)
297. M. Graf et al., Appl. Phys. Lett. **84**, 475 (2004)
298. A. Majumdar, K.K. Choi, J.L. Reno, D.C. Tsui, Appl. Phys. Lett. **83**, 5130(2003)
299. M.P. Touse et al., Appl. Phys. Lett. **86**, 093501 (2005)
300. S.V. Bandara et al., Appl. Phys. Lett. **86**, 151104 (2005)
301. J. Li et al., Appl. Phys. Lett. **89**, 081128 (2006)
302. W. Liu, D.H. Zhang, Z.M. Huang, W.J. Fan, J. Appl. Phys. **101**, 033114(2007)
303. H.C. Liu, E. Dupont, M. Ershow, J. Nonlinear Opt. Phys. Mater. **11**, 433 (2002)
304. T. Maier et al., Appl. Phys. Lett. **84**, 5162 (2004)
305. H. Schneider et al., Opt. Lett. **30**, 287 (2005)
306. T. Maier et al., Appl. Phys. Lett. **88**, 051117 (2006)
307. H.C. Liu, C.Y. Song, A.J. SpringThorpe, J.C. Cao, Appl. Phys. Lett. **84**, 4068(2004)
308. H. Luo, H.C. Liu, C.Y. Song, Z.R. Wasilewski, Appl. Phys. Lett. **86**, 231103 (2005)
309. F. Castellano, R.C. Iotti, F. Rossi, Appl. Phys. Lett. **88**, 182111 (2006)
310. F. Castellano, R.C. Iotti, F. Rossi, Appl. Phys. Lett. **92**, 091108 (2008)
311. F. Castellano, R.C. Iotti, F. Rossi, J. Appl. Phys. **104**, 123104 (2008)
312. L. Esaki, R. Tsu, IBM J. Res. Develop. **14**, 61 (1970)
313. R.F. Kazarinov, R.A. Suris, Sov. Phys. Semicond. **5**, 707 (1971)
314. C. Gmachl, F. Capasso, D.L. Sivco, A.Y. Cho, Rep. Prog. Phys. **64**, 1533 (2001)
315. J. Faist et al., Science **264**, 553 (1994)
316. J. Faist et al., Phys. Rev. Lett. **76**, 411 (1996)
317. G. Scamarcio et al., Science **276**, 773 (1997)
318. J. Faist et al., Nature **387**, 777 (1997)
319. C. Gmachl et al., Science **280**, 1556 (1998)
320. A. Tredicucci et al., Nature **396**, 350 (1998)
321. C. Gmachl et al., Science **286**, 749 (1999)
322. J. Faist, Science **290**, 1713 (2000)
323. R. Paiella et al., Science **290**, 1739 (2000)
324. G. Dehlinger et al., Science **290**, 2277 (2000)
325. R.C. Iotti, F. Rossi, Phys. Rev. Lett. **87**, 146603 (2001)
326. M. Beck et al., Science **295**, 301 (2002)
327. C. Gmachl et al., Nature **415**, 883 (2002)
328. R. Köhler et al., Nature **417**, 156 (2002)
329. F. Eickemeyer et al., Phys. Rev. Lett. **89**, 047402 (2002)
330. A. Tredicucci, Science **302**, 1346 (2003)
331. R. Colombelli et al., Science **302**, 1374 (2003)
332. N. Owschimikow et al., Phys. Rev. Lett. **90**, 043902 (2003)
333. G. Scalari et al., Phys. Rev. Lett. **93**, 237403 (2004)
334. M. Troccoli et al., Nature **433**, 845 (2005)
335. R.F. Service, Science **307**, 1551 (2005)
336. I.S. Osborne, Science **311**, 579 (2006)
337. M. Lee, M.C. Wanke, Science **316**, 64 (2007)
338. R. Terazzi et al., Nat. Phys. **3**, 329 (2007)

339. M.A. Belkin et al., Nat. Photonics **1**, 288 (2007)
340. B.S. Williams, Nat. Photonics **1**, 517 (2007)
341. I. Waldmueller, M.C. Wanke, W.W. Chow, Phys. Rev. Lett. **99**, 117401 (2007)
342. H. Thienpont, Nat. Photonics **2**, 132 (2008)
343. A. Wade et al., Nat. Photonics **3**, 41 (2008)
344. K.J. Franz et al., Nat. Photonics **3**, 50 (2008)
345. J. Faist, Nat. Photonics **3**, 11 (2008)
346. C. Sirtori, Nat. Photonics **3**, 13 (2008)
347. Y. Bonetti, J. Faist, Nat. Photonics **3**, 32 (2008)
348. H. Choi et al., Phys. Rev. Lett. **100**, 167401 (2008)
349. Y. Chassagneux et al., Nature **457**, 174 (2009)
350. N. Yu et al., Phys. Rev. Lett. **102**, 013901 (2009)
351. C. Sirtori et al., Appl. Phys. Lett. **73**, 3486 (1998)
352. R.C. Iotti, F. Rossi, Phys. Stat. Sol. B **238**, 462 (2003)
353. R.C. Iotti, F. Rossi, Appl. Phys. Lett. **76**, 2265 (2000)
354. R.C. Iotti, F. Rossi, Appl. Phys. Lett. **78**, 2902 (2001)
355. F. Compagnone, M. Manenti, A. Di Carlo, P. Lugli, Physica B Condens. Matter **314**, 336 (2002)
356. S.M. Goodnick, P. Lugli, Phys. Rev. B **37**, 2578 (1988)
357. V. Spagnolo, G. Scamarcio, H. Page, C. Sirtori, Appl. Phys. Lett. **84**, 3690 (2004)
358. C. Sirtori et al., IEEE J. Quantum Electron. **34**, 1722 (1998)
359. R.C. Iotti, F. Rossi, J. Comput. Electron. **2**, 191 (2003)
360. M. Sargent III, M.O. Scully, W.E. Lamb, *Laser Physics* (Addison Wesley, London, 1974)
361. W.W. Chow, M. Sargent III, S.W. Koch, *Semiconductor-Laser Physics* (Springer, Berlin Heidelberg, 1994)
362. W.W. Chow, S.W. Koch, *Semiconductor-Laser Fundamentals* (Springer, Berlin Heidelberg, 1999)
363. R. Köhler, R.C. Iotti, A. Tredicucci, F. Rossi, Appl. Phys. Lett. **79**, 3920 (2001)
364. A. Tredicucci et al., Appl. Phys. Lett. **74**, 638 (1999)
365. A. Tredicucci et al., Appl. Phys. Lett. **77**, 2286 (2000)
366. R. Colombelli et al., Appl. Phys. Lett. **78**, 2620 (2001)
367. R.C. Iotti, F. Rossi, AIP Conf. Proc. **772**, 1569 (2005)
368. R.C. Iotti et al., Appl. Phys. Lett. **97**, 033110 (2010)
369. M.S. Vitiello et al., Appl. Phys. Lett. **86**, 111115 (2005)
370. M.S. Vitiello et al., Appl. Phys. Lett. **89**, 131114 (2006)
371. M.S. Vitiello et al., Appl. Phys. Lett. **90**, 121109 (2007)
372. V.A. Shchukin, D. Bimberg, Rev. Mod. Phys. **71**, 1125 (1999)
373. Y. Alhassid, Rev. Mod. Phys. **72**, 895 (2000)
374. S.M. Reimann, M. Manninen, Rev. Mod. Phys. **74**, 1283 (2002)
375. W.G. van der Wiel et al., Rev. Mod. Phys. **75**, 1 (2003)
376. R. Hanson et al., Rev. Mod. Phys. **79**, 1217 (2007)
377. R. Shankar, Rev. Mod. Phys. **80**, 379 (2008)
378. E. Biolatti, R.C. Iotti, P. Zanardi, F. Rossi, Phys. Rev. Lett. **85**, 5647 (2000)
379. E. Biolatti, I. D'Amico, P. Zanardi, F. Rossi, Phys. Rev. B **65**, 075306 (2002)
380. S. De Rinaldis et al., Phys. Rev. B **65**, 081309 (2002)
381. F. Widmann et al., Phys. Rev. B **58**, 15989 (1998)

382. M. Arlery et al., Appl. Phys. Lett. **74**, 3287 (1999)
383. A.D. Andreev, E.P. O'Reilly, Phys. Rev. B **62**, 15851 (2000)
384. D.P. Williams, A.D. Andreev, E.P. O'Reilly, D.A. Faux, Phys. Rev. B **72**, 235318 (2005)
385. T. Nakaoka, S. Kako, Y. Arakawa, Phys. Rev. B **73**, 121305 (2006)
386. D.P. Williams, A.D. Andreev, E.P. O'Reilly, Phys. Rev. B **73**, 241301 (2006)
387. S. Schulz, S. Schumacher, G. Czycholl, Phys. Rev. B **73**, 245327 (2006)
388. M. Winkelnkemper, A. Schliwa, D. Bimberg, Phys. Rev. B **74**, 155322 (2006)
389. A.F. Jarjour et al., Phys. Rev. Lett. **99**, 197403 (2007)
390. N. Baer et al., Phys. Rev. B **76**, 075310 (2007)
391. D. Simeonov, A. Dussaigne, R. Butté, N. Grandjean, Phys. Rev. B **77**, 075306 (2008)
392. O. Marquardt et al., Phys. Rev. B **78**, 235302 (2008)
393. S. Schulz, A. Berube, E.P. O'Reilly, Phys. Rev. B **79**, 081401 (2009)
394. R. Cingolani et al., Phys. Rev. B **61**, 2711 (2000)
395. S. De Rinaldis, I. D'Amico, F. Rossi, Appl. Phys. Lett. **81**, 4236 (2002)
396. I. D'Amico, F. Rossi, Appl. Phys. Lett. **81**, 5213 (2002)
397. S. De Rinaldis, I. D'Amico, F. Rossi, Phys. Rev. B **69**, 235316 (2004)
398. I. D'Amico, F. Rossi, Appl. Phys. Lett. **79**, 1676 (2001)
399. E. Pazy, I. D'Amico, P. Zanardi, F. Rossi, Phys. Rev. B **64**, 195320 (2001)
400. W. Yourgrau, S. Mandelstam, *Variational Principles in Dynamics and Quantum Theory*, 3rd edn. (Dover Publications, New York, NY, 2009)
401. S.A. Gurvitz, Phys. Rev. B **56**, 15215 (1997)
402. A.N. Korotkov, Phys. Rev. B **60**, 5737 (1999)
403. A.N. Korotkov, Phys. Rev. B **63**, 115403 (2001)
404. A. Shnirman, G. Schön, Phys. Rev. B **57**, 15400 (1998)
405. U. Hohenester et al., Appl. Phys. Lett. **77**, 1864 (2000)
406. K. Bergmann, H. Theuer, B.W. Shore, Rev. Mod. Phys. **70**, 1003 (1998)
407. M. Elk, Phys. Rev. A **52**, 4017 (1995)
408. T.A. Laine, S. Stenholm, Phys. Rev. A **53**, 2501 (1996)
409. J.J. Finley et al., Appl. Phys. Lett. **73**, 2618 (1998)
410. J.I. Cirac, P. Zoller, Phys. Rev. Lett. **74**, 4091 (1995)
411. T. Pellizzari, S.A. Gardiner, J.I. Cirac, P. Zoller, Phys. Rev. Lett. **75**, 3788 (1995)
412. C. Monroe et al., Phys. Rev. Lett. **75**, 4714 (1995)
413. A. Sorensen, K. Molmer, Phys. Rev. Lett. **82**, 1971 (1999)
414. Q.A. Turchette et al., Phys. Rev. Lett. **75**, 4710 (1995)
415. A. Imamoglu et al., Phys. Rev. Lett. **83**, 4204 (1999)
416. N. Gershenfeld, I. Chuang, Science **275**, 350 (1997)
417. *Focus issue on Solid State Quantum Information*, ed. by R. Fazio, New J. Phys. 7 (2005)
418. J.C. Gallop, *Superconducting Quantum Interference Devices, the Josephson Effects and Superconducting Electronics* (Taylor & Francis, London, 1991)
419. *Principles and Applications of Superconducting Quantum Interference Devices*, ed. by A. Barone (World Scientific Publishing, Singapore, 1992)
420. J.C. Gallop, J. MacFarlane, L. Hao, *Superconducting Electronics and Devices* (Taylor & Francis, London, 2009)
421. A.O. Niskanen et al., Science **316**, 723 (2007)

422. J.H. Plantenberg, P.C. de Groot, C.J.P.M. Harmans, J.E. Mooij, Nature **447**, 836 (2007)
423. B.E. Kane, Nature **393**, 133 (1998)
424. M.V. Gurudev Dutt et al., Science **316**, 1312 (2007)
425. J.J.L. Morton et al., Nature **455**, 1085 (2008)
426. K.C. Nowack, F.H.L. Koppens, Y.V. Nazarov, L.M.K. Vandersypen, Science **318**, 1430 (2007)
427. N.H. Bonadeo et al., Science **282**, 1473 (1998)
428. G. Chen et al., Science **289**, 1906 (2000)
429. X. Li et al., Science **301**, 809 (2003)
430. E.A. Stinaff et al., Science **311**, 636 (2006)
431. X. Xu et al., Science **317**, 929 (2007)
432. L. Robledo et al., Science **320**, 772 (2008)
433. E. Pazy et al., Europhys. Lett. **62**, 175 (2003)
434. D. Loss, D.P. DiVincenzo, Phys. Rev. A **57**, 120 (1998)
435. P. Zanardi, F. Rossi, Phys. Rev. Lett. **81**, 4752 (1998)
436. P. Zanardi, F. Rossi, Phys. Rev. B **59**, 8170 (1999)
437. F. Troiani, U. Hohenester, E. Molinari, Phys. Rev. B **62**, 2263 (2000)
438. F. Troiani, E. Molinari, U. Hohenester, Phys. Rev. Lett. **90**, 206802 (2003)
439. P. Solinas, P. Zanardi, N. Zanghí, F. Rossi, Phys. Rev. B **67**, 121307 (2003)
440. P. Solinas, P. Zanardi, N. Zanghí, F. Rossi, Phys. Rev. A **67**, 052309 (2003)
441. P. Solinas, P. Zanardi, N. Zanghí, F. Rossi, Phys. Rev. A **67**, 062315 (2003)
442. M. Feng, I. D'Amico, P. Zanardi, F. Rossi, Phys. Rev. A **67**, 014306 (2003)
443. S. Lloyd, Science **261**, 1569 (1993)
444. P. Borri et al., Phys. Rev. Lett. **87**, 157401 (2001)
445. S. Hameau et al., Phys. Rev. Lett. **83**, 4152 (1999)
446. O. Verzelen, R. Ferreira, G. Bastard, Phys. Rev. B **62**, 4809 (2000)
447. O. Verzelen, R. Ferreira, G. Bastard, Phys. Rev. B **64**, 075315 (2001)
448. O. Verzelen, R. Ferreira, G. Bastard, Phys. Rev. Lett. **88**, 146803 (2002)
449. A. Vasanelli, R. Ferreira, G. Bastard, Phys. Rev. Lett. **89**, 216804 (2002)
450. S. Hameau et al., Phys. Rev. B **65**, 085316 (2002)
451. O. Verzelen, G. Bastard, R. Ferreira, Phys. Rev. B **66**, 081308 (2002)
452. I. Favero et al., Phys. Rev. B **68**, 233301 (2003)
453. R. Oulton et al., Phys. Rev. B **68**, 235301 (2003)
454. E.A. Zibik et al., Phys. Rev. B **70**, 161305 (2004)
455. T. Grange, R. Ferreira, G. Bastard, Phys. Rev. B **76**, 241304 (2007)
456. E.A. Zibik et al., Phys. Rev. B **77**, 041307 (2008)
457. B. Krummheuer, V.M. Axt, T. Kuhn, Phys. Rev. B **65**, 195313 (2002)
458. A. Vagov, V.M. Axt, T. Kuhn, Phys. Rev. B **66**, 165312 (2002)
459. A. Vagov, V.M. Axt, T. Kuhn, Phys. Rev. B **67**, 115338 (2003)
460. A. Vagov et al., Phys. Rev. B **70**, 201305 (2004)
461. V.M. Axt, P. Machnikowski, T. Kuhn, Phys. Rev. B **71**, 155305 (2005)
462. K. Roszak, A. Grodecka, P. Machnikowski, T. Kuhn, Phys. Rev. B **71**, 195333 (2005)
463. B. Krummheuer et al., Phys. Rev. B **71**, 235329 (2005)
464. M. Glanemann, V.M. Axt, T. Kuhn, Phys. Rev. B **72**, 045354 (2005)
465. V.M. Axt, T. Kuhn, A. Vagov, F.M. Peeters, Phys. Rev. B **72**, 125309 (2005)
466. B. Krummheuer, V.M. Axt, T. Kuhn, Phys. Rev. B **72**, 245336 (2005)
467. A. Krügel, V.M. Axt, T. Kuhn, Phys. Rev. B **73**, 035302 (2006)

468. D. Reiter, M. Glanemann, V.M. Axt, T. Kuhn, Phys. Rev. B **73**, 125334 (2006)
469. A. Vagov et al., Phys. Rev. Lett. **98**, 227403 (2007)
470. P. Machnikowski, V.M. Axt, T. Kuhn, Phys. Rev. A **75**, 052330 (2007)
471. D. Reiter, M. Glanemann, V.M. Axt, T. Kuhn, Phys. Rev. B **75**, 205327 (2007)
472. A. Krügel, A. Vagov, V.M. Axt, T. Kuhn, Phys. Rev. B **76**, 195302 (2007)
473. K. Roszak, V.M. Axt, T. Kuhn, P. Machnikowski, Phys. Rev. B **76**, 195324 (2007)
474. J. Huneke et al., Phys. Rev. B **78**, 085316 (2008)
475. J.M. Kikkawa, I.P. Smorchkova, N. Samarth, D.D. Awschalom, Science **277**, 1284 (1997)
476. J. Berezovsky et al., Science **320**, 349 (2008)
477. A. Bertoni et al., Phys. Rev. Lett. **84**, 5912 (2000)
478. R. Ionicioiu, P. Zanardi, F. Rossi, Phys. Rev. A **63**, 050101 (2001)
479. A. Yacoby, M. Heiblum, D. Mahalu, H. Shtrikman, Phys. Rev. Lett. **74**, 4047 (1995)
480. F. Alsina et al., Phys. Rev. B **66**, 165330 (2002)
481. F. Alsina et al., Phys. Rev. B **67**, 161305 (2003)
482. G. Gumbs, Y. Abranyos, Phys. Rev. A **70**, 050302 (2004)
483. G. Giavaras et al., Phys. Rev. B **74**, 195341 (2006)
484. J.A. del Alamo, C.C. Eugster, Appl. Phys. Lett. **56**, 78 (1990)
485. M. Kitagawa, M. Ueda, Phys. Rev. Lett. **67**, 1852 (1991)
486. A.E. Popescu, R. Ionicioiu, Phys. Rev. B **69**, 245422 (2004)
487. T. Zibold, P. Vogl, A. Bertoni, Phys. Rev. B **76**, 195301 (2007)
488. Z. Chen et al., Science **311**, 1735 (2006)
489. A. Einstein, B. Podolsky, N. Rosen, Phys. Rev. **47**, 777 (1935)
490. D.M. Greenberger, M.A. Horne, A. Shimony, A. Zeilinger, Am. J. Phys. **58**, 1131 (1990)
491. A. Aspect, J. Dalibard, G. Roger, Phys. Rev. Lett. **49**, 1804 (1982)
492. C. Monroe, D.M. Meekhof, B.E. King, D.J.Wineland, Science **272**, 1131 (1996)
493. D. Bouwmeester et al., Phys. Rev. Lett. **82**, 1345 (1999)
494. L. Quiroga, N.F. Johnson, Phys. Rev. Lett. **83**, 2270 (1999)
495. J.S. Bell, Physics **1**, 195 (1964)
496. J.F. Clauser, M.A. Horne, A. Shimony, R.A. Holt, Phys. Rev. Lett. **23**, 880 (1969)
497. K. Kwok, J.C. Ellenbogen, Mater. Today **5**, 28 (2002)
498. A. Jalabert, A. Amara, F. Clermidy, *Molecular Electronics Materials, Devices and Applications* (Springer, Dordrecht, 2008)
499. M.A. Reed et al., Science **278**, 252 (1997)
500. N.S. Hush, Ann. N.Y. Acad. Sci. **1006**, 1 (2003)
501. A. Aviram, M.A. Ratner, Chem. Phys. Lett. **29**, 277 (1974)
502. J. McGinness, P. Corry, P.H. Proctor, Science **183**, 853 (1974)
503. H. Akamatsu, H. Inokuchi, Y. Matsunaga, Nature **173**, 168 (1954)
504. B.A. Bolto, R. McNeill, D.E. Weiss, Aust. J. Chem. **16**, 1090 (1963)
505. H.W. Kroto et al., Nature **318**, 162 (1985)
506. H.W. Kroto, Nature **329**, 529 (1987)
507. Y. Meir, N.S. Wingreen, Phys. Rev. Lett. **68**, 2512 (1992)
508. C.H. Joachim, J.K. Gimzewski, R.R. Schlittler, C. Chavy, Phys. Rev. Lett. **74**, 2102 (1995)
509. A. Aviram, C. Joachim, M. Pomerantz, Chem. Phys. Lett. **146**, 490 (1988)

510. C. Joachim, J.K. Gimzewski, Chem. Phys. Lett. **265**, 353 (1997)
511. S.J. Tans et al., Nature **386**, 474 (1997)
512. G.J. Ashwell, D.S. Gandolfo, J. Mater. Chem. **3**, 411 (2002)
513. I. Zutic, J. Fabian, S. Das Sarma, Rev. Mod. Phys. **76**, 323 (2004)
514. S.A. Wolf, A.Y. Chtchelkanova, D.M. Treger, IBM J. Res. Dev. **50**, 101 (2006)
515. A. Fert, Rev. Mod. Phys. **80**, 1517 (2008)
516. S. Bandyopadhyay, *Intro to Spintronics* (Taylor & Francis, London, 2008)
517. M. Johnson, R.H. Silsbee, Phys. Rev. Lett. **55**, 1790 (1985)
518. M.N. Baibich et al., Phys. Rev. Lett. **61**, 2472 (1988)
519. G. Binasch, P. Grünberg, F. Saurenbach, W. Zinn, Phys. Rev. B **39**, 4828 (1989)
520. S. Datta, B. Das, Appl. Phys. Lett. **56**, 665(1990)
521. Y.M. Lee et al., Appl. Phys. Lett. **90**, 212507 (2007)
522. B.T. Jonker et al., Phys. Rev. B **62**, 8180 (2000)
523. A.T. Hanbicki, B.T. Jonker, Appl. Phys. Lett. **80**, 1240(2002)
524. X. Jiang et al., Phys. Rev. Lett. **90**, 256603 (2003)
525. J.M. Kikkawa, D.D. Awschalom, Phys. Rev. Lett. **80**, 4313 (1998)
526. B.T. Jonker, Polarized optical emission due to decay or recombination of spin-polarized injected carriers, U.S. Patent 5,874,749, 1999
527. X. Lou et al., Nat. Phys. **3**, 197 (2007)
528. I. Appelbaum, B. Huang, Nature **447**, 295 (2007)
529. H. Dery, P. Dalal, L. Cywinski, L.J. Sham, Nature **447**, 573 (2007)
530. M. Holub, J. Shin, D. Saha, P. Bhattacharya, Phys. Rev. Lett. **98**, 146603 (2007)

Index